Exploration
of the Universe

Exploration of the Universe
Updated Brief Edition

George Abell
University of California, Los Angeles

Holt, Rinehart and Winston
New York   Chicago   San Francisco   Atlanta
Dallas   Montreal   Toronto   London   Sydney

Acknowledgments

Cover photograph: The Dumbbell nebula (M27) in Valpecula
(Courtesy Hale Observatories.)
Half-title photograph: M8, Lagoon Nebula in Sagittarius.
(Courtesy Hale Observatories.)
Frontpiece photograph: The "Whirlpool" Galaxy, M51.
(U.S. Naval Observatory; Flagstaff Station.)

Copyright © 1964, 1969, 1973 by Holt, Rinehart and Winston, Inc.
All rights reserved
Library of Congress Catalog Card Number: 69-13567
SBN: 03-075950-1
Printed in the United States of America
 3456 41 987654321

Preface

"When we try to pick out any thing by itself," wrote John Muir, "We find it hitched to everything else in the Universe." It is a purpose of science, and particularly of astronomy, to learn the precise way in which the parts are hitched together, and more important, the rules that govern the machinery of the Universe. Science is, in brief, a specific method for making models that describe the behavior of natural phenomena, and that enable us to understand nature better. Evidently the method of science is fruitful: not only have we learned many of the rules of the game Nature plays, but we have learned to apply them to our technology to make our world a more comfortable and interesting — at least a more complex — place to live.

Astronomy has had an enormous impact on our way of life. Every student of science or engineering first studies mechanics and gravitational theory, developed by Kepler, Galileo, Newton and others largely as a result of observations of the motions of the planets. Other astronomical investigations led Einstein to relativity theory, and the realization of the equivalence of mass and energy — the basis of nuclear power. Optics, spectroscopy, radiation theory, much of quantum theory, and other modern disciplines — indeed, even mathematics — have developed in large part for the purpose of or as a consequence of astronomical studies.

This is an enormously exciting time for astronomy. We are studying parts of the moon brought to earth by man himself, and the planets and the rest of the solar system with advanced space probes.

We now know the surface of Mars through photographic mapping far more precisely than we knew the geography of our own planet two or three centuries ago. Today we observe the Universe in the light of x-rays, ultraviolet and infrared radiation, and radio waves. We have discovered hundreds of quasars, the greatest energy emitters yet known, and dozens of pulsars, probably superdense stars made of neutrons. We may be observing gravitational waves from the collapse of matter near the center of our Galaxy. Most cosmologists now believe that our radio telescopes are detecting radiation from the primeval fireball that started the expansion of the Universe itself. We cannot begin to guess yet how today's discoveries will affect our technology of tomorrow, but we can be quite certain that the science of our time will continue to lead to a technological reshaping of our future, whether we like it or not. Scientists of many fields, expecting that in astronomy lie the most fundamental discoveries and advances to be made in the near future, are turning to astronomy from other widely varied disciplines.

Of course astronomy is "relevant" to all of us — to use a word that holds a special meaning in the vocabulary of our time. We must not, however, err in assuming that the only value of astronomy is in its direct and indirect applications to our physical needs. From ancient times astronomy *did*, to be sure, find practical applications; as Glaucon said (Plato's *Republic*, Book VII), "The observation of the seasons and of months and years is as essential to the general as it is to the farmer or sailor." But to this Socrates replied, "I am amused, Glaucon, at your fear of insisting on useless studies; in every man there is an eye of the soul which, by them, is purified and re-illumined, and is more precious by far than ten thousand bodily eyes, for it alone sees truth." Should it not move our spirits to consider that the most remote galaxies in the Universe are made of atoms identical to those that compose our bodies — and more so, that they behave according to the same natural laws? Perhaps a knowledge of the uncanny order of the Universe is astronomy's greatest contribution.

I believe that a study of astronomy is one of the best ways to bridge the dichotomy of human knowledge and thought between the sciences, on the one hand, and the humanities and social sciences on the other. Most students of the latter disciplines have not learned the rules of scientific thought, nor the most rudimentary vocabulary of the various scientific languages. There was a time when a man was not considered educated who was ignorant of natural philosophy, but unfortunately, as science and technology matured the natural philosophers, that is, the scientists, had to develop new techniques and languages to cope with the problems they encountered at the expanding frontier of knowledge. It is, in fact, usually impossible even for scientists to become expert in those sciences that lie outside their own increasingly narrow disciplines. A person generally requires years of post-graduate study even to understand thoroughly the work at the frontier, let alone to contribute to it.

Yet all of us can learn enough to understand some of the laws of nature and to appreciate the significance of many new discoveries. Astronomy encompasses principles of nearly all of the physical sciences, many of which find astronomical applications at an elementary enough level to be understood by any interested student. Moreover, by its very nature astronomy, perhaps more than any other science, excites great general interest, and introductory courses in the subject have become increasingly popular among students not majoring in science or mathematics. It is for such courses that this text has been written. It is not intended primarily as a text for a *first* course, but rather a *terminal* course, for it is not expected that most students will have an opportunity (nor, probably, the mathematical preparation) to study astronomy further. For many it may be their only science course.

This text, an updated version of the *Brief Edition* of the author's *Exploration of the Universe,* is intended for a one-quarter or one-semester course at a junior or liberal-arts college. Mathematics has been avoided almost entirely. Although there are a few simple equations scattered through the text, they need not be understood to appreciate the gist of the book and can be skipped over entirely if desired, without substantial loss of continuity. It seems to me to be not particularly important that a student of, say, history know how to calculate the sidereal period of a planet of known synodic period, or his weight on a hypothetical planet of twice the radius and five times the mass of the Earth. On the other hand, there are three things that I hope he *does* learn from the course (and text): (1) a general overview of the Universe, with an idea of how it is made up, of the kinds of objects in it, and of their distances and dimensions; (2) an appreciation of the beauty of the natural laws that apply so universally throughout the cosmos and how we can use our knowledge of these laws, not only to expand our technology but to investigate further the Universe itself; and (3) a realization that while openmindedness is a virtue, it should not be confused with gullibility, and that all ideas are not equally valid simply because experts may disagree with one another on details at the frontier of knowledge; that is to say, I hope the student learns a healthy skepticism and does not fall for all crank ideas and mystic nonsense. Beyond this, I hope the student is "turned on" by science in general and astronomy in particular. I hope that his course is so taught and that this book is so written that he will thoroughly enjoy his introduction to the cosmos.

The present *Updated Brief Edition* differs from its predecessor in the inclusion of a good deal of new material. New photographs and improved star charts are provided. There are accounts of recent findings by the Apollo missions and by the Mars probe, Mariner 9. There is new discussion of long-baseline interferometry and recent radio observations of interstellar molecules. Descriptions of both the theory and observational basis for neutron stars and black holes are much expanded. There is new discussion of the origin and future evolution of the solar system. The chapters on galaxies and cosmology are updated, and a description of Weber's experiments to detect gravitational waves is included.

There are numerous other changes as well, but despite our best efforts, no text can remain completely timely, even for the several months it takes to print it, in a field so rapidly changing as astronomy. It is the responsibility of the instructor to insure that the students are apprised of the most recent developments, and it is also his responsibility to present such advances in their proper context and with appropriate emphasis.

Every effort has been made to avoid and eliminate errors in fact and minor printing errors. Numerous corrections to the original *Brief Edition,* as well as to the full edition of *Exploration of the Universe* have been incorporated, and I particularly want to thank those correspondents who have called the need for them to my attention. Still, all kinds of errors doubtless remain, and I will continue to be most grateful to be informed of such slips.

I am most indebted to those who read part or all of the manuscript for this text, and who made many valuable and helpful suggestions. These include Mr. Paul Wylie, and Drs. Grant Athay, William Bidelman, Owen Gingerich, Louis Henyey, William Kaula, Dimitri Mihalas, Thornton Page, Stanton Peale, Daniel Popper, Paul Routly, John Schopp, Hyron Spinrad, Victor Thoren, Harold Weaver, George Wetherill, and Harold Zirin.

Los Angeles, California   G. O. A.
20 November 1972

Contents

Exploration
of the Universe

CHAPTER 1

Prologue

Two thousand years ago, the earth was believed to occupy a very special place in the cosmos. To the ancient observer the earth seemed vast and immobile — the center of the universe. Indeed, the entire realm of the universe, the whole scope of his knowledge and imagination, was the *realm of the earth*. He regarded the sky as a huge inverted bowl overhead, or actually, as a great hollow sphere completely surrounding the earth. This celestial sphere turned daily, carrying around the sky, the sun, moon, stars, and planets that seemed to be affixed to its inner surface. It was natural, of course, for early man to attach such central importance to his earth. His experience had not prepared him to accept the sky as virtually empty space extending to such unimaginable — perhaps infinite — depths that the earth, in comparison, becomes a mere point in the cosmos.

1.1 DEVELOPMENT OF ASTRONOMICAL THOUGHT

Certain original and imaginative individuals, however, began to make systematic observations of the positions and motions of the celestial objects, to formulate more precise theories to explain their behavior, and to subject such theories to critical tests to ascertain their validity. Thus, with the development of the scientific procedure, man's concept of the universe began to change until, a few centuries ago, he had come to the realization that the earth was one of several planets that revolve in precise paths and in regular periods about the sun.

The sun, man learned, is a great, luminous, gaseous sphere with a diameter more than a hundred times that of the earth. The earth, on the other hand, is more or less like the rest of the planets — cold and solid. Further, man learned that there are many lesser objects as well in the solar system. Some are minor planets, tiny planets, most of them not over a few miles across, circling the sun by the thousands. Others, in unknown numbers, are comets, loose swarms of solid particles embedded in tenuous gases, revolving about the sun in elongated orbits that carry them billions of miles into space and eventually back to the vicinity of the sun in long, but regular, periods. Still others are meteoroids, cluttering space by

FIG. 1-1   The Sun. *(Hale Observatories.)*

FIG. 1-2 The planet Mars, photographed in orange light in August, 1956. *(Hale Observatories.)*

the countless trillions, particles that range in size down to microscopic dimensions.

The earth, in other words, is only one member of a large family of objects in the solar system, all of which derive their light and heat from that central and by far most massive body, the sun. Until recent times, however, it was the planets, not the sun, that commanded the principal attention of astronomers. These were other worlds, in some respects like the earth. It was natural to investigate them, to wonder and speculate about them. Was it possible that some of them might also maintain life? Man's concept of space had moved out to encompass the solar system — the *realm of the planets.*

Human curiosity, however, does not long remain inattentive to what lies beyond. As early as 1600, the monk Giordano Bruno was burned alive in Rome for writing such heresies as that the stars themselves are suns. But in 1838, Bessel's detection of the parallax of the star 61 Cygni (a slight angu-

lar displacement due to the earth's orbital motion) made possible the first direct measurement of a stellar distance. The investigations that followed proved conclusively that the stars are suns, more or less like our own sun, but at such immense distances that they appear as mere points. Light, which travels with a speed of 186,000 mi/sec, requires minutes or hours to reach us from the planets but years to travel from and between the stars. Once more man's concept of space had enlarged, now to encompass the *realm of the stars.*

It had been suspected in the middle eighteenth century, and demonstrated convincingly by William Herschel in 1787, that the stars are not strewn over the sky at random but rather make up a huge assemblage, or *galaxy.* Herschel described the assemblage as shaped like a "grindstone." The Milky Way, a luminous band completely circling the sky, is the light from the myriads of stars that are projected into our line of sight as we look edge on through the grindstone. In addition to individual stars, it was found that the Galaxy contains clusters of stars and diffuse luminous clouds, or *nebulae.* Although the disklike character of the Galaxy was understood throughout the nineteenth century, astronomers had no clear idea of its extent until well into the twentieth century. It is a system of about 100 billion stars — suns like our own — stretching across an expanse of space that light may require 100,000 years to traverse. The scope of astronomy had now extended to include the *realm of the Galaxy.*

Immanuel Kant had speculated in 1755 that the sun was a part of a system or great cluster of stars, and that there were other such clusters, or "island universes," distributed through space far beyond the borders of our own system. From time to time in the decades to follow, the *island universe* theory gained the attention of astronomers. It seemed possible that many of the nebulae were composed of stars and might actually lie out beyond the stars of our Galaxy — "candidates" for

FIG. 1-3 Star clouds and dark obscuring matter in the Milky Way in the region of Sagittarius. *(Yerkes Observatory.)*

such outposts of space. By the close of the nineteenth century, however, some of the nebulae had been shown to be of a definitely gaseous nature, and even as late as 1920 the weight of astronomical opinion did not favor the notion that any of the nebulae could be *island universes* — galaxies in their own right.

The controversy over the theory continued, however, until 1924, when it was finally settled. Kant had been right. Cepheid variables were discovered in some of the "nebulae." These variables are stars that can be recognized because of the peculiar and characteristic manner in which they vary in their brightness. The Cepheid variables in our own galaxy are known to be very luminous stars; yet in the nebulae they appear very faint, which indicates that these Cepheids, and the nebulae in which they reside, must lie at distances far greater than the farthest stars in our own galaxy.

Subsequent investigations have shown that most of the other nebulae — in fact, all but the comparatively few that are glowing gases between the stars of the Milky Way — *are* galaxies. They speckle space to the depths of the observable universe — to distances requiring light billions of years to cross. The number of observable galaxies must be close to a billion. Most of these appear as insignificant, slightly irregular specks, barely visible on photographs made over long time exposures with the world's greatest telescopes. The number of remote galaxies too faint to observe cannot even be estimated with any degree of confidence. Yet each galaxy is a system of billions of stars, each star a sun like our own. Even in our galaxy, which is probably only one of many billions, the nearest other suns are so remote as to appear as points of light. Man's concept of the universe has enlarged to encompass the *realm of the galaxies.*

Since World War II, new techniques have been developed which make it possible to receive and record radio radiation emitted by celestial objects. Still more recently, instruments carried by rockets, artificial satellites, and space probes have permitted the detection of energy in the far ultraviolet, x-rays, and even gamma rays — adding new data to that obtained from ground-based observatories. This new information has raised the inevitable new problems — for example, the source of x-rays from certain peculiar stars and of radio waves from peculiar galaxies. Especially perplexing is the nature of the quasi-stellar objects or "quasars" — stellar-appearing objects that emit much of their energy at radio wavelengths and which may be the most remote things yet observed. Many other sources of radio waves may be galaxies too remote to be visible through even our greatest optical telescopes. Some of the feeble radio radiation coming from all directions in space is now thought to be energy

FIG. 1-4   The spiral galaxy M81 in Ursa Major.
(Lick Observatory.)

astronomer, of necessity, must examine his subjects from afar; the whole of the cosmos is his laboratory. Except for his limited exploration of the moon and neighboring planets, made possible by the new space technology, and for his observations of cosmic rays and of fallen meteorites, he has no direct contact with the objects of his investigations. In this respect astronomy differs from the other sciences.

The astronomer must call upon all his resources of ingenuity and employ more clever devices than the most astute detective to solve the problems of his business. In this respect astronomy is representative of the other sciences.

Man, by nature, is curious about his surroundings, particularly about unexplained phenomena. To satisfy his curiosity, he may employ one of two entirely dissimilar approaches. The first is to invent, to make up an explanation out of the imagination. The procedure is extremely "efficient" — it answers all the big questions. Here, of course, is the root of myth, superstition, and primitive religion. One such "religion" that grew rather

generated billions of years ago from a primeval state of the universe, and may be an immensely important clue to our investigations in the *realm of cosmology.*

Today, in the new age of space technology, man stands at the "threshold of space." Yet, he is not really at the threshold of the astronomical universe. Even if man should go beyond the moon, to the planets, or even if, in some unforeseen, unimaginable manner, he should succeed in traveling at nearly the speed of light and reaching the nearest stars — even then, he would scarcely have begun on a journey through *space* in the astronomical sense of the word.

1.2 ASTRONOMY AS A SCIENCE

Astronomy is the science of the universe. The

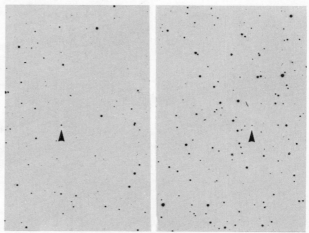

FIG. 1-5   Quasistellar objects BSO-1 and 3C-9; 200-inch telescope photograph. *(Hale Observatories.)*

naturally from a very early and invented idea of the nature of the universe is astrology.

The other approach is the *scientific method.* One gathers clues by making observations or performing experiments. With these clues at hand, one formulates tentative hypotheses to explain the phenomena or the experimental results. Then — the crucial step — one tests these hypotheses by predicting from them new phenomena or the results of new experiments. If the tests fail, the hypotheses must be discarded in favor of better ones.

The purpose of science is to find order in the chaos of natural phenomena. Science attempts to *represent* nature as simply and accurately as possible with *natural laws* — descriptions of how nature behaves. Note that science *describes the how* but does not *explain the why* of nature; it makes no attempt to establish the true and absolute "nature of things." This latter activity belongs in the province of religion. Science, then, is a *method,* not a *subject.* It is a *method* for the organized investigation of nature.

The scientific method is the most successful means we have found for achieving the aims of science. It has made possible our modern technology with its skyscrapers, airplanes, television, and rockets. It does not, however, yield absolutes of knowledge. Rather it allows us to describe how nature has behaved in the past and how we might expect it to behave in the future. The *laws* of

FIG. 1-6   Stonehenge, an assemblage of stones in Western England, believed to have been an astronomical shrine of early peoples. *(Rapho Guillumette Pictures.)*

science are the rules of the game that nature seems to play. Some of these rules are extremely well established and appear never to be violated. We cannot, of course, consider them absolute and binding on nature; yet they are so consistent that we believe in them almost religiously, and predictions based on them are, to all intents and purposes, "established facts." For example, if we raise a stone and drop it, we can be quite certain that it will fall to the ground. Similarly, all known phenomena are so consistent with the hypothesis that the earth goes around the sun, and not vice versa, that we regard the earth's motion as "fact," not "theory." Science, however, deals with other postulates that are by no means so firmly established. We shall come across many of these in the chapters to follow.

Although science itself does not ascribe an absolutism to its laws, it does not follow that one theory is as good as another. We hear much from cranks and charlatans proposing new (and not so new) theories in the name of science. They dwell on the scientific errors of the past, on "academic prejudice" (against new ideas that conflict with established theory), and on their inability, against such persecution, to obtain general acceptance of their new revelations, or even to be heard. They are not, however, always so unheard. The public invests millions of dollars a year in the purchase of crank literature. In 1950 a best-selling book, from a highly reputable publisher, described the eruption from Jupiter of a comet, which stopped or slowed the rotation of the earth, and eventually became the planet Venus. The most elementary knowledge of the natures of comets and planets, and of the simplest and best-established physical laws, is sufficient to reveal the absolute absurdity of the idea. The author, in common with nearly all authors of such hypotheses, acknowledged that his theory is counter to established laws of science, and that if it is to be accepted those laws must be rejected. To the general public, uneducated

in even elementary science, and used to many technical miracles, the scientific-sounding jargon and forceful arguments of cranks are often convincing.

Of course, scientists *do* make mistakes — and frequently. Even the most eminent of them often disagree. But wrong tracks and conflicting views are almost always at the frontiers of knowledge, where progress is made by the trial and error process of the scientific method. Moreover, "established" ideas are under continual reevaluation and, if necessary, revision; but they are not likely to be rejected entirely. Newton's laws have not been abandoned because they fail to apply to the motions of electrons in atoms or to bodies moving with speeds near that of light (for which the quantum theory and relativity, respectively, are superior descriptions); they are as valid as ever within the domain in which they have been shown to apply. If our most basic rules of science are wrong, then it is hard to understand how repeated tests continue to verify them, let alone how our modern technology ever came to be. In any case, the burden rests on new theories to predict the results of tests or observations more accurately than the old ones.

One of the most important, and perhaps most difficult, tasks for a student taking a course in science is to learn to evaluate what he hears and reads in terms of basic and well-tested natural laws and to tell the crank from the legitimate investigator. The task is not always easy for the scientist himself, and there exists a middle ground where the distinction itself is at best vague. The student should learn to develop at least a critical and objective attitude and to distinguish open-mindedness from gullibility.

Astronomy is probably the oldest of the sciences; it is also one of the most exciting and rapidly developing fields of knowledge today. What we have learned of the universe may well stagger the imagination, but no less impressive is the list of unsolved problems. A text in astronomy must therefore be somewhat of a "progress report" to the student; in no other field do textbooks go out of date so rapidly. Astronomy is an excellent subject with which to demonstrate the methods of the detectives known as scientists. In scientific investigation, however, unlike detective fiction, the case is never closed. Certain facets may be "cleared up" from time to time, but new "developments" continually complicate the picture.

1.3 CONTRIBUTIONS OF ASTRONOMY

Astronomy, as well as serving as an excellent example of the scientific approach, is important for the effect its contributions have had on our everyday lives. Observations of the motions of the planets led Newton to his great discovery of the laws of motion and gravitation — the basis of much of our modern technology. Further, many mathematical techniques used daily by our engineers were first developed to solve astronomical problems. The science of optics was initiated and developed largely by astronomers for the investigation of astronomical problems. Yet it has led to the invention not only of the telescope, but also of the microscope, so vital to medicine and other fields unrelated to astronomy. The field of spectroscopy, which had an astronomical origin and is still widely applied to astronomical problems, has proved invaluable to physics, chemistry, metallurgy, biology, and other areas of knowledge that bear directly on our daily living. Of course, timekeeping, navigation, and geodesy are closely related to astronomy. The vastly important field of physics, in fact, developed largely as an offshoot of astronomy.

Perhaps the most spectacular contribution of astronomy in the twentieth century has been the realization that thermonuclear energy is the source of energy of the sun and other stars. This realization has led to experiments in nuclear physics, and to man's effort to duplicate the source of stellar energy. It is, perhaps, a dubious honor to astronomy that man first succeeded in this venture with the production of nuclear bombs.

Now let us begin the story of man's exploration of the universe. We shall see how astronomers dabble in their laboratory — the universe itself.

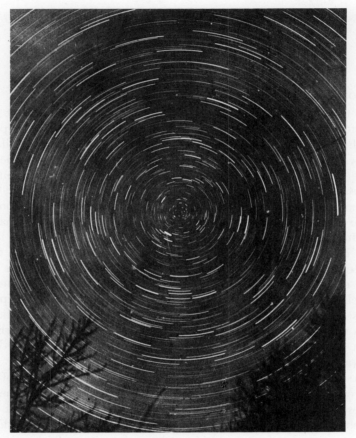

CHAPTER 2

Beginnings of Astronomy

2.1 EARLIEST ASTRONOMERS

Primitive speculations on the nature of the universe must date from prehistoric times. It is difficult to state definitely when the earliest observations of a more or less quantitative sort were made, or when astronomy as a science began. Certainly, in many of the ancient civilizations the regularity of the motions of celestial bodies was recognized and attempts were made to keep track of and predict celestial phenomena. In particular, the invention of and keeping of a calendar requires at least some knowledge of astronomy — the basic units of the calendar being the day, the month (originally, the 29- and 30-day cycling of the moon's phases), and the year of seasons (the *tropical year*).

The Chinese had a working calendar at least as early as the thirteenth or fourteenth century B.C., and probably much earlier. About 350 B.C., the astronomer Shih Shen prepared what was probably the earliest star catalogue, containing about 800 entries. The Chinese also kept rather accurate records of comets, meteors, and fallen meteorites from 700 B.C. Records were made of sunspots visible to the naked eye and of what the Chinese called "guest stars," stars that are normally too faint to be seen but suddenly flare up to become visible for a few weeks or months (such a star is now called a *nova*). The most significant of the Chinese observations of nova outbursts was of the great nova of 1054 A.D. in the constellation of Taurus. Today's remnant of that cosmic explosion is believed to be the Crab nebula, a chaotic, expanding mass of gas (see Chapters 18 and 20).

The Babylonians and Assyrians also knew the approximate length of the year several centuries B.C. In pre-Christian Egypt the astronomers were the priests. Although Egyptian astronomers had divided the sky into constellations and had named these star groups for deities in their mythology, their main astronomical function was in keeping the calendar. An important contribution of the Egyptians was the development of the rudiments of civil engineering. They made measurements with ropes and surveyed their land. Apparently they had developed some of the concepts of arithmetic, and primitive geometry has been found on papyri prepared between the nineteenth and sixteenth centuries B.C. Both the Egyptians and Babylonians had learned to construct fairly accurate sundials for time keeping. The earliest sundial still preserved is an Egyptian one dating from the eighth century B.C.

Most of the peoples of antiquity believed in and practiced astrology. The origin of this religion was probably in Babylonia. The heavenly bodies were quite naturally regarded as different in substance from terrestrial matter, and were generally associated with deities. These gods, although not omnipotent, at least had influence on earthly and human affairs. Thus the particular configuration of the sun, moon, and planets in the sky at the time and place of one's birth was believed to be of significance to that person's character and destiny. Particularly important was the object in the ascendancy — that is, just rising — for that planet's influence must be strong in the air inhaled by the newborn child's first gasp. Astrology was greatly developed and expanded by the ancient Greeks; indeed, the earliest known systematic exposition of the subject is Ptolemy's *Tetrabiblos* in the second century A.D. Astrology is still widely accepted in some countries, and even in the Western nations of Judeo-Christian heritage it has a surprisingly large number of adherents.

The science of astronomy owes a great debt to astrology, because its practice required a detailed knowledge of the motions of the celestial bodies. This knowledge was needed not only because the destiny of a person was believed to be under the continual influence of the planets and their configurations, but in particular to construct his *horoscope* — a chart of the planets in the sky at the time and place of his birth, when their influence was at its greatest. Consequently, astronomers from the time of antiquity to the time of Galileo, even some who may not have believed in astrology per se, devoted most of their energy to observing carefully the positions and motions of the heavenly bodies, or to constructing models or schemes that would allow the accurate calculation of the planetary positions at arbitrary times in the future. Both Tycho Brahe and Kepler, great astronomers in the Shakespearian era, prepared horoscopes as part of their duties. Had they not been employed to do so, the development of the laws of planetary motion, and ultimately our modern technology, would certainly not have come when it did.

2.2 EARLY GREEK ASTRONOMY

The high point in ancient science was Greek culture from 600 B.C. to 400 A.D. The earliest Greeks were not scientists in the modern sense. In general, their research took the form of solving academic problems, reasoning on a purely abstract level from axioms that were "obvious" to every Greek scholar. To make observations, perform experiments, or otherwise make contact with the physical world was not considered a "proper" form of investigation. Yet, as we shall see, in that great Greek reservoir of ideas and inspiration, some experiments were performed and some observations were carried out, with the result that science in general, and astronomy in particular, was raised to a level unsurpassed until the sixteenth century.

(a) Early Concepts of the Sky

THE CELESTIAL SPHERE

If we gaze upward at the sky on a clear night, we cannot avoid the impression that the sky is a great hollow spherical shell with the earth at the center. The early Greeks regarded the sky as just such a *celestial sphere;* some apparently thought of it as an actual sphere of a crystalline material, with the stars embedded in it like tiny jewels. The sphere, they reasoned, must be of very great size, for if its surface were close to the earth, as one moved from place to place he would see an apparent angular displacement in the directions of the stars.

Of course, at any one time we see only a hemisphere overhead, but with the smallest effort of imagination we can envision the remaining hemisphere, that part of the sky that lies below the horizon. The fact that the sky appears as a full hemisphere and not merely part of one, convinced the ancients that the celestial sphere must be extremely large — infinite as far as the eye can tell — compared to the earth. If we watch the sky for several hours, we see that the celestial sphere is gradually and continually changing its orientation. The effect is caused simply by the rotation of the earth, which carries us under successively different portions of the sphere. Following along with us must be our *horizon,* that line in the distance at which the ground seems to dip out of sight, providing a demarcation between earth and sky. (The horizon may, of course, be hidden from view by mountains, trees, buildings, or in large cities, smog.) As our horizon tips down in the direction that the earth's rotation carries us, stars hitherto

FIG. 2-1   Time exposure showing trails left by stars as a consequence of the apparent rotation of the celestial sphere. *(Lick Observatory.)*

hidden beyond it appear to rise above it. In the opposite direction the horizon tips up, and stars hitherto visible appear to set behind it. Analogously, as we round a curve in a mountain road, new scenery comes into view while old scenery disappears behind us.

The direction around the sky toward which the earth's rotation carries us is *east;* the opposite direction is *west.* The Greeks, unaware of the earth's rotation, imagined that the celestial sphere rotated about an axis that passed through the earth. As it turned, it carried the stars up in the east, across the sky, and down in the west.

CELESTIAL POLES

A careful observer will notice that some stars do not rise or set. As seen from the Northern Hemisphere, there is a point in the sky some distance above the northern horizon about which the whole celestial sphere appears to turn. As stars circle about that point, those close enough to it can pass beneath it without dipping below the northern horizon. A star exactly at the point would appear motionless in the sky. Today *Polaris* (the North Star) is within 1° of this pivot point of the heavens.

The Greeks regarded that pivot point as one end of the axis about which the celestial sphere rotates. We know today that it is the earth that spins about an axis through its North and South Poles. An extension of the axis would appear to intersect the sky at points in line with the North and South Poles of the earth but, because of the nearly infinite size of the celestial sphere, immensely far away. As the earth rotates about its polar axis, the sky appears to turn in the opposite direction about those *north* and *south celestial poles.*

An observer at the North Pole of the earth would see the north celestial pole directly overhead (at his *zenith*). The stars would all appear to circle about the sky parallel to the horizon, none rising or setting. An observer at the earth's equator, on the other hand, would see the celestial poles at the north and south points on his horizon. As the sky apparently turned about these points, all the stars would appear to rise straight up in the east and set straight down in the west. For an observer at an arbitrary place in the Northern Hemisphere (for example, in Greece) the north celestial pole would appear at a point between the zenith and the north point on his horizon, its location depending on his relative distances from the equator and North Pole of the earth (see Chapter 5). The stars that were not always above the horizon would rise at an oblique angle in the east, arc across the sky in a slanting path, and set obliquely in the west.

RISING AND SETTING OF THE SUN

The sun is always present at some position on the celestial sphere. When the apparent rotation of the sphere carries the sun above the horizon, the brilliant sunlight scattered about by the molecules of the earth's atmosphere produces the blue sky that hides the stars that are also above the horizon. The early Greeks were aware that the stars were there during the day as well as at night.

ANNUAL MOTION OF THE SUN

The Greeks were also aware, as were the Chinese, Babylonians, and Egyptians before them, that the sun gradually changes its position on the celestial sphere, moving each day about 1° to the east among the stars. Of course, the daily westward rotation of the celestial sphere (or eastward rotation of the earth) carries the sun, like everything else in the heavens, to the west across the sky. Each day, however, the sun rises, on the average, about 4 minutes later with respect to the stars; the celestial sphere (or earth) must make just a bit more than one complete rotation to bring the sun up again. The sun, in other words, has an independent motion of its own in the sky, quite apart from the daily apparent rotation of the celestial sphere.

In the course of 1 year the sun completes a circuit of the celestial sphere. The early peoples mapped the sun's eastward journey among the

FIG. 2-2   Time exposure showing star trails in the region of the north celestial pole. The bright trail near the center was made by Polaris (the North Star). *(Yerkes Observatory.)*

stars. This apparent path of the sun is called the *ecliptic* (because eclipses can occur only when the moon is on or near it — see Chapter 6). The sun's motion on the ecliptic is in fact merely an illusion produced by another motion of the earth — its annual revolution about the sun. As we look at the sun from different places in our orbit, we see it projected against different stars in the background, or we would, at least, if we could see the stars in the daytime; in practice, we must deduce what stars lie behind and beyond the sun by observing the stars visible in the opposite direction at night. After a year, when we have completed one trip around the sun, it has apparently completed one circuit of the sky along the ecliptic. We have an analogous experience if we walk around a campfire at night; we see the flames appear successively in front of each of the people seated about the fire.

It was also noted by the ancients that the ecliptic does not lie in a plane perpendicular to the line between the celestial poles, but is inclined (today) at an angle of about 23½° to that plane. This angle is called the *obliquity* of the ecliptic, and was measured surprisingly accurately by several ancient observers. The obliquity of the ecliptic, as we shall see, is responsible for the seasons (Chapter 5) and also for the invariable tilt in the axes of terrestrial globe maps.

FIXED AND WANDERING STARS

The sun is not the only moving object among the stars. The moon and each of the five planets visible to the unaided eye — Mercury, Venus, Mars, Jupiter, and Saturn — change their positions in the sky from day to day. The moon, being the earth's nearest celestial neighbor, has the fastest apparent motion; it completes a trip around the sky in about 1 month. We are referring now to the independent motions of these objects among the stars; these motions are superimposed upon the daily rotation of the celestial sphere. The ancient

Greeks distinguished between what they called the *fixed stars*, the real stars that appeared to maintain fixed patterns among themselves throughout many generations, and the *wandering stars* or *planets.* The word *planet* means "wanderer." Today, we do not regard the sun and moon as planets, but the Greeks applied the term to all seven of the moving objects in the sky. Because the seven "planets" were of great astrological significance, much of ancient astronomy was devoted to observing and predicting their motions. In fact, they give us the names for the seven days of our week; Sunday is the sun's day, Monday the moon's day, and Saturday is Saturn's day. We have only to look at the names of the other days of the week in the Romance Languages to see that they are named for the remaining planets.

THE ZODIAC

The individual paths of the moon and planets in the sky all lie close to the ecliptic, although not exactly on it. The reason is that the paths of the planets about the sun, and of the moon about the earth, are all in nearly the same plane, as if they were marbles rolling about on the top of a table. The planets and moon are always found in the sky within a narrow belt centered on the ecliptic, called the *zodiac* (the "zone of the animals"). The zodiac was divided into twelve parts or *signs:* Aries, Taurus, Gemini, Cancer, Leo, Virgo, Libra, Scorpio, Sagittarius, Capricornus, Aquarius, and Pisces. The apparent motions of the planets in the sky result from a combination of their actual motions and the motion of the earth about the sun, and consequently they are somewhat complex.

CONSTELLATIONS

The backdrop for the motions of the "wanderers" in the sky is the canopy of stars themselves. Like the Chinese and the Egyptians, the Greeks had divided the sky into *constellations,* apparent configurations of stars. Modern astronomers still make use of these constellations to denote approximate locations in the sky, much as geographers use political areas to denote the locations of places on the earth. The boundaries between the modern constellations are imaginary lines in the sky running north-south and east-west, so that every point in the sky falls in one constellation or another. The *signs* of the zodiac, incidentally, are evenly spaced regions which do *not* correspond to the constellations bearing the same names. There was a rough correspondence some 2000 years ago, but that has long since changed due to precession (Chapter 4). Many of the 88 recognized constellations are of Greek origin and bear names which are Latin translations of those given them by the Greeks. Today, the lay person is often puzzled because the constellations seldom resemble the people or animals for which they were named. In all likelihood the Greeks themselves did not name groupings of stars because they resembled actual people or objects, but rather named sections of the sky in *honor* of the characters in their mythology, and then fitted the configurations of stars to the animals and people as best they could.

(b) The Earliest Astronomers

The earliest Greek scientists were the Ionians, who lived in what is now Asia Minor.

Pythagoras (died about 497 B.C.) was originally an Ionian, but he later founded a school of his own in southern Italy. He pictured a series of concentric spheres, in which each of the seven moving objects — the planets, the sun, and the moon — was carried by a separate sphere from the one that carried the stars, so that the motions of the planets resulted from independent rotations of the different spheres about the earth. The friction between them gave rise to harmonious sounds, the *music of the spheres,* which only the most gifted ear could hear.

Pythagoras also believed that the earth, moon, and other heavenly bodies are spherical. It is doubtful that he had a sound reason for this belief, but it may have stemmed from the realization that the moon shines only by reflected sunlight, and that the moon's sphericity is indicated by the curved shape of the *terminator,* the demarcation line between its illuminated and dark portions. If he had so reasoned that the moon is round, the sphericity of the earth might have seemed to follow by analogy. The belief that the earth is round never disappeared from Greek thought.

Another member of the Pythagorean school was Philolaus, who lived in the following century. He was the first, as far as we know, to introduce the concept that the earth is in motion. Apparently he held that it is too base to occupy the center of the universe, and assigned a *central fire* to that position. About the fire revolved the earth and other planets. The earth and seven planets or "wandering stars" made eight moving objects, and the sphere of "fixed" stars made a ninth object. Philolaus, however, as a confirmed Pythagorean, believed 10 to be the most perfect number because it is the sum of 1, 2, 3, and 4 (and perhaps also because he had 10 fingers); therefore, he believed there must be another body. This, he proposed, was a *counter earth,* or *antichthon,* which revolved around the central fire exactly between it and the earth — thus hiding the fire from view from any place on earth. (The central fire itself seems not to have been counted in the numbering of celestial objects.) The period of revolution of earth and antichthon was 1 day, and the earth rotated as it revolved about the fire so as to keep Greece always turned away from it and the antichthon. Philolaus regarded the celestial sphere as motionless and its apparent rotation as the result of the revolution and rotation of the earth. He proposed that the sun, moon, and planets moved in their respective spheres outside the orbit of the earth. It was an imaginative concept based entirely on fancy and superstition and cannot be regarded as a forerunner of the heliocentric theory.

Nevertheless, the concept of a moving earth

had been introduced, although in a completely erroneous manner. It was a bold idea that may have had some influence on later Greek thought. Other Greek philosophers of the sixth to fourth centuries B.C. who are said to have believed in a moving earth are Hicetas, Heracleides, and Ecphantus. Centuries later Copernicus, in his *De Revolutionibus,* quoted the Pythagoreans as authorities for his own doctrines.

(c) "To Save the Phenomena"

Some Greek philosophers invented complicated schemes to describe the apparent motions of the planets. Eudoxus of Cnidus (408–355? B.C), for example, represented the motions of celestial bodies by a complex combination of rotating spheres pivoted on each other. In their invention of cosmological schemes, the Greeks did not always necessarily attempt to describe what they regarded as reality. A knowledge of reality, in the eyes of many of them, was available only to the gods; at any rate, mortal man could hardly expect to understand the real nature of the universe. Rather, Greek philosophers were often trying to find a scheme — a model — that would *describe the phenomena* and would predict events (eclipses, configurations of the planets, and so on). It is difficult to say to what extent this philosophy applies to all the Greek astronomical theories. Quite likely, however, the spheres of Eudoxus were intended as a mere mathematical representation of the motions of the planets. It was a scheme that "saved the phenomena" better than ones before it, and in this respect it was successful. The epicycles of Ptolemy, developed later, may similarly be regarded as mathematical representations not intended to describe reality.

Modern science does no more. The laws of nature "discovered" by science are merely mathematical or mechanical models that describe *how* nature behaves, not *why,* nor what nature "actually" is. Science strives to find representations that accurately describe nature, not absolute truths. This fact distinguishes science from religion.

(d) Aristotle

Aristotle (384–322 B.C.), most famous of the Greek philosophers, wrote encyclopedic treatises on nearly every field of human endeavor. Although his writings well summarized the total of Greek knowledge to that date, the greatest of the Greek accomplishments in the fields of physics and astronomy were yet to come. Aristotle, however, was accepted as the ultimate authority during the medieval period, and his views were upheld by authority even much later, when research had begun to make them untenable.

THE MOON'S PHASES

In spite of their errors, the writings of Aristotle contain several discussions that represent a substantial advance. First, we find among them clear and correct explanations of the phases of the moon and of eclipses, although these phenomena were almost certainly understood at an earlier date (the explanation is often attributed to Anaxagoras, sometime before 430 B.C.).

Aristotle recognized that the progression of the moon's phases — that is, its changing shape during the month — results from the fact that the moon is not itself luminous but is illuminated by sunlight. Because of its sphericity, only half of the moon is illuminated, that is, having daylight, at one time — the half turned toward the sun. The apparent shape of the moon in the sky depends simply on how much of its daylight hemisphere is turned to our view.

It was generally accepted by Aristotle and earlier Greek astronomers that the sun is more distant than the moon. This was surmised from the sun's slower apparent motion among the stars on the celestial sphere and also from the fact that the moon occasionally passes exactly between the earth and sun and temporarily hides the sun from view (*solar eclipse*). Thus when the moon is in the same general direction from earth as the sun (position *A* in

Figure 2-3), its daylight side is turned away from the earth. Because its night side — the side turned toward us — is dark and invisible, we do not see the moon in that position. The phase of the moon is then *new*. (Perhaps it would seem more reasonable to call it "no moon" instead of "new moon," for we do not see any moon at all.) To appear silhouetted in front of the sun, producing a solar eclipse, the moon must be at the new phase and must also lie on the line joining the earth and sun (see Chapter 6). A solar eclipse does not occur at every new moon because the plane of the moon's orbit is inclined slightly to the plane of the ecliptic, so that the new moon usually lies above or below the earth-sun line.

A few days after new moon, the moon reaches position *B*, and from the earth we see a small part of its daylight hemisphere. The illuminated crescent increases in size on successive days as the moon moves farther and farther around the sky away from the direction of the sun. During these days the moon is in the *waxing crescent* phase. About a week after new moon, the moon is one quarter of the way around the sky from the sun (position *C*) and is at the *first quarter* phase. Here the line from the earth to the moon is at right angles to the line from the earth to the sun and half of the moon's daylight side is visible — it appears as a half moon. (The moon is seen as half full at this point because the sun is very much farther away than the moon, and the sun's rays that illuminate the moon and earth are essentially parallel.)

During the week after the first quarter phase we see more and more of the moon's illuminated hemisphere, and the moon is in the *waxing gibbous* phase (position *D*). Finally, about 2 weeks after new moon, the moon (at *E*) and the sun are opposite each other in the sky; the side of the moon turned toward the sun is also turned toward the earth; we have *full moon*. During the next 2 weeks the moon goes through the same phases again in reverse order — through *waning gibbous, third* (or *last*) *quarter,* and *waning crescent.* Occasionally the full moon passes through the earth's shadow, which of course extends outward in space in the direction opposite the sun. This is a *lunar eclipse*.

(If the student finds difficulty in picturing the phases of the moon from this verbal account, he should try a simple experiment: He should stand about 6 feet in front of a bright electric light outdoors at night and hold in his hand a small round object such as a tennis ball or an orange. If the object is then viewed from various sides, the portions of its illuminated hemisphere that are visible will represent the analogous phases of the moon.)

The lunar phases are thus easily understood. Nevertheless, this understanding was a notable accomplishment for observers at the time of Aristotle and before, who had no a priori knowledge

that the moon was spherical, that it was not self-luminous, much less that it derived its light from the sun, and who had no original information whatsoever regarding the relative distances of the sun and moon.

THE SPHERICAL SHAPE OF THE EARTH

A second important topic discussed by Aristotle is the shape of the earth. He cited two convincing arguments for the earth's sphericity. First is the fact that during a lunar eclipse as the moon enters or emerges from the earth's shadow, the shape of the shadow seen on the moon is always round (see Figure 2-4). Only a spherical object always produces a round shadow. If the earth were a disk, for example, there would be some occasions when the

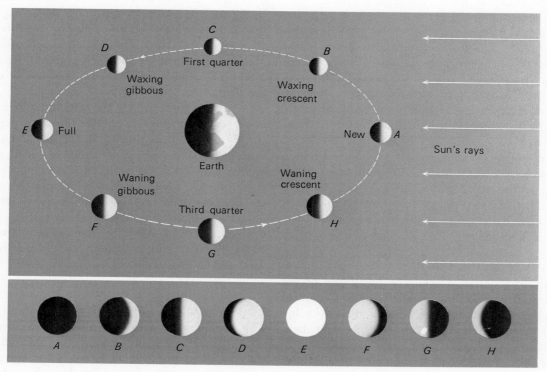

FIG. 2-3 Phases of the moon. The moon's orbit is viewed obliquely. Below, the appearance of the moon as seen from the earth is shown for several positions.

FIG. 2-4　Partially eclipsed moon moving out of the earth's shadow.
Note the curved shape of the shadow. *(Yerkes Observatory.)*

sunlight would be striking the disk edge on, and the shadow on the moon would be a line.

As a second argument, Aristotle explained that northbound travelers observed hitherto invisible stars to appear above the northern horizon and other stars to disappear behind the southern horizon. Southbound travelers observed the opposite effect. The only possible explanation is that the travelers' horizons had tipped to the north or south, respectively, which indicates that they must have moved over a curved surface of the earth. As a third piece of evidence that the earth is round, Aristotle mentioned that elephants had been observed to the east in India and also to the west in Morocco; evidently, those two places must not be far apart! He also listed a number of other proofs of an even less objective nature.

THE MOTION OF THE EARTH

Finally, Aristotle considered the possibility of the motion of the earth. The apparent daily motion of the sky, he pointed out, can be explained by a hypothesis of the rotation of either the celestial sphere or the earth. He rejected the latter explanation.

He also considered the possibility that the earth revolves about the sun rather than the sun

about the earth. He discarded this *heliocentric* hypothesis in the light of an argument that has been used many times since. Aristotle explained that if the earth moved about the sun we would be observing the stars from successively different places along our orbit and their apparent directions in the sky would then change continually during the year.

Any apparent shift in the direction of an object as a result of motion of the observer is called *parallax*. An annual shifting in the apparent directions of the stars that results from the earth's orbital motion is called *stellar parallax*. For the nearer stars it is observable with modern telescopes (see Chapter 13), but it is impossible to measure with the naked eye because of the great distances of even the nearest stars. Aristotle, thus, was unaware of stellar parallax and incorrectly concluded that the earth is stationary.

2.3 LATER GREEK ASTRONOMY

The early Greeks, as we have seen, were aware of, and to some extent understood, the phenomena of the sky. Real progress in Greek astronomy, however, waited until after Aristotle, when the first quantitative astronomical measurements were made. We turn now to the school of astronomers centered in Alexandria, where Greek science attained its greatest heights.

(a) Aristarchus of Samos

Aristarchus of Samos (310–230? B.C.), the first famous astronomer of the Alexandrian school, devised a most ingenious method to find the relative distances from the earth to the sun and moon. He assumed that the moon goes about the earth in a perfectly circular orbit at constant speed, and that the sun, although more distant from the earth than the moon, is near enough so that its rays travel along diverging paths to different parts of the moon's orbit. As it happens, these assumptions are incorrect. The moon's orbit is not a circle, but an ellipse — a slightly "flattened" circle — and both the moon's distance and speed vary. Furthermore, the sun is about 400 times as distant as the moon, and thus light from the sun strikes all parts of the moon's orbit along virtually parallel lines. At the

time, however, Aristarchus' assumptions seemed reasonable or at least compatible with any observations yet made.

Aristarchus then reasoned as follows: The moon appears exactly *half full* (first and last quarters) when the terminator — the line dividing the light and dark halves — is a perfectly straight line as viewed from the earth. But the moon is spherical, and the terminator, being a line upon its surface, must be curved. Thus, the only way it can appear straight is for us to view it exactly edge on. That is, the plane of the terminator must contain the line of sight from the earth to the center of the moon. If that is true, the line from the moon to the earth must be at right angles to the line from the moon to the sun. In Figure 2-5 these right angles are *EMS* and *EM′S*. Now we see that because the sun is not infinitely far away, by Aristarchus' assumption, the points *M′*, *E*, and *M* do not lie along a straight line. Hence the moon, moving at a uniform rate, should require a shorter time to go from *M′* to *M* than from *M* to *M′*. The difference between

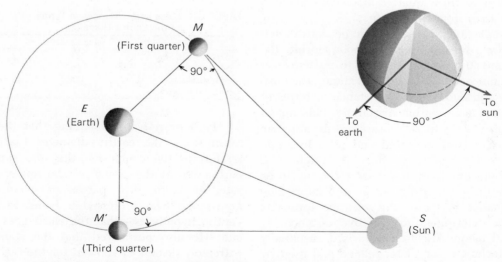

FIG. 2-5 Aristarchus's method of measuring the relative distances of the sun and moon.

these intervals from third quarter to first quarter moon and from first quarter to third quarter determines the angle $M'EM$. For example, if the periods were equal, $M'EM$ would be a straight line, and the sun would have to be infinitely far away; if the period from M to M' were, say, twice that from M' to M, the angle $M'EM$ would be a third of a circle, or 120°.

Aristarchus then attempted to observe the exact instants of first and third quarter moons and determine the difference between the two intervals. Even with our modern equipment of the mid-twentieth century we could not observe the exact instants of quarter moon with sufficient accuracy to measure the angle $M'EM$, because of the sun's great distance. At the time Aristarchus made his observations, however, the moon's elliptical orbit may have been so oriented with respect to the sun that the distance the moon had to move between the first and third quarter phases, combined with the effects of the moon's variable speed, resulted in an observable difference between the periods. In any case, Aristarchus determined that the interval from first to third quarter moon was about 1 day longer than from third to first.

The angle $M'EM$ could then be constructed inside a circle representing the moon's orbit; the lines MS and $M'S$, drawn tangent to the circle at M and M', intersect as S, thus determining the position of the sun and hence its distance in terms of the size of the moon's orbit. Aristarchus determined that the sun was 18 to 20 times more distant than the moon. This was accepted and used by other investigators for centuries after. Although it is about 20 times too small, it did show the sun to be much more distant than the moon, and its method of determination was a worthy scientific approach.

Having determined the relative distances to the sun and moon, Aristarchus devised an equally ingenious technique, one later refined and used by Hipparchus, to determine the relative sizes of the sun, moon, and earth. As Aristotle had noted, the shape of the earth's shadow on the moon during a lunar eclipse is curved. During such an eclipse, from the time it took for the moon to cross the earth's shadow, Aristarchus estimated that the shadow, out there at the distance of the moon, is eight thirds the size of the moon itself.

The relative distances of the moon and sun, the (approximately) equal angular sizes of those objects, and the relative sizes of the moon and earth's shadow are enough data to construct geometrically a scale drawing of the earth, moon, and sun. On such a drawing, we need only measure with a ruler to obtain their relative sizes and distances. Although the data obtained and used by Aristarchus were inaccurate and his results somewhat far from the truth, at least they were better than any available before; indeed, they were the first objective measures of astronomical dimensions, and they were based on logical reasoning. The approximate values found by Aristarchus, and the actual values, all in units of the earth's diameter, are given in Table 2.1.

TABLE 2.1 Sizes and Distances in Terms of the Earth's Diameter

	ARISTARCHUS	MODERN
Moon's distance	10	30
Moon's diameter	$\frac{1}{3}$	0.272
Sun's distance	200	11,700
Sun's diameter	7	109

Perhaps it was his finding that the sun was seven times the earth's diameter that led Aristarchus to the conclusion that the sun, not the earth, was at the center of the universe. At any rate, he is the first person of whom we have knowledge who professed a belief in the heliocentric hypothesis — that the earth goes about the sun. He also postulated that the stars must be extremely distant to account for the fact that their parallaxes could not be observed.

Jupiter. *(Lick Observatory.)*

Saturn. *(Lick Observatory.)*

Mars. *(Lick Observatory.)*

Orion Nebula. *(Hale Observatories).*

(b) Measurement of the Earth by Eratosthenes

Aristarchus had measured some astronomical dimensions but only in terms of the size of the earth. The latter was not measured by him. The first fairly accurate determination of the earth's diameter was made by Eratosthenes (276–195 or 196 B.C.), who was another astronomer of the Alexandrian school.

To appreciate Eratosthenes' technique for measuring the earth, which is in principle the same as many modern methods, we must understand that the sun is so distant from the earth compared to its size, even by Aristarchus' estimate, that the sun's rays intercepted by all parts of the earth approach it along sensibly parallel lines. Imagine a light source near the earth, say at position A in Figure 2-6. Its rays strike different parts of the earth along diverging paths. From a light source at B, or at C, still farther away, the angle between rays that strike extreme parts of the earth is smaller. The more distant the source, the smaller the angle between the rays. For a source *infinitely* distant, the rays travel along parallel paths. The sun is not, of course, infinitely far away, but light rays striking the earth from a point on the sun diverge from each other by at most an angle of less than ⅓ of a minute of arc ('), far too small to be observed with the unaided eye. As a consequence, if people all over the earth who could see the sun were to point at it, their fingers would all be pointing in the same direction — they would all be parallel to each other.

The concept that rays of light from the sun, planets, and stars approach the earth along parallel lines is vital to the art of celestial navigation — the determination of position at sea.

Eratosthenes noticed that at Syene, Egypt (vicinity of the modern Aswân), on the first day of summer, sunlight struck the bottom of a vertical well at noon, which indicated that Syene was on a direct line from the center of the earth to the sun. At the corresponding time and date in Alexandria, 5000 stadia north of Syene (the *stadium* was a Greek unit of length), he observed that the sun was not directly overhead but slightly south of the zenith, so that its rays made an angle with the vertical equal to 1/50 of a circle (about 7°). Yet the sun's rays striking the two cities are parallel to each other. Therefore (see Figure 2-7), Alexandria must be 1/50 of the earth's circumference north of Syene, and the earth's circumference must be 50 × 5000, or 250,000, stadia. The figure was later revised to 252,000, so that each degree on the earth's surface would have exactly 700 stadia.

It is not possible to evaluate precisely the accuracy of Eratosthenes' solution because there is doubt as to which of the various kinds of Greek stadia he used. If it was the common Olympic stadium, his result was about 20 percent too large. According to another interpretation, he used a stadium equal to about ¹⁄₁₀ mi, in which case his figure was within 1 percent of the correct value of 24,900

FIG. 2-6   The more distant an object, the more nearly parallel are the rays of light coming from it.

FIG. 2-7   Eratosthenes' method of measuring the size of the earth.

mi. The diameter of the earth is found from the circumference, of course, by dividing the latter by π.

(c) Hipparchus

The greatest astronomer of pre-Christian antiquity was Hipparchus, who was born in Nicaea in Bithynia. The dates of his life are not accurately known, but he carried out his work at Rhodes, and possibly also at Alexandria, in the period 160–127 B.C. Hipparchus either invented, or at least highly developed, trigonometry, which makes it possible to solve for the elements of a triangle without geometrical construction — a process that is not feasible with the long, "skinny" triangles encountered in astronomical applications. He also made extensive and systematic observations with the greatest accuracy that his instruments would permit, and he made systematic and critical use of old observations, comparing them with later observations of his own to detect changes too slow to observe in a single lifetime. Many of his own observations, especially of the planets, were designed not for his own analysis but for use by astronomers in future generations. Finally, he devised a geometrical representation, involving *eccentrics* and *epicycles,* that described the motions of the sun and moon more precisely than had any previous scheme.

HIPPARCHUS' STAR CATALOGUE

Hipparchus erected an observatory on the island of Rhodes and built instruments with which he measured as accurately as possible the directions of objects in the sky. He compiled a star catalogue of about 850 entries. He designated for each star its celestial coordinates, that is, quantities analogous to latitude and longitude that specify its position (direction) in the sky. He also divided the stars according to their apparent brightness into six categories, or *magnitudes,* and specified the magnitude of each star. In the course of his observations of the stars, and in comparing his data with older observations, he made one of his most remarkable discoveries — that the position in the sky of the north celestial pole had altered over the previous century and a half. Hipparchus correctly deduced that the direction of the axis about which the celestial sphere appears to rotate continually changes. The real explanation for the phenomenon is that the direction of the earth's rotational axis changes slowly because of the gravitational influence of the moon and sun, much as a top's axis describes a conical path as the earth's gravitation tries to tumble the top over. This variation of the earth's axis, called *precession,* requires about 26,000 years for one cycle (Chapter 4).

OTHER MEASUREMENTS

Hipparchus, refining the technique first applied by Aristarchus, also obtained a good estimate of the moon's size and distance. Unlike Aristarchus, he used the correct value of ½° for the angular diameters of the sun and moon. He found the moon's distance to be 29½ times the earth's diameter; the correct number is 30.

He determined the length of the year to within 6 minutes, and even analyzed his possible errors, estimating that he could not be farther off than about 15 minutes. He also carefully observed the motions of the sun, moon, and planets, and found a method by which he could predict the position of the sun on any date of the year with an accuracy equal to the best observations and the position of the moon with somewhat less accuracy. His work made possible the reliable prediction of eclipses, and with the information he left, any astronomer thereafter could predict a lunar eclipse to within an hour or so.

(d) Ptolemy

The last great Greek astronomer of antiquity was Claudius Ptolemy (or Ptolemeus), who flourished about 140 A.D. He compiled a series of 13 volumes on astronomy known as the *Almagest*. All of the *Almagest* does not deal with Ptolemy's own work, for it includes a compilation of the astronomical achievements of the past, principally of Hippar-chus. In fact, it is our main source of information about Greek astronomy. The *Almagest* also contains accounts of the contributions of Ptolemy himself.

One of Ptolemy's accomplishments was the measurement of the distance to the moon by a technique essentially identical to one used today. The method he used, the principle of which is illustrated in Figure 2-8, makes use of the moon's parallax. Suppose we could observe the moon directly overhead. We would have to be, then, at position A on the earth, on a line between the center of the earth E, and the center of the moon M. Suppose that at the same time someone else at position B were to observe the angle ZBM between the moon's direction and the point directly over his head, Z. The angle MBE would then be determined in the triangle MBE (it is $180° -$ angle ZBM). The distance from A to B determines the angle BEM. For example, if A is one twelfth of the way around the earth from B, the angle BEM is 30°. The side BE is of course the radius of the earth. We therefore know two angles and an included side in the triangle MBE. It is now possible to determine, either by trigonometry or geometrical construction, the distance EM between the centers of the earth and moon.

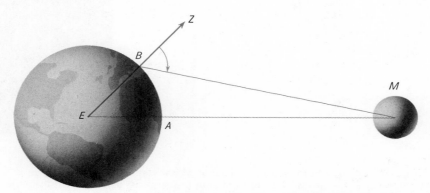

FIG. 2-8   Ptolemy's method of measuring the distance to the moon.

This is an example of the principle of *surveying*. We shall discuss it further later.

In practice, we do not need another observer at *B*, for the rotation of the earth will carry us over there in a few hours anyway, and we can observe the angle *ZBM* then. We shall have to correct, however, for the motion of the moon in its orbit during the interval between our two observations; the moon's motion being known, the correction is a detail easily accomplished. Using the principle described, Ptolemy determined the moon's distance to be 59 times the radius of the earth or $29\frac{1}{2}$ times the earth's diameter — very nearly the correct value.

PTOLEMY'S SCHEME OF COSMOLOGY

Ptolemy's most important original contribution was a geometrical representation of the solar system that predicted the motions of the planets with considerable accuracy. Hipparchus, having determined by observation that earlier theories of the motions of the planets did not fit their actual behavior, and not having enough data on hand to solve the problem himself, instead massed observational material for posterity to use. Ptolemy supplemented the material with a few observations of his own and with it produced a cosmological hypothesis that endured until the time of Copernicus.

The complicating factor in the analysis of the planetary motions is that their apparent wanderings in the sky result from the combination of their own motions and the earth's orbital revolution. Notice, in Figure 2-9, the orbit of the earth and the orbit of a hypothetical planet farther from the sun than the earth. The earth travels around the sun in the same direction as the planet and in

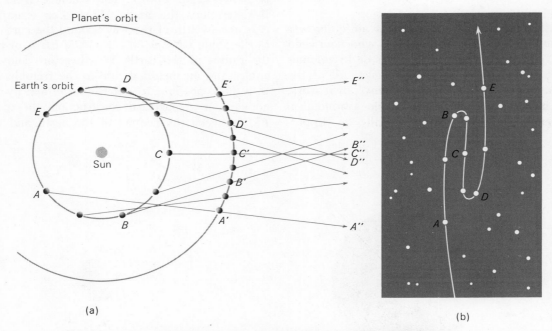

(a)

(b)

FIG. 2-9   Retrograde motion of a superior planet. (a) Actual positions of the planet and earth; (b) the apparent path of the planet as seen from the earth, against the background of distant stars.

nearly the same plane, but has a higher orbital speed. Consequently, it periodically overtakes the planet, like a faster race car on the inside track. The apparent directions of the planet, seen from the earth, are shown at successive intervals of time along lines *AA'A''*, *BB'B''*, and so on. In the right side of the figure we see the resultant apparent path of the planet among the stars. From positions *B* to *D*, as the earth passes the planet, it appears to drift backward, to the *west* in the sky, even though it is actually moving to the *east*. Similarly, a slowly moving car appears to drift backward with respect to the distant scenery when we pass it in a faster-moving car. As the earth rounds its orbit toward position *E*, the planet again takes up its usual eastward motion in the sky. The temporary westward motion of a planet as the earth swings between it and the sun is called *retrograde* motion. (During and after its retrograde motion the planet's apparent path in the sky does not trace exactly over itself because of the slight inclinations between the orbits of the earth and other planets. Thus, the retrograde path is shown as an open loop in Figure 2-9.) Obviously, it is difficult to find an explanation for retrograde motion on the hypothesis that the planet is revolving about the earth.

Ptolemy solved the problem by having a planet *P* (Figure 2-10) move on the circumference of a small portable circle called an *epicycle,* whose center, in turn, revolves about the earth in a larger circle called the *deferent.* When the planet is at position *x,* it is moving in its epicyclic orbit in the same direction as the point *C* moves about the earth, and the planet appears to be moving eastward. When the planet is at *y,* however, its epicyclic motion is in the opposite direction to the motion of *C.* By choosing the right combination of speeds and distances, Ptolemy succeeded in having the planet moving westward at the right speed at *y* and for the correct interval of time. However, because the planets, like the earth, travel about

the sun in elliptical orbits, their actual behavior cannot be represented accurately by so simple a scheme of uniform circular motions. Consequently, Ptolemy made the deferent an *eccentric* circle, centered not on the earth, but slightly away from the earth at *A.* Furthermore, he had the center of the epicycle, *C,* move at a uniform angular rate, not around *A,* or *E,* but a point *B,* called the *equant,* on the opposite side of *A* from the earth.

It is a tribute to the genius of Ptolemy as a mathematician that he was able to conceive such a complex system to account successfully for the observations. His hypothesis, with some modifications, was accepted as absolute authority throughout the Middle Ages, until it finally gave way to the heliocentric theory in the seventeenth century. In the *Almagest,* however, Ptolemy made no claim that his cosmological model described reality. He intended his scheme rather as a mathematical representation to predict the positions of the planets at any time. Modern astronomers do the same thing with algebraic formulas. Our modern mathematical methods were not available to Ptolemy; he had to use geometry.

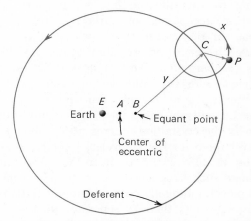

FIG. 2-10   Ptolemy's system of deferent, epicycle, eccentric, and equant.

2.4 INDIAN AND ARABIAN ASTRONOMY

In the next 13 centuries after Ptolemy the only significant astronomical investigations were carried out by the Hindus and Arabs. References have been made to Indian observations as far back as 1200 B.C., but there was apparently no substantial development as a science in India until about 300 A.D., after the influence of Greek and Babylonian science had reached there. In the period following, the Hindus made some creditable observations, among them the determination of the moon's distance as $64\frac{1}{2}$ earth radii (the actual figure is nearer 60). Their most important contribution to the science of our culture is very likely our system of numbers and place counting.

The Arabs brought the Hindu system of numbers to Europe and developed algebra. They also had access to some of the records of the Greek astronomers. Their greatest contribution was probably to provide some continuity between ancient astronomy, on the one hand, and the development of modern astronomy in the Renaissance.

2.5 MEDIEVAL ASTRONOMY

In medieval Europe no new astronomical investigations of importance were made. Rather than turning to scientific inquiry, the medieval mind rested in acceptance of authority and absolute dogma. In the realm of cosmology, this dogma involved belief in the crystalline spheres of Pythagoras (as perpetuated by Aristotle). To reconcile this view with the epicycles necessary to Ptolemaic technical astronomy, Purbach (1423–1461) revived a scheme, going back to Ptolemy himself, whereby the crystalline spheres were hollowed out to make room for the epicycles! Jerome Fracastor, writing in 1538, went even further. Desiring to eliminate such irregularities as eccentrics, equants, and epicycles altogether, he invented a system that required no less than 79 separate spheres.

Such was the state of affairs when Copernicus' *De Revolutionibus* was published in 1543.

FIG. 2-11   Nicolas Copernicus. *(Yerkes Observatory.)*

2.6 COPERNICUS

Nicolas Copernicus (also Coppernicus, or Koppernig, 1473–1543) was born in Thorn on the Vistula in Poland, although since then the region often has been in the domain of Prussia. His training was in law and medicine, but Copernicus' main interest was in astronomy and mathematics. By the time he had reached middle age, he was well known as an authority on astronomy.

Copernicus was not principally an observer; his forte was mathematics. His most important contribution to science was a critical reappraisal of the existing theories of cosmology and the development of a new model of the solar system. His unorthodox and heretical idea that the sun, not the earth, is the center of the solar system had become known by 1530, chiefly through a manuscript circulated by him and his friends.

His ideas were set forth in full detail in his great book, *De Revolutionibus,* published in the year of his death, 1543. Supervision over the publication of the book fell into the hands of one Osiander, a Lutheran preacher, who was probably responsible for the augmented title of the work — *De Revolutionibus Orbium Celestium* (On the Revolutions of the Celestial Spheres). Osiander wrote a preface, which he neglected to sign, expressing the (modern) view that science presented only an abstract mathematical hypothesis, and implying that the theory set forth in the book was only a convenient calculating scheme. The preface is almost certainly in contradiction to Copernicus' own feelings.

In *De Revolutionibus,* Copernicus sets forth certain postulates from which he derives his system of planetary motions. His postulates include the assumptions that the universe is spherical and that the motions of the heavenly bodies must be made up of combinations of uniform circular motions. We see, thus, that the work of Copernicus was not free from all preconceived notions.

(a) The Motion of the Earth

Copernicus argued that the apparent annual motion of the sun about the earth could be equally well represented by a motion of the earth about the sun. By a simple extension of the idea, Copernicus showed that the apparent rotation of the celestial sphere could be accounted for by assuming that the earth rotates about an axis and that the celestial sphere is fixed.

Contrary to popular belief, Copernicus did not *prove* that the earth is in motion, but he did offer substantial evidence to show that the hypothesis of a moving earth was simpler than that of a stationary one. To the objection that if the earth rotated about an axis it would fly into pieces, Copernicus answered that if such motion would tear the earth apart, the even faster motion (because of its greater size) of the celestial sphere required by the alternative hypothesis would be even more devastating to it.

(b) The Planetary Motions According to Copernicus

The most important concept set forth by Coper-

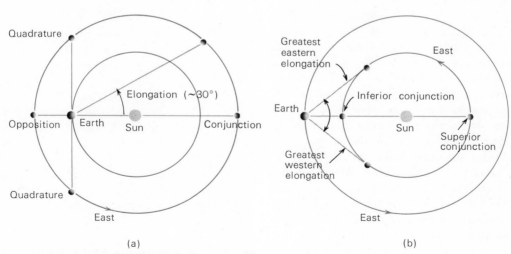

(a)   (b)

FIG. 2-12 (a) Configurations of a superior planet; (b) configurations of an inferior planet.

nicus in *De Revolutionibus* is that the earth is but one of six (then known) planets that revolve about the sun. Through this principle, he was able to work out the correct general picture of the solar system. He placed the planets, starting nearest the sun, in the order Mercury, Venus, Earth, Mars, Jupiter, and Saturn. Further, he deduced that the nearer a planet is to the sun, the greater is its orbital speed. Thus the retrograde motions of the planets (Section 2.3d) were easily understood without the necessity for epicycles. It will be helpful here to define a few terms that describe the positions of planets in their orbits. These are illustrated in Figure 2-12.

A *superior planet* is any planet whose orbit is larger than that of the earth, that is, a planet that is farther from the sun than the earth (Mars, Jupiter, and Saturn). An *inferior planet* is a planet closer to the sun than the earth (Venus and Mercury).

Every now and then, the earth passes between a superior planet and the sun. Then that planet appears in exactly the opposite direction in the sky from the sun — or at least as nearly opposite as is allowed by the slight differences of inclination among the planes of the orbits of the other planets and the earth. At such time, the planet rises at sunset, is above the horizon all night long, and sets at sunrise. We look one way to see the sun, and in the opposite direction to see the planet. The planet is then said to be in *opposition*.

On other occasions, a superior planet is on the other side of the sun from the earth. It is then in the same direction from the earth as the sun is, and of course is not visible. At such time, the planet is said to be in *conjunction*.

In between these extremes (but not halfway between), a superior planet may appear 90° away from the sun in the sky, so that a line from the earth to the sun makes a right angle with the line from the earth to the planet. Then the planet is said to be at *quadrature*. At quadrature, a planet rises or sets at either noon or midnight.

The angle formed at the earth between the earth-planet direction and the earth-sun direction is called the planet's *elongation*. In other words, the elongation of a planet is its angular distance from the sun as seen from the earth. At conjunction, a planet has an elongation of 0°, at opposition 180°, and at quadrature 90°.

An inferior planet can never be at opposition, for its orbit lies entirely within that of the earth. The greatest angular distance from the sun, on either the east or west side, that the inferior planet can attain is called its *greatest eastern elongation* or *greatest western elongation*.

When an inferior planet passes between the earth and sun, it is in the same direction from earth as the sun and is said to be in *inferior conjunction*. When it passes on the far side of the sun from the earth, and is again in the same direction as the sun, it is said to be at *superior conjunction*.

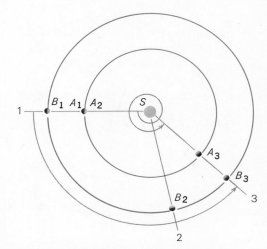

FIG. 2-13 Relation between the sidereal and synodic periods of a planet.

Copernicus recognized the distinction between the *sidereal period* of a planet — that is its actual period of revolution about the sun with respect to the fixed stars — and its *synodic period,* its apparent period of revolution about the sky with respect to the sun. The synodic period is also the time required for it to return to the same configuration, such as the time from opposition to opposition or from conjunction to conjunction.

Consider two planets, A and B, A moving faster in the smaller orbit (Figure 2-13). At position (1), planet A passes between B and the sun S. Planet B is at opposition as seen from A, and A is at an inferior conjunction as seen from B. When A has made one revolution about the sun and has returned to position (1), B has, in the meantime, moved on to position (2). In fact, A does not catch up with B until both planets reach position (3). Now planet A has gained one full lap on B. Planet A has revolved in its orbit through 360° *plus* the angle that B has described in traveling from position (1) to position (3) in its orbit. The time required for the faster-moving planet to gain a lap on the slower-moving one is the synodic period of one with respect to the other. If B is the earth and A an inferior planet, the synodic period of A is the time required for the inferior planet to gain a lap on the earth; if A is the earth and B a superior planet, the synodic period of B is the time for the earth to gain a lap on the superior planet.

What is observed from the earth is the synodic, not the sidereal period of a planet. By reasoning along the lines outlined in the last paragraph, however, Copernicus was able to deduce the sidereal periods of the planets from their synodic periods.

Copernicus was also able to find their relative distances from the sun. The problem is particularly simple for the inferior planets. When an inferior planet is at greatest elongation (Figure 2-14), the line of sight from the earth to the planet, EP, must be tangent to the orbit of the planet, and hence perpendicular to the line from the planet to the sun, PS. We have, therefore, a right triangle, EPS. The angle PES is observed (it is the greatest elongation) and the side ES is the earth's distance from the sun. The planet's distance from the sun can then be found, in terms of the earth's distance, by geometrical construction, or by trigonometric calculation.

The values obtained by Copernicus for the distances of the various planets from the sun, in units of the earth's distance, are summarized in

FIG. 2-14 Determination of the relative distance of an inferior planet from the sun.

Table 2.2. Also given are the values determined by modern measurement.

TABLE 2.2 Distances of Planets from the Sun

PLANET	COPERNICUS	MODERN
Mercury	0.38	0.387
Venus	0.72	0.723
Earth	1.00	1.00
Mars	1.52	1.52
Jupiter	5.22	5.20
Saturn	9.17	9.54

So far, we have discussed the Copernican theory as though Copernicus regarded the planets as having circular orbits centered on the sun. However, we recall that Hipparchus and Ptolemy had introduced, centuries ago, epicycles, eccentrics, and equants to account for those minor irregularities that arise because of deviations from uniform circular motion (actually because the true orbits of planets are ellipses). Copernicus rejected the equants of Ptolemy as unworthy of the perfection of heavenly bodies and instead introduced a system of eccentrics and small epicycles to take care of the irregularities. In all, his system required 34 circles: 4 for the moon, 3 for the earth, 7 for Mercury, and 5 each for the other planets.

His excellence as a mathematician enabled Copernicus to work out the details of his planetary system so that it accounted for the apparent motions of the sun and planets with fair accuracy. The Copernican system is certainly closer to our modern view of the solar system than those that preceded it. On the other hand, Copernicus' use of epicycles, apparently intended as a literal description of nature, and his acceptance of received principles lead us to wonder whether his system of logic was any closer to that of modern science than were those of Hipparchus and Ptolemy.

2.7 TYCHO BRAHE

Three years after the publication of *De Revolutionibus*, Tycho Brahe (1546–1601) was born of

FIG. 2-15 Tycho Brahe. *(Yerkes Observatory.)*

a family of Danish nobility. Tycho (as he is generally known) developed an early interest in astronomy and as a young man made significant astronomical observations.

In 1572 he observed a nova or "new star" (now believed to be a supernova — see Chapter 18) that rivaled the planet Venus in brilliance. Tycho observed the star for 16 months until it disappeared from naked-eye visibility. Now the moon exhibits a diurnal parallax, or apparent displacement in direction, because of the rotation of the earth, which constantly shifts our position of observation. The effect is the same whether we regard it as being caused by the earth's rotation or a rotation of the celestial sphere carrying the moon about us. Tycho, despite the most careful observations, was unable to detect any parallax of his nova and accordingly concluded that it must be more distant than the moon. This conclusion was of the utmost importance, for it showed that changes can occur in the celestial sphere, generally regarded as perfect and unchanging, apart from the regular motions of the planets.

The reputation of the young Tycho Brahe as an astronomer gained him the patronage of Frederick II, and in 1576 Tycho was able to establish a fine astronomical observatory on the Danish island of Hveen. The chief building of the observatory was named *Uraniborg*. The facilities at Hveen included a library, laboratory, living quarters, workshops, a printing press, and even a jail. There, for 20 years, Tycho and his assistants carried out the most complete and accurate astronomical observations yet made.

Unfortunately, Tycho was both arrogant and extravagant, and after Frederick II died, the new king, Christian IV, lost patience with the astronomer and eventually discontinued his support. Thus, in 1597 Tycho was forced to leave Denmark. He took up residence near Prague, taking with him some of his instruments and most of his records. There Tycho Brahe spent the remaining years of his life analyzing the data accumulated over 20 years of observation. In 1600, the year before his death, he secured the assistance of a most able young mathematician, Johannes Kepler.

(a) Tycho's Observations

Tycho, like others of his time and before him, believed that comets were luminous vapors in the earth's atmosphere. In 1577, however, a bright comet appeared for which he could observe no parallax. Tycho concluded that the comet was at least three times more distant than the moon, and guessed that it probably revolved about the sun, in contradiction to his earlier beliefs. Other comets

FIG. 2-16 Tycho Brahe at Uraniborg, observing with his great mural quadrant. *(Yerkes Observatory.)*

were observed by him or his students in 1580, 1582, 1585, 1590, 1593, and 1596.

Tycho is most famous for his very accurate observations of the positions of the stars and planets. With instruments of his own design he was able to make observations accurate to the very limit of vision with the naked eye. The positions of the nine fundamental stars in his excellent star catalogue were accurate in most cases to within 1'. Only in one case was he off by as much as 2', and this was a star whose position was distorted by atmospheric refraction (see Chapter 8).

Tycho's observations included a continuous record of the positions of the sun, moon, and planets. His daily observations of the sun, extending over years and comprising thousands of individual sightings, led to solar tables that were good to within 1′. He reevaluated nearly every astronomical constant and determined the length of the year to within 1 second. His extensive and precise observations of planetary positions enabled him to note that there were variations in the positions of the planets from those given in published tables, and he even noted regularities in the variations.

(b) Tycho's Cosmology

Tycho rejected the Copernican heliocentric hypothesis on what seemed at the time to be very sound grounds. First, he found it difficult to reconcile a moving earth with certain Biblical statements; nor could he even imagine an object as heavy and "sluggish" as the earth to be in motion. The fact that he could not detect a parallax for even a single star, moreover, meant that the stars would have to be enormously distant if the earth revolved around the sun.

Tycho did, however, suggest an original system of cosmology, although it was not worked out in full detail. He envisioned the earth in the center, with the sun revolving about the earth each year, and with the other planets revolving about the sun in the order Mercury, Venus, Mars, Jupiter, and Saturn.

Exercises

1. Which (if any) of the following can never appear at opposition? At conjunction? At quadrature?
   (a) Mars
   (b) A star
   (c) Sun
   (d) Earth
   (e) Jupiter
   (f) Moon
   (g) Venus
   (h) Mercury

2. Suppose you are on a strange planet, and observe, at night, that the stars do not rise and set, but circle parallel to the horizon. Now you walk in a constant direction for 8000 miles, and at your new location on the planet you find that all stars rise straight up in the east and set straight down in the west, perpendicularly to the horizon.
   (a) How could you determine the circumference of the planet without any further observations?
   (b) What evidence is there that the Greeks could have done what you suggest?
   (c) What *is* the circumference of the planet, in miles?

3. On March 13, there will be a total eclipse of the moon. About what time will the moon rise that night?

4. During the middle ages people believed that the planets went around the earth in epicyclic orbits. Suppose you had a good telescope and auxiliary equipment so that you could accurately measure the angular diameter of the apparent disk of a particular planet. How could you prove that the epicycle theory was incorrect?

5. Suppose you observed that an inferior planet never reached a greater elongation than 30°. Assume that the planet and earth both go around the sun in perfectly circular orbits. Explain how you could find the planet's distance from the sun in astronomical units, and give that distance.

6. Where on earth:

(a) are all stars at some time visible above the horizon?

(b) is only half the sky ever above the horizon?

7. Why can an eclipse of the sun never occur on the day following an eclipse of the moon?

8. Suppose Aristarchus had found that the interval from third quarter to first quarter moon was one week, and that from first quarter to third was two weeks. What then would he have found for the sun's distance in terms of the moon's?

9. Give some everyday examples of parallax.

10. Suppose Eratosthenes had found that at Alexandria at noon on the first day of summer the line to the sun makes an angle of 45° with the vertical. What then would he have found for the earth's circumference?

View to the northeast in a mosaic of photographs made with the television camera on Surveyor VII, from a position about 18 miles north of the crater Tycho. *(NASA.)*

CHAPTER 3

The Birth of
Modern Astronomy

We come now to the boundary between the old astronomy and the new. Brahe was of the older school; Galileo was certainly of the newer. Between them stood the man who was a little of each — half mystic philosopher, half modern scientist. That man was Kepler.

3.1 KEPLER

Johannes Kepler (1571–1630) was born in Weil-der-Stadt, Württemberg (southwestern Germany).

FIG. 3-1 Johannes Kepler. *(Yerkes Observatory.)*

He studied for a theological career, and attended college at Tübingen, where he learned the principles of the Copernican system. He became an early convert to the heliocentric hypothesis and defended it in arguments with his fellow scholars.

In 1594, because of his facility as a mathematician, he was offered a position teaching mathematics and astronomy at the high school at Gratz. As part of his duties at Gratz, he prepared almanacs that gave astronomical and astrological data. Eventually, however, the power of the Catholic church in Gratz grew to the point where Kepler, a Protestant, was forced to quit his post. Accordingly, he went to Prague to serve as an assistant to Tycho Brahe.

Tycho set Kepler to work trying to find a satisfactory theory of planetary motion —one that was compatible with the long series of observations made at Hveen. After Tycho's death, Kepler succeeded him as mathematician to the Emperor Rudolph and obtained possession of the majority of Tycho's records. Their study occupied most of Kepler's time for the next 25 years.

(a) The Investigation of Mars

Kepler's most detailed study was of Mars, for which the observational data were the most extensive. He published the first results of his work in 1609 in *The New Astronomy*, or *Commentaries on the Motions of Mars*. He had spent nearly a decade trying to fit various combinations of circular motion, including eccentrics and equants, to the observed motion of Mars, but without success. At one point he found a hypothesis that agreed with observations to within 8′ (about one quarter the diameter of the full moon), but he believed that Tycho's observations could not have been in error by even this small amount, and so, with characteristic integrity, discarded the hypothesis. Finally, Kepler tried to represent the orbit of Mars with an oval, and soon discovered that the orbit could be fitted extremely well by a curve known as an *ellipse*.

PROPERTIES OF THE ELLIPSE

Next to the circle, the ellipse is the simplest kind of closed curve. It belongs to a family of curves known as *conic sections* (Figure 3-2). A conic section is simply the curve of intersection between a hollow cone and a plane that cuts through it. If the plane is perpendicular to the axis of the cone (or parallel to its base), the intersection is a circle. If the plane is inclined at an arbitrary angle, but still cuts completely through the surface of the cone, the resulting curve is an ellipse. If the plane is parallel to a line in the surface of the cone, it never quite cuts all the way through the cone, and the curve of intersection is open at one end. Such a curve is called a *parabola*. If the plane is inclined at an even smaller angle to the axis of the cone, an open curve results that is called a *hyperbola*. The ellipse, then, ranges from a circle at one extreme to a parabola at the other. The parabola separates the family of ellipses from the family of hyperbolas.

An interesting and important property of an ellipse is that from *any point* on the curve, the sum of the distances to two points inside the ellipse, called the *foci* of the ellipse, is the same. This property suggests a simple way to draw an ellipse. The ends of a length of string are tied to two tacks pushed through a sheet of paper into a drawing board, so that the string is slack. If a pencil is then pushed against the string, so that the string is held taut, and then slid against the string around the tacks (Figure 3-3), the curve that results is an ellipse; at any point where the pencil may be, the sum of the distances from the

FIG. 3-2 Conic sections.

pencil to the two tacks is a constant length — the length of the string. The tacks, of course, are at the two foci of the ellipse.

The maximum diameter of the ellipse is called its *major axis*. Half this distance, that is, the distance from the center of the ellipse to one end, is the *semimajor axis*. The *size* of an ellipse depends on the length of the major axis. The *shape* of an ellipse depends on how close together the two foci are compared to the major axis. The ratio of the distance between the foci to the major axis

FIG. 3-3 Drawing an ellipse.

is called the *eccentricity* of the ellipse. If an ellipse is drawn as described above, the length of the major axis is the length of the string, and the eccentricity is the distance between the tacks divided by the length of the string (Figure 3-4). If the foci (or tacks) coincide, the ellipse is a circle; a circle is, then, an ellipse of eccentricity zero. Ellipses of various shapes are obtained by varying the spacing of the tacks (as long as they are not farther apart than the length of the string). If one tack is removed to an infinite distance, and if enough string is available, "our end" of the resulting, infinitely long ellipse is a parabola. A parabola has an eccentricity of 1. An ellipse is completely specified by its major axis and its eccentricity.

Kepler found that Mars has an orbit that is an ellipse and that the sun is at one focus (the other focus is empty). The eccentricity of the orbit of Mars is only about $\frac{1}{10}$; the orbit, drawn to scale, would be practically indistinguishable from a circle. It is a tribute to Tycho's observations and to Kepler's perseverance that he was able to determine that the orbit was an ellipse at all.†

THE VARYING SPEED OF MARS
Before he saw that the orbit of Mars could be

---

†Tycho did not die in Hveen.

represented accurately by an ellipse, Kepler had already investigated the manner in which the planet's orbital speed varied. After some calculation, he found that Mars speeds up as it comes closer to the sun and slows down as it pulls away from the sun. Kepler expressed this relation by imagining that the sun and Mars are connected by a straight, elastic line. As Mars travels in its elliptical orbit around the sun, in equal intervals of time the areas swept out in space by this imaginary line are always equal (Figure 3-5). This relation is commonly called the *law of areas*.

KEPLER'S FIRST TWO LAWS OF PLANETARY MOTION SUMMARIZED
We may summarize the most important contributions in *The New Astronomy*, or *Commentaries on the Motions of Mars*, by stating what are now known as Kepler's first two *laws of planetary motion:*

KEPLER'S FIRST LAW: *Each planet moves about the sun in an orbit that is an ellipse, with the sun at one focus of the ellipse.*

KEPLER'S SECOND LAW (THE LAW OF AREAS): *The straight line joining a planet and the sun sweeps out equal areas in space in equal intervals of time.*

At the time of publication of the *New Astronomy* (1609), Kepler appears to have demonstrated the validity of these two laws only for the case of Mars. However, he expressed the opinion that they held also for the other planets.

(b) The Harmony of the Worlds
Although an excellent mathematician, Kepler was also a mystic, and he indulged freely in wild speculation and the occult. In his endeavor to find an underlying harmony in nature, he constantly

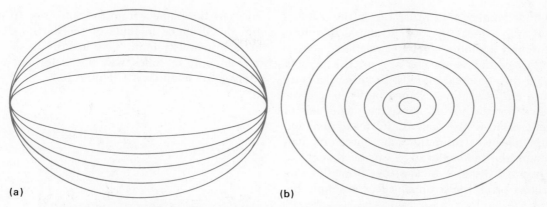

(a)   (b)

FIG. 3-4   (a) Ellipses of the same major axis but various eccentricities;
(b) ellipses of the same eccentricity but various major axes.

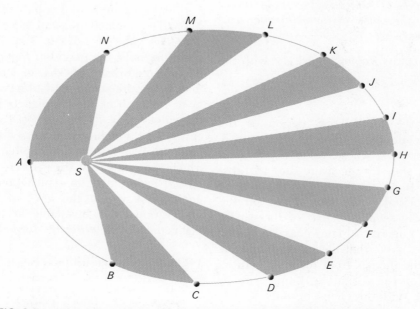

FIG. 3-5   Law of equal areas. A planet moves most rapidly on its elliptical
orbit when it is at position *A*, nearest the focus of the ellipse, *S*, where the sun
is. The planet's orbital speed varies in such a way that in equal intervals of
time it moves distances *AB*, *BC*, *CD*, and so on, so that regions swept out by
the line connecting it and the sun (shaded and clear zones) are always
the same in area.

searched for numerological relations in the celestial realm. It was a great personal triumph, therefore, that he found a simple algebraic relation between the lengths of the semimajor axes of the planets' orbits and their sidereal periods. Because planetary orbits are elliptical, the distance between a given planet and the sun varies. Now, the major axis of a planet's orbit is the sum of its maximum and minimum distances from the sun. Therefore, half of this sum, the semimajor axis, can be thought of as the *average* distance of a planet from the sun. In a circular orbit, the semimajor axis is simply the radius of the circle.

Kepler published his discovery in 1619 in *The Harmony of the Worlds*. The relation is now known as his *third*, or *harmonic, law*.

KEPLER'S THIRD LAW: *The squares of the sidereal periods of the planets are in direct proportion to the cubes of the semimajor axes of their orbits.*

It is simplest to express Kepler's third law with the algebraic equation.

$$P^2 = Ka^3,$$

where P represents the sidereal period of the planet, a is the semimajor axis of its orbit, and K is a numerical constant whose value depends on the kinds of units chosen to measure time and distance. It is convenient to choose for the unit of time the earth's period — the year — and for the unit of distance the semimajor axis of the earth's orbit. This latter quantity is known as the *astronomical unit* (AU). Note that the values found for the distances of the planets from the sun, both as determined by Copernicus and by Kepler, are in astronomical units. With this choice of units, $K = 1$, and Kepler's third law can be written

$$P^2 = a^3.$$

We see that to arrive at his third law it was not necessary for Kepler to know the *actual* dis-

tances of the planets from the sun (say, in miles), only the distance in units of the earth's distance, the astronomical unit. The length of the astronomical unit in miles was not determined accurately until later (see Chapter 13).

FIG. 3-6 Notes written by Kepler, representing the music "sung" by the planets (from his *Harmony of the Worlds*).

As an example of Kepler's third law, consider Mars. The semimajor axis of Mars' orbit, a, is 1.524 AU. The cube of 1.524 is 3.54. According to the above formula, the period of Mars, in years, should be the square root of 3.54, or 1.88 years, a result that is in agreement with observations. Table 3.1 gives for each of the six planets known to Kepler the values of a, P, a^3, and P^2. To the limit of accuracy of the data given, we see that Kepler's law holds exactly, except for Jupiter and Saturn, for which there are very slight discrepancies. Decades later, Newton gave an explanation for the discrepancies, but within the limit of accuracy of the observational data available in 1619, Kepler was justified in considering his formula to be exact.

TABLE 3.1 Observational Test of Kepler's Third Law

PLANET	SEMIMAJOR AXIS OF ORBIT, a (AU)	SIDEREAL PERIOD (YEARS)	a^3	P^2
Mercury	0.387	0.241	0.058	0.058
Venus	0.723	0.615	0.378	0.378
Earth	1.000	1.000	1.000	1.000
Mars	1.524	1.881	3.537	3.537
Jupiter	5.203	11.862	140.8	140.7
Saturn	9.534	29.456	867.9	867.7

The harmonic law is the most valuable part of *Harmony of the Worlds*. Most of the book deals with Kepler's theories on the harmony of the planets; he even derives notes of music they play as they move harmoniously in their orbits. The earth, for example, plays the notes *mi, fa, mi*, which he took to symbolize the *"miseria* (misery), *fames* (famine), *miseria"* of our planet.

(c) The Epitome

In 1618, 1620, and 1621, Kepler published (in installments) his text on astronomy, the *Epitome of the Copernican Astronomy*. The book includes accounts of discoveries, both by himself and by Galileo (see Section 3.2), and firmly supports the Copernican view. Here, for the first time, Kepler implies that his first two laws had been tested and found valid for the other planets besides Mars (including the earth) and for the moon. Also, he states that the harmonic (third) law applies to the motions of the four newly discovered satellites of Jupiter as well as to the motions of the planets about the sun.

3.2 GALILEO

Galileo Galilei (1564–1642), the great Italian contemporary of Kepler, was born in Pisa. Galileo, like Copernicus, began training for a medical career, but he had little interest in the subject and later switched to mathematics. In school he incurred the wrath of his professors by refusing to accept on faith dogmatic statements based solely on the authority of great writers of the past. From his classmates he gained the nickname "Wrangler."

For financial reasons, Galileo was never able to complete his formal university training. Nevertheless, his exceptional ability as a mathematician gained him the post, in 1589, of professor of mathematics and astronomy at the university at Pisa. In 1592 he obtained a far better position at the university at Padua, where he remained until 1610,

FIG. 3-7   Galileo Galilei. *(Yerkes Observatory.)*

when he left to become mathematician to the Grand Duke of Tuscany. While at Padua he became famous throughout Europe as a brilliant lecturer and as a foremost scientific investigator.

(a) Galileo's Experiments in Mechanics

Galileo's greatest contributions were in the field of mechanics. In his time the principles of mechanics outlined by Aristotle were still regarded as absolutely authoritative. Most of Aristotle's notions, however, were based upon intellectual assumptions rather than actual experiment. Although the seeds of experimental science had been sown by certain of the later Greek scholars, notably Archimedes, the concept of performing experiments to learn physical laws was virtually unheard of in

Galileo's time; scholars believed that all that was to be known could be found within the pages of Aristotle.

Galileo experimented with pendulums, with balls rolling down inclined planes, with light and mirrors, with falling bodies, and many other objects. Aristotle had said that heavy objects fall faster than lighter ones. Galileo showed that if a heavy and light object were dropped together, even from a great height, both would hit the ground at practically the same time. What little difference there is could easily be accounted for by the resistance of the air.

LAWS OF MOTION

In the course of his experiments, Galileo discovered laws that invariably described the behavior of physical objects. The most far-reaching of these is the *law of inertia* (now known universally as Newton's first law). The inertia of a body is that property of the body that resists any change of motion. It was familiar to all persons then as it is to us now that if a body is at rest it tends to remain at rest, and requires some outside influence to start it in motion. Rest was thus generally regarded as the *natural state of matter*. Galileo showed, however, that rest was no more natural than motion. If an object is slid along a rough horizontal floor, it soon comes to rest, because friction between it and the floor acts as a retarding force. However, if the floor and object are both highly polished, the body, given the same initial speed, will slide farther before coming to rest. On a smooth layer of ice, it will slide farther still. Galileo noted that the less the retarding force, the less the body's tendency to slow down, and he reasoned that if all resisting effects could be removed (for example, the friction of the floor or ground, and of the air) the body would continue in a steady state of motion indefinitely. In fact, he argued, not only is a force required to start an object moving from rest, but a force is also required to slow down, stop, speed

up, or change the direction of a moving object.

Galileo also studied the way bodies accelerated, that is, changed their speed, as they fell freely, or rolled down inclined planes. He found that such bodies accelerate uniformly, that is, in equal intervals of time they gain equal increments in speed. Galileo formulated these newly found laws in precise mathematical terms that enabled one to predict, in future experiments, how far and how fast bodies would move in various lengths of time. It remained for Newton to incorporate and generalize Galileo's principles into a few simple laws so fundamental that they have become the basis of a great part of our modern technology.

The great contribution of Galileo, then, was to lay the foundations of experimental science and to bring about a renaissance of the scientific method — a philosophy that was all but lost since its infancy in ancient Greece.

(b) Galileo's Astronomical Contributions

Sometime in the 1590's Galileo adopted the Copernican hypothesis of the solar system. In Roman Catholic Italy, this was not a popular philosophy, for the Church authorities still upheld as absolute truth the ideas of Aristotle and Ptolemy. It was primarily because of Galileo that in 1616 the Church issued a prohibition decree which stated that the Copernican doctrine was "false and absurd" and was not to be held or defended.

The prevailing notion of the time was that the celestial bodies belonged to the realm of the heavens where all is perfect, unchanging, and incorruptible. Perpetual circular motion, being the "perfect" kind of motion, was regarded as the natural state of affairs for those heavenly bodies. Once Galileo had established the principle of inertia — that on the earth bodies in undisturbed motion remain in motion — it was no longer necessary to ascribe any special status to the fact that the planets remain perpetually in orbit. By the same token, even the earth could continue to move, once started. What *does* need to be explained is why the planets move in curved paths around the sun rather than in straight lines. Evidently, Galileo was sufficiently imbued with Aristotelian concepts that he accepted uniform circular celestial motion without subjecting the planets to the same objective scrutiny that he applied in his terrestrial experiments.

In answer to the common objection that objects could not remain on the earth if it were in motion, Galileo noted that if a stone is dropped from the masthead of a moving ship it does not fall behind the ship and land in the water beyond its stern, but rather lands at the foot of the mast, for the stone already has a forward inertia gained from its common motion with the ship before it is dropped. In an analogous way, objects on the earth would not be swept off and left behind if the earth were moving, for they share the earth's forward motion.

GALILEO'S TELESCOPES

It is not certain when the principle was first conceived of combining two or more pieces of glass to produce an instrument that enlarged distant objects, making them appear nearer. Claims for the discovery exist as early as the time of Roger Bacon (thirteenth century). At any rate, the first telescopes that attracted much notice were made by the Dutch spectacle-maker Hans Lippershey in 1608. Galileo heard of the discovery in 1609, and without ever having seen an assembled telescope he constructed one of his own with a three-power magnification — that is, that made distant objects appear three times nearer and larger. He quickly built other instruments, his best with a magnification of about 30 diameters.

SIDEREAL MESSENGER

It was a fairly obvious step to apply the newly invented telescope to celestial observations. The idea may have occurred to others about the same time as it did to Galileo, or possibly sooner. Galileo,

FIG. 3-8 Galileo's drawings of Jupiter and its satellites. *(Yerkes Observatory.)*

however, rightly deserves the honor of having been the first to make significant astronomical telescopic observations, for he realized the importance of careful and persistent study of the objects he viewed. In 1610 he startled the world by publishing a list of his remarkable discoveries in a small book, *Sidereal Messenger* (*Sidereus Nuncius*).

Galileo found that many stars too faint to be seen with the naked eye became visible with his telescope. In particular, he found that some nebulous blurs resolved into many stars (for example, the Praesepe in Cancer) and that the Milky Way was made up of multitudes of individual stars. He found that Jupiter had four satellites or moons revolving about it with periods ranging from just under 2 days to about 17 days (eight other satellites of Jupiter have been found since). This discovery was particularly important because it showed that there could be centers of motion that in turn are in motion. It had been argued that if the earth were in motion the moon would be left behind, because it could hardly keep up with a rapidly moving planet. Yet here were Jupiter's satellites doing exactly that!

PHASES OF VENUS

Another important telescopic discovery that strongly supported the Copernican view was the fact that Venus goes through phases like the moon. In the Ptolemaic system, Venus is always closer to the earth than the sun, and thus, because Venus never has more than about 40° elongation, it would never be able to turn its fully illuminated surface to our view — it would always appear as a crescent. Galileo, however, saw that Venus went through both crescent and gibbous phases, and concluded that it must travel around the sun, passing at times behind and beyond it, rather than revolving directly around the earth (Figure 3-9). Mercury also goes through all phases.

IRREGULARITIES IN THE HEAVENS

Galileo's observations revealed much about our nearest neighbor, the moon. He saw craters, mountain ranges, valleys, and flat dark areas that he guessed might be water (the dark *maria,* or "seas," on the moon were thought to be water until long after Galileo's time). Not only did these discoveries show that the heavenly bodies, regarded as perfect, smooth, and incorruptible, do indeed have irregularities, as does the earth, but they showed the moon to be not so very dissimilar to the earth, which suggested that the earth, too, could belong to the realm of celestial bodies.

Of course, Galileo's conclusions did not go uncriticized. Ludovico delle Colombe, one of his critics, argued that the valleys and mountains on the moon were actually submerged beneath an invisible sea of a crystalline material whose outer surface was perfectly smooth. Galileo answered that this was a splendid suggestion with which he completely agreed, and that furthermore there were mountains of this same invisible substance rising

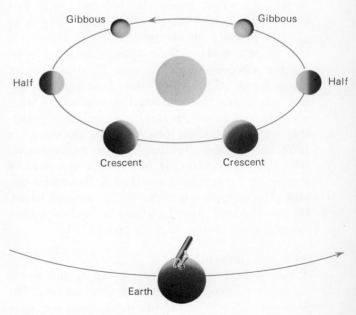

FIG. 3-9   Phases of an inferior planet.

FIG. 3-11 Galileo's drawings of sunspots. *(Yerkes Observatory.)*

FIG. 3-10 Galileo's drawings of the moon. *(Yerkes Observatory.)*

10 times higher than any mountains that could be seen on the moon!

Galileo also found that Saturn did not appear round, although his telescope was not good enough to show the true nature of the planet. It was not until 1655 that Huygens described the magnificent system of rings about Saturn (Chapter 11).

One of Galileo's most disturbing observations, to his contemporaries, was of spots on the sun, showing that this body also had "blemishes." Sunspots are now known to be large, comparatively cool areas on the sun that appear dark because of their contrast with the brighter and hotter solar surface (Chapter 17). Sunspots are temporary, lasting usually only a few weeks to a few months. Large sunspots actually had been observed before, with the unaided eye, but were generally regarded either as something in the earth's atmosphere or as planets between the earth and sun silhouetting themselves against the sun's disk in the sky. In fact, some of Galileo's critics attempted to explain the spots as satellites revolving about the sun.

Galileo observed the spots to move, day by day, across the disk of the sun. He also noted that they moved most rapidly when near the center of the sun's disk and increasingly slowly as they approached the limb (the sun's limb is its apparent "edge" as we see it in the sky). Often, after about 2 weeks, the same spots would reappear on the opposite limb, and move slowly at first, then more rapidly toward the center of the disk. Galileo explained that the spots must be either on the surface of the sun or very close to it and that they were carried around the sun by its own rotation. Their variable speed, he showed, is an effect of foreshortening; when near the center of the sun's disk they are being carried directly across our line of sight, but near the limb most of their motion is either toward or away from us. He determined the sun's period of rotation to be a little under a month.

(c) Dialogue on the Great World Systems

As we have seen, Galileo had accumulated a great deal of evidence to support the Copernican system. By the decree of 1616 he was forbidden to "hold or defend" the odious hypothesis, but he still hoped to convert his countrymen to the heliocentric view — to help them to "see the light." He finally prevailed upon Pope Urban VIII, an old friend, to allow him to publish a book that explained fully all arguments for and against the Copernican system, not for the purpose of extolling it, but merely to examine it, and to show those of other nationalities that Italians were not ignorant of new theories.

The book appeared in 1632 under the title, *Dialogue on the Great World Systems* (*Dialogo dei Massimi Sistemi*). The *Dialogue* is written in Italian (not Latin) to reach a larger audience, and is a magnificent and unanswerable argument for Copernican astronomy. It is in the form of a conversation, lasting 4 days, among three philosophers:

Salviati, the most brilliant and the one through whom Galileo generally expresses his own views; Sagredo, who is usually quick to see the truth of Salviati's arguments; and Simplicio, an Aristotelian philosopher who brings up all the usual objections to the Copernican system, which Salviati promptly shows to be absurd.

It is pointed out in the preface to the *Dialogue* that the arguments to follow are merely a mathematical fantasy, and that divine knowledge assures us of the immobility of the earth. This was thinly cloaked irony, however, and Galileo's enemies acted quickly to build a case against him. He was called before the Roman Inquisition on the charge of believing and holding doctrines that are false and contrary to the Divine Scriptures. Urban VIII, apparently won over by Galileo's enemies, refused to come to his aid. Galileo was forced to plead guilty and deny his own doctrines. His sentence was very light, for that time, consisting of confinement to his own home, under close watch, for the last 10 years of his life. He was then nearly seventy.

The *Dialogue*, meanwhile, took its place along with Copernicus' *De Revolutionibus* and Kepler's *Epitome* on the *Index of Prohibited Books*, where it remained until 1835.

3.3 NEWTON

Isaac Newton (1643–1727) was born at Woolsthrope, in Lincolnshire, England, almost exactly 1 year after the death of Galileo. (Newton was born on Christmas Day, 1642, according to the calendar in use at his time, but by the modern Gregorian calendar his birth date was January 4, 1643.) Newton's scientific successor, Lagrange, described him as "the greatest genius that ever existed." It was Newton who formulated the basic laws of modern mechanics and showed them to be universal throughout the solar system, applying to the motions of the celestial objects as well as to objects on the earth. Furthermore, Newton's greatest contributions in the fields of astronomy, mechanics, optics, and mathematics were conceived by the time he had reached the age of 24.

Newton entered Trinity College at Cambridge in 1661 and 8 years later was appointed Lucasion Professor of Mathematics, a post that he held during most of his productive career. As a young man in college, he became interested in natural philosophy. He had originally worked out many of his ideas on mechanics and optics during the plague years of 1665 and 1666. Unfortunately, however, he did not publish the results of his investigations at that time, partially because his interests were easily diverted from one subject to another and partially because of outstanding problems and inconsistencies that he had not yet solved. As a result, others from time to time worked out and published the solutions to some of the problems that Newton had actually solved first.

FIG. 3-12 Isaac Newton. *(Yerkes Observatory.)*

Eventually Newton's friend, Edmond Halley, prevailed on him to collect and publish the results of his investigations in mechanics and gravitation. The result was Newton's *Philosophiae Naturalis Principia Mathematica*. The *Principia,* as the book is generally known, was published at Halley's expense in 1687.

As in most great contributions, some of the ideas in the *Principia* were developed independently by others at about the same time, or even earlier. In the *Principia,* however, for the first time a very few natural laws are shown to be general, and to apply to a wide range of phenomena and experiences. Much of the immense variety of the universe became unified under these extraordinarily simple principles.

3.4 NEWTON'S LAWS OF MOTION

Newton's entire system is based on three propositions, now known as his three laws of motion. The laws of motion were not entirely new; Galileo had stated the first law and, in a less general form, the second. In the *Principia,* however, the laws are stated as a fundamental basis for all mechanics, and we find there, too, the first clear discussion of the concepts of *mass* and *force.*

(a) The First Law — Inertia

Newton's first law of motion states that in the absence of any outside influence an object at rest tends to remain at rest, and an object in motion tends to remain in motion, in a straight line, and at a constant speed. This property of a body that resists a change in motion is called its *inertia.* The first law of motion is therefore often called the law of inertia. Newton postulated that the law is general and holds for all objects, whether on earth or in space.

MOMENTUM

A measure of the inertia of an object is its *mo-*

mentum. Newton defined the momentum of a body as being proportional to its velocity. The constant of proportionality is that property of the body that gives it its inertial characteristics — that enables it to resist a change in its state of motion. This property is called *mass.*† The momentum of a body, then, is the product of its mass and velocity; that is,

$$momentum = mass \times velocity.$$

Momentum involves two important concepts: *velocity* and *mass.*

VELOCITY

Velocity is a description of the speed a body moves and also its *direction* of motion. The velocity of a body changes if either its speed or direction changes; thus a racing car traveling at a uniform rate around a circular track constantly changes its velocity.

Velocity is often represented graphically by a *vector.* A vector that describes velocity is a line whose length is proportional to the speed and whose direction represents the direction of motion (see Figure 3-13).

Any change in the velocity of an object, either to start it, stop it, speed it up, slow it down, or change its direction, is *acceleration.*

Frequently the direction of motion of a body may not be relevant to the problem at hand. In such a case the velocity often is specified simply by the speed of the object. For example, we might say that the orbital velocity of the earth is 18 mi/sec and not trouble to indicate its direction of motion. Strictly speaking, therefore, we have not really specified completely the earth's velocity; properly we should have said that its speed is 18 mi/sec.

MASS

Mass is a concept first clearly defined by Newton.

†Strictly, this is called *inertial mass,* as opposed to *gravitational mass.* (Section 3.6a).

It is important that we thoroughly understand at the outset what is meant by "mass."

We all have an intuitive notion of mass. Whenever we speak of the *weight* of an object, we almost always mean its *mass.* There is an important distinction between weight and mass. The weight of an object is the *force of attraction* between that object and the earth (or some other body if the object is not on the earth). When we consider weight later in the chapter, we shall find that its value depends upon the location of the object in question. On the surface of the moon, a person would weigh only about one sixth of what he does on earth; far off in interstellar space he would be virtually weightless. When we speak of weight in everyday conversation, however, we generally regard it as a property of an object itself; we should say "mass."

Mass is a property that characterizes the actual amount of material of a body. That property, its total content of matter, depends in no way upon where it is. If a 300-lb man goes to the moon, he may only *weigh* 50 lb there, but he has not changed his mass — he is as obese as ever. We might think of the mass of something as a measure of the total number of subatomic particles that compose it. Changing its shape, vaporizing it, freezing it, or sending it to the moon will not change its mass. By convention, we take the mass of an object to be numerically equal to its weight at sea level on the earth. Thus, 10 lb of lead weighs 10 lb here and nothing in remote space, but it has a mass of 10 lb everywhere.

In reality, of course, we have no direct means of counting the total number of subatomic particles in an object. To speak of mass as a measure of something we must define it in such a way that it can be experimentally determined. In the context of Newton's laws of motion, we can define mass as *that property of a body that gives it inertia,* that is, that *resists acceleration.* This is the *inertial* definition of mass. It will be clear how mass can

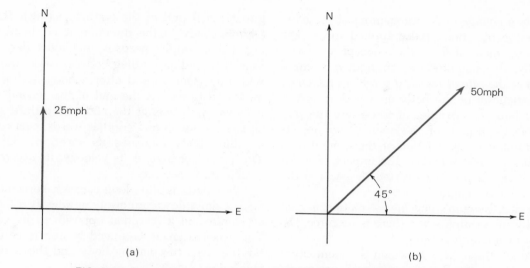

FIG. 3-13 Two vectors, representing velocities of 25 and 50 mi/hr, respectively, the first to the north, the second to the northeast.

actually be measured when we discuss Newton's second and third laws. The most common units of mass are the *gram* (the mass of 1 cubic centimeter of water) and the *pound*.

MASS, VOLUME, AND DENSITY

It is important not to confuse mass, volume, and density. *Volume* is simply a measure of the physical space occupied by a body, say in cubic inches, cubic feet, liters, and so on. In short, the volume is the "size" of an object — it has nothing to do with its mass. A lady's wristwatch and an inflated balloon may both have the same mass, but they have very different volumes.

The watch and balloon are also very different in *density*, which is a measure of how much mass is contained within a given volume. Specifically, it is the ratio of mass to volume:

$$\text{density} = \frac{\text{mass}}{\text{volume}}.$$

The units of density could be pounds per cubic inch (lb/in.^{3}), grams per cubic centimeter (gm/cm^3), ounces per cubic mile, and so on. Sometimes density is given in terms of the density of water (62 lb/ft^3 or 1 gm/cm^3), in which case it is called *specific gravity*. Iron has a specific gravity of 7.9 or a density of 7.9 gm/cm^3; gold has a specific gravity of 19.3. To sum up, then, *mass* is "how much," *volume* is "how big," and *density* is "how tightly packed."

THE FIRST LAW — SUMMARY

With the concepts of mass and momentum firmly established, we can now state Newton's first law of motion as follows: In the absence of any outside influence, the momentum of any body remains constant — that is, is *conserved*.

(b) The Second Law — Force

The second law of motion deals with changes in momentum. It states that if a force acts on a body it produces a *change* in the momentum of the body that is in the direction of the applied force. The second law, then, defines the concept of *force*. Specifically, the magnitude or strength of a force is defined as the *rate at which it produces a change in the momentum of the body on which it acts.*

Some familiar examples of forces are the pull of the earth, the friction of air slowing down objects moving through it, the friction of the ground or a floor similarly slowing bodies, the impact of a bat on a baseball, the pressure exerted by air, and the thrust of a rocket engine.

Note that Newton's first law of motion is consistent with his second; when there is no force, the change in momentum is zero.

There are three ways in which the momentum of a body can change. Its velocity can change, or its mass, or both. Most often the mass of a body does not change when a force acts upon it; a change in momentum usually results from a change in velocity. Thus, in the vast majority of examples, the second law can be written as the simple formula

$$\text{force} = \text{mass} \times \text{acceleration},$$

because acceleration is the rate at which velocity changes. Let us consider acceleration in more detail.

We recall that acceleration occurs whenever the velocity of a body changes, and that the body's velocity changes when either the speed or the direction of its motion is altered. If the acceleration occurs in the same direction as the velocity, the body simply speeds up; if the acceleration occurs in the opposite direction to the velocity, the body slows down. If acceleration occurs exactly at *right angles* to the velocity, only the direction of motion of the body, and not its speed, changes.

Galileo had investigated falling bodies indirectly by experimenting with bodies rolling down inclined planes. The acceleration of falling bodies is downward (in the direction toward which the gravitational pull of the earth is acting). It accelerates a body in the direction it is already moving, and so simply speeds it up. Galileo determined that the speed of a falling body increases *uniformly* with time. One second after a body is dropped it moves 32 ft/sec. At the end of the second second it moves 64 ft/sec; at the end of the third second, 96 ft/sec; and so on. In other words, every second a falling body increases its speed by 32 ft/sec. The rate of change of its velocity, its *acceleration*, is then 32 ft/sec/sec.

If a body is slid along a rough horizontal surface, it slows down uniformly in time. It is therefore accelerated in a direction opposite to its velocity. The acceleration is produced by the force of friction between the moving body and the rough surface.

Of course, a body accelerates, that is, its momentum changes, only while a force acts upon it. The force of gravity is constantly pulling falling bodies downward; thus, they continually accelerate. On the other hand, once a baseball ceases to be in actual contact with the bat, it continues forward in a straight line at a constant speed and would do so forever if air friction did not slow it, the earth pull it down, or an eager outfielder intercept it.

Whenever the mass of an object is subject to change, we cannot assume that force = mass × acceleration. The variation of mass must be taken into account in calculating the acceleration produced by a force. An important and familiar example is a rocket. Exhaust gases are ejected from the back to accelerate the rocket forward, with its remaining fuel. Thus, as the rocket accelerates, its mass continually diminishes.

Two equal accelerations may correspond to entirely different forces. Consider the forces required to accelerate an automobile and a bicycle each to a speed of 20 mi/hr in 20 seconds. Clearly, because of the car's greater mass, a proportionately greater force will be required to produce the necessary acceleration. Similarly, once the bodies are both moving at that speed, a far greater force is needed to *stop* the automobile as quickly as the bicycle.

COMBINATIONS OF FORCES

Newton showed how to compute the effect of two or more forces acting at once on an object. We shall illustrate the procedure when there are two forces, f_1 and f_2, acting together on a body at O. We can represent the forces by vectors just as we can velocities (Figure 3-14). The directions of the vectors show the directions of the forces and the lengths show the magnitudes of the forces (say, so many dynes to the inch). The arrows representing the vectors both start from the object on which the two forces are acting. We note now that the two arrows define two sides of a parallelogram, Of_1ff_2. The diagonal of the parallelogram, Of, is the resultant force acting on the body; the body accelerates as if acted upon by just that one force. The figure shown is called a *parallelogram of forces.*

Of special interest is the situation in which two or more forces act upon a body in such a way as to cancel each other. Then the body is said to be

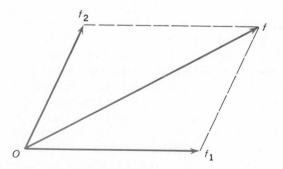

FIG. 3-14 Parallelogram of forces.

in a state of *static equilibrium,* and does not accelerate. An example is a ball hanging from a rubber band — the rubber stretches until its tension exactly balances the force of gravity pulling the ball downward. Another example is the column of mercury in a barometer; the mercury stands in static equilibrium between its own weight and the pressure of the air.

(c) The Third Law — Reaction

Newton's third law of motion was a new idea. It states that all forces occur as *pairs* of forces that are mutually equal to and opposite each other. If a force is exerted upon an object, it must be exerted by something else, and the object will exert an equal and opposite force back upon that something. All forces, in other words, must be mutual forces acting *between* two objects or things.

Suppose a boy jumps off a table down to the ground. The force pulling him down is a mutual gravitational force between him and the earth. Both he and the earth suffer the same total change of momentum because of the influence of this mutual force. Of course, the boy does most of the moving; because of the greater mass of the earth, it can experience the same change of momentum by accelerating only a negligible amount.

A more obvious manifestation of the mutual nature of forces between objects is familiar to all who have played baseball. The recoil of the bat shows that the ball exerts a force on the bat during the impact, just as the bat does on the ball. The momentum imparted to the bat by the ball is transmitted through the batter to the earth, so the acceleration produced is far less than that suffered by the ball. Similarly, when a rifle is discharged, the force pushing the bullet out the muzzle is equal to that pushing backward upon the gun and marksman.

If a swimmer dives from the side of a rowboat, the force he exerts upon himself to spring out away from the boat is equal to a force that accelerates the craft backward away from him. Here, in fact, is the principle of rockets — the force that discharges the exhaust gases from the rear of the rocket is accompanied by a force that shoves the rocket forward. The exhaust gases need not push against air or the earth; a rocket operates best of all in a vacuum. Incidentally, passengers on a rocket ship would have to take care how they disposed of waste material. For example, if garbage were discharged through a port in the missile, the reaction force would accelerate the rocket slightly off course.

In all the cases considered above, a mutual force acts upon the two objects concerned; each object always experiences the same total change of momentum, but in opposite directions. Because momentum is the product of velocity and mass, the object of lesser mass will end up with proportionately greater velocity. Today we are developing atomic rocket engines. Atoms are shot from the rear of such engines. The masses of the atoms are extremely small, but they are propelled outward with speeds near that of light, so they carry momentum that is not completely negligible, and if enough atoms are shot out, the engine can measurably accelerate a rocket.

GENERALITY OF THE THIRD LAW

Newton phrased his third law by stating that for every action there is an equal and opposite reaction. In fact, a system may be far more complex than just a pair of bodies exerting mutual forces on each other. Suppose, for example, a rocket in space were to explode into thousands of pieces. It is then pointless to speak of two bodies exerting mutual forces on each other; we generalize by saying that each of the remnants of the explosion has exerted a force on each other in such a way that all the forces balance each other. Each particle has suffered a change in momentum, but the sum total of all changes in momentum of particles

accelerated one way will be equal and opposite to the total change in momentum associated with all the particles going the opposite direction. All these changes in momentum balance each other, so that the momentum of the entire system, as long as it has not been acted on by *external* forces, is the same as before the explosion.

In other words, Newton's third law can be thought of as a generalization of his first. The first law states that in the absence of a force, a body's momentum is conserved. The third law means that if we isolate an entire system from outside forces, the total momentum of the system is conserved. *Internal* forces in the system may result in changes of momentum within it, but these are always accompanied by equal and opposite changes. The total momentum of a rocket, for example, does not change, as long as we always include the momentum of its exhaust gases and any other object (such as the atmosphere or the earth) that may have come in contact with these gases.

MEASURING MASS

We are now in a position to see how we can measure mass. Initially, some object must be adopted as a standard and said to have *unit mass* — for example, 1 cubic centimeter of water, or perhaps the mass of some king's favorite gemstone. Then the mass of any other object can be compared to it by measuring the relative accelerations produced when the same force acts on each of the two. We assure that each is subjected to the same force by isolating them from other objects and producing equal and opposite changes in their momenta with an *internal* force — say, by separating them with a compressed spring placed between them and released by an internal triggering device, or by detonating a small charge between them. Thus, the third law permits an *operational definition* of mass.

Having found a way to measure mass, we can now express the value of a force numerically. The most common units of force are *dynes* and

The launching of Surveyor I by an Atlas-Centaur rocket. *(NASA.)*

Astronaut Edward H. White II performing a space walk during the third orbit of Gemini IV in May, 1965. *(NASA.)*

The Moon during a lunar eclipse.
(William Close.)

The Milky Way in the direction of the galactic center.
(U.S. Naval Observatory; Flagstaff Station.)

poundals. A dyne is the force needed to give a mass of 1 gm an acceleration of 1 cm/sec/sec; a poundal is the force needed to give a mass of 1 lb an acceleration of 1 ft/sec/sec. At the surface of the earth the force due to gravity acting on a 1-lb mass is 32 poundals.

(d) The Laws of Motion Summarized

For convenience, we now summarize Newton's three laws of motion:

> NEWTON'S FIRST LAW: *In the absence of outside forces, the momentum of a system remains constant.*

> NEWTON'S SECOND LAW: *If a force acts upon a body, the body accelerates in the direction of the force, its momentum changing at a rate numerically equal to that force.*

> NEWTON'S THIRD LAW: *Forces are always mutual; thus if a force is exerted upon a body, that body reacts with an equal and opposite force upon whatever exerts the force upon it.*

Newton's three laws, as we have seen, contain, respectively, definitions of the three concepts, momentum, force, and mass.

3.5 ACCELERATION IN A CIRCULAR ORBIT

It might be assumed that some force or power is required to keep the planets in motion. However, Galileo argued from the principle of inertia (Newton's first law of motion) that once started, the planets would remain in motion — that the state of motion for planets was as natural as for terrestrial objects. What does require explanation, however, is why the planets move in nearly circular orbits rather than in straight lines (the latter motion would eventually carry them away from the vicinity of the solar system). Galileo had not considered this problem.

By Newton's time, a number of investigators had considered the problem of circular motion. The correct solution to the problem was first published (in 1673) by the Dutch physicist Christian Huygens (1629–1695). However, Newton had found the solution independently in 1666.

Clearly, for a body to move in a circular path rather than in a straight line, it must continually suffer an acceleration toward the center of the circle. Such an acceleration is called *centripetal acceleration.* The central force that produces the centripetal acceleration (*centripetal force*) is, for a planet, an attraction between the planet and the sun. For a stone whirled about at the end of a string, the centripetal force is the tension in the string. With the help of Newton's laws of motion and some elementary mathematics it can be shown that if a particle of mass m moves with a speed v on the circumference of a circle of radius r, the centripetal force on it is given by the formula

$$F = \frac{mv^2}{r}.$$

3.6 UNIVERSAL GRAVITATION

It is obvious that the earth exerts a force of attraction upon all objects at its surface. This is a mutual force; a falling apple and the earth are pulling on each other. Newton wondered whether this force of attraction between the earth and objects on or near its surface might extend as far as to the moon and produce the centripetal acceleration required to keep the moon in its orbit. Newton wondered further whether there might be a general force of attraction between *all* material bodies. If so, the attractive force between the sun and each of the planets could provide the centripetal acceleration necessary to keep each in its respective orbit. Thus Newton hypothesized that there is a universal attraction between all bodies everywhere in space. It remained for him to determine the mathematical nature of the attraction and test the hypothesis by using it to predict *new* phenomena.

Newton derived the mathematical form of his law of universal gravitation from the formula for centripetal force and from Kepler's laws. For the purpose of illustration, suppose the planets to move on circular orbits (Newton did not make this restrictive assumption). Then the semimajor axis of a planet's orbit is just its radius, r, and its speed is constant, namely the circumference of the orbit ($2\pi r$) divided by the planet's period of revolution. With Kepler's third law, the period can be expressed in terms of r ($P \propto r^{3/2}$); then the gravitational force on the planet, the centripetal force that makes it move on a circular orbit, can be expressed in terms of r. It is found to be proportional to the planet's mass and to the reciprocal of r^2. In order that the force of attraction be mutual between the sun and planet (to be consistent with the third law of motion), Newton reasoned that it must actually be proportional to both the mass of the planet and the mass of the sun. Therefore, the attractive force between the two has the mathematical form

$$\text{force} \propto \frac{m_s m_p}{r^2},$$

where m_p and m_s are the masses of the planet and sun, respectively. (Newton, of course, took the elliptical shapes of the planet's orbits into account, but derived exactly the same formula for gravitation.) Both the sun and a planet revolving around it experience the same change of momentum as a result of this mutual force between them. The fact that the sun is observed to remain more or less at the center of the solar system while the planets revolve around it is evidence that the sun's mass must be enormously greater than that of the planets. Therefore its acceleration is relatively small; actually the sun's acceleration is observable.

For Newton's hypothesis of universal attraction to be correct, there must be an attractive force be- tween all pairs of objects everywhere whose value is given by the same mathematical formula as that above for the force between the sun and a planet. Thus the force F between two bodies of masses m_1 and m_2, and separated by a distance d, is

$$F = G \frac{m_1 m_2}{d^2}.$$

Here G, the constant of proportionality in the equation, is a number called the *constant of gravitation*, whose value depends on the units of mass, distance, and force used. The actual value of G has to be determined by laboratory measurement of the attractive force between two material bodies (see Section 3.6c). If metric units are used (grams for mass, centimeters for distance, and dynes for force), G has the numerical value 6.67×10^{-8}.[†]

The above equation expresses Newton's law of universal gravitation, which is stated as follows:

Between any two objects anywhere in space there exists a force of attraction that is in proportion to the product of the masses of the objects and in inverse proportion to the square of the distance between them.

Not only is there a force between the sun and each planet, but also between any two planets. Because of the sun's far greater mass, the dominant force felt by any planet is that between it and the sun. The attractive forces between the planets have relatively little influence. Similarly, there is a gravitational attraction between any two objects on earth (for example, between two flying airplanes, or between the kitchen sink and a tree outside the house), but this force is insignificant compared to the force between each of them and the very massive earth.

---

† The notation 6.67×10^{-8} means 0.0000000667; see Appendix 3.

FIG. 3-15 Attraction of a sphere is as though all its mass
were concentrated at its center.

Before we see how Newton tested his law of gravitation, let us investigate some of its other consequences.

(a) Weight

Newton hypothesized a force of attraction between all pairs of bodies. The earth is a large spherical mass, however, that can be thought of as being composed of a large number of component parts. An object, say a man, on the surface of the earth feels the simultaneous attractions of the many parts of the earth pulling on him from many different directions. Exactly what is the resultant gravitational effect of the many parts of a sphere, each pulling independently upon a mass outside the surface of the sphere? Here was a difficult problem to which Newton had to find a solution before he could test his law of gravitation.

To solve the problem, Newton had to calculate the force between an object on the surface of the earth and each infinitesimal piece of the earth, and then calculate how all of these forces combined.

It was necessary for him to invent and use a new method of mathematics which he called *inverse fluxions* (today we call it *integral calculus*). Fortunately, the solution to the problem gives a beautifully simple result. A spherical body acts gravitationally as though all its mass were concentrated at a point at its center (Figure 3-15).†
This meant that Newton could consider the earth, moon, sun, and planets as geometrical points as far as their gravitational influences are concerned.

The gravitational force, then, between an object of mass m on the earth and the earth itself, of mass M, is equal to the constant of gravitation times the product of the masses m and M, divided by the square of the distance from the object to the center of the earth. The latter distance is just the radius of the earth, R. This gravitational force between the earth and a body on its surface is the

†Strictly, the statement is correct only if the density distribution within the body is spherically symmetrical.

body's *weight.* Algebraically, the weight W of a body is given by

$$W = G\frac{mM}{R^2}.$$

We see that the weight of a body is proportional to its mass. This circumstance gives us another method of measuring mass, namely, by measuring the weight of a body. Whenever the mass of an object is determined by its gravitational influence, as by measuring its weight, the mass so determined is defined as a *gravitational mass.* So far, the results of all experiments indicate that gravitational mass is exactly equivalent to *inertial mass.*

Because the gravitational attraction between two bodies decreases as the square of their separation, an object weighs less if it is lifted above the surface of the earth. A person actually weighs less at the top of a step ladder than on the ground. Careful laboratory experiments are able to detect such subtle changes in gravitational force. At 4000 mi above the earth's surface, a body is twice as far from the earth's center as it would be at the surface, so an object there would weigh $\frac{1}{2}^2$, or $\frac{1}{4}$, times as much as at the surface of the earth. Far out in space, a person's weight drops virtually to zero. In the vicinity of another gravitating body, his weight is determined by the attraction between him and that body.

If an object is dropped from a height, the downward acceleration is equal to the force acting on it, that is, its weight, divided by its mass:

$$\text{acceleration} = \frac{W}{m} = G\frac{M}{R^2}.$$

Thus, although the gravitational force between the earth and a massive object is greater than between the earth and a less-massive one, all objects experience the same acceleration and fall at the same rate, as Galileo had found.

(b) Test of Gravitation: The Apple and the Moon
Suppose an apple is dropped from a height above the surface of the earth. We have seen that it accelerates 32 ft/sec/sec. If the hypothesis of gravitation is correct, the accelerations toward the earth of the apple (at the earth's surface) and of the moon (239,000 mi away) should both be given by the equation

$$\text{acceleration} = G\frac{M}{D^2},$$

where M is the mass of the earth and D is the distance of the object in question from the center of the earth. The acceleration, in other words, should be inversely proportional to the square of the distance from the earth's center. The apple's distance is about 4000 mi, and the moon's distance is 239,000 mi, about 60 times as far. Thus the acceleration of the moon should be $\frac{1}{60}^2$, or $\frac{1}{3600}$, times that of the apple.

If we assume that the moon's orbit about the earth is a perfect circle (its orbit is, indeed, very nearly circular), we can use the formula for centripetal acceleration given in Section 3.5 to calculate the moon's acceleration:

$$\text{acceleration} = \frac{v^2}{D}.$$

The moon's distance (the radius of its orbit) is 239,000 mi, or about 1.26×10^9 ft. Its orbital speed is about 3350 ft/sec. If we substitute these numbers in the above formula we find

$$\text{moon's acceleration} = \frac{(3350)^2}{1.26 \times 10^9}$$
$$= 0.0089 \text{ ft/sec}^2$$
$$= 0.107 \text{ in./sec}^2.$$

The acceleration predicted by the law of gravitation is

$$\text{moon's acceleration} = \frac{32}{(60)^2}$$
$$= 0.0089 \text{ ft/sec}^2$$
$$= 0.107 \text{ in./sec}^2.$$

The law of gravitation predicts that because the moon is 60 times as far from the center of the earth as the apple is, its acceleration should be $\frac{1}{60}^2$ times as much; this is exactly the acceleration that we observe for the moon. The test gives results that are consistent with Newton's law of universal gravitation.

(c) The Mass of the Earth and the Determination of G

Note that in the formula for the weight of an object at the surface of the earth, or for its acceleration toward the earth, the constant of gravitation, G, always occurs multiplied by the mass of the earth. If the latter were known, G could be evaluated at once from the known acceleration of gravity.

There are various methods of finding the mass of the earth. All depend on a comparison of the gravitational attraction between some object of known mass and the earth, with the attraction between two objects of known mass.

An accurate modern method of "weighing" the earth is with a torsion balance. In a torsion balance two spheres are connected by a horizontal light rod that is suspended in the middle by a thin quartz fiber. Then another large mass is placed near one of the spheres. The gravitational force between the large mass and the sphere causes the sphere to move toward it, twisting the quartz fiber. The amount of twist, or torsion, in the fiber is an accurate measure of the force between the two gravitating masses. This force is then compared to that between the large mass and the earth — that is, the *weight* of the former — to obtain the earth's mass.

The results of the best determinations give for the mass of the earth 5.98×10^{27} gm, or 6.6×10^{21} English tons.

3.7 LIMITATIONS TO NEWTON'S LAWS

There are many examples to demonstrate the generality of Newton's laws. During the eighteenth and nineteenth centuries, Newton's laws seemed so powerful that they were generally believed to apply to *all* natural phenomena. However, by the twentieth century, certain experiments had been performed which yielded results that did not conform to Newtonian mechanics. From these experiments grew two new branches of physics, based on new postulates. These new disciplines provide descriptions of physical phenomena within realms in which Newton's laws are found not to apply. We describe them only briefly.

(a) Relativity

Particularly perplexing were the results of an experiment carried out by Michelson and Morley in 1887, which was intended to reveal the absolute speed of the earth in space by measuring the velocity of light in different directions. Much to the surprise of all, the experiment gave null results — it completely failed to reveal any motion of the earth at all! Other similar experiments gave similar results.

In 1905, Albert Einstein (1879–1955) published a paper in which he outlined his *special theory of relativity*. The theory was based on two postulates: (1) there is no absolute reference system with respect to which we can measure "absolute" motions in space; the best we can hope to measure is the *relative* motion of one object with respect to another; and (2) the speed of light, with respect to all observers, is always the same. There is much experimental evidence to support the latter postulate, particularly evidence obtained from the orbits of double stars. The first postulate seems a perfectly reasonable one.

From these two postulates, Einstein showed that two observers, moving with respect to each other, will disagree on their measurements of length, time, velocity, and mass. Einstein derived equations that showed by exactly how much the two observers would disagree. Not only did the theory seem sound and reasonable, but it com-

FIG. 3-16   Albert Einstein. *(Yerkes Observatory.)*

pletely explained the results of perplexing experiments such as those of Michelson and Morley. Today the results of a vast number of observations and experiments are in complete accord with special relativity. Among these is the equivalence of mass and energy predicted by the special theory through the famous equation $E = mc^2$.

In 1916 Einstein generalized the theory to include the effects of accelerations between various observers. This more far-reaching theory is the *general theory of relativity*. The general theory predicts new laws of motion and a new law of gravitation. Einstein proposed three astronomical tests for this new law of gravitation. They are: (1) the orbit of Mercury should undergo an observable perturbation — the orientation of its major axis should slowly change in space at a rate that is different from that predicted by the perturbation theory of Newtonian mechanics (Chapter 11). (2) Light passing near the surface of the sun should be deflected slightly in direction. The effect should be observable during a total solar eclipse, when the sun's brilliance is blotted out and stars appearing in nearly the same direction as the sun can be observed (Chapter 6). (3) The wavelengths of light leaving certain extremely dense stars (white dwarfs) should be slightly lengthened (Chapter 16). All of these effects have been observed, as well as other phenomena predicted since Einstein's time. Now, whereas the results of tests made so far appear to support Einstein's relativity theory, they are also compatible with certain other theories. However, most modern physicists accept the general theory of relativity, as well as the special theory, as an accurate description of nature.

The predictions of special relativity differ from those of Newtonian mechanics only when speeds are encountered that are very near the speed of light. Einstein's more general laws predict results identical to Newton's when the relative velocities encountered are within our ordinary range of experience.

(b) The Quantum Theory

All matter is composed of atoms. Atoms, in turn, are composed of still smaller subatomic particles — protons, electrons, neutrons, and so on. Experiments concerned with the behavior of these subatomic particles and the nature of radiation show that Newtonian mechanics is not adequate to describe their motions. There gradually developed in the present century a theory based on new physical principles that describes the behavior of these tiny particles to the degree of accuracy of present-day observations. This is the *quantum theory*. Only when it is applied to these tiny particles does the quantum theory predict results that are measurably different from those expected from Newton's laws. When applied to problems encountered in everyday experience, the quantum theory, like relativity, predicts the same results that we obtain from Newtonian theory.

Later, especially in Chapter 7, we shall have occasion to look more closely at some of the consequences of the quantum theory.

(c) Newton's Laws Wrong?

In view of these limitations to Newton's laws, we might be inclined to say that Newtonian theory is in error. If we do, we are forgetting the significance of natural laws. Laws are the means by which science attempts to *describe* nature, not *explain* it. The mistake is only to apply them to situations that are outside their range of validity. The very simple principles of Newtonian mechanics, well within the grasp of any high school student, give an amazingly accurate representation of a great many natural phenomena. Among these are the motions of the planets and other bodies in the solar system, and even of stars and systems of stars. The application of Newtonian mechanics to the motions of celestial bodies is *celestial mechanics,* the subject of Chapter 4.

Exercises

1. A friend tells you that the earth cannot be rotating, as the scientists claim, for if a point on the equator were being carried eastward at about 1000 mi/hr, as claimed, a baseball pitcher could throw a ball straight up in the air, and the earth and pitcher would move to the east out from under the ball, which would land some distance behind (to the west) of the pitcher. How might you straighten out this stupid friend of yours?

2. You weigh 100 pounds at sea level. What would you weigh if:

    (a) the earth were suddenly to double in radius, its mass remaining the same?

    (b) the earth were suddenly to double in mass, its radius remaining the same?

3. Suppose you tie a string to a stone and whirl it around over your head at the end of the string. Now you hold up a razor blade directly in front of your face so that you cut the string just as the stone is out at the end of the string directly in front of you. Ignore the pull of the earth's gravity and the friction of the air.

    (a) Just after the string is cut, what forces are acting on the stone?

    (b) Which way does the stone move after the string is cut? Why?

4. Why is it nonsense to speak of "force of forward motion"?

5. What is the major axis of a circle of radius 13 in.?

6. The distance from the sun of a hypothetical planet varies from 60,000,000 mi. to 90,000,000 mi. What is the eccentricity of the planet's orbit?

    Answer: 0.2

7. What is the distance, in astronomical units, of a planet from the sun if the planet has a period of 8 years?

8. Draw a diagram showing Galileo's argument in favor of sunspots being on or very near the solar surface.

9. How much would a 200-pound man weigh 4000 mi. above the surface of the earth? How much would he weigh 16,000 mi. above the surface of the earth?

  Answer to Second Part: about 8 pounds

10. Can a quantity of lead be more massive than an inflated balloon? Can it be less massive? Can it have a greater volume? A smaller volume? A greater density? A smaller density? Give examples.

11. How many accelerators are there on a standard passenger car? Explain.

12. What does a pan balance measure? A spring balance?

13. Did Newton need to know the exact size of the earth to test his theory of gravitation on the apple and moon? Why or why not?

14. Calculate the gravitational attraction between a 160-pound man and a 120-pound woman, and compare it to the attraction between the man and the earth. Assume that the man and woman are 50 feet apart, and that each can be assumed to act gravitationally like a point mass.

CHAPTER 4

Motions of
Celestial Bodies

Newton's laws of motion and gravitation enable us to predict the motions of bodies under the influence of their mutual gravitation. We shall first consider the interactions of two bodies, both of which are either point masses or spherically symmetrical, so that they act (gravitationally) as point masses. The subject is called the *two-body problem*.

4.1 CENTER OF MASS

According to Newton's third law, the total momentum of an isolated system is conserved; that is, all changes of momentum within it are balanced. We can, therefore, define a point within the system that remains fixed (or moves uniformly) as if the entire mass of the system were concentrated at that point. It is called the *center of mass*. It can be shown that the center of mass of a complex body (which is a collection of point masses joined rigidly together) is that point at which the body balances when placed near a gravitating body; thus it is also often called the *center of gravity*.

The center of mass (or gravity) for two bodies is given a special name: the *barycenter*. It must lie on a line connecting the centers of the bodies. With elementary mathematics it can be shown that the barycenter is located so that the distance of each body from it is in inverse proportion to its mass.

The center of mass of a system of revolving bodies consisting of the sun and a planet, because of the sun's far greater mass, lies very close to the center of the sun — in most cases, within its surface.

(a) Relative Orbits

As two bodies revolve mutually about their bary-center they must always maintain the same relative distances from it; if one body doubles its distance so must the other, and if one comes closer, the other must lessen its distance by a proportional amount. We see, then, that the orbits of the two bodies are exactly similar to each other (that is, are the same shape), the sizes of the orbits being in inverse proportion to the masses of the bodies.

The similarity of the two orbits makes it easy to refer to the orbit of one object relative to the other. It is conventional to choose the center of the more massive body as reference. Of course neither body is actually fixed — both move about the bary-center — but it is often convenient to *regard* one (the more massive) as origin, and consider the motion of the other with respect to it. This is called the *relative orbit*. The earth and moon, for example, revolve about their barycenter on a line between their centers. The earth is *not* at the center of the moon's orbit, but we can and often do speak of the relative orbit of the moon about the earth's center.

The size of the relative orbit is the sum of the sizes of the individual orbits about the barycenter. If the bodies move in circular paths (Figure 4-1), the relative orbit is a circle with a radius equal to the sum of the distances of the two bodies from the barycenter (it is also equal to the distance between the bodies). If the orbits are ellipses, the relative orbit is an ellipse of major axis equal to the sum of the major axes of the elliptical orbits of the two bodies; all three ellipses have the same eccentricity.

We now turn to the motions of planets, satellites, and other celestial objects.

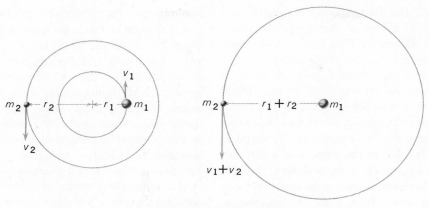

FIG. 4-1   Relative orbit.

4.2 ORBITAL MOTION EXPLAINED

If an object is dropped toward the ground, at the end of 1 second it has accelerated to a speed of 32 ft/sec. Its average speed during that second is 16 ft/sec. Thus it drops, in 1 second, through a distance of 16 ft. The moon, which accelerates only 1/3600 as much, drops toward the earth only about 16/3600 ft in 1 second, or about $\frac{1}{20}$ in. In other words, as the moon moves forward in its orbit for 1 second and travels 3350 ft (about ⅔ mi) it falls $\frac{1}{20}$ in. toward the earth. However, because of the earth's curvature, the ground has fallen away under the moon by that same distance of $\frac{1}{20}$ in., so the moon is still the same distance from the earth. In this way the moon literally "falls around the earth." In the period of about 1 month it has "fallen" through one complete circuit of the earth and is back to its starting point (Figure 4-2).

Orbital motion is thus easily understood in terms of Newton's laws of motion and gravitation. At any given instant, the orbital speed of the moon would tend to carry it off in space in a straight line tangent to its orbit. The gravitational attraction between the earth and moon provides the proper centripetal force to accelerate the moon into its nearly circular path. Consequently, the moon continually falls toward the earth without getting any

closer to it. The orbital motions of the planets are similarly explained.

(a) The Fable of "Centrifugal Force"

Orbital motion is often incorrectly described as being a balance between two forces, the force of

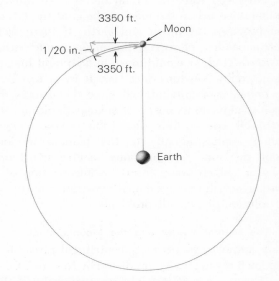

FIG. 4-2   Velocity and acceleration of the moon.

gravitation pulling the moon toward the earth, and an outward "centrifugal force" that keeps the moon from falling into the earth. This explanation, however, is more confusing than explanatory.

If one whirls a rock around over his head at the end of a string, he feels a "tug" in the string. This tug is the tension in the string. It is the result of the mutual pull between his hand and the stone. He provides, through the string, the centripetal force needed to accelerate the stone into a circular path. The stone tugs with a mutual force on the person's hand. However, there is no outside force pulling the stone away from him. If there were, and the string were to break, removing the central or centripetal force, the stone would move radially away from the person. Actually, if the string broke, the stone would fly off in a straight line in a direction *tangent* to its former circular path, continuing in the direction it was moving at the instant the string broke, in accord with Newton's first law of motion.

In short, there is no such thing as centrifugal (or outward) force in orbital motion. No force is pulling outward on the moon to balance the attraction between the moon and earth. If there were such a balance of forces, the moon would either not move at all or would move in a straight line, as predicted by Newton's first law. It is, in fact, just the *unbalanced* gravitational force that causes the moon to move in its nearly circular path around the earth. Of course, this discussion of the moon's motion applies equally to the planets' motions about the sun. (A "fictitious" centrifugal force, however, which arises from the introduction of a rotating coordinate system, is sometimes useful in the solution of difficult problems.)

(b) The Mutual Forces and the Moon's Mass
Many authors explain that "centrifugal force" is a reaction force to gravity, predicted by Newton's third law. This too is incorrect. The reaction to the earth's gravitational force on the moon is the moon's equal but opposite gravitational attraction for the earth. In other words, both pull mutually and centrally *toward* each other; there is no *outward* "reaction" force. The earth and moon suffer the same change of momentum as they revolve around each other, but the earth, being much more massive than the moon, accelerates far less. The center of mass about which they mutually revolve can be located by making careful observations of the other planets — especially of Mars — and, more recently, by tracking space probes. Small monthly periodic variations in the apparent motion of Mars result from the earth's monthly revolution about the barycenter. From the size of those variations we find that the barycenter is about 3000 mi from the center of the earth, or about 1000 mi below its surface. The moon is about 81 times as far from the barycenter and so is correspondingly less massive than the earth (Chapter 6).

4.3 NEWTON'S DERIVATION OF KEPLER'S LAWS
Kepler's laws of planetary motion are *empirical* laws; that is, they describe the way the planets are *observed* to behave. Kepler himself did not succeed in finding more fundamental laws or relationships from which his three laws of planetary motion would follow. On the other hand, Newton's laws of motion and gravitation were proposed by him as the basis of all mechanics. Thus it should be possible to derive Kepler's laws from them. Newton did, in fact, show that the motions of the planets, as described by Kepler, followed from his fundamental postulates.

To investigate the relative motion of two bodies under the influence of their mutual gravitation, Newton had to invent a new branch of mathematics, which he called *fluxions* (we call it *differential calculus*). Newton found that the motion of one body with respect to the other did not have to be an ellipse, but did have to be some kind of conic section — an ellipse, circle, hyperbola, or parabola. Which conic section describes the relative orbit depends on the initial separation and relative speed of the bodies.

Circular and parabolic orbits require theoretically precise speeds that would not be expected to occur in nature; thus we would not expect to find a planet (or other object) with *exactly* a circular or parabolic orbit. The latter divides the family of elliptical (closed) from the family of hyperbolic (open) orbits that actually do occur. The planets, of course, do not have hyperbolic orbits or they would long since have receded into interstellar space; their orbits, then, must be elliptical, as found by Kepler.

Newton also showed that the law of areas must hold, not only in closed (elliptical) orbits, but between *any* two bodies moving under the influence of an attractive force between their centers. The law of areas happens to be the geometrical manifestation of the principle of mechanics known as the *conservation of angular momentum.*

Finally, Newton derived the period of revolution of two bodies, of masses m_1 and m_2, revolving about each other under the attraction of their mutual gravitation. If the bodies have a relative orbit of semimajor axis a, their period of mutual revolution, P, is related to a by the equation:

$$(m_1 + m_2)P^2 = \frac{4\pi^2}{G} a^3.$$

Then the above equation looks the same as the formula we gave in Chapter 3 for Kepler's third law, except for the factor $m_1 + m_2$ and the factor $4\pi^2/G$. The latter is simply a constant of proportionality. If the proper units are chosen for distance, time, and mass (astronomical units, years, and solar masses, for example, or the moon's distance and period and the earth's mass), G will take on such a value that $4\pi^2/G$ is equal to unity.

Newton derived his equation, not only for the planets moving about the sun, but also for any pair of mutually revolving bodies — two stars, a planet and a satellite, or even a plate and a spoon revolving about each other in space.

Newton's version of Kepler's third law differs from the original in that it contains as a factor the sum of the masses of the two revolving bodies. It will become clear why Kepler was not aware of that term if we note that we can consider the sun and earth to be a pair of mutually revolving bodies. We shall see in Chapter 15 that the sun has a mass of about 330,000 times that of the earth. Thus the combined mass of the sun and the earth is, to all intents and purposes, the mass of the sun itself, the earth's mass being negligible in comparison. Suppose we choose the mass of the sun for our unit of mass. Then, in the earth-sun system, $m_1 + m_2 = 1$. Furthermore, the sum of the masses of the sun and any other planet is also very nearly unity. Even Jupiter, the most massive planet, has only 1/1000 of the mass of the sun; for the sun and Jupiter, $m_1 + m_2 = 1.001$, a number so nearly equal to 1.000 that Kepler was unable to detect the difference from Tycho's observations. (However, the fact that the masses of Jupiter and Saturn are not completely negligible compared to the sun accounts, in part, for the slight discrepancies in Kepler's version of his third law as applied to Jupiter and Saturn; see Table 3.1.) Thus, if we apply the equation Newton derived to the mutual revolution of the sun and a planet, and choose years and astronomical units for the units of time and distance, and the solar mass for the unit of mass, Newton's equation reduces to

$$(m_1 + m_2)P^2 = (1)P^2 = P^2 = a^3,$$

in agreement with Kepler's formulation of the law.

(a) Kepler's Laws Restated — Summary

We now restate Kepler's three laws of planetary motion in their more general form, as they were derived by Newton:

KEPLER'S FIRST LAW: *If two bodies interact gravitationally, each will describe an orbit that is a conic section about the common center of mass of the pair. In particular, if the bodies are permanently associated, their orbits will be ellipses. If they are not permanently associated, their orbits will be hyperbolas.*

KEPLER'S SECOND LAW: *If two bodies revolve about each other under the influence of a central force (whether or not in a closed elliptical orbit), a line joining them sweeps out equal areas in the orbital plane in equal intervals of time.*

KEPLER'S THIRD LAW: *If two bodies revolve mutually about each other, the sum of their masses times the square of their period of mutual revolution is in proportion to the cube of the semimajor axis of the relative orbit of one about the other.*

4.4 MASSES OF PLANETS AND STARS

Our only means of measuring the masses of astronomical bodies is to study the way in which they react gravitationally with other bodies. Newton's derivation of Kepler's third law, which includes a term involving the sum of the masses of the revolving bodies, is most useful for this purpose.

Consider a planet, such as Jupiter, that has one or more satellites revolving about it. We can select one of those satellites, and regard it and its parent planet as a pair of mutually revolving bodies. We measure the period of revolution of the satellite (say, in sidereal months), and the distance of the satellite from the planet (in terms of the distance of the moon from the earth), and insert those values into the equation

$$m_1 + m_2 = \frac{a^3}{P^2}.$$

Since both a and P are observed, we can immediately calculate the combined mass of the planet

and its satellite. Obviously most of this mass belongs to the planet, its satellites all being very small compared to it. Thus $m_1 + m_2$ is, essentially, the mass of the planet in terms of the mass of the earth.

As a numerical example, Deimos, the outermost satellite of Mars, has a sidereal period of 1.262 days and a mean distance from the center of Mars of 14,600 mi. In sidereal months, the period of the satellite is $1.262/27.3 = 0.0463$. In units of the distance of the moon from the earth, Deimos has a distance from the center of Mars of $14,600/239,000 = 0.0611$. Thus the mass of Mars plus the mass of Deimos is given by

$$m_{\text{Mars}} + m_{\text{Deimos}} = \frac{(0.0611)^3}{(0.0463)^2} = \frac{2.28 \times 10^{-4}}{2.14 \times 10^{-3}}$$

$$= 0.11 \text{ earth mass.}$$

Since Deimos is a very tiny satellite (only about 5 mi across), its mass can be neglected compared to that of Mars, and we find that Mars has a mass of just over one tenth that of the earth.

In Chapter 15 we shall see that we use the same mathematical technique to determine the masses of stars that are members of binary-star systems (a binary star is a pair of stars that revolve around each other). In fact, we can use Newton's version of Kepler's third law to estimate the mass of our entire Galaxy (Chapter 19), or even of other galaxies (Chapter 21).

4.5 ORBITS OF PLANETS

A classical problem in celestial mechanics is to compute the orbit of a planet (or minor planet) from observations of its directions at various times as seen from the earth. Kepler determined the orbit of Mars geometrically from Tycho Brahe's data. With the additional knowledge of Newton's laws of motion and gravitation it is possible to find the orbit of a planet with far fewer observational data. In 1801 Karl Friedrich Gauss (1777–

1855) invented a method of determining the orbit of an object moving around the sun from observations of it extending over only a few weeks. Since Gauss's time various mathematical techniques have been developed to handle the same problem. The complication is that we observe the other planets from a moving earth. Accurate positions in the sky of the object on each of at least three different dates, preferably separated by more than a week, must be known to calculate its detailed path around the sun. The orbit, when finally determined, is usually designated by a set of quantities known as its elements.

Two of the orbital elements, the semimajor axis and the eccentricity, determine the size and shape of the orbit, as we have seen. Another element is the inclination of the orbital plane of the object to the plane of the earth's orbit. Two other elements indicate the orientations of the line of intersection of the two planes and of the long axis of the object's orbit in its orbital plane. A final element gives one of the exact times that the object passed perihelion (nearest approach to the sun), so that its position in its orbit can be calculated for any time. There are various other combinations of data that also specify the orbit uniquely, but these are the ones most used in astronomical tradition.

4.6 ARTIFICIAL SATELLITES

An artificial satellite is a manmade object that is in orbit around the earth. It is an astronomical body in its own right. If some of the artificial satellites that have been launched are temporary astronomical objects, it is because they dip into the atmosphere of the earth during some portions of their revolutions. The friction of the air causes a satellite to lose energy so that eventually it spirals into the denser part of the atmosphere where friction heats it until it burns up completely. If an artificial satellite is launched so that its entire orbit

is outside the earth's atmosphere, it will remain in orbit indefinitely as an astronomical body.

To illustrate how a satellite can be launched, imagine a man on top of a high mountain, firing a rifle in a direction exactly parallel to the surface of the earth (Figure 4-3 — adapted from a similar one in Newton's *Principia*). Imagine, further, that the friction of the air could be removed, and that all hindering objects, such as other mountains, buildings, and so on, were absent. Then the only force that acts on the bullet after it leaves the muzzle of the rifle is the gravitational force between the bullet and earth.

If the bullet is fired with muzzle velocity v_a, it will continue to have that forward speed, but meanwhile the gravitational force acting upon it will accelerate it downward so that it strikes the ground at a. However, if it is given a higher muzzle velocity v_b, its higher forward speed will carry it farther before it hits the ground, for, regardless of its forward speed, its downward gravitational acceleration is the same. Thus this faster-moving bullet

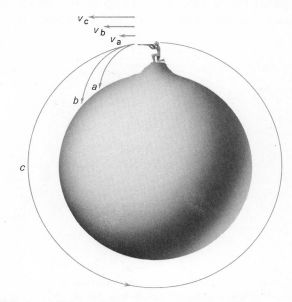

FIG. 4-3   Firing a bullet into a satellite orbit.

will strike the ground at *b*. If the bullet is given a high enough muzzle velocity, v_c, as it accelerates toward the ground, the curved surface of the earth will cause the ground to tip out from under it so that it remains the same distance above the ground, and "falls around" the earth in a complete circle. This is another way of saying that at a critical speed v_c the gravitational force between the bullet and earth is just sufficient to produce the centripetal acceleration needed for a circular orbit about the earth. The speed v_c, the *circular satellite velocity* at the surface of the earth, is about 5 mi/sec, or 18,000 mi/hr.

Novelist Jules Verne anticipated earth satellites long ago. In one of his stories an enemy force was planning to bomb a city with a gigantic cannon ball. However, the cannon ball was propelled with too great a speed — in fact, the circular satellite velocity — so it passed harmlessly over the city and on into a circular orbit around the earth.

(a) Possible Satellite Orbits*

Suppose that a missile is shot up to an altitude of a few hundred miles, then turned so that it is

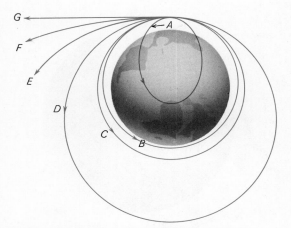

FIG. 4-4   Various satellite orbits that result from different burnout velocities but that are all parallel to the earth's surface.

moving horizontally, and finally given a forward horizontal thrust. It will proceed in an orbit the size and shape of which depend critically on the exact direction and speed of the missile at the instant of its "burnout," that is, the instant when the thrust supplied by its fuel is shut off. First, suppose that it is moving exactly horizontally, or parallel to the ground, at burnout. The possible kinds of orbits it can enter are shown in Figure 4-4.

If the missile's burnout speed is less than the circular satellite velocity, its orbit will be an ellipse, with the center of the earth at one focus of the ellipse. The *apogee* point of the orbit, that point that is *farthest* from the center of the earth, will be the point of burnout; the *perigee* point (closest approach to the center of the earth) will be halfway around the orbit from burnout.

If the burnout speed is substantially below the circular satellite velocity, most of its elliptical orbit will lie beneath the surface of the earth (orbit *A*), where, of course, the satellite cannot travel; consequently, it will traverse only a small section of its orbit before colliding with the surface of the earth (or more likely, burning up in the dense lower atmosphere of the earth). If the burnout speed is just slightly below the circular satellite velocity, the missile may clear the surface of the earth (orbit *B*), although its orbit will probably lie too low in the atmosphere for the satellite to be long-lived.

If the burnout speed were exactly the circular satellite velocity, a circular orbit centered on the center of the earth would result (orbit *C*). It is extraordinarily unlikely that a missile could be given so accurate a direction and speed that a perfectly circular orbit could be achieved. A slightly greater burnout speed will produce an elliptical orbit with *perigee* at burnout point and apogee halfway around the orbit (orbit *D*).

A burnout speed equal to the velocity of escape from the earth's surface, that is, the parabolic velocity (about 7 mi/sec), will put the missile into a parabolic orbit that will just enable the vehicle to escape from the earth into space (orbit *E*). A still higher burnout speed will produce a hyperbolic orbit in which the missile escapes the earth with energy to spare (orbit *F*). The higher the burnout speed, the nearer will the orbit be to a straight line (orbit *G*).

(b) First Artificial Satellites

For the International Geophysical Year 1957–1958, it was proposed to launch small earth satellites that would carry instruments to investigate the conditions just outside the earth's atmosphere.

The first successful launching of an artificial earth satellite, an important milestone in the history of human technology, was accomplished by the Soviet Union on October 4, 1957. This first Soviet satellite, called *Sputnik I* (the Russian word for "satellite"), had an overall weight of about 4 tons, and a scientific instrumentation package weighing 184 lb. The instrument package and launching rocket traveled about the earth in a period of 96 minutes in an elliptical orbit that ranged from 142 to 588 mi above the earth's surface. Since Sputnik I was never completely outside the earth's atmosphere, it gradually lost energy, and finally burned up in the denser lower atmosphere on January 4, 1958. On November 3, almost a month after Sputnik I was put in orbit, the Russians launched a second satellite that survived in orbit until April 14, 1958. This second satellite was very similar to the first except that it had a heavier instrumentation payload that included a live dog.

The first successful American satellite was launched by an Army Jupiter C missile on January 31, 1958. By this time, the "space age" was well under way. Satellites carry instruments that radio back to earth data about our immediate space environment. Ballistic rockets (which are satellites in a sense, as we have explained) carry instruments 100 or more miles above the earth's surface. One advantage of rockets that return to the ground is

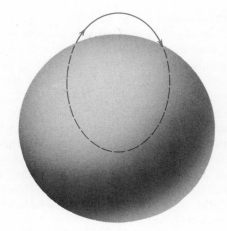

FIG. 4-5   Orbit of a ballistic missile.

that instrument payloads can sometimes be recovered intact, so that photographs can be obtained of the earth, the sun, the solar spectrum, and so on.

Among the kinds of information obtained by various rockets and space probes are data on the density and temperature of the earth's outer atmosphere, the density of micrometeorites in space, cosmic radiation, the density of charged particles near the earth, the nature of the earth's outer magnetic field, the nature of the solar radiation in the ultraviolet, the direct appearance of the sun in ultraviolet light, the appearance of the back side of the moon, the gravitational field of the earth, and ultraviolet and x-ray radiation from beyond the solar system. We shall discuss these data where appropriate in the following chapters. Information is so rapidly accumulating from satellite and rocket research, however, that we cannot be completely up to date.

4.7 INTERPLANETARY PROBES AND TRAVEL

We have now learned the principles of space travel. Rockets, once they have left the earth, are astronomical bodies. They obey the same laws of celestial mechanics as the planets and natural and artificial satellites. In other words, rockets or space probes travel in orbits. If the space vehicles carry auxiliary rocket engines and extra fuel, it may be possible to alter their orbits at will, but the principles remain the same.

There are many kinds of orbits a space vehicle can follow to reach, say, one of the planets. Those that require the expenditure of the least energy as the rocket leaves the earth are, of course, the most economical of fuel. The orbits of the successful United States *Mariner* Venus and Mars probes, and of the similar Soviet probes, were nearly of this type.

Suppose, for simplicity, that the orbits of Venus, Earth, and Mars are circles centered on the sun (when the slight ellipticity of planetary orbits is taken into account, the problem is similar but slightly more complicated). The least-energy orbit that will take us to Mars is an ellipse tangent to the earth's orbit at the space vehicle's

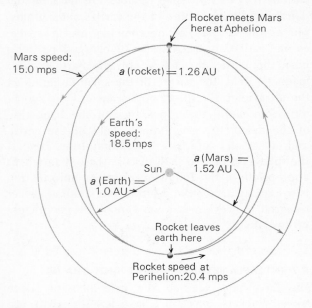

FIG. 4-6 Least-energy orbit to Mars.

perihelion (closest approach to the sun) and tangent to the orbit of Mars at the vehicle's *aphelion* (farthest from the sun) (see Figure 4-6).

The earth is traveling around the sun at the right speed for a circular orbit. For us on the earth to enter the elliptical orbit to Mars, we must achieve a speed, in the same direction as the earth is moving, that is slightly greater than the earth's circular velocity (which is about 18.5 mi/sec). The required speed turns out to be slightly over 20 mi/sec. Since the earth is already moving 18.5 mi/sec, we need to leave the earth with the proper speed and direction so that when we are far enough from it that its gravitational influence on us is negligible compared to the sun's, we are still moving in the same direction as the earth with a speed relative to it of just under 2 mi/sec.

We have now entered an orbit that will carry us out to the orbit of Mars. The time required for the trip can be found from Kepler's third law, because our spaceship is a planet. The period required to traverse the entire orbit is 1.41 years. The time required to reach the aphelion point (Mars' orbit) is half of this, or about 8½ months. The trip will have to be planned very carefully so that when we reach the aphelion point of the least-energy orbit, Mars will be there at the same time. Space probes generally carry rocket engines that can be activated by radio command from the earth, so that minor corrections in their trajectories can be made, as necessary, for them to achieve their missions.

The return trip from Mars to the earth is of the same type as the trip from the earth to Venus. The least energy orbit to Venus is similar to that to Mars, except now it is at the *aphelion* point that the trajectory ellipse is tangent to the earth's orbit, and at the *perihelion* point that it is tangent to the orbit of Venus (Figure 4-7). The space vehicle would have to leave the earth, as before, with enough speed so that when it has left the earth's vicinity it has a speed with respect

a (rocket) =
0.86 AU

Rocket leaves earth
here at Aphelion;
speed = 17 mps

Venus speed:
21.8 mps

a (Earth) =
1.0 AU

Sun

a (Venus) =
0.72 AU

Earth's speed:
18.5 mps

Rocket meets Venus
here at Perihelion

FIG. 4-7   Least energy orbit to Venus.

to the earth of 1.5 mi/sec but in a direction *oppo-site* that of the earth's motion. Then, relative to the sun, the vehicle is moving at 17 mi/sec and will reach the orbit of Venus along the desired elliptical orbit. The travel time to Venus, found as before from Kepler's third law, is about 5 months. Returning from Venus to earth is similar to traveling from earth to Mars.

From the foregoing discussion it is obvious that the earth and the planet to be visited must be at a critical configuration at the time of launch, in order that the space vehicle meet the planet at the other end of the vehicle's heliocentric orbit. These critical configurations occur at intervals equal to the synodic period of the planet. In practice it is not necessary, and is seldom feasible, to launch the rocket at exactly the proper instant to achieve the least-energy orbit. However, there is a short range of time (typically a few weeks) during which a *nearly* least-energy orbit can be achieved. The length of this time period, called a "window" in space jargon, depends on the thrust capabilities of the available rockets (that is, on

how much energy, above the least possible needed, can be supplied by the rocket). "Windows" for Mars journeys occur at intervals of about 780 days; those for Venus trips at intervals of about 584 days.

4.8 THE PROBLEM OF MORE THAN TWO BODIES
Until now we have considered the sun and a planet, or a planet and one of its satellites, as a pair of mutually revolving bodies. Actually, the planets (and different satellites of a planet) exert gravitational forces upon each other as well. These interplanetary attractions cause slight variations from the orbits that would be expected if the gravitational forces between planets were neglected. Unfortunately, the problem of treating the motion of a body that is under the gravitational

FIG. 4-8   The Jet Propulsion Laboratory's Mariner II spacecraft which made a successful flight to the vicinity of Venus in 1962.
(Jet Propulsion Laboratory.)

influence of two or more other bodies is very complicated, and is called the *multibody* or *n-body problem*.

The *n-body* problem is, in general, that of describing the motion of any body in a collection (or cluster) of many objects all interacting under the influence of their mutual gravitation. Unlike the case in the two-body problem, for 3 or more bodies it is not possible to write an equation that describes the trajectories (or orbits) for all time. Consequently, the *n*-body problem is often said to have no solution. Actually, by numerical calculation many problems can be solved to a desired precision, although for some problems of importance, such as the evolution of the solar system, even the biggest electronic computer is not adequate. In principle, however, the *n*-body problem is not solvable only in the sense that a single equation does not describe the motion.

Fortunately, the many-body problem can be solved rather accurately when a given body feels predominantly the gravitational force of one other mass. The motion of a planet around the sun, for example, is determined mainly by the gravitational force between it and the sun, the force between it and any of the other planets being very small in comparison. Thus the influences of the other planets can be regarded as small corrections to be applied to the two-body solution; these corrections are called *perturbations*.

4.9 THE GRAVITATIONAL EFFECTS OF NONSPHERICAL BODIES

Bodies with spherical symmetry act, gravitationally, as point masses, for which the gravitational influences are easily calculated. In nature, however, most bodies are not exactly spherical, and the simple two-body theory does not give precise results. If the shape of a body deviates only slightly from a sphere, we usually approximate its gravitational influence by that produced by a point mass and treat the small effects of its asphericity as perturbations. A common cause of the deformation of a star or planet from a perfect sphere is its rotation. Rapidly rotating planets, such as Jupiter, are noticeably flattened. The rotational flattening of the earth is slight but is important.

(a) The Shape of the Earth

Because of the earth's rotation, the inertia of its constituent parts tends to make them fly off tangentially into space. Therefore, each particle on and in the earth must be undergoing a constant *centripetal* acceleration to keep it in place. Of course, it is the earth's gravitation that provides this acceleration. Only a small part of the gravitational force on an object at the earth's surface is required to provide the centripetal acceleration that keeps it on the ground; the remainder is the object's actual *weight*. In other words, the rotation of the earth slightly reduces the weights of objects on the earth's surface (and the interior parts of the earth itself as well), because some part of the gravitational force is used up in providing the centripetal acceleration.

On the other hand, a body exactly on the axis of the earth has zero speed and suffers no centripetal acceleration. Therefore, at the poles of the earth (they are the ends of its rotational axis) the full force of gravity goes into the weight of objects. As a person travels away from a pole, his weight gradually diminishes and is least at the equator, where the centripetal acceleration is greatest. The effect is slight; a 300-lb man would lose only about 1 lb this way.

Newton showed that the result of this effect should be a distortion of the earth from a purely spherical shape. Suppose the earth were spherical (Figure 4-9). The direction of the gravitational pull upon a constituent rock in the earth near its surface is toward the center of the earth. This force must produce the combined effects of the centripetal acceleration on the rock and its weight. Now the centripetal force is not directed toward the

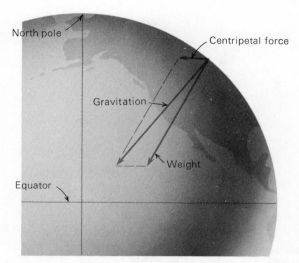

FIG. 4-9   Direction of the weight of a body on a spherical earth.

center of the earth (except for objects at its equator) but rather is toward the center of the path on which the rock is moving. Thus the centripetal force is directed perpendicularly toward the earth's axis of rotation. If we once again apply the principle of the parallelogram of forces, we see that the centripetal force must represent one side, the gravitational force the diagonal, or resultant, and the weight of a body the other side of the parallelogram. In other words, the weight of the rock in question is *not directed toward the earth's center* and is not in a direction perpendicular to the surface of the earth. Thus the surface rocks will slide toward the equator, adjusting their distribution until the weight of each *is* perpendicular to the surface of the earth at its location.

Analysis shows that the resulting shape of the earth should be nearly that of an *oblate spheroid,* a slightly flattened sphere that has an elliptical cross section. That is, the earth should be flattened at the poles, whereas the equatorial regions should be slightly bulged out. Consequently, a person at

one of the poles would be nearer the earth's center than he would be at the equator. Here is an additional effect that causes the weight of an object to be less at the equator. Both effects together, the rotation of the earth and the earth's oblate shape, reduce the weight of an object at the equator compared to its weight at a pole by 1 part in 190.

The actual oblateness of the earth is small. Its diameter from pole to pole is only 27 mi less than through the equator, about 1 part in 298.

(b) Perturbations on Earth Satellites

The equatorial bulge of the earth is responsible for a deformation in the earth's gravitational field from that which would be produced by a point mass or spherically symmetrical earth. These deformations are especially important near the surface of the earth and produce conspicuous perturbations in the orbits of low-altitude earth satellites.

Since the motions of earth satellites depend on the precise shape of the earth, careful studies of perturbations on satellite orbits should enable us to derive the earth's shape rather accurately.

Analyses of satellite perturbation, combined with sensitive measures of surface gravity over the earth, lead to an accurate knowledge of the form of the earth and permit the preparation of maps like the one in Figure 4-10, by William Kaula at UCLA. The earth is first represented by an oblate spheroid (a flattened sphere with elliptical cross section) in which the equatorial diameter exceeds the polar diameter by only 1 part in 298.25. The figures on the map are the heights, in meters, of the geoid above (positive numbers) or below (negative numbers) the surface of that oblate spheroid. The geoid is the same as mean sea level over the oceans; it is a surface everywhere perpendicular to the direction of gravity. The *topography* — the ups and downs of continents, mountains, and other crustal features — affects the geoid only indirectly through slight gravitational attraction.

FIG. 4-10 Height, in meters, of the mean surface of the earth above (positive numbers) and below (negative numbers) the surface of an oblate spheroid with a flattening of 1 part in 298.25. *(Courtesy of William Kaula, UCLA.)*

4.10 TIDES

Two perfectly spherical bodies, as has been explained, attract each other as if they were point masses, located at their own centers. If however, a body deviates from perfect sphericity, even slightly, it no longer acts as a point mass, and we have seen in Section 4.9 that satellite orbits are consequently perturbed. The shape of a planet can be affected by rotation. Another contribution to asphericity is supplied by the *differential gravitational forces* that two neighboring bodies exert on each other. These forces, in turn, result in such phenomena as *tides* and *precession*.

Early in history it was realized that tides were related to the moon, because the daily delay in high tide ("high water") is the same as the daily delay in successive transits of the moon across the local meridian. A satisfactory explanation of the tides, however, awaited the theory of gravitation, supplied by Newton.

First, we shall consider the effects of the moon's attraction on the solid earth. For the moment, we ignore the flattening of the earth due to its rotation. Our planet can be regarded as being composed of a large number of particles, each of unit mass, all bound together by their mutual gravitational attraction and cohesive forces. The gravitational forces exerted by the moon at several arbitrarily selected places in the earth are illustrated in Figure 4-11. These forces differ slightly from each other because of the earth's finite size; all parts are not equally distant from the moon, nor are they all in exactly the same direction from the moon. If the earth retained a perfectly spherical shape, the resultant of all these forces would be that of the force on a point mass, equal to the mass of the earth, and located at the earth's center. Such is approximately true, because the earth is nearly spherical, and it is this resultant force on the earth that causes it to accelerate each month in an elliptical orbit about the barycenter of the earth-moon system.

The earth, however, is not *perfectly* rigid. Consequently, the differential or *tidal* force (that is, the difference in forces) of the moon's attraction on different parts of the earth causes the earth to distort slightly. The side of the earth nearest the moon is attracted toward the moon

FIG. 4-11   Moon's attraction on different parts of the earth.

more strongly than is the center of the earth, which, in turn, is attracted more strongly than is the side of the earth opposite the moon. Thus, the differential force tends to "stretch" the earth slightly into a *prolate spheroid* with its major axis pointed toward the moon. That is, the earth takes on a shape such that a cross section whose plane contains the line between the centers of the earth and moon is an ellipse with its major axis in the earth-moon direction.

If the earth were fluid, like water, it would distort until it took on a shape such that all places at its surface were in equilibrium between the tidal forces and the gravitational pull of the earth itself. Measures have been made to investigate the actual deformation of the earth. It is found that the solid earth does distort, as would a liquid, but only about one third as much, because of the high rigidity of the earth's interior. The maximum tidal distortion of the solid earth amounts at its greatest to only about nine inches. As the earth rotates, different parts of it are continually being carried under the moon, so the direction and magnitude of the tidal force acting at any given place on the earth's surface are constantly changing. Direct observation shows that the earth readjusts its shape under the influence of the changing tidal force almost instantaneously.

Rotation, of course, also distorts the earth's shape. The equatorial bulge of the earth that re-

sults from the tidal distortions described above is superimposed on its rotation. This polar flattening of the earth is, of course, a very much greater distortion than the distortion due to the tides.

As we have seen, if the earth were fluid it would take on a shape such that points on its surface would be in equilibrium and feel no horizontal force. However, the earth is sufficiently rigid to be distorted from a sphere by only about one third of the amount required to remove all unbalanced forces. Consequently, objects at the surface of the earth that are not restrained from horizontal motion, for example, the waters in the oceans, are free to flow in the direction of the horizontal components of the tidal forces. Let us suppose, first, that the earth were covered uniformly by a deep ocean, and investigate the nature of the tides produced in it.

Water on the lunar side of the earth would be drawn toward the sublunar point (the point on the earth where the moon appears in the zenith), piling up water to greater depths on that side of the earth, with the greatest depths at the sublunar point. On the opposite side of the earth, water is drawn in the *opposite* direction, producing a tidal bulge on the side of the earth opposite the moon (see Figure 4-12). The tidal bulges in the oceans do not result from the moon compressing or expanding the water, nor from the moon lifting the water "away from the earth." Rather,

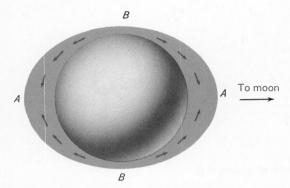

FIG. 4-12 Tidal bulges in the "ideal" oceans.

the tidal bulges result from an actual flow of water over the earth's surface, toward the regions below and opposite the moon, causing the water to pile up to greater depths at those places.

The tidal bulge on the side of the earth *opposite* the moon often seems mysterious to students who picture the tides as being formed by the moon "lifting the water away from the earth." What actually happens, of course, is that the differential gravitational force of the moon on the earth tends to stretch the earth, elongating it slightly toward the moon. The solid earth distorts slightly, but because of its high rigidity, not enough to reach complete equilibrium with the tidal forces. Consequently, the ocean, moving freely over the earth's surface, flows in such a way as to increase the elongation and piles up at points under and opposite the moon.

So far, we have regarded the earth as though its ocean waters were distributed uniformly over its surface. In this idealized picture, only approximately realized even in the largest oceans, the tides would cause the depths of the ocean to range through only a few feet. The rotation of the earth would carry an observer at any given place alternately into regions of deeper and shallower water. As he was being carried toward the regions under or opposite the moon where the water was deepest, he would say, "the tide is

coming in"; when carried away from those regions, he would say, "the tide is going out." During a day, he would be carried through two tidal bulges (one on each side of the earth) and so would experience two "high tides" and two "low tides" each day.

The sun also produces tides on the earth, although the sun is less than half as effective a tide-raising agent as the moon. Actually, the gravitational attraction between the sun and the earth is about 150 times as great as that between the earth and the moon. We recall, however, that the tidal force is the differential gravitational force of a body on the earth. The sun's attraction for the earth is much greater than the moon's, but the sun is so distant that it attracts all parts of the earth with almost equal strength. The moon, on the other hand, is close enough for its attraction on the near side of the earth to be substantially greater than its attraction on the far side. In other words, its *differential gravitational pull* on the earth is greater than the sun's, even though its total gravitational attraction is less. If there were no moon, the tides produced by the sun would be all we would experience, and the tides would be less than half as great as those we now have. The moon's tides, therefore, dominate.

The "simple" theory of tides, described in the preceding paragraphs, would be sufficient if the earth were completely surrounded by very deep oceans. However, the presence of land masses stopping the flow of water, the friction in the oceans and between oceans and the ocean floors, the rotation of the earth, the variable depth of the ocean, winds, and so on, all complicate the picture.

Both the times and the heights of high tide vary considerably from place to place on the earth. The tidal flow of waters over shallow parts of the ocean, and into irregular coastal regions, causes the tides there to range generally far more than the 2 or 3 ft that they range in the deep oceans.

In particular, where the tidal flow of water is funneled shoreward by a V-shaped inlet that narrows back away from the sea, the difference between high and low water may be especially large. Such a place is the Bay of Fundy between New Brunswick and Nova Scotia in Southeast Canada, where the tidal range sometimes exceeds 50 ft.

The prediction of the actual time of high tide at any given place is even more complicated. The earth's rapid rotation causes the tide-raising forces within a given mass of water to vary too rapidly for the water to adjust completely to them. These forces, however, recurring periodically, set up forced oscillations in the ocean surfaces, so that the water over a large area rises and lowers in step. Consequently, the highest water does not necessarily occur when the moon is on the meridian (or in the lower half of the meridian, extended below the horizon through the nadir), but rather when the oscillations of the ocean, produced by the tidal forces acting upon it, pile up the water to its greatest depth at that location. The latter depends critically upon the shape and depth of the adjacent ocean basin. The United States Coast and Geodetic Survey prepares and publishes each year the *Tide Tables,* which give the times and heights of tides at principal ports throughout the world.

4.11 PRECESSION

The earth, because of its rapid rotation, is not perfectly spherical but has taken on the approximate shape of an oblate spheroid, its equatorial diameter being 27 mi greater than its polar diameter. As we have seen, the plane of the earth's equator, and thus of its equatorial bulge, is inclined at about 23½° to the plane of the ecliptic, which, in turn, is inclined at 5° to the plane of the moon's orbit. The differential gravitational forces of the sun and moon upon the earth not only cause the tides but also attempt to pull the equatorial bulge of the earth into coincidence with the ecliptic.

The latter should be clear from inspection of Figure 4-14. The solid arrows are vectors that represent the attraction of the moon on representa-

FIG. 4-13   Bay of Fundy at high and low tides. *(National Film Board of Canada.)*

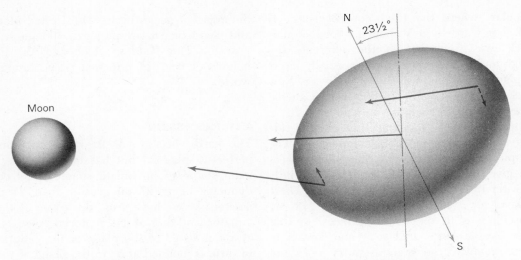

FIG. 4-14   Differential force of the moon on the oblate earth tends to "erect" its axis.

tive parts of the earth. The part of the earth's equatorial bulge nearest the moon is pulled more strongly than the part farthest from the moon, and the earth's center is pulled with an intermediate force. The dashed arrows show the differential forces with respect to the earth's center. Note how they tend not only to "stretch" the earth toward the moon, but also to pull the equatorial bulge into the plane

of the ecliptic. The differential force of the sun, although less than half as effective, does the same thing. Thus, the gravitational attractions of the sun and moon upon the earth act in such a way as to attempt to *change the direction of the earth's axis of rotation,* so that it would stand perpendicular to the orbital plane of the earth. To understand what actually takes place, we must digress for a moment to consider what happens when a similar force acts upon a top or gyroscope.

(a) Precession of a Gyroscope

Consider the top (a simple form of gyroscope) pictured in Figure 4-15. If the top's axis is not perfectly vertical, its weight (the force of gravity between it and the earth) tends to topple it over. The actual force that acts to change the orientation of the axis of rotation of the top is that component of the top's weight that is perpendicular to its axis. We know from watching a top spin that the axis of the top does not fall toward the horizontal, but rather moves off in a direction *perpendicular to the plane defined by the axis and the force tending to change its orientation.* Until the spin of the

Component of weight tending to "topple" top

Weight of top

FIG. 4-15   Precession of a top.

top is slowed down by friction the axis does not change its angle of inclination to the vertical (or to the floor), but rather describes a conical motion (a cone about the vertical line passing through the pivot point of the top). This conical motion of the top's axis is called *precession*. The phenomenon of precession may seem surprising, but it is predicted in detail by Newton's laws of motion.

(b) Precession of the Earth

The differential gravitational force of the sun on the earth tends to pull the earth's equatorial bulge into the plane of the ecliptic, and that of the moon tends to pull the bulge into the plane of the moon's orbit, which is nearly in the ecliptic. These forces, in other words, tend to pull the earth's axis into a direction approximately perpendicular to the ecliptic plane. Like a top, however, the earth's axis does not yield in the direction of these forces, but precesses. The obliquity of the ecliptic

FIG. 4-16   Precession of the earth.

remains approximately $23\frac{1}{2}°$. The earth's axis slides along the surface of an imaginary cone, perpendicular to the ecliptic, and with a half-angle at its apex of $23\frac{1}{2}°$ (see Figure 4-16). The precessional motion is exceedingly slow; one complete cycle of the axis about the cone requires about 26,000 years.

Precession is this motion of the axis of the earth. It must not be confused with *variation in latitude* (Chapter 5) which is caused by a slight wandering of the terrestrial poles with respect to the earth's surface. Precession does not affect the cardinal directions on the earth or the positions of geographical places that are measured with respect to the earth's rotational axis, but only the orientation of the axis with respect to the celestial sphere.

Precession does, however, affect the positions among the stars of the celestial poles, those points where extensions of the earth's axis intersect the celestial sphere. In the twentieth century, for example, the north celestial pole is very near Polaris. This was not always so. In the course of 26,000 years, the north celestial pole will move on the celestial sphere along an approximate circle of about $23\frac{1}{2}°$ radius, centered on the pole of the ecliptic (where the perpendicular to the earth's orbit intersects the celestial sphere). This motion of the pole is shown in Figure 4-17. In about 12,000 years, the celestial pole will be fairly close to the bright star Vega.

As the positions of the poles change on the celestial sphere, so do the regions of the sky that are circumpolar; that is, that are perpetually above (or below) the horizon for an observer at any particular place on earth. The Little Dipper, for example, will not always be circumpolar as seen from north temperate latitudes. Moreover, 2000 years ago, the Southern Cross was sometimes visible from parts of the United States. It was by noting the very gradual changes in the positions of stars with respect to the celestial poles that Hipparchus discovered precession in the second century B.C. (Section 2.3c).

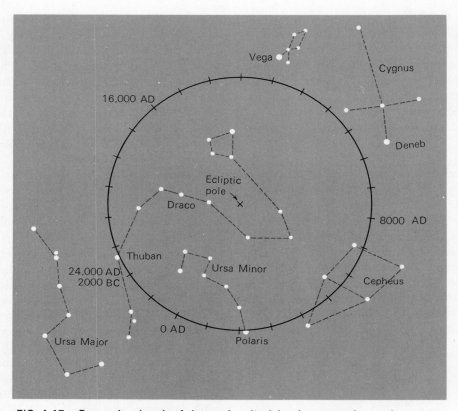

FIG. 4-17   Precessional path of the north celestial pole among the northern stars.

Exercises

1. Two bodies 5 feet apart have masses of 10 lbs. and 40 lbs., respectively. How far is their mutual center of mass from the 10-pound body?

2. Jupiter has a mass of about 0.001 that of the sun. Its distance from the sun is about 780,000,000 km. Roughly, how far is the center of mass of the sun-Jupiter system from the sun's center?

3. Two stars revolve about each other in a period of 2 years and in a relative orbit of semimajor axis 2 astronomical units. What is their combined mass in units of the solar mass?

4. A hypothetical planet has a satellite about 240,000 miles away from it, and the satellite revolves about the planet in about 14 days. Is the planet's mass greater than, the same, or less than the earth's? Why?

5. Would it be feasible to launch an artificial satellite that was intended to revolve about the earth in an orbit of semimajor axis 3 astronomical units? Explain.

6. A ballistic missile is a missile that is given an initial thrust and then allowed to coast to its target. Describe how it can be regarded as an earth satellite. What is its orbit like? Where are the foci? What are some complications that must be taken into account during launch?

7. Give some examples of cases where perturbation theory might be applied.

8. Describe how we would have to launch a rocket in order to make it hit the sun. After it is far from the earth, but still essentially the same distance from the sun that the earth is, what are its speed and direction with respect to the earth? Why should it be difficult to send a very large payload to the sun?

9. Strictly speaking, should it be a 24-hour period during which there are two "high tides?" If not, what should the interval be?

10. Does a bicycle offer another example of precession? Explain. *Hint*: Consider how a rider can steer by leaning to one side.

Following page: View of earth from Apollo 8 spacecraft. South America is conspicuous near the center. *(NASA.)*

CHAPTER 5

Earth and Sky

In the preceding chapter we were concerned with the mechanics that dictate the motions of celestial bodies. Now we turn our attention to the motion of the earth, and the relation between earth and sky. We shall consider the physical properties of the earth as a planet in Chapter 10.

The most apparent of the earth's motions are its *rotation* and *revolution*. In astronomy "rotation" is used to describe the turning of a body about an axis running through it and "revolution" to describe the motion of a body around an exterior point. Thus the earth rotates on its axis and revolves around the sun.

5.1 ROTATION OF THE EARTH

We have seen that the apparent rotation of the celestial sphere could be accounted for either by a daily rotation of the sky around the earth, or by the rotation of the earth itself. Copernicus, Kepler, Galileo, and Newton had piled up convincing circumstantial evidence in favor of a rotating earth, but it was not until the nineteenth century that simple direct proofs of the rotation of the earth were devised.

(a) The Foucault Pendulum

In 1851 the French physicist Jean Foucault suspended a 200-ft pendulum weighing 60 lb from the domed ceiling of the Pantheon in Paris. He started the pendulum swinging evenly by drawing it to one side with a cord and then burning the cord. The direction of swing of the pendulum was recorded on a ring of sand placed on a table beneath its point of suspension. At the end of each swing a pointed stylus attached to the bottom of the bob cut a notch in the sand. Foucault had taken great care to avoid air currents and other influences that would disturb the direction of swing of the pendulum. Yet, after a few moments

FIG. 5-1   A Foucault pendulum. *(Griffith Observatory.)*

Lake Titicaca in the Andes. *(Gemini IX photograph; NASA.)*

The Southern Cross. *(John B. Irwin.)*

Moon and clouds over the western Pacific. *(Gemini VII photograph; NASA.)*

Start of solar eclipse produced by the earth, photographed from the moon by Surveyor III, April 24, 1967. The color is obtained by reconstruction from three black and white television photographs made through different filters and transmitted to earth. *(NASA.)*

Same eclipse, 38 minutes later. The bright ring is sunlight refracted into the earth's shadow and onto the moon by the earth's atmosphere. *(NASA.)*

it became apparent that the plane of oscillation of the pendulum was slowly changing with respect to the ring of sand, and hence with respect to the earth.

The only force acting upon the pendulum was that of gravity between it and the earth, and, of course, this force was in a downward direction. If the earth were stationary, there would be no force that could cause the plane of oscillation of the pendulum to alter, and, in accord with Newton's first law, the pendulum should continue to swing in the same direction. The fact that the pendulum slowly changed its direction of swing with respect to the earth is proof that the earth rotates.

It is comparatively easy to visualize a Foucault pendulum experiment at the North Pole. Here we can imagine the plane of swing of the pendulum maintaining a fixed direction in space with respect to the stars, while the earth turns under it. Thus, at the North (or South) Pole, a pendulum would *appear* to rotate its plane of oscillation completely in one day. At other places than the poles, the problem is complicated because the pendulum must always swing in a vertical plane that passes through the center of the earth. That plane of oscillation obviously must change with respect to the stars.† However, the fact that the pendulum appears to change at all proves that the earth rotates — or at least that its rotation is the only alternative consistent with Newton's laws.

5.2 RELATION OF EARTH AND SKY

(a) Positions on the Earth

To denote positions of places on the earth, we must set up a system of coordinates on the earth's surface. The earth's axis of rotation (that is, the locations of its North and South Poles) is the basis for such a system.

A *great circle* is any circle on the surface of a sphere whose center is at the center of the sphere.

The earth's *equator* is a great circle on the earth's surface halfway between the North and South Poles. We can also imagine a series of great circles that pass *through* the North and South Poles. These circles are called *meridians;* they intersect the equator at right angles.

A meridian can be imagined passing through an arbitrary point on the surface of the earth (see Figure 5-2). This meridian specifies the east-west location of that place. The longitude of the place is the number of degrees, minutes, and seconds of

---

†The following is a brief derivation of the period of a Foucault pendulum: Let the angular velocity of the earth about its polar axis be ω (see the figure). Then the component of this angular velocity about a radius vector through a point on the earth's surface at latitude ϕ is $\omega \sin \phi$. The time required for one full rotation about this radius vector is $360°/\omega \sin \phi$. Since $\omega = 360°/(23$ hours 56 minutes$)$, the period of a pendulum at latitude ϕ is

$$P = \frac{23^h 56^m}{\sin \phi}.$$

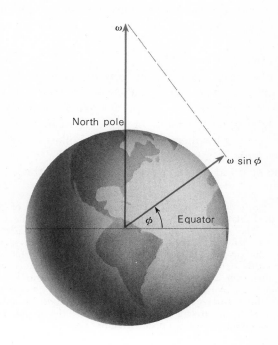

arc along the equator between the meridian passing through the place and the one passing through Greenwich, England, the site of the old Royal Observatory. Longitudes are measured either to the east or west of the Greenwich meridian from 0 to 180°. The convention of referring longitudes to the Greenwich meridian is of course completely arbitrary. As an example, the longitude of the bench mark in the clock house of the Naval Observatory in Washington D.C., 77°03′56″.7 W. Note in Figure 5-2 that the number of degrees along the equator between the meridians of Greenwich and Washington is also the angle at which the planes of those two meridians intersect at the earth's axis.

The latitude of a place is the number of degrees, minutes, and seconds of arc measured along its meridian to the place from the equator. Latitudes are measured either to the north or south of the equator from 0 to 90°. As an example, the latitude of the above-mentioned bench mark is 38°55′14″.0 N. Note that the latitude of Washington is also the angular distance between it and the equator as seen from the center of the earth.†

(b) Positions in the Sky

In denoting positions of objects in the sky, it is often convenient to make use of the fictitious *celestial sphere,* a concept, we recall, that many early peoples accepted literally. We can think of the celestial sphere as being a hollow shell of extremely large radius, centered on the observer. The celestial objects appear to be set in the inner

---

†Strictly, this is *geocentric* latitude. Because of the earth's oblate shape, there are several ways to define latitude. The *geodetic* (or *geographical*) latitude commonly used is defined by the angle between the equatorial plane and the perpendicular to the mean "sea-level" surface of the earth at the place in question. It may differ by several minutes from geocentric latitude. *Astronomical* latitude and longitude, obtained directly from astronomical observations, may differ from geodetic latitude and longitude by a few seconds of arc, owing to the deflection of the plumb bob by mountains or other crustal irregularities.

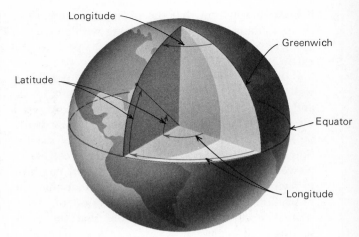

FIG. 5-2   Latitude and longitude of Washington, D.C.

surface of this sphere, so we can speak of their positions *on* the celestial sphere. We can devise coordinate systems, analogous to latitude and longitude, to designate these positions. Of course we are really only denoting their *directions* in the sky.

The point on the celestial sphere directly overhead an observer (defined as opposite to the direction of a plumb bob) is his *zenith*. Straight down, 180° from his zenith, is the observer's *nadir*. Halfway between, and 90° from each, is his *horizon*. (This is the *celestial* horizon and will not necessarily coincide with the apparent horizon, which may be interrupted with such things as mountains, buildings, and trees.) Note that observers at different places have different zeniths, nadirs, and horizons.

CELESTIAL EQUATOR AND POLES

The apparent rotation of the sky takes place about an extension of the earth's axis of rotation. That is, the sky appears to rotate about points directly in line with the North and South Poles of the earth — the *north celestial pole* and the *south celestial pole*. Halfway between the celestial poles, and thus 90° from each, is the *celestial equator,* a great circle on the celestial sphere that is in the same plane as the earth's equator; it would appear to pass directly through the zenith of a person on the equator of the earth. Great circles passing through the celestial poles and intersecting the celestial equator at right angles (analogous to meridians on the earth) are called *hour circles.*

THE CELESTIAL MERIDIAN

The great circle passing through the celestial poles and the zenith (and also through the nadir) is called the observer's *celestial meridian.* It coincides with the projection of his terrestrial meridian, as seen from the earth's center, onto the celestial sphere. The celestial meridian intercepts the horizon at the *north* and *south* points on the horizon. Halfway between these north and south points are the *east* and *west* points on the horizon.

As the earth turns, the observer's terrestrial meridian moves under the celestial sphere; therefore, his celestial meridian sweeps continuously around the celestial sphere. Another way of putting it is to say that as the sky turns around the earth, the stars pass by the observer's stationary celestial meridian.

It helps to visualize these circles in the sky if we imagine that the earth is a hollow transparent

spherical shell with the terrestrial coordinates (latitude and longitude) painted on it. Then we imagine ourselves at the center of the earth, looking out through its transparent surface to the sky. The terrestrial poles, equator, and meridians will be superimposed upon the celestial ones.

ALTITUDE AND AZIMUTH

The most obvious coordinate system is based on the horizon and zenith of the observer. Great circles passing through the zenith (*vertical circles*) intersect the horizon at right angles. Imagine a vertical circle through a particular star (Figure 5-3). The *altitude* of that star is the number of degrees along this circle from the horizon up to the star. It is also the angular "height" of the star as seen by the observer.

The *azimuth* is the number of degrees along the horizon to the vertical circle of the star from some reference point on the horizon. In astronomical tradition, azimuth formerly was measured from the south point on the observer's horizon, but in modern practice azimuth is measured from the north point, in conformity with the convention of navigators and engineers. In any case, azimuth is measured to the east (clockwise to one looking down from the sky) along the horizon from 0 to 360°.

RIGHT ASCENSION AND DECLINATION

The principal disadvantage of the altitude and azimuth system (the *horizon system*) is that as the earth turns, the coordinates of the celestial objects are constantly changing. It is desirable, therefore, to devise a coordinate system that is attached to the celestial sphere itself, just as the system of latitude and longitude is permanently attached to the earth. Then the positions of the stars remain fixed rather than changing rapidly as the earth's rotation causes the sky to rotate. A system which comes close to meeting these requirements is *right ascension* and *declination,* or the *equator system.*

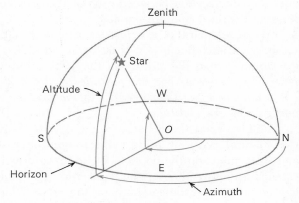

FIG. 5-3 Altitude and azimuth.

This and several other celestial coordinate systems are in common use. Each has its advantages for special purposes. It is important for astronomers and navigators to understand these systems, but their value to students taking a survey course in astronomy may not be great unless practical exercises or sessions at the telescope are included. We shall not, therefore, describe astronomical coordinate systems here, but rather refer the reader to Appendix 7.

(c) The Orientation of the Celestial Sphere

The next step is to determine the orientation of the celestial sphere with respect to the zenith and horizon of a particular observer on the earth. At the North (or South) Pole, the problem is very simple indeed. The north celestial pole, directly over the earth's North Pole, appears at the zenith. The celestial equator, 90° from the celestial poles,

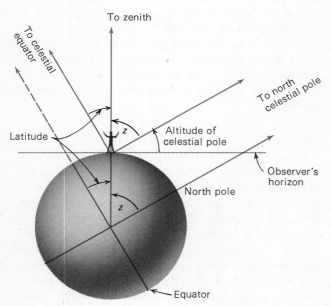

FIG. 5-4 The altitude of the celestial pole equals the observer's latitude.

lies along the horizon. An observer at one of the terrestrial poles would always see the same half of the sky.

At the equator the problem is almost as simple. The celestial equator, in the same plane as the earth's equator, passes through the zenith, and since it runs east and west, it intersects the horizon at the east and west points. The celestial poles, being 90° from the celestial equator, must be at the north and south poles on the horizon. Evidently at points on the earth between the equator and poles, one of the celestial poles must be a certain distance above the horizon.

Now consider an observer at an arbitrary latitude on the earth. In Figure 5-4 the latitude is shown as north of the equator, but for a southern latitude the case would be exactly analogous. Since the terrestrial North Pole is on the observer's terrestrial meridian, the north celestial pole will have to be on his celestial meridian, at some altitude above the north point on the horizon. Suppose the angle from the observer's zenith down to the north celestial pole is z. Being on the celestial sphere, the north celestial pole is so distant that an observer at the center of the earth would see it in the same direction as our observer. Thus the angle at the center of the earth between the observer and the north celestial poles is also z. (See Figure 5-4; the angles of intersection between each of two parallel lines and a third line are equal.)

We recognize that z is just 90° *minus* the observer's latitude. But also z is 90° (the altitude of the zenith) *minus* the altitude of the north celestial pole. Thus we see that *the altitude of the north (or south) celestial pole is equal to the observer's north (or south) latitude.*

Finally, since the celestial equator is 90° from the celestial poles, it must cut through the east and west points on the horizon, tilt southward as it extends up above the horizon, and cross the celestial meridian a distance south of the zenith that is also equal to the observer's latitude.

(d) The Motion of the Sky As Seen From Different Places On Earth

Imagine an observer at the earth's North Pole. The celestial north pole is at his zenith and the celestial equator along his horizon. As the earth rotates, the sky turns about a point directly overhead. The stars neither rise nor set; they circle parallel to the horizon. Only that half of the sky that is north of the celestial equator is ever visible to this observer. Similarly, an observer at the South Pole would only see the southern half of the sky (see Figure 5-5).

The situation is very different for an observer at the equator (Figure 5-6). There the celestial poles, the points about which the sky turns, lie at the north and south points on the horizon. All stars rise and set; they move straight up from the east side of the horizon and set straight down on the west side. During a 24-hour period, all stars are above the horizon exactly half the time.

For an observer between the equator and North Pole, say at 34° north latitude, the situation

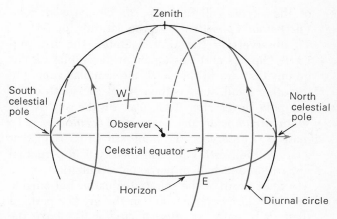

FIG. 5-6   Sky from the equator.

is as depicted in Figure 5-7. Here the north celestial pole is 34° above the observer's northern horizon. The south celestial pole is 34° *below* the southern horizon. As the earth turns, the stars appear to circle around and around, parallel to the celestial equator. The whole sky seems to pivot about the north celestial pole. For this observer, stars within 34° of the North Pole can never set. They are always above the horizon, day and night. This part

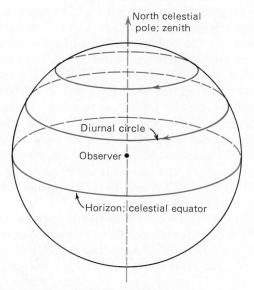

FIG. 5-5   Sky from the North Pole.

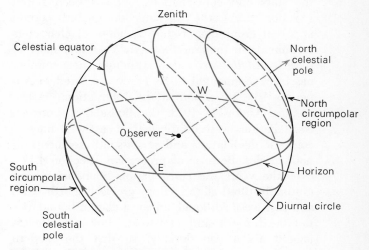

FIG. 5-7   Sky from latitude 34° N.

of the sky is called the *north circumpolar zone* (*perpetual apparition*) at the latitude of 34° N. To observers in the United States, the Big and Little Dippers and Cassiopeia are examples of star groups that are in the north circumpolar zone. On the other hand, stars within 34° of the south celestial pole never rise. That part of the sky is the *south circumpolar zone* (*perpetual occultation*). To most U.S. observers, the Southern Cross is in that zone. At the North Pole, half the sky is in perpetual apparition and half in perpetual occultation.

Stars north of the celestial equator, but outside the north circumpolar zone, in the greater parts of their daily paths, or *diurnal circles,* lie above the horizon; hence they are up more than half the time. Stars *on* the celestial equator are up exactly half the time, for their diurnal circle is the celestial equator. Because it is a great circle, exactly half of it must be above the horizon. Stars south of the celestial equator, but outside the south circumpolar zone, are up less than half the time.

(e) Nomenclature of Stars

While we are discussing the celestial sphere, it is in order to describe briefly the system for naming the stars and constellations. As has been stated (Section 2.2a) the ancients designated certain apparent groupings of stars in honor of characters or animals in their mythology. We retain most of these *constellations* today, although their number has been augmented to 88 (they are listed in Appendix 16). By action of the International Astronomical Union in 1928, the boundaries between constellations were established as east-west lines of constant declination and north-south segments of hour circles. Because of precession (Section 4.11), over the years the constellation boundaries have gradually tilted slightly from precisely north-south and east-west. Although all constellation boundaries ran north-south and east-west, they nevertheless jogged about considerably, so that the modern constellations still contain most or all of the brighter stars assigned to them by the ancients. Consequently, the boundaries often delineate highly irregular regions of the sky, reminding one of the boundaries between Congressional districts that result from the gerrymandering practices of many state legislatures. At any rate, because every position on the celestial sphere (or every direction in the sky) lies in one or another constellation, we commonly use constellations today to designate the places of stars or other celestial objects.

Many of the brighter stars have proper names. Often these are Arabic names that describe the positions of the stars in the imagined figures that the Greek constellations represented. For example, *Deneb,* is Arabic for "tail" and is the star that marks the tail of *Cygnus* (the Swan), and *Denebola* is the star at the tail of *Leo* (the Lion). Some star names, however, are of modern origin; for example, *Polaris,* the "pole star."

A superior designation of stars was introduced by a Bavarian, J. Bayer, in his *Atlas* of the constellations, published in 1603. He assigned successive letters of the Greek alphabet to the more conspicuous stars in each constellation, in approximate order of decreasing brightness. The full star designation is the Greek letter, followed by the genitive form of the constellation name. Thus Deneb, the brightest star in Cygnus, is α *Cygni,* and Denebola, the second brightest in Leo, is β *Leonis.* Bayer's ordering of stars by brightness was not always correct, and on occasion he deviated from the scheme altogether and assigned letters to stars of comparable brightness according to their geometrical arrangement in the constellation figure, for example, in the Big Dipper. Bayer's nomenclature is still in wide use today.

Fainter stars were subsequently given number designations, with the numbers increasing in order of the stars' right ascensions. The majority of stars, however, too faint to see without a telescope, are designated, if at all, only by their numbers in various catalogues. Data for the nearest stars, listed in Appendix 12, are seen, from the variety of nomenclature, to have been selected from several different catalogues. Many stars, of course, are listed in more than one catalogue and thus bear various names — a circumstance that has sometimes confused even astronomers. The vast majority of stars, however, too faint and numerous to measure and catalogue, remain nameless.

5.3 TIME OF DAY

The measurement of time is based on the rotation of the earth. As the earth turns, objects in the sky appear to move around us, crossing the meridian each day. Time is determined by the position in the sky, with respect to the local meridian, of some reference object (for example, the sun) on the celestial sphere. The interval between successive meridian crossings or *transits* of that object is defined as a *day*. The actual length of a day depends on the reference object chosen; several different kinds of days, corresponding to different reference objects, are defined. Each kind of day is divided into 24 equal parts, called *hours*.

(a) The Passage of Time; Hour Angle

Time is reckoned by the angular distance around the sky that the reference object has moved since it last crossed the meridian. The motion of that point around the sky is analogous to the motion of the hour hand on a 24-hour clock. The angle measured to the west along the celestial equator from the local meridian to the hour circle passing through any object (for example, a star) is that object's *hour angle*. (An hour circle is a great circle on the celestial sphere running north and south through the celestial poles.) *Time* can be defined as the *hour angle of the reference object*.

As an example, suppose that the star *Rigel* is chosen as the reference for time. Then when Rigel is on the meridian it is $0^h0^m0^s$, "Rigel time." Twelve *Rigel hours* later, Rigel is halfway around the sky,

at an hour angle of 180°, and the Rigel time is $12^h0^m0^s$. When Rigel is only 1° east of the meridian, and one *Rigel day* is nearly gone, the star is at an hour angle of 359°, and the Rigel time is $23^h56^m0^s$.

Because of the relation between hour angle and time, it is often convenient to measure angles in time units. In this notation, 24 hours corresponds to a full circle of 360°, 12 hours to 180°, 6 hours to 90°, and so on. One hour equals 15°, and 1° is 4 minutes of time.

Here we must distinguish between minutes and seconds of *time* (subdivisions of an hour), denoted ^{m} and ^{s}, respectively, and minutes and seconds of *arc* (subdivisions of a degree), denoted ′ and ″, respectively. The conversion between units of time and arc is given in Table 5.1.

TABLE 5.1 Conversion Between Units of Time and Arc

TIME UNITS	ARC UNITS
24^h	360°
1^h	15°
4^m	1°
1^m	15′
4^s	1′
1^s	15″

Because we base time on the rotation of the earth, it might seem that the earth's rotation with respect to the stars would be the nearest we could come to defining the "true rotation rate" of the earth. We could then measure the passage of time by the hour angle of some fixed point on the celestial sphere. The interval between two successive meridian transits of this point would define such a "stellar day." This is the kind of rotation period of the earth that would be measured, say, by the Foucault pendulum. (Actually, of course, the stars are in motion and do not define a perfectly fixed reference system; however, they are so distant that over moderate periods of time they remain sensibly fixed on the celestial sphere.)

We could measure time by such a scheme, but it turns out that other kinds of time are more convenient. The most common kinds of time in use are *sidereal time* and *solar time,* which are based on the sidereal day and the solar day.

(b) The Sidereal and Solar Day

The *sidereal* day is defined as the interval between successive transits of the *vernal equinox,* the point on the celestial sphere where the sun, in its apparent annual path around the sky, crosses the celestial equator to the north on the first day of spring. Because of *precession* (Section 4.11), the vernal equinox is not quite a fixed point on the celestial sphere, but slowly shifts its position. Technically, therefore, the term "sidereal day" is a misnomer, for its length is not based strictly on the earth's rotation with respect to the stars. However, because the motion of the vernal equinox in the sky is very slow, a sidereal day is very close to the true period of rotation of the earth with respect to the stars — within one hundredth of a second.

The *solar day* is the period of the earth's rotation with respect to the sun. The solar and sidereal days are *not* the same, as a study of Figure 5-8 will show. Suppose we start counting a day when the earth is at *A,* with the sun on the meridian of an observer at point *O* on the earth. The direction from the earth to the sun, *AS,* if extended, points in the direction *C* among stars on the celestial sphere. After the earth has made one rotation with respect to the stars, the same stars in direction *C* will again be on the local meridian to the observer at *O.* However, because the earth has moved from *A* to *B* in its orbit about the sun during its rotation, the sun has not yet returned to the meridian of the observer but is still slightly to the east. The vernal equinox is so nearly fixed among the stars that the earth has completed, essentially, one *sidereal* day, but to complete a *solar* day it must turn a little more to bring the sun back to the meridian.

In other words, a solar day is slightly *longer* than a sidereal day, or one complete rotation of

the earth. There are about 365 days in a year and 360° in a circle; thus the daily motion of the earth in its orbit is about 1°. This 1° angle, *ASB,* is nearly the same as the additional angle over and above 360° through which the earth must turn to complete a solar day. It takes the earth about 4 minutes to turn through 1°. A solar day, therefore, is about 4 minutes longer than a sidereal day.

Each kind of day is subdivided into hours, minutes, and seconds. A unit of solar time (hour, minute, or second) is longer than the corresponding unit of sidereal time by about 1 part in 365. In

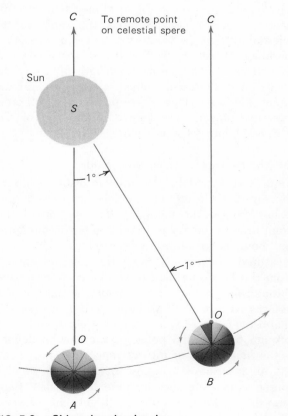

FIG. 5-8   Sidereal and solar day.

units of solar time, one sidereal day is $23^h56^m4^s.091$. The period of the earth's rotation with respect to the stars is $23^h56^m4^s.099$ in solar time units.

(c) Sidereal Time

Sidereal time is based on the sidereal day with its subdivisions of sidereal hours, minutes, and seconds. It is defined as the *hour angle of the vernal equinox*. The sidereal day begins at sidereal *noon*, $0^h0^m0^s$ sidereal time, with the vernal equinox on the meridian.

Sidereal time is very useful to astronomy and navigation. The common coordinate system used to denote positions of stars and planets on the celestial sphere (right ascension and declination) is referred to the celestial equator and the vernal equinox, much as latitude and longitude on the earth are referred to the earth's equator and the meridian of Greenwich, England (see Appendix 7). Therefore, the position of a star in the sky with respect to the observer's meridian is directly related to the sidereal time. Every observatory maintains clocks that are rated to keep accurate sidereal time.

We regulate our everyday lives, however, by the sun, not the vernal equinox; for example, the working day is usually determined according to the daylight hours. It is far more desirable to use solar time for ordinary purposes.

(d) Apparent Solar Time

Just as sidereal time is reckoned by the hour angle of the vernal equinox, so *apparent solar time* is determined by the hour angle of the sun. At midday, apparent solar time, the sun is on the meridian. The hour angle of the sun is the time *past midday* (*post meridiem,* or P.M.). It is convenient to start the day not at noon, but at midnight. Therefore the elapsed apparent solar time since the beginning of a day is the hour angle of the sun *plus* 12 hours. During the first half of the day, the sun has not yet reached the meridian. We designate those hours

before midday (ante meridiem, or A.M.). We customarily start numbering the hours after noon over again, and designate them P.M., to distinguish them from the morning hours (A.M.). On the other hand, it is often useful to number the hours from 0 to 24, starting from the beginning of the day at midnight. For example, in various conventions, 7:46 P.M. may be written as 19^h46^m, 19:46, or simply 1946.

Apparent solar time, defined as the hour angle of the sun plus 12 hours, is the most obvious and direct kind of solar time. It is the time that is kept by a sundial. In a sundial, a raised marker, or *gnomon,* casts a shadow whose direction indicates the hour angle of the sun. Apparent solar time was the time kept by man through many centuries.

The exact length of an apparent solar day, however, varies slightly during the year. Recall that the difference between an apparent solar day and a sidereal day, if time is counted from noon on one day, is the extra time required, after one rotation of the earth with respect to the vernal equinox, to bring the sun back to the meridian. The length of this extra time depends on how far *east* of the meridian the sun is after the completion of 1 sidereal day. The earth rotates to the east at a very constant rate of 1° every 4 sidereal minutes. Thus, if the sun were exactly 1° east of the meridian, about 4 sidereal minutes would be needed to bring it to the meridian. (Actually, it is just over 4 sideral minutes, because the earth is still advancing in its orbit during that period, which moves the sun another 10″ to the east along the ecliptic.) If the sun were more or less than 1° east of the meridian, the extra time required would be a little more or less than 4 sidereal minutes.

The length of the apparent solar day would be constant if the eastward progress of the sun, in its apparent annual journey around the sky, were precisely constant. However, because of the slightly varying speed of the earth in its orbit, and the fact that the earth's axis of rotation is not perpendicular to the plane of its revolution, the amount by which the sun appears to shift to the east is not exactly the same every day of the year.

The apparent solar day is always *about* 4 minutes longer than a sidereal day, but because of the sun's variable progress to the east, the precise interval varies by up to ½ minute one way or the other. The variation can accumulate after a number of days to several minutes. After the invention of clocks that could run at a uniform rate, it became necessary to abandon the apparent solar day as the fundamental unit of time. Otherwise, all clocks would have to be adjusted to run at a different rate each day.

(e) Mean Solar Time

Mean solar time is based on the *mean solar day,* which has a duration equal to the *average* length of an apparent solar day. Mean solar time is defined as the hour angle of the mean sun plus 12 hours, where the *mean sun* is a fictitious point in the sky that moves uniformly to the east along the *celestial equator,* with the same average eastern rate as the true sun. In other words, mean solar time is just apparent solar time averaged to be uniform.

Although mean solar time has the advantage of progressing at a uniform rate, it is still inconvenient for practical use. Recall that it is defined as the hour angle of the mean sun. But hour angle is referred to the local celestial meridian, which is different for every longitude on earth. Thus, observers on different north-south lines on the earth have a different hour angle of the mean sun and hence a different mean solar time. If mean solar time were strictly observed, a person traveling east or west would have to reset his watch continually as his longitude changed, if it were always to read the local mean time correctly. For instance, a commuter traveling from Oyster Bay to New York City would have to adjust his watch as he rode through the East River tunnels, because Oyster

Bay time is actually a few seconds more advanced than that of Manhattan.

(f) Standard and Zone Time

Until near the end of the last century, every city and town in the United States kept its own local mean time. With the development of railroads and telegraph, however, the need for some kind of standardization became evident. In 1883 the nation was divided into four time zones. Within each zone, all places keep the same time, the local mean solar time of a standard meridian running more or less through the middle of each zone. Now a traveler resets his watch only when the time change has amounted to a full hour. For local convenience, the boundaries between the four time zones are chosen to correspond, as much as possible, to divisions between states. Mean solar time, so standardized, is called *standard time.* The standard time zones in the United States (not including Alaska and Hawaii) are Eastern Standard Time (EST), Central Standard Time (CST), Mountain Standard Time (MST), and Pacific Standard Time (PST), which respectively keep the mean times of the meridians of 75, 90, 105, and 120° west longitude. Hawaii and Alaska both keep the time of the meridian 150° west longitude, 2 hours less advanced than Pacific Standard Time.

(g) Daylight Saving Time

To take advantage of the maximum amount of sunlight during waking hours, most states in this country, as well as many foreign nations, keep what is called *daylight saving time* during the spring and summer. Daylight saving time is simply the local standard or zone time of the place *plus* 1 hour. Thus, on a summer evening, when it would ordinarily grow dark about 8:00 P.M. standard time, it is light until 9:00 P.M. daylight saving time.

It might seem that daylight saving time is needed more in the winter than the summer. It is

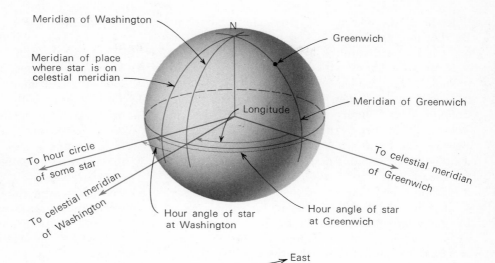

Meridian of Washington

Meridian of place where star is on celestial meridian

Greenwich

N

Meridian of Greenwich

Longitude

To hour circle of some star

To celestial meridian of Greenwich

To celestial meridian of Washington

Hour angle of star at Washington

Hour angle of star at Greenwich

East

FIG. 5-9   A difference in hour angle equals the difference in longitude.

not practical in the winter, however, for if clocks were set an hour ahead in December, in many parts of the country it would still be dark at 7:30 A.M. while people are on their way to work.†

(h) Time Around the World; The International Date Line

The direction of the earth's rotation is to the east. Therefore, places to the east of us must always have a time *more advanced* than ours. The celestial meridian of New York sweeps under the sun on the celestial sphere about 3 hours earlier than does the celestial meridian of San Francisco. Thus in New York the hour angle of the sun, the local time there, is 3 hours ahead of San Francisco time. The times of places halfway around the world from each other differ by 12 hours.

In general, the difference in the local time of any two places on earth is equal to their difference in longitude, if longitude is measured in time units

rather than degrees. The more easterly place always has the more advanced time. The statement is true of any kind of time, sidereal, apparent solar, mean solar, and so on, because any kind of local time is defined as the hour angle of some reference object. The difference in the hour angles of the same object seen at two different places is just the difference in the directions of the local meridians of the two places, that is, the difference in their longitudes. Thus, the determination of longitude is equivalent to finding the difference between local time and the time at some known place on earth — say the time at Greenwich (Figure 5-9).

The local mean solar time of the meridian running through Greenwich, England, is called *universal time.* Data are usually given in navigational tables, such as the *Nautical Almanac,* for various intervals of universal time.

The fact that time is always more advanced to the east presents a problem. Suppose a man moving eastward travels around the world. He passes into a new time zone, on the average, for about every 15° of longitude he travels and, each

†Daylight saving time has been compared to the action of a woman who cut 1 ft off the top of her blanket and sewed it on to the bottom to make it longer.

time, he dutifully sets his watch ahead an hour. By the time he has completed his trip, he has set his watch ahead through a full 24 hours, and has thus *gained a day* over those who stayed home. Let us look at the problem another way. It is 3 hours later in New York than in Berkeley, California. In London, it is 8 hours later than in Berkeley; in Tokyo, still farther to the east, it is 17 hours later. Because San Francisco is about 100° east of Tokyo, it follows that in San Francisco, a few miles to the west across the San Francisco Bay from Berkeley, it is 7 hours later than at Tokyo, or 24 hours − 1 day − later than at Berkeley! One might suppose that by going around the world to the east or west enough times, we could go into the future or past as far as we desired, creating the equivalent of a time machine.

The solution to the dilemma is the *international date line,* set by international agreement along the 180° meridian of longitude. The date line runs about down the middle of the Pacific Ocean, although it jogs a bit in a few places to avoid cutting through groups of islands and through Alaska. By convention, at the date line, the date of the calendar is changed by 1 day. If a person crosses the date line from west to east, so that he is advancing his time, he compensates by decreasing his date; if he crosses from east to west, he increases his date by 1 day.

If an ocean liner crosses the international date line from west to east just after its passengers have been served a Christmas dinner, the crew can begin the preparation of a second feast, for the next day will be Christmas again. On the other hand, one could skip Christmas altogether by crossing the date line from east to west.

(i) Summary of Time

We may briefly summarize the story of time as follows:

(1) The ordinary day is based on a rotation of the earth with respect to the sun (not the stars).

(2) The length of the apparent solar day is slightly variable because of the earth's variable orbital speed and the obliquity of the ecliptic.

(3) Therefore, the average length of an apparent solar day is defined as the mean solar day.

(4) Time based on the mean solar day is mean solar time. It is defined as the local hour angle of the mean sun plus 12 hours. The mean sun is an imaginary sun that revolves annually on the celestial equator at a perfectly uniform eastward rate.

(5) The mean solar time at any one place, called *local mean time,* varies continuously with longitude, so that two places a few miles east and west of each other have slightly different times.

(6) Therefore time is standardized, so that each place in a certain region or zone keeps the same time − the local mean time of the standard meridian in that zone.

(7) In many localities, standard time is advanced by 1 hour in spring and summer to take advantage of the maximum number of hours of sunshine during the waking hours. This is daylight saving time.

(j) Variations in the Earth's Rotation

Since the innovation of quartz crystals, time standards have been accurate enough to measure irregularities in the rotation of the earth. Slight expansions and contractions in the earth can change its rotation rate enough to produce accumulated clock errors of $\frac{1}{2}$ minute or more. Furthermore, slight seasonal variations have been detected in the length of the day; these probably result from the shifting of ice and snow deposits and air masses over the earth. Superposed upon these slight fluctuations of the earth's rotation rate is a gradual slowing down of the rotation in the amount of about $\frac{1}{1000}$ second per century. This gradual slowing is generally attributed to friction in the tides.

Because of the slight irregularities in the earth's rotation rate, data in astronomical tables are now given for intervals of *ephemeris time,* a recently-invented time that runs at a theoretically precise rate, namely the rate that mean solar time had at the beginning of the year 1900. The slight difference between ephemeris and universal (Greenwich) time (just over a half minute in the late 1960's) is continually reevaluated from accurate observations of the motions of the moon and planets.

5.4 THE REVOLUTION OF THE EARTH

We have seen that the earth's revolution about the sun produces an apparent annual motion of the sun around the sky. The sun's apparent eastward journey among the stars on the celestial sphere is along the *ecliptic.* However, we have also seen that the sun's apparent revolution about the earth can be explained either by a motion of the sun or of the earth. What evidence is there that it is the earth, not the sun, that moves?

(a) Proofs of the Earth's Revolution

As in discussing the earth's rotation, we can only *prove* that the earth revolves if we are willing to accept certain postulates. If we adopt Newton's laws of motion, it follows simply and directly that the earth must revolve about the sun and not vice versa.

It is obvious that *either* the earth goes around the sun or the sun around the earth. Thus we have a system of *two mutually revolving bodies.* The problem is simply to determine where the common center of revolution is. In Chapter 15 we shall see how the mass of the sun is determined. It is calculated that the sun is about 330,000 times as massive as the earth. Thus, we find that the common center of mass of the earth-sun system must be less than 1/300,000 of the distance from the center of the sun to the center of the earth.

This puts it well inside the surface of the sun. Essentially, then, the earth revolves around the sun.

5.5 THE SEASONS

The earth's orbit around the sun is an ellipse, its distance from the sun varying by about 3 percent. However, the changing distance of the earth from the sun is *not* the cause of the seasons. The seasons result because the plane in which the earth revolves is not coincident with the plane of the earth's equator. The planes of the equator and ecliptic are inclined to each other by about 23½°. This angle of 23½° is called the *obliquity of the ecliptic.*

Globes of the earth are usually mounted with the earth's axis tilted from the vertical. This tilt is the same angle of 23½°, for that is the angle the earth's axis must make with the perpendicular to the plane of its orbit around the sun. The result of the obliquity of the ecliptic is that the Northern Hemisphere is inclined *toward* the sun in June and away from it in December.

(a) The Seasons and Sunshine

Figure 5-10 shows the earth's path around the sun. The line *EE'* is in the plane of the celestial equator. In the figure the earth appears to pass alternately above and below this plane, but the celestial sphere is so large, and the celestial equator so far away, that a line from the center of the earth through the earth's equator always points to the celestial equator.

We see in the figure that on about June 22 (the date of the *summer solstice*), the sun shines down most directly upon the northern hemisphere of the earth. It appears 23½° *north* of the equator and thus on that date passes through the zenith of places on the earth that are at 23½° north latitude. The situation is shown in detail in Figure 5-11, which shows the earth on the date of the summer solstice. To an observer on the equator, the sun appears 23½° north of the zenith at noon. To a

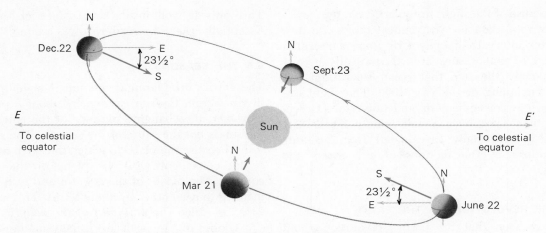

FIG. 5-10   The seasons are caused by the inclination of the plane of the earth's orbit to the plane of the equator.

person at a latitude 23½° N, the sun is overhead at noon. This latitude on the earth, at which the sun can appear at the zenith at noon on the first day of summer is called the *Tropic of Cancer*. We see also in Figure 5-11 that the sun's rays shine down past the North Pole; in fact, all places within 23½° of the pole, that is, at a latitude greater than 66½° N, have sunshine for 24 hours on the first day of summer. The sun is as far north on this date as it can get; thus, 66½° is the southernmost latitude where

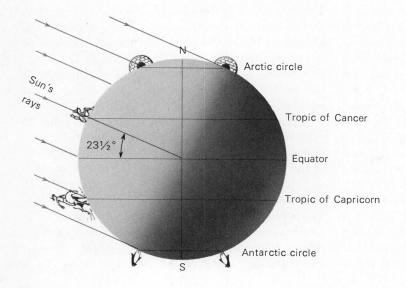

FIG. 5-11
The earth on June 22.

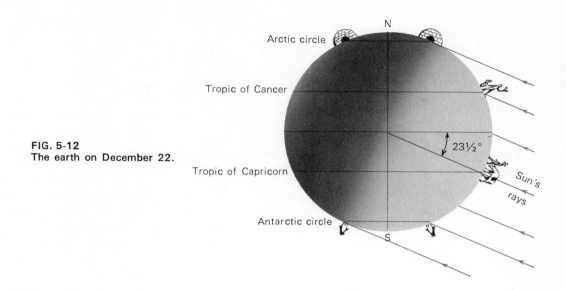

FIG. 5-12
The earth on December 22.

the sun can ever be seen for a full 24-hour period (the *midnight sun*); that circle of latitude is called the *Arctic Circle*.

On the other hand, the sun's rays shine very obliquely on the Southern Hemisphere. In fact, all places within 23½° of the South Pole — that is, south of latitude 66½° S (the *Antarctic Circle*) —

have no sight of the sun for the entire 24-hour period.

The situation is reversed 6 months later, about December 22 (the date of the *winter solstice*), as is shown in Figure 5-12. Now it is the Arctic Circle that has a 24-hour night and the Antarctic Circle that has the midnight sun. At latitude 23½° S, the

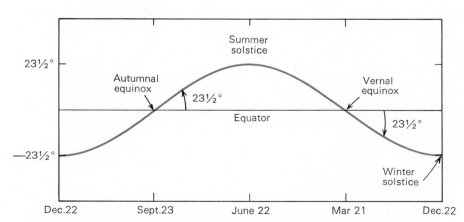

FIG. 5-13   Plot of the ecliptic around the celestial equator.

Tropic of Capricorn, the sun passes through the zenith at noon. It is winter in the Northern Hemisphere, summer in the Southern.

Finally, we see in Figure 5-10 that on about March 21 and September 23 the sun appears to be in the direction of the celestial equator, and, on these dates, the equator itself is the diurnal circle for the sun. Every place on the earth then receives exactly 12 hours of sunshine and 12 hours of night. These points, where the sun crosses the celestial equator, are called the *vernal* (spring) *equinox* and *autumnal* (fall) *equinox. Equinox* means "equal night."

Figure 5-13 is a map in which the sky is shown flattened out, as in a Mercator projection of the earth. The equator runs along the middle of the map, and the ecliptic is shown as a wavy line crossing the equator at the two equinoxes. Both equator and ecliptic are, of course, great circles, but they cannot both be shown as straight lines on a flat surface. Notice that the ecliptic intersects the equator at an angle of 23½°, and that its northernmost extent is 23½° north of the equator (the summer solstice) and its southernmost extent is 23½° south of the equator (the winter solstice).

Figure 5-14 shows the aspect of the sky at a typical latitude in the United States. During the spring and summer, the sun is north of the equator and is thus up more than half the time. A typical spot in the United States, on the first day of summer (about June 22), receives about 14 or 15 hours of sunshine. Also, notice that the sun appears *high* in the sky, and so in these seasons the sunlight is more direct, and thus more effective in heating than in the fall and winter when the sun appears at a lower altitude in the sky.

In the fall and winter the sun is south of the equator, where most of its diurnal circle is below the horizon, and so it is up less than half the time. On about December 22, a typical city at, say 30 to 40° north latitude receives only 9 or 10 hours of sunshine. Also, the sun is low in the sky; a bundle

of its rays is spread out over a larger area on the ground (Figure 5-15) than in summer; because the energy is spread out over a larger area, there is less for each square foot, and so the sun at low altitudes is less effective in heating the ground.

(b) The Seasons at Different Latitudes

At the equator all seasons are much the same. Every day of the year, the sun is up half the time, so there are always 12 hours of sunshine at the equator. About June 22, the sun crosses the meridian 23½° north of the zenith, and about December 22, 23½° south of the zenith.

The seasons become more severe as one travels north or south of the equator. At the Tropic of Cancer, on the date of the summer solstice, the sun is at the zenith at noon. On the date of the winter solstice, the sun crosses the meridian 47° south of the zenith. At the Arctic Circle, on the first day of summer the sun never sets, but at midnight can be seen just skimming the north point on the horizon. About December 22, the sun does not quite rise at the Arctic Circle, but just gets up to the south point on the horizon at noon. Between

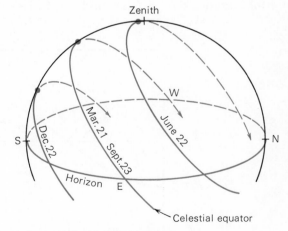

FIG. 5-14 Diurnal paths of the sun for various dates at a typical place in the United States.

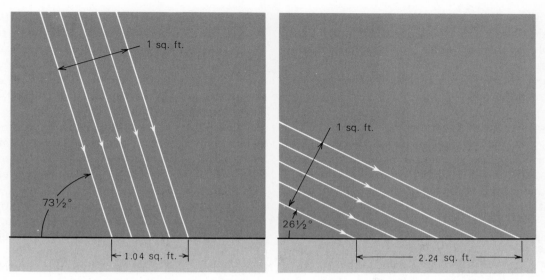

FIG. 5-15 Effect of the sun's altitude. When the sun is low in the sky, its rays are more oblique to the ground and are spread over a larger area than when the sun is high in the sky.

the Tropic of Cancer and the Arctic Circle, the number of hours of sunshine and the noon altitude of the sun range between these two extremes.

We recall that at the North Pole, all celestial objects that are north of the celestial equator are always above the horizon and, as the earth turns, circle around parallel to it. The sun is north of the celestial equator from about March 21 to September 23, and so at the North Pole the sun rises when it reaches the vernal equinox and sets when it reaches the autumnal equinox. There are 6 months of sunshine at the Pole. The sun has its maximum altitude of 23½° about June 22; before that date it climbs gradually higher each day, and after that it drops gradually lower. A navigator can easily tell when he is at the North Pole, for there the sun circles around the sky parallel to the horizon, getting no higher or lower (except gradually as the days go by).†

In the Southern Hemisphere, the seasons are similar, except that they are reversed from those in the north. While we are having summer in the United States, in Australia it is winter. Furthermore, in the Southern Hemisphere, the sun crosses the meridian generally to the *north* of the zenith.

The earth, in its elliptical orbit, reaches its closest approach to the sun about January 4. It is then said to be at *perihelion*. It is farthest from the sun, at *aphelion,* about July 5. We see, then, that the earth is closest to the sun when it is winter in the north. However, it is summer in the Southern Hemisphere when the earth is at perihelion, and the earth is farthest from the sun during the Southern Hemisphere's winter. Therefore, we might expect the seasons to be somewhat more severe in

---

†It is said that one botanist considered the North Pole to be an excellent place to raise sunflowers, because there were so many hours of sunshine during the summer months. He accordingly planted some there, and they did quite well for a while. However, sunflowers like to *face* the sun, and as they followed the sun around and around the sky, they ended by wringing their own necks!

the Southern Hemisphere than in the Northern. However, there is more ocean area in the Southern Hemisphere; this and other topographical factors are more important in their influence on the seasons on the earth than is the earth's changing distance from the sun.

(c) Precession of the Equinoxes

As the earth's axis precesses in its conical motion (Section 4.11), the equatorial plane retains (approximately) its 23½° inclination to the ecliptic plane; that is, the obliquity of the ecliptic remains constant. However, the intersections of the celestial equator and the ecliptic (the equinoxes) must always be 90° from the celestial poles (because all points on the celestial equator are 90° from the celestial poles). Thus, as the poles move because of precession, the equinoxes slide around the sky, moving westward along the ecliptic. This motion is called the *precession of the equinoxes*. The angle through which the equinoxes move each year, the *annual precession,* is 1/26,000 of 360°, or about 50″. Each year as the sun completes its *eastward* revolution about the sky with respect to, say, the vernal equinox, that equinox has moved *westward,* to meet the sun, about 50″. Since it takes the sun about 20 minutes to move 50″ along the ecliptic (or, more accurately, because it takes the earth that long to move through an angle of 50″ in its orbit about the sun), a *tropical year,* measured with respect to the equinoxes, is 20 minutes shorter than a *sidereal* year, measured with respect to the stars.

Precession has no important effect on the seasons. The earth's axis retains its inclination to the ecliptic, so the Northern Hemisphere is still tipped toward the sun during one part of the year and away from it during the other. Our calendar year is based on the beginnings of the seasons (the times when the sun reaches the equinoxes and solstices), so spring still begins in March (in the Northern Hemisphere), summer in June, and so on. The only effect is that as the precessional cycle goes on, a given season will occur when the earth is in gradually different places in its orbit with respect to the stars. In the twentieth century, for example, Orion is a *winter constellation;* we look out at night, away from the sun, and see Orion in the sky during the winter months. In 13,000 years, half a precessional cycle later, it will be summertime when we look out in the same direction, away from the sun, and see Orion. Similarly, Scorpio is a summer constellation now, whereas in the year 15,000 it will be a winter constellation (see Figure 5-16).

The vernal equinox is sometimes called the *first point of Aries,* because about 2000 years ago, when it received that name, it lay in the constellation of Aries. Now, because of precession, the vernal equinox has slid westward into the constellation of Pisces.

(d) Twilight

We all know that the sky does not immediately darken when the sun sets. Even after the sun is no longer visible from the ground, the upper atmosphere of the earth can catch some of the rays of the setting sun and scatter them helter-skelter, illuminating the sky. Gases in the earth's atmosphere are dense enough to scatter appreciable sunlight up to altitudes of about 200 mi. The sun must be at least 18° below the horizon for all traces of this postsunset or presunrise sky light *(twilight)* to be absent. At latitudes near the equator, where the sun rises and sets nearly vertically to the horizon, twilight lasts only a little over 1 hour. However, at far northern and southern latitudes, the sun rises and sets in a much more oblique direction and takes correspondingly longer to reach a point 18° below the horizon, so twilight may last for 2 hours or more. In the far northern countries twilight lasts all night in the summertime. At the North Pole there are 6 weeks of twilight in the late winter before sunrise and again in the early fall after sunset.

FIG. 5-16 Summer and winter constellations change due to precession.

5.6 THE DATE OF THE YEAR

The natural units of the calendar are the day, based on the period of rotation of the earth; the month, based on the period of revolution of the moon about the earth, and the year, based on the period of revolution of the earth about the sun. The difficulties in the calendar have resulted from the fact that these three periods are not commensurable — that is, one does not divide evenly into any of the others.

The period of revolution of the moon with respect to the stars, the *sidereal month,* is about 27⅓ days. However, the interval between corresponding phases of the moon, the more obvious kind of month, is the moon's period of revolution with respect to the sun, the *synodic month,* which has about 29½ days.

The period of revolution of the earth with respect to the vernal equinox, that is, with respect to the beginnings of the various seasons, is as we have seen, the *tropical* year. Its length is 365.-242199 mean solar days, or $365^d5^h48^m46^s$. Our calendar, to keep in step with the seasons, is based on the tropical year. Because of precession (Section 5.5c), the tropical year is slightly shorter than the sidereal year.

(a) The Roman Republican Calendar

The roots of our modern calendar go back to the Roman republican calendar, which derives from earlier Roman and Greek calendars dating from at least the eighth century B.C. The earliest Roman calendar probably had 10 months, the last 4 of which have given us the names of our months, September, October, November, and December. But by the first century B.C., 2 additional months, January and February, had been added.

The original Roman calendar was a lunar one in that the months were based on the moon's synodic period; each month began with a new moon. To give the months an average length of 29½ days (the lunar synodic period) the months had 29 and 30 days alternately. The difficulty with the lunar calendar is that 12 lunar months add up to only 354 days, whereas the tropical year has about 365¼ days. After about 3 years, the difference accumulates to a whole month. To keep their year in step with the year of the seasons, the ancients adopted the policy of intercalation — that is, they simply inserted a thirteenth month every third year or so. The normal 12-month years were "empty years" and the 13-month years were "full years."

Unfortunately, the management of the calendar was left to the discretion of the priests, who greatly abused their authority by declaring as full years those in which their friends were in public office. The intercalation process became such a political football that by the time of the reign of Julius Caesar, a Roman traveler going from town to town could find himself going from year to year! Thus, in 46 B.C., Julius Caesar instigated a calendar reform.

(b) The Julian Calendar
At the advice of the Alexandrian astronomer Sosigenes, Caesar adopted a new calendar, which had 12 more or less equal months averaging about 30½ days in length rather than 29½. The features of the Julian calendar reform of 46 B.C. were as follows:

(1) The lunar synodic month was abandoned as a basic unit in the calendar. Instead, each year contained 12 months, which contained a total of 365 days.

(2) The calendar was to be based on the tropical year, whose length had at that time been determined to be 365¼ days. Of course, one fourth of a day could not be "tacked on" to the end of the calendar year. Therefore, *common* years were to contain only 365 days. However, after 4 years, this ¼ day per year adds up to 1 full day. Thus every fourth year was to have 366 days (a *leap year*), the extra day being added to February. The *average* length of the year would then be 365¼ days. Note that the process of leap year is analogous to the intercalation of extra months in "empty years" to make them "full years" of 13 months.

(3) To bring the date of the vernal equinox, which had fallen badly out of place in the Roman republican calendar, back to its traditional date of March 25, Caesar intercalated 3 extra months in the year 46 B.C., bringing its length to 445 days. Forty-six B.C. was known as the "year of confusion." The Julian Calendar was introduced, then, on January 1, 45 B.C.

After Caesar's death in 44 B.C., the month Quintilis (the fifth month in the original Roman calendar) was renamed in his honor (thus our name, July). Later, the Roman senate did some further juggling with the Julian calendar. Sextilis (originally the sixth month) was renamed in honor of Augustus Caesar, successor to Julius, and the present number of days for each month resulted.

(c) The Council of Nicaea
The dates of observance of Easter and certain other religious holidays were fixed by order of the Council of Nicaea (Nice) in 325 A.D. Easter, according to the rule adopted, falls on the first Sunday after the fourteenth day of the moon (almost full moon) that occurs on or after March 21. At that time March 21 was the date of the vernal equinox. (The Sunday *after* full moon was specified intentionally to avoid the possibility of an occasional coincidence with the Jewish Passover.)

Note that between 45 B.C. and 325 A.D., the date of the vernal equinox had slipped back from March 25 to March 21. This was because the Julian year, with an average length of 365¼ days, is $11^m 14^s$ longer than the tropical year of $365^d 5^h 48^m 46^s$. The slight discrepancy had accumulated to just over 3 days in those four centuries.

(d) The Gregorian Calendar

By 1582, that 11 minutes and 14 seconds per year had added up to another 10 days, so that the first day of spring was occurring on March 11. If the trend were allowed to continue, eventually Easter and the related days of observance would be occurring in early winter. Therefore, Pope Gregory XIII instituted a further calendar reform.

The Gregorian calendar reform consisted of two steps. First, 10 days had to be dropped out of the calendar to bring the vernal equinox back to March 21, where it was at the time of the Council of Nicaea. This step was expediently accomplished. By proclamation the day following October 4, 1582, became October 15.

The second feature of the new Gregorian calendar was that the rule for leap year was changed so that the average length of the year would more closely approximate the tropical year. In the Julian calendar, every year divisible by four was a leap year, so that the average year was 365.250000 mean solar days in length. The error between this and the tropical year of 365.242199 mean solar days accumulates to a full day every 128 years. Ideally, therefore, one leap year should be made a common year, thus dropping 1 day, every 128 years. Such a rule, however, is cumbersome.

Instead, Gregory decreed that three out of every four century years, all leap years under the Julian calendar, would be common years henceforth. The rule was that only century years divisible by 400 should be leap years. Thus, 1700, 1800, and 1900, all divisible by 4, and thus leap years in the old Julian calendar, were *not* leap years in the Gregorian calendar. On the other hand, the years 1600, and 2000, both divisible by 400, are leap years under both systems. The average length of this Gregorian year was 365.2425 mean solar days, and was correct to about 1 day in 3300 years.

The Catholic countries immediately put the Gregorian reform into effect, but countries under control of the Eastern Church and most Protestant countries did not adopt it until much later. It was 1752 when England and the American colonies finally made the change. The year 1700 had been a leap year in the Julian calendar but not the Gregorian; thus the discrepancy between the two systems had become 11 days. By parliamentary decree, September 2, 1752, was followed by September 14. Although special laws were passed to prevent such breaches of justice as landlords collecting a full month's rent for September, there were still riots, and people demanded their 11 days back. To make matters worse, in England it had been customary to follow the ancient practice of starting the year on March 25, originally the date of the vernal equinox. In 1752, however, the start of the year was moved back to January 1, so in England and the Colonies 1751 had no months of January and February and had lost 24 days of March! We celebrate George Washington's Birthday on February 22, 1732, but at the time of his birth, a calendar would have read February 11, 1731. Russia did not abandon the Julian calendar until the time of the Bolshevik revolution. The Russians then had to omit 13 days to come into step with the rest of the world.

The Gregorian calendar has now been modified slightly to come into better conformity with the tropical year: the years 4000, 8000, 12,000, and so on, all leap years in the original Gregorian calendar, are now common years. The calendar is thus accurate to 1 day in about 20,000 years.

5.7 THE MANY MOTIONS OF THE EARTH

In this chapter we have discussed in detail two of the earth's motions: rotation and revolution. However, there are many other motions of the earth, dealt with in other sections. Here, for completeness, these motions are summarized:

(1) The earth *rotates* daily on its axis.

(2) The earth periodically shifts slightly with respect to its axis of rotation *(variation in latitude)*. The actual migrations of the terrestrial poles along the earth's surface are spread over only about 50 ft or less.

(3) The earth *revolves* about the sun.

(4) The gravitational pull of the sun and moon on the earth's equatorial bulge causes a very slow change in orientation of the axis of the earth called *precession* (Section 4.11).

(5) Because the moon's orbit is not quite in the plane of the ecliptic, and because of a slow change in orientation of the moon's orbit, there is a small periodic motion of the earth's axis superimposed upon precession, called *nutation.*

(6) Actually, it is the center of mass of the earth-moon system, or *barycenter,* that revolves about the sun in an elliptical orbit. Each month the center of the earth revolves about the barycenter (Section 4.2b).

(7) The earth shares the motion of the sun and the entire solar system among its neighboring stars. This *solar motion* is about 12 mi/sec (Chapter 13).

(8) The sun, with its neighboring stars, shares in the general *rotation of the Galaxy.* Our motion about the center of the Galaxy is about 150 mi/sec (Chapter 19).

(9) All other galaxies are observed to be in motion. Therefore, our *Galaxy is in motion* with respect to other galaxies in the universe (Chapter 22).

In view of the many motions of the earth, one might wonder what the absolute speed of the earth is in the universe. To determine the absolute velocity of the earth was the object of the Michelson-Morley experiment, which we have seen (Section 3.7a) gave null results. A fundamental postulate of Einstein's special theory of relativity is that it is not possible to define an absolute coordinate system with respect to which the absolute speed of an object in space can be determined.

Exercises

1. If a star rises in the northeast, in what direction does it set?
2. What is the latitude of:
   (a) The North Pole?
   (b) The South Pole?
   (c) A point two-thirds of the way from the equator to the South Pole?
3. Why has longitude no meaning at the South Pole?
4. Tell where you are on earth from the following descriptions:
   (a) The stars rise and set perpendicular to the horizon.
   (b) The stars circle the sky parallel to the horizon.
   (c) The celestial equator passes through the zenith.
   (d) In the course of a year, all stars are visible.
   (e) The sun rises on September 23 and does not set until March 21.
5. How many more sidereal than solar days are there in one year? Why?
6. If a star rises at 8:30 P.M. tonight, approximately what time will it rise two months from now?

7. If the local mean time is 2:30 P.M. and the Universal Time is 10:30 A.M., what is the longitude?

8. Where on earth is it possible for the ecliptic to lie along the horizon?

9. If the earth were to speed up in its orbit so that a tropical year contained 365 days 2 hours, how often would we need to have a leap year?

10. In far northern countries the winter months tend to be so cloudy that astronomical observations are nearly impossible. Why is it that good observations cannot be made in those places during the summer months?

11. If it is 1:20 AM. on April 13 at a longitude 160° W, what are the time and date at longitude 160° E?

12. What is the altitude of the sun at noon on December 22 as seen from a place on the Tropic of Cancer?

Following page: Earthrise from the moon, photographed by Apollo 8 astronauts, December 1968. *(NASA.)*

CHAPTER 6

Aspects and Motions of the Moon; Eclipses

The moon, the earth's nearest celestial neighbor and its only known natural satellite, accompanies the earth in its annual revolution about the sun. Although the moon shines only by reflected sunlight, it is nevertheless the second most brilliant object in the sky, and is one of the three celestial objects (the sun, the moon, and Venus) that can be seen in broad daylight.

Having examined the motions of the earth and how those motions affect the appearance of the sky, we now turn our attention to the apparent and real motions of the moon and certain other of its aspects. The physical nature of the moon itself, however, is discussed in Chapter 10.

6.1 ASPECTS OF THE MOON

The moon, because of its proximity, appears to move more rapidly in the sky than any other natural astronomical object, except meteors, which are within the earth's atmosphere. The moon appears the same size as the sun in the sky, subtending ½° of arc. Its larger surface features are easily visible to the unaided eye and form the facial markings of the "man in the moon." As it travels about the earth each month, it displays different parts of its daylight hemisphere to our view and progresses through its cycle of phases.

(a) The Moon's Apparent Path in the Sky

If the moon's position among the stars on the celestial sphere is carefully noted night after night, it is seen that the moon changes its position rather rapidly, moving, on the average, about 13° to the east per day. In fact, during a single evening the moon can be seen visibly creeping eastward among the stars. The moon's apparent path around the celestial sphere is a great circle (or very nearly so) that intersects the sun's path, the ecliptic, at an angle of about 5°.

The moon's path intersects the ecliptic at two points on opposite sides of the celestial sphere. These points are called the *nodes* of the moon's orbit. The node at which the moon crosses the ecliptic while moving northward is called the *ascending node,* and the node at which the moon crosses the ecliptic moving southward is the *descending node.*

The moon's orbit is constantly and gradually changing because of perturbations, just as the orbits of artificial satellites change. The most important of the perturbations are caused by the gravitational attraction of the sun. One of the effects of the perturbations on the moon's orbit is that the nodes slide westward along the ecliptic, completing one trip around the celestial sphere in about 18.6 years. This motion is called the moon's *regression of the nodes.* Perturbations also cause the inclination of the moon's orbit to the ecliptic to vary from 4°57′ to 5°20′; the average inclination is 5°9′.

If it were not for the regression of the nodes, the moon's orbit would maintain a nearly fixed angle to the celestial equator. It maintains a nearly fixed angle of about 5° to the ecliptic, but because the nodes constantly shift, its angle of inclination

FIG. 6-1   Phases of the moon and the time of day. (The outer series of figures shows the moon at various phases as seen in the sky from the earth's surface.)

to the equator varies from 23½ + 5°, or about 28½°, to 23½ − 5°, or about 18½° (23½° is the inclination of the ecliptic to the celestial equator).

(b) Delay in Moonrise from Day to Day

We have seen that the moon's average eastward motion with respect to the stars is about 13° per day. The sun, on the other hand, apparently moves to the east about 1° per day. With respect to the sun, therefore, the moon moves eastward about 12° per day. As the earth turns on its axis, the moon, like other celestial objects, appears to rise in the east, move across the sky, and set in the west. But because of its daily eastward motion on the celestial sphere, it crosses the local meridian each day about 50 minutes later, on the average, than on the previous day.

(c) The Progression of the Moon's Phases

The phases of the moon (explained in Section 2.2d) were thoroughly understood by the ancients. The relation between any phase of the moon and the moon's corresponding position in the sky at any time of day should now be clear. Except at extreme northern or southern latitudes, one can easily tell where to look for the moon in the sky, if he knows its phase, from a consideration of Figure 6-1.

In Figure 6-1 we imagine ourselves looking down upon the earth and the moon's orbit from the north. The moon is shown in several positions in its monthly circuit of the earth: 1, 2, 3, and so on. The sun is off to the right of the figure at a distance so great that its rays approach the earth and all parts of the moon's orbit along essentially parallel paths. The daylight sides of the earth and moon — the sides of those bodies turned toward the sun —

are indicated. For each position of the moon, its phase, that is, its appearance *as viewed from the earth,* is shown just outside its orbit. Several observers are at various places on the earth, *A, B, C,* and so on. The time of day is indicated for each observer.

For person *A* it is 3:00 P.M. If he sees the moon on his meridian, it must be in the waxing crescent phase (the moon can be seen easily at noon when in this phase). If the moon is in the waning crescent phase it is setting, for it lies on his western horizon. West is the direction away from which the turning earth carries the observer, and his horizon lies in a plane tangent to the surface of the earth at the point where he is standing. If the moon is new it is in about the same direction as the sun in the western sky, and if it is at first quarter it is in the eastern sky. If the moon is rising it must be in the waxing gibbous phase.

For person *B* it is 6:00 P.M. (Person *B* could be person *A* 3 hours later. During a period of even a full day the moon does not move enough for its phase to change appreciably.) For *B* the moon is setting if new, and rising if full. If it is a waxing crescent or gibbous it is in the western or eastern sky, respectively. If it is in the first quarter phase, it appears on the meridian.

For person *D* it is midnight. If the moon is full, it rose at sunset, and is now on the meridian. The first or third quarter moon is just setting or rising, respectively. The waxing or waning gibbous moons must appear in the western or eastern sky.

By studying Figure 6-1 we can see that as the moon goes about the earth during the month and gradually changes its phase, its approximate position in the sky is completely determined for any time of day or night. For example, the full moon rises at sunset and sets at sunrise. The first quarter

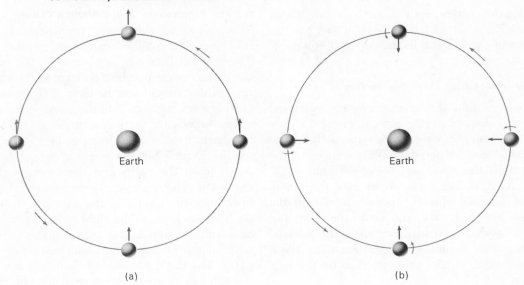

FIG. 6-2   If the moon did not rotate, (a) it would turn all its sides to our view; actually it does rotate (b) in the same period as it revolves, so we always see the same side.

(a)   (b)

moon rises at noon and sets at midnight, and so on.

(d) The Rotation of the Moon

Even naked-eye observation is sufficient to determine that as the moon goes about the earth it keeps the same side toward the earth. The same facial characteristics of the "man in the moon" are always turned to our view. Because the moon always presents the same side to us, it is sometimes said not to rotate on its axis. However, this statement is incorrect. In Figure 6-2 the arrow on the moon represents some lunar feature. If the moon did not rotate, as in (a), we would see that feature part of the time, and part of the time we would see the other side of the moon. Actually the moon rotates on its axis with respect to the stars in the same period as it revolves about the earth, and so always turns the same side toward us (b).

The coincidence of the periods of the moon's rotation and revolution can hardly be accidental. It is believed to have come about as a result of the earth's tidal forces on the moon.

We often hear the back side of the moon (the side we do not see) called the "dark side." Of course, the back side is dark no more frequently than the front side. Since the moon rotates, the sun rises and sets on all sides of the moon. The back side of the moon is receiving full daylight at new moon; the dark side is then turned toward the earth.

Actually, we do see a little more than one hemisphere of the moon during a month, because: (1) while the moon rotates at a constant rate its orbital speed varies, so that the rotation and revolution are not always exactly in step with each other; (2) its axis of rotation is not quite perpendicular to its orbit plane, so we can see sometimes a little over its north pole and sometimes a little over its south pole; and (3) the rotation of the earth carries us around a circle of some thousands of miles in radius, thus constantly changing the vantage point from which we observe the moon. These three effects are called *librations*. Because

of them, about 59 percent of the lunar surface is sometime or other exposed to our view.

6.2 THE MOON'S DISTANCE AND SIZE

The moon's distance is only about 30 times the diameter of the earth; consequently, the direction of the moon differs slightly as seen from various places on the earth. We have seen (Section 2.3d) how Ptolemy made use of this principle to determine geometrically the distance of the moon. Modern methods of geometrically determining the moon's distance are essentially the same as Ptolemy's. Today, however, the moon's distance can be determined more accurately by radar.

(a) Radar Measures of the Moon's Distance

The technique of radar is that of transmitting radio waves to a distant object and then detecting those waves that are reflected by that body back to the original source. The direction from which the reflected waves come indicates the direction of the body that reflects them. The time the waves spend making the round trip (the speed of transmission of the radiation being known) indicates the distance of that object.

We know the moon's direction from optical observations, and so are interested only in the time required for the radar signals to travel to the moon and back. The radar signals, being radio waves (a form of electromagnetic radiation), travel with the speed of light — an accurately known quantity. The interval of time between the instant that a wave is broadcast to the moon and the instant it returns can be measured electronically to within a few millionths of a second. The distance to the moon is, of course, the speed of the waves multiplied by half the time required for the round trip.

The first radar contact with the moon was achieved by the United States Army Signal Corps in 1946. A very accurate determination of the moon's distance by this technique was made by the Naval Research Laboratory in 1957. The latter

238,857 miles

Moon's diameter

Hypothetical observer
at center of earth

31'5" = 1865"

FIG. 6-3   Measuring the moon's diameter.

measure gives for the mean distance from the center of the earth to the center of the moon the value 384,403 km, or 238,857 mi. The NRL scientists consider their measure correct to within about 1 mi.

(b) The Moon's Diameter

The mean angular diameter of the moon is 31' 5". From its angular size and its distance, the linear size of the moon can easily be found. Because the method is the same as that applied to measure the diameters of planets and other astronomical objects that subtend a measurable angle, we shall explain the procedure in detail.

Notice in Figure 6-3 that because the moon's angular size is relatively small, its linear diameter is essentially a small arc of a circle, with the observer as center and with a radius equal to the moon's distance. Obviously, the moon's diameter is the same fraction of a complete circle as the angle subtended by the moon is of 360°. A complete circle contains 1,296,000" (there are 60" per minute, 60' per degree, and 360° in a circle). As seen from the center of the earth, the moon's mean angular diameter of 31'5", or 1865", is thus $\frac{1}{695}$ of a circle. The moon's diameter, therefore, is $\frac{1}{695}$ of the circumference of a circle of radius 238,857 mi. Since the circumference of a circle is 2π times its radius, we have

$$\text{diameter of moon} = \frac{2\pi(238,857)}{695} = 2160 \text{ mi.}$$

This type of calculation can be generalized if we note that the distance along the arc of a circle of radius R subtended by 1" is $2\pi R/1,296,000 = R/206,265$. Thus, if an object, say a planet, subtends an angle of α seconds, and has a distance D, its linear diameter d is given by

$$d = \frac{\alpha D}{206,265}.$$

Accurate calculations of the moon's diameter give as a result 3475.9 km, or 2159.86 mi, with an uncertainty of a few hundredths of a mile. This is the moon's equatorial diameter in a direction perpendicular to its direction from the earth. The moon's diameter is a little over a quarter that of the earth's and its volume is about $\frac{1}{49}$ of the earth's.

In practice, of course, we do not observe the moon from the center of the earth but from its surface. Both the angular size and direction of the moon differ for different observers on the earth. However, because the moon's distance and the size of the earth are known, the measured angular size of the moon can easily be corrected to the value it would have if the moon were viewed from the earth's center.

6.3 THE TRUE ORBIT OF THE MOON

When the moon's distance is measured during different times of the month, it is found to vary by more than 10 percent. In obedience to Kepler's first law, the moon's orbit, basically, is an ellipse with the earth at one focus. The sun, however, produces strong perturbations on the moon, and the elements of its orbit can be stated only in an average sense. Indeed, the moon's orbit changes so rapidly that if the moon's positions over a month, relative to the center of the earth, are plotted carefully, even on a sheet of standard typing paper, the orbit is seen to not close on itself.

Because of the perturbations on the moon, the eccentricity of its orbit changes even during 1 month. The mean value, over many years, is 0.0549. The moon, however, may come as close as 221,463 mi from the earth's center and pass as far from it as 252,710 mi. At its farthest, however, the moon is still only about $\frac{1}{370}$ as far as the sun, and its orbital speed of about $\frac{3}{5}$ mi/sec is only about $\frac{1}{30}$ of the earth's orbital speed about the sun. Thus, if the moon's actual path with respect to the sun is plotted, it is seen to vary only minutely from the earth's orbit; in fact, the moon's orbit is always concave to the sun.

We saw in Section 4.2b how the moon's mass can be found from the location of the barycenter of the earth-moon system. The best modern determinations of the moon's mass, however, are obtained by measuring the accelerations the moon produces on space probes — either those sent to the moon itself, or that pass it by on interplanetary missions. Analysis of the motion of the Mariner V probe, which passed about 6000 mi from Venus on October 19, 1967, gives for the earth-moon mass ratio the value 81.3004 ± 0.0007; this is the best determination at the time of writing.

6.4 SHADOWS AND ECLIPSES

Important phenomena associated with the motions of the earth-moon system are eclipses. Eclipses occur whenever any part of either the earth or the moon enters the shadow of the other. When the moon's shadow strikes the earth, people on earth within that shadow see the sun covered at least partially by the moon; that is, they witness a *solar eclipse*. When the moon passes into the shadow of the earth, people on the night side of the earth see the moon darken — a *lunar eclipse*.

Because the sun is not a point source of light, the shadow cast in sunlight by the moon (or earth) has two parts: the *umbra*, the cone of darkness within which all of the sun is hidden; and the *penumbra*, the region surrounding the umbra from which part, but not all, of the sun appears hidden behind the eclipsing body.

(a) Eclipse Seasons

For the moon to appear to cover the sun and thus to produce a solar eclipse, it must be in the same direction as the sun in the sky, that is, it must be at the *new* phase. For the moon to enter the earth's shadow and produce a lunar eclipse, it must be opposite the sun; that is, it must be at the *full* phase. Eclipses occur, therefore, only at new moon and at full moon. If the orbit of the moon about the earth lay exactly in the plane of the earth's orbit about the sun — in the ecliptic — an eclipse of the sun would occur at every new moon and a lunar eclipse at every full moon. However, because the moon's orbit is inclined at about 5° to the ecliptic, the new moon, in most cases, is not *exactly* in line with the sun, but is a little to the north or to the south of the sun in the sky. Similarly, the full moon usually passes a little south or north of the earth's shadow.

However, if full or new moon occurs when the moon is at or near one of the *nodes* of its orbit (where its orbit intercepts the ecliptic), an eclipse can occur. The line through the center of the earth that connects the nodes of the moon's orbit is called the *line of nodes*. If the direction of the sun lies along, or nearly along, the line of nodes, new or full moon occurs when the moon is near a node, and an eclipse results. The situation is illustrated in

FIG. 6-4   Eclipses occur only when the sun is along, or nearly along, the line of nodes.

Figure 6-4. The orientation of the moon's orbit, and the line of nodes, *nn′*, remains relatively fixed during a revolution of the earth about the sun. There are, therefore, just two places in the earth's orbit, points *A* and *B,* where the sun's direction lies along the line of nodes. It is only during the times in the year, roughly 6 months apart, when the earth-sun line is approximately along the line of nodes that eclipses can occur. These times are called *eclipse seasons.*

Because of perturbations of the moon's orbit, the line of nodes is gradually moving westward in the ecliptic, making one complete circuit in 18.6 years. Therefore, the eclipse seasons occur earlier each year by about 20 days. In 1970 the eclipse seasons are near February and August; in 1975 they are near November and May.

6.5 ECLIPSES OF THE SUN

One of the most surprising coincidences of nature is that the two most prominent astronomical objects, the sun and the moon, have so nearly the same apparent size in the sky. Although the sun is about 400 times as large in diameter as the moon, it is also about 400 times farther away, so both the sun and moon subtend about the same angle — about ½°.

The apparent or angular sizes of both sun and moon vary slightly from time to time, as their respective distances from the earth vary. The average angular diameter of the sun (as seen from the center of the earth) is 31′ 59″, and the average angular diameter of the moon is slightly less, 31′ 5″. However, the sun's apparent size can vary from the mean by about 1.7 percent and the moon's by 7 percent. The maximum apparent size of the moon

FIG. 6-5   Geometry of a total solar eclipse (not to scale).

is 33′ 16″, larger than the sun's apparent size, even at its largest. Therefore, if an eclipse of the sun occurs when the moon is somewhat nearer than its average distance, the moon can completely hide the sun, producing a *total solar eclipse*. An equivalent way of stating it is to say that a total eclipse of the sun occurs whenever the umbra of the moon's shadow reaches the surface of the earth.

(a) Geometry of a Total Solar Eclipse

The geometry of a total solar eclipse is illustrated in Figure 6-5. The earth must be at a position in its orbit such that the direction of the sun is nearly along the line of nodes of the moon's orbit. Furthermore, the moon must be at a distance from the surface of the earth that is less than the length of the umbra of the moon's shadow. Then, at new moon, the moon's umbra intersects the ground at a point X on the earth's surface. Anyone on the earth at X (within the small area covered by the moon's umbra) will not see the sun and will witness a total eclipse. The moon's penumbra, on the other hand, covers a larger area of the earth's surface. Any person within the penumbra will see part but not all of the sun eclipsed by the moon — a partial solar eclipse.

As the moon moves eastward in its orbit at about 2400 mi/hr, its shadow sweeps eastward across the earth at the same speed. The earth, however, is rotating eastward at the same time, so the speed of the shadow with respect to a particular place on earth is less than 2400 mi/hr. At the equator, where the rotation of the earth carries places eastward at about 1040 mi/hr, the shadow moves relative to the earth with a speed of about 1060 mi/hr. In higher latitudes the speed is greater. In any case, the tip of the truncated cone of the umbra of the moon's shadow sweeps along a thin band across the surface of the earth, and the total solar eclipse is observed successively along this band (refer to Figure 6-5). This path across the earth within which a total solar eclipse is visible (weather permitting) is called the *eclipse path*. Within a zone about 2000 mi on either side of the eclipse path, a partial solar eclipse is visible — the observer, inside this limit, being located in the penumbra of the shadow.

Because the moon's umbra just barely reaches the earth, the width of the eclipse path, within which a total eclipse can be seen, is very small. Under the most favorable conditions, the path is only 167 mi wide in regions near the earth's equator. At far northern or southern latitudes, because the moon's shadow falls obliquely on the ground, it can cover a path somewhat more than 167 mi wide.

It does not take long for the moon's umbra to sweep past a given point on earth. The duration of totality may be only a brief instant. It can never exceed about 7½ minutes.

FIG. 6-6   Eclipse path of total solar eclipses occurring between 1963 and 1984. *(Yerkes Observatory.)*

sun is covered by the moon. *Second contact* occurs from 1 to 2 hours after first contact, at the instant when the sun becomes completely hidden behind the moon. In the few minutes before second contact (the beginning of totality) the sky noticeably darkens; some flowers close up, and chickens may go to roost. Because the diminished light that reaches the earth must come solely from the edge of the sun's disk, and consequently from the higher layers in its atmosphere (see Chapter 17), the sky and landscape take on strange colors. In the last instant before totality, the only parts of the sun that are visible are those that shine through the lower valleys in the moon's irregular profile and line up along the periphery of the advancing edge of the moon — a phenomenon called *Baily's beads*. During totality, the sky is quite dark and the brighter stars are visible.

As Baily's beads disappear and the bright disk of the sun becomes entirely hidden behind the

(b) Appearance of a Total Solar Eclipse

A total solar eclipse is one of the most spectacular of natural phenomena. If a person is anywhere near the path of totality of a solar eclipse, it is well worth his while to move into the eclipse path so that he may witness this rare and impressive event.

The very beginning of a solar eclipse is the *first contact*, when the moon just begins to silhouette itself against the edge of the sun's disk. The *partial phase* of the eclipse is the period following first contact, during which more and more of the

FIG. 6-7   The corona of the sun, photographed during the eclipse of June 8, 1918. *(Hale Observatories.)*

moon, the *corona* flashes into view. The corona is the sun's outer tenuous atmosphere, consisting of sparse gases that extend for millions of miles in all directions from the apparent surface of the sun. It is ordinarily not visible because the light of the corona is very feeble compared to that from the underlying layers of the sun that radiate most of the solar energy into space. Only when the brilliant glare from the sun's visible disk is blotted out by the moon during a total eclipse is the pearly white corona, the sun's outer extension, visible. Recently, however, it has become possible to photograph the inner, brighter, part of the corona with an instrument called a coronagraph, a telescope in which a black disk in the telescope's focal plane produces an artificial eclipse, enabling at least the brighter part of the corona to be studied at any time.

Also, during a total solar eclipse, the chromosphere can be observed — the layer of gases just above the sun's visible surface. Prominences, great jets of gas extending above the sun's surface, are sometimes viewed. These outer parts of the sun's atmosphere are discussed more completely in Chapter 17.

The total phase of the eclipse ends, as abruptly as it began, with *third contact,* when the moon begins to uncover the sun. Gradually the partial phases of the eclipse repeat themselves, in reverse order. At *last contact* the moon has completely uncovered the sun.

In addition to being beautiful to watch, total eclipses of the sun have considerable astronomical value. Many data are obtained during eclipses that are otherwise not accessible. For example, during an eclipse we can determine the exact relative positions of the sun and moon by timing the instants of the four contacts. We can take direct photographs and make spectrographic observations of the sun's outer atmosphere and prominences. We can measure the light and heat emitted by the corona. We can determine how meteorological conditions are affected by solar eclipses and can learn something about the light-scattering properties of the earth's atmosphere. Measurements of positions of stars near the sun's direction during eclipses have also been made to check a prediction of general relativity (Section 3.7a).

(c) Annular Eclipses of the Sun

More than half the time the moon does not appear large enough in the sky to cover the sun completely, which means that the umbra of its shadow does not reach all the way to the surface of the earth. The geometry of the situation is illustrated in

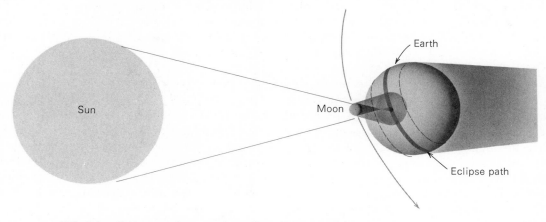

FIG. 6-8   Geometry of an annular eclipse (not to scale).

Figure 6-8. When the moon's shadow cone does not reach the earth, a total eclipse is not possible. However, if the boundaries of the umbra of the shadow are extended until they intersect the earth's surface, they define a region on the ground within which the moon can be seen completely silhouetted against the sun's disk, with a ring of sunlight showing around the moon. This kind of eclipse is called an *annular eclipse,* from the Latin word *annulus,* meaning "ring." The extension of the moon's umbra, within which an annular eclipse is visible, sweeps across the ground in a path much like the path of totality of a total eclipse. As is true for total eclipses, a partial eclipse is visible within a region of 2000 mi or more on either side of the annular eclipse path.

An annular eclipse begins and ends like a total eclipse. However, because the sun is never completely covered by the moon, the corona is not visible, and although the sky may darken somewhat, it does not get dark enough for stars to be seen. An annular eclipse is not so spectacular, nor has it the scientific value of a total eclipse.

(d) Partial Eclipses of the Sun

A *partial eclipse* of the sun is one in which only the penumbra of the moon's shadow strikes the earth. During such an eclipse, the moon's umbra passes north or south of the earth, and from nowhere can the sun appear to be covered completely by the moon. Also, a total or annular eclipse appears partial from regions outside the eclipse path but within the zone of the earth that is intercepted by the moon's penumbra. Few people, therefore, have seen total or annular solar eclipses, whereas most have had the opportunity to see the sun partially eclipsed. The moon seems to "skim" across the northern or southern part of the sun. How much of the sun can appear covered depends, of course, on how close the observer is to the path of totality or annularity. Partial eclipses are interesting but not spectacular. The progress of the

FIG. 6-9   An annular eclipse of the sun. Bright spots are sunlight streaming through lunar valleys. *(Lick Observatory.)*

eclipse can be observed conveniently through heavily smoked glass or densely exposed photographic film. Only if the observer is within a few hundred miles of the eclipse path will he see the sky darken appreciably.

6.6 ECLIPSES OF THE MOON

A lunar eclipse occurs when the moon, at the full phase, enters the shadow of the earth. There are three kinds of lunar eclipses: *total, partial,* and *penumbral.*

(a) Geometry of a Lunar Eclipse

The geometry of lunar eclipses is shown in Figures 6-10 and 6-11. Unlike a solar eclipse, which is visible only in certain local areas on the earth, a lunar eclipse is visible to everyone who can see the moon. Weather permitting, a lunar eclipse can be seen from the entire night side of the earth, including those sections of the earth that are carried into the earth's umbra while the eclipse is in progress. A lunar eclipse, therefore, is observed far more fre-

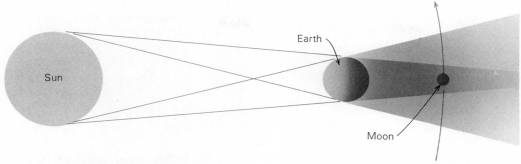

FIG. 6-10   Geometry of a lunar
eclipse (not to scale).

quently from a given place on earth than is a solar
eclipse.

Figure 6-11 shows the cross section of the
earth's shadow at the moon's distance. The umbra
there averages about 5700 mi in diameter, and the
penumbra about 10,000 mi across. Shown are four
of the many possible paths of the moon through
the earth's shadow. A total lunar eclipse occurs
when the moon passes completely into the umbra
(path *A*). A *partial eclipse* occurs if only part of

the moon skims through the umbra (path *B*), and
a *penumbral eclipse* occurs if the moon passes
through the penumbra, or partially through the
penumbra, but does not come into contact with
the umbra (paths *C* and *D*).

(b) Appearance of Lunar Eclipses

Penumbral eclipses usually go unnoticed even
by astronomers. Only within about 700 mi of the
umbra is the penumbra dark enough to produce a

FIG. 6-11   Different kinds of
lunar eclipses.

9:04 P.M.

11:32 P.M.

9:38 P.M.

12:04 A.M.

10.01 P.M.

12:23 A.M.

FIG. 6-12   Sequence of photographs of the total lunar eclipse of November 17/18, 1956. *(Photographed by Paul Roques, Griffith Observatory.)*

noticeable darkening on the moon. However, the diminished illumination on the moon's surface can be detected photometrically.

Every total or partial lunar eclipse must begin with a penumbral phase. About 20 minutes or so before the moon reaches the shadow cone of the earth, the side nearest the umbra begins to darken somewhat. At the moment called *first contact*, the limb of the moon (the "edge" of its apparent disk in the sky) begins to dip into the umbra of the earth. As the moon moves farther and farther into the umbra, the curved shape of the earth's shadow upon it is very apparent. In fact, Aristotle listed the round shape of the earth's shadow as one of the earliest proofs of the fact that the earth is spherical (Section 2.2d).

If the eclipse is a partial one, the moon never gets completely into the umbra of the earth's shadow but passes on by, part of it remaining in the penumbra, where it still receives some sunlight. At *last contact* the moon emerges from the umbra.

On the other hand, if the eclipse is a total one, at the instant of *second contact* the moon is completely inside the umbra, and the total phase of the eclipse begins. Even when totally eclipsed, the moon is still faintly visible, usually appearing a dull coppery red. Kepler explained this phenomenon in his treatise *Epitome*. The illumination on the eclipsed moon is sunlight that has passed through the earth's atmosphere and been refracted by the air into the earth's shadow (Figure 6-13).

The eclipse is darkest if the center of the lunar disk passes near the center of the umbra. The darkness of the lunar eclipse depends also on weather conditions around the terminator of the earth. It is only here, on the line between day and night on the earth, that sunlight passing through the atmosphere can be refracted into the shadow. Heavy cloudiness in that critical region will allow less light to pass through. The light striking the eclipsed moon inside the umbra is reddish because red light of longer wavelengths penetrates through the long path of the earth's atmosphere most easily.

Totality ends at *third contact*, when the moon begins to leave the umbra. It passes through its partial phases to *last contact*, and finally emerges

FIG. 6-13   Illumination of the moon during a total eclipse by sunlight refracted by the earth's atmosphere into the earth's shadow.

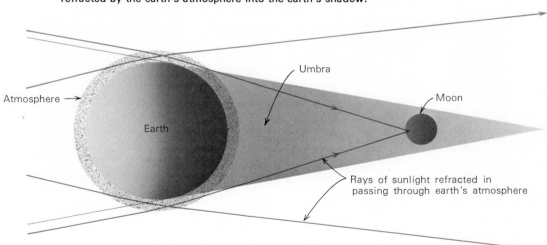

Atmosphere →

Umbra

Moon

Earth

Rays of sunlight refracted in passing through earth's atmosphere completely from the penumbra. The total duration of the eclipse depends on how closely the moon's path approaches the axis of the shadow during the eclipse. The moon's speed with respect to the shadow is about 2100 mi/hr; if it passes through the center of the shadow, therefore, about 6 hours will elapse from the time the moon starts to enter the penumbra until it finally leaves it. The penumbral phases at the beginning and end of the eclipse last about 1 hour each, and each partial phase consumes at least 1 hour. The total phase itself can last as long as 1 hour 40 minutes if the eclipse is central, or, of course, a shorter time if the moon does not pass through the center of the umbra.

6.7 FREQUENCY OF ECLIPSES

We have seen that solar eclipses occur only when the moon is new and lunar eclipses when the moon is full. Furthermore, eclipses occur only when the new or full moon is near a node. If the sun, earth, and moon were geometrical points, the new or full moon would have to occur *exactly* at a node if there were to be an eclipse. Because of the finite sizes of the three bodies, however, part of the sun may appear covered by part of the moon for observers at certain places on earth even though at new moon the moon is not located *exactly* at the node. Similarly, in a lunar opposition, it is possible that, because of the large size of the earth's shadow, the

TABLE 6.1 Important Total Solar Eclipses in the Second Half of the Twentieth Century

DATE	DURATION OF TOTALITY (MIN)	WHERE VISIBLE
1952 Feb. 25	3.0	Africa, Asia
1954 June 30	2.5	North-Central U.S. (Great Lakes), Canada, Scandinavia, U.S.S.R., Central Asia
1955 June 20	7.2	Southeast Asia
1958 Oct. 12	5.2	Pacific, Chile, Argentina
1959 Oct. 2	3.0	Northern and Central Africa
1961 Feb. 15	2.6	Southern Europe
1962 Feb. 5	4.1	Indonesia
1963 July 20	1.7	Japan, Alaska, Canada, Maine
1965 May 30	5.3	Pacific Ocean, Peru
1966 Nov. 12	1.9	South America
1970 March 7	3.3	Mexico, Florida, parts of U.S. Atlantic coastline
1972 July 10	2.7	Alaska, Northern Canada
1973 June 30	7.2	Atlantic Ocean, Africa
1974 June 20	5.3	Australia
1976 Oct. 23	4.9	Africa, Indian Ocean, Australia
1977 Oct. 12	2.8	Northern South America
1979 Feb. 26	2.7	Northwest U.S., Canada
1980 Feb. 16	4.3	Central Africa, India
1981 July 31	2.2	Siberia
1983 June 11	5.4	Indonesia
1984 Nov. 22	2.1	Indonesia, South America
1987 March 29	0.3	Central Africa
1988 March 18	4.0	Philippines; Indonesia
1990 July 22	2.6	Finland, Arctic Regions
1991 July 11	7.1	Hawaii, Central America, Brazil
1992 June 30	5.4	South Atlantic
1994 Nov. 3	4.6	South America
1995 Oct. 24	2.4	South Asia
1997 March 9	2.8	Siberia, Arctic
1998 Feb. 26	4.4	Central America
1999 Aug. 11	2.6	Central Europe, Central Asia

full moon can pass partially through it or even pass completely into it, even if the full moon is not exactly at the node. Thus, the requirement that the moon be at a node can be relaxed slightly; there is some leeway within which eclipses can occur.

The leeway is great enough that it is impossible to go through an eclipse season without some kind of solar eclipse. There are, therefore, at least two solar eclipses visible from someplace on earth each year. It is possible to have as many as 5 solar eclipses in a calendar year. The number of lunar eclipses ranges from none to 3. The maximum number of eclipses of all kinds in any one year is 7.

The significant solar eclipses in the second half of the 20th century are listed in Table 6.1.

6.8 PHENOMENA RELATED TO ECLIPSES

There are other phenomena with geometrical properties similar to those of eclipses that involve other celestial bodies. Examples are *occultations* and *transits.*

The moon often passes between the earth and a star; the phenomenon is called an *occultation.* During an occultation, a star suddenly disappears as the eastern limb of the moon crosses the line between the star and observer. If the moon is at a phase between new and full, the eastern limb will not be illuminated and the star may appear to vanish mysteriously as the dark edge of the moon covers it. Because the moon moves through an angle about equal to its own diameter every hour, the longest time that an occultation can last is about 1 hour. It can have a much shorter duration if the occultation is not central. Geometrically, occultations are equivalent to total solar eclipses, except that they are total eclipses of stars other than the sun.

Occultations of the brighter stars are listed in advance in various astronomical publications. The times and durations of the occultations and the places on earth from which they are visible are given. Also listed are the comparatively rare occultations of planets by the moon, and of stars by planets.

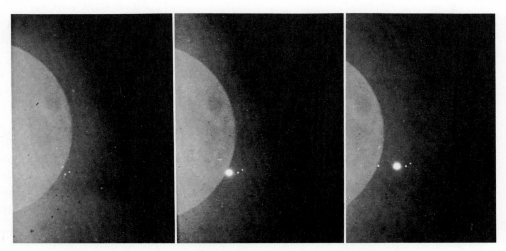

FIG. 6-14   Series of photographs showing *(left to right)* the emergence of Jupiter and three of its satellites after their occultation by the moon. *(Photographed by Paul Roques, Griffith Observatory.)*

A *transit* is a passage of an inferior planet (Mercury or Venus) across the front of the sun's disk at the time of inferior conjunction. Usually each of these planets appears to pass north or south of the sun, but it can pass in front of the sun if inferior conjunction occurs when the planet is near one of the nodes of its orbit — the points where its orbit crosses the ecliptic.

The sun passes the nodes of Mercury's orbit in May and November; these are the seasons for transits of Mercury. There are, on the average, about 13 transits of Mercury per century. Transits of Venus are more rare than those of Mercury. At present, transits of Venus occur in pairs separated by intervals of about 8 years. The last pair of Venusian transits were those of 1874 and 1882. The next pair will be on June 8, 2004, and June 6, 2012.

Transits are analogous to annular eclipses of the sun. The planets Venus and Mercury, seen from earth, are too small to cover the sun completely; their shadow cones fall far short of reaching the surface of the earth. The appearance of a transit is that of a black dot slowly crossing the disk of the sun from east to west. The silhouette of Mercury against the sun is too small to see without a telescope. That of Venus can be barely observed, without optical aid, if the sun is properly viewed by projection, or through dense filters to protect the eye.

Exercises

1. When earthshine is brightest on the moon, what must be the phase of the earth as seen from the moon?

2. About what time does the moon rise when it is at each of the following phases:
   (a) New?   (d) Two days past new?
   (b) Full?   (e) Three days past full?
   (c) Third quarter?   (f) First quarter?

3. Suppose you lived on a plain on the moon near the center of the moon's disk as seen from earth.
   (a) How often would you see the sun rise?
   (b) How often would you see the earth set?
   (c) Over what fraction of the time would the vernal equinox be above your horizon?

4. Use data in the appendix to calculate how long radar waves (which travel with the speed of light) require to make the round trip from earth to moon and back.

5. The red army staged a surprise attack on the blue army on the night of February 10 during the relative darkness of a total eclipse of the moon. If you were the general of the blue army, when would you plan a return attack if you wanted to be sure of a complete night of no moonlight?

6. Sketch the shadow of a pencil cast by an electric lamp. Indicate the umbra and penumbra.

7. If a solar eclipse were annular over part of the eclipse path and total over part, would the total eclipse be of long or short duration? Why?

8. Describe what an observer on the moon would observe during a total lunar eclipse. Assume that he is on the side of the moon turned toward the earth.

9. Describe what the same observer would see during a total solar eclipse (total solar eclipse as viewed from the earth).

10. Where is the moon in the sky under each of the following conditions:

(a) Full moon; local time is 2:00 P.M.?

(b) Moon at first quarter; local time is midnight?

(c) Waning crescent phase of the moon; local time is 5:00 A.M.?

(d) Sun is totally eclipsed; local time is 4:00 P.M.?

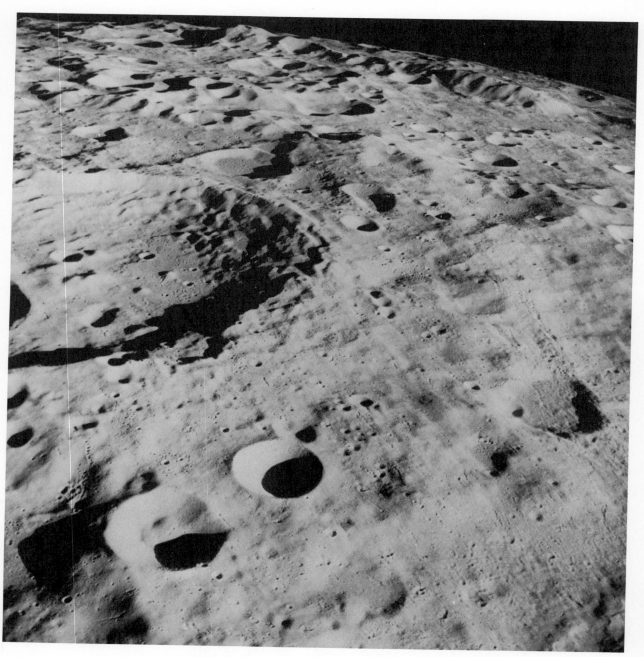

Looking south on the lunar farside from the Apollo 8 spacecraft. *(NASA.)*

CHAPTER 7

Atoms and Light

Light is one form of *electromagnetic energy,* so called because the energy is carried by electric and magnetic waves through empty space. Actually, visible light constitutes only a small part of a wide range of different kinds of electromagnetic radiation, which differ only in *wavelength* but are called by different names: gamma rays, x-rays, ultraviolet, light, infrared, and radio waves.

7.1 NATURE AND PROPERTIES OF ELECTROMAGNETIC RADIATION

By its omnipresence, light is familiar to us all. Yet, to describe its nature, or to represent it with pictorial conceptions, is extremely difficult. The best we can do is to describe its properties with mathematical models. Most of the characteristics of electromagnetic radiation can be described adequately if it is represented as energy propagated in waves, although no medium is required to transmit them.

Any wave motion can be characterized by a *wavelength.* Ocean waves provide an analogy. The wavelength is simply the distance separating successive wave crests. Various forms of electromagnetic energy differ from each other only in their wavelengths. Those with the longest waves, ranging up to miles in length, are called *radio waves.* Forms of electromagnetic energy of successively shorter wavelengths are called, respectively, infrared radiation, light, ultraviolet radiation, x-rays, and finally the very short wave gamma rays. Short radio waves and long infrared waves are sometimes referred to as *radiant heat.* All these forms of radiation are the same basic kind of energy and could be thought of as different kinds of light. In this book, however,

we shall reserve the term "light" to describe those wavelengths of electromagnetic radiation that, by their action upon the organs of vision, enable them to perform the functions of sight. Light is also called *luminous energy.*

Not all the properties of electromagnetic radiation can be described by this simple wave model. Experiments with electrons have shown that radiation of any given wavelength is always absorbed or emitted in quantities of energy that are whole multiples of some basic tiny quantity of radiant energy. It is as if electromagnetic energy were composed of many discrete "packets" of energy. These units of radiant energy are called *photons.* Photons must not be thought of as particles, for, as we have said, electromagnetic energy travels with a wavelike motion. Photons can be regarded as individual wave trains propagating through space, each spreading out in all directions from its source. This picture, however, describes only how electromagnetic radiation behaves, not what it actually "is."

Light, with other forms of electromagnetic radiation, is thus regarded both as a wave form of energy, and as composed of great numbers of photons, or packets of energy. The wave picture of light is adequate to describe many of the properties of light, including those principles of optics that are relevant to the design of telescopes. For some purposes, however, we shall have to call upon the photon representation of light. Different kinds of photons, corresponding to electromagnetic radiation of different wavelengths, differ from each other in their energies and are detected by different means. Before we investigate the differences among photons, however, we should investigate those properties that they have in common.

(a) The Inverse-Square Law

An important property of the propagation of electromagnetic energy is the *inverse-square law*, a property that also belongs to the propagation of other kinds of energy, for example, sound. The amount of energy that would be picked up by a telescope, or an eye, or any other detecting device of fixed area, located at any given distance d from a light source O, is proportional to the amount of energy, or the number of photons, crossing each square inch of the surface of an imaginary sphere of radius d (Figure 7-1). A certain finite amount of energy, or number of photons, is emitted by the source in a given time interval. That energy spreads out over a larger and larger area as it moves away from the source. When it has moved out a distance d, it is spread over a sphere of area $4\pi d^2$ ($4\pi d^2$ is the surface area of a sphere of radius d). If E is the amount of energy in question, an amount $E/4\pi d^2$

passes through each square inch (if d is measured in inches).

We see, then, that the amount of energy passing through a unit area *decreases with the square of the distance from the source*. At distances d_1 and d_2 from a light source, the amounts of energy received by a telescope (or other detecting device), l_1 and l_2, are in the proportion:

$$\frac{l_1}{l_2} = \frac{4\pi d_2^2}{4\pi d_1^2} = \left(\frac{d_2}{d_1}\right)^2.$$

The above relation is known as the inverse-square law of light propagation. Figure 7-2 also illustrates the inverse-square law of propagation.

(b) Speed of Light

All electromagnetic energy propagates in a vacuum with a speed of 2.997925×10^{10} cm/sec, or about 186,000 mi/sec. Both theory and all available evidence demonstrate that this speed of light is the highest possible speed, and furthermore that an observer anywhere in space would observe light to move with the same speed with respect to him. In fact, this universality of the observed speed of light is one of the postulates of the special theory of relativity.

(c) Different Kinds of Electromagnetic Energy

All forms of electromagnetic energy have certain characteristics in common: their wavelike method of propagation and their speed. Nevertheless, the various kinds of radiant energy, differing from each other in wavelength, are detected by very different means.

Radio waves have the longest wavelength, ranging up to miles in length. Those used in short-wave communication and in television have wavelengths ranging from inches to yards. When these waves pass a conductor, such as a radio antenna, they induce in it a feeble current of electricity, which can be amplified until it can drive a loudspeaker or recording apparatus.

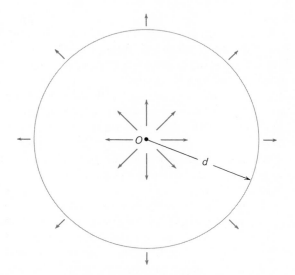

FIG. 7-1   Inverse-square law of light propagation.

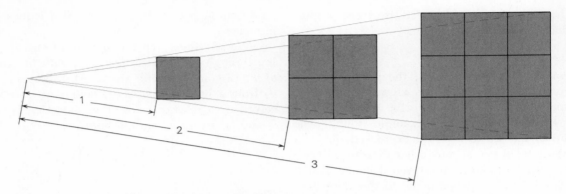

FIG. 7-2   As light energy radiates away from its source, it
spreads out, so that the energy passing through unit area decreases as
the square of the distance from the source.

The shortest wavelengths of radio radiation, less than about one millimeter ($\frac{1}{25}$ in.), merge into infrared radiation. Infrared photons can be detected with thermocouples, and lead sulfide cells. These are devices that generate small currents when they absorb radiation of wavelengths to which they are sensitive. Infrared radiation of wavelength less than about 1/1000 in. can be photographed with special photographic emulsions. A simple device for detecting infrared radiation is a container of water. As the water absorbs the radiation its temperature rises.

Electromagnetic radiation with a wavelength in the range 0.000016 to 0.000028 in. comprises visible light. An appropriate unit to express the wavelength of visible light is the *angstrom*. One angstrom (abbreviated Å) is 1 hundred-millionth of a centimeter; 1 cm is about $\frac{2}{5}$ in. Visible light, then, has wavelengths that range from about 4000 to 7000 Å. The exact wavelength of visible light determines its *color*. Radiation with a wavelength in the range 4000 to 4500 Å or so gives the impression on the retina of the eye of the color violet. Radiations of successively longer wavelengths give the impression of the colors blue, green, yellow, orange, and red, respectively. The array of colors of visible light is called the *spectrum*. A mixture of light of all wavelengths, in about the same relative proportions as are found in the light emitted into space by the sun, gives the impression of white light.

Radiation of wavelengths too short to be visible to the eye is called *ultraviolet*. Radiations with wavelengths less than about a millionth of an inch comprise x-rays. Both ultraviolet radiation and x-rays, like visible light, can be detected photographically.

Electromagnetic radiation of the shortest wavelength, to less than a billionth of an inch, is called *gamma radiation*. Gamma rays are often emitted in the course of nuclear reactions and by radioactive elements. Gamma radiation is generated in the deep interiors of stars; it is gradually degraded into visible light by repeated absorption and reemission by the gases that comprise stars.

The array of radiation of all wavelengths, from radio waves to gamma rays, is called the *electromagnetic spectrum*.

(At one time, *cosmic rays* were thought to be very short-wavelength electromagnetic radiation. It is now known, however, that cosmic rays are mostly nucleii of atoms from outer space, bombarding the earth's upper atmosphere with extremely high speeds — near that of light. They are thought to originate from the Galaxy or beyond.)

(d) Relation Between Wavelength, Frequency, and Speed

For any kind of wave motion we can derive a simple relation between wavelength and *frequency*. The frequency of light (or sound) is the rate at which wave crests pass a given point, that is, the number of wave crests that pass per second. Imagine a long train of waves moving to the right, past point O (Figure 7-3), at a speed c. If c is measured, say, in centimeters per second, we can measure back to a distance of c centimeters to the left of O and find the point P along the wave train that will just reach the point O after a period of 1 second. The frequency f of the wave train — that is, the number of waves that pass O during that second — is obviously the number of waves between P and O. That number of waves, times the length of each, λ, is equal to the distance c. Thus we see that for any wave motion, the speed of propagation equals the frequency times the wavelength, or, symbolically,

$$c = f \times \lambda.$$

(e) The Energy of Photons

According to modern quantum theory, each photon carries a certain discrete amount of energy that depends only on the frequency of the radiation it comprises. Specifically, the energy of a photon is proportional to the frequency; the constant of proportionality, denoted h, is called *Planck's constant,* named for Max Planck (1858–1947), the great German physicist who was one of the originators of the quantum theory. If metric units are used (that is, if energy is measured in ergs and frequency in cycles or waves per second), Planck's constant has the value $h = 6.62 \times 10^{-27}$ erg·sec. The energy of a photon is thus

$$E = h \times f.$$

Since we have seen in Section 7.1d that the frequency times the wavelength is equal to the speed of light, we also note that the energy of a photon is inversely proportional to the wavelength; that is,

$$E = \frac{hc}{\lambda}.$$

Photons of violet and blue light are thus of higher energy than those of red light. The highest energy photons of all are gamma rays; those of lowest energy are radio waves.

7.2 LAWS OF GEOMETRICAL OPTICS

Having discussed the properties of electromagnetic radiation in general, we now turn our attention to those of visible light. The properties of light that are most important to the design and construction of telescopes and other astronomical instruments can be summarized simply in three laws of *geometrical optics* — the laws of *reflection, refraction,* and *dispersion.*

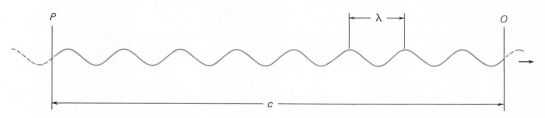

FIG. 7-3   Relation between wavelength, frequency, and the speed of radiation.

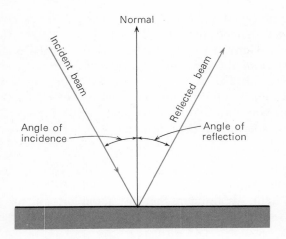

FIG. 7-4   Law of reflection.

normal to the surface at the point where it strikes. That angle is the *angle of incidence*. The angle that the reflected beam of light makes with the normal is called the *angle of reflection*. The law of reflection states that the angle of reflection is equal to the angle of incidence and that the reflected beam lies in the plane formed by the normal and the incident beam (see Figure 7-4).

(b) The Law of Refraction

The law of refraction deals with the deflection of light when it passes from one kind of transparent medium into another. Every transparent substance can be characterized by its *index of refraction*, a measure of the degree to which the speed of light is diminished in passing through it. Specifically, the index of refraction is the ratio of the speed of light in a vacuum to that in the substance. Whenever light passes from a medium having one index of refraction into a medium having another index, such as from air into glass or from glass into air, it is always bent, or refracted, at the interface between the two media. Usually, media of higher densities have higher indices of refraction. Thus if light passes from air into glass, it passes from a

(a) The Law of Reflection

The law of reflection describes the manner in which light is reflected from a smooth, shiny surface. We shall speak frequently of the normal to a surface. The *normal* to a surface at some point is simply a line or direction perpendicular to that surface at that point. If light strikes a shiny surface, its direction must make a certain angle with the

FIG. 7-5   Refraction of light between glass and air: (a) when light passes from air to glass; (b) when light passes from glass to air.

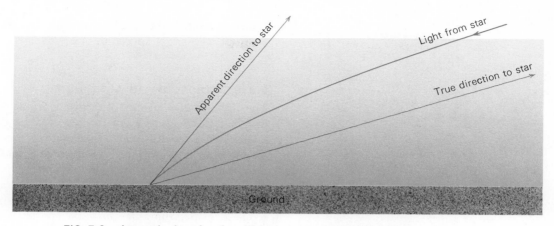

FIG. 7-6 Atmospheric refraction. Light is bent upon entering the earth's atmosphere in such way as to make stars appear at a higher altitude than they actually are. (The effect is grossly exaggerated in this figure.)

medium of lesser into a medium of higher index of refraction.

The law of refraction states that if light passes from one medium into a second one of a different index of refraction, the angle the light beam makes with the normal to the interface between the two substances is always *less* in the medium of higher index. Thus, if light goes from air into glass or water, it is bent *toward* the normal to the interface, while if it goes from water or glass into air, it is bent *away* from the normal. The law of refraction is expressed mathematically by Snell's law.†

Water has an index of refraction of about $1\frac{1}{3}$; crown glass and flint glass have indices of about 1.5 and 1.6, respectively. Diamond has the very high index of refraction of 2.4.

In a vacuum, the index of refraction is taken, by convention, to be exactly 1.0. The index of air at sea level is about 1.00029. The light from stars, planets, the sun, and the moon, is bent, upon enter-

ing the earth's atmosphere, in such a way as to make the object appear to be at a greater altitude above the horizon than it actually is. Atmospheric refraction is greatest for objects near the horizon. It raises the apparent altitude of objects on the horizon by about $\frac{1}{2}°$. Refraction of light, both by glass and by the earth's atmosphere, is illustrated in Figures 7-5 and 7-6.

(c) The Law of Dispersion

Dispersion of light is the manner in which white light, a mixture of all wavelengths of visible light, can be decomposed into its constituent wavelengths or colors when it passes from one medium into another. The phenomenon of dispersion occurs because the index of refraction of a transparent medium is different for light of different wavelengths. In general, the index of refraction is greater for light of shorter wavelengths. Thus, whenever light is refracted in passing from one medium into another, the violet and blue light, of shorter wavelengths, is bent more than the orange and red light of longer wavelengths.

Figure 7-7 shows the way in which light of different wavelengths is bent different amounts in

---

†*Snell's law* is
$$n \sin \alpha = n' \sin \alpha',$$
where α and α' are the angles between a light beam and the normal to the interface between two media of indices of refraction n and n', respectively.

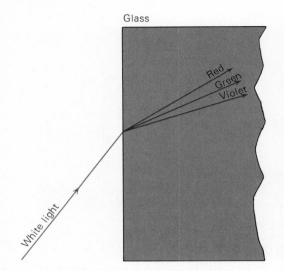

FIG. 7-7   Dispersion at the interface of two media.

passing from, say, air into glass. Figure 7-8 shows a prism, a triangular piece of glass. Upon entering one face of the prism, light is refracted once, the violet light more than the red, and upon leaving the opposite face, the light is bent again, and so is further dispersed. Even greater dispersion can be obtained by passing the light through a series of prisms. If the light leaving a prism is focused upon a screen, the different wavelengths or colors that compose white light are lined up side by side. The array of colors is a spectrum.

7.3 SPECTROSCOPY IN ASTRONOMY

The light from a star or other luminous astronomical body, as well as from the sun, can be decomposed into its constituent wavelengths, producing a spectrum. A device used to observe visually the spectrum of a light source is a *spectroscope;* one used to photograph a spectrum is a *spectrograph.* In Chapter 8 it will be explained how a spectroscope (or spectrograph) may be constructed and how it can be attached to a telescope so that we may study the spectra of astronomical objects.

(a) The Value of Stellar Spectra

If the spectrum of the white light from the sun and stars were simply a continuous rainbow of colors, astronomers would not have such intensive interest in the study of stellar spectra. To Newton, the solar spectrum did appear as just a continuous band of colors. However, in 1802, William Wollaston (1766–1828) observed several dark lines running across the solar spectrum. He attributed these lines to natural boundaries between the colors. Later, in 1814–1815, the German physicist Joseph

FIG. 7-8   Dispersion by a prism.

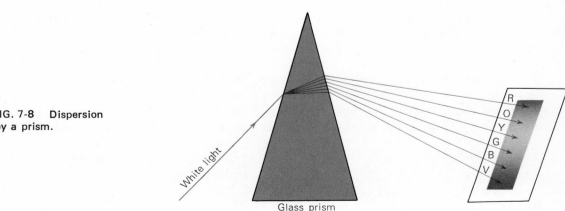

Fraunhofer (1787-1826), upon a more careful examination of the solar spectrum, found about 600 such dark lines. He noted the specific positions in the spectrum, or the wavelengths, of 324 of these lines. To the more conspicuous lines he assigned letters of the alphabet, with the letters increasing from the red to the violet end of the spectrum. Today we still refer to several of these lines in the solar spectra by the letters assigned to them by Fraunhofer.

Subsequently, it was found that such dark spectral lines could be produced in the spectra of artificial light sources by passing their light through various transparent substances. On the other hand, the spectra of the light emitted by certain glowing gases were observed to consist of several separate bright lines. A preliminary explanation of these phenomena (although grossly oversimplified in light of modern knowledge) was provided in 1859 by Gustav Kirchhoff (1824–1887) of Heidelberg. Kirchhoff's explanation is often given in the form of *three laws of spectral analysis:*

FIRST LAW: *A luminous solid or liquid emits light of all wavelengths, thus producing a continuous spectrum.*

SECOND LAW: *A rarefied luminous gas emits light whose spectrum shows bright lines, and sometimes a faint superposed continuous spectrum.*

THIRD LAW: *If the white light from a luminous source is passed through a gas, the gas may abstract certain wavelengths from the continuous spectrum so that those wavelengths will be missing or diminished in its spectrum, thus producing dark lines.*

We distinguish, then, among three types of spectra. A *continuous* spectrum is an array of all wavelengths or colors of the rainbow. A *bright line* or *emission* spectrum appears as a pattern or series of bright lines; it is formed from light in which only certain discrete wavelengths are present. A *dark line* or *absorption spectrum* consists of a series or pattern of dark lines — missing wavelengths — superposed upon the continuous spectrum of a source of white light (see Figure 7-9).

The great significance of Kirchhoff's laws is that each particular chemical element or compound, when in the *gaseous form*, produces its own characteristic pattern of dark or bright lines. In other words, each particular gas can absorb or emit only

FIG. 7-9   Three kinds of spectra: (a) continuous; (b) bright line; (c) dark line. (The spectra are shown as they appear on photographic negatives.)

FIG. 7-10   The solar spectrum. Labels indicate the elements in the sun's
photosphere that cause some of the dark lines. The wavelengths,
in Angstroms, and the colors of the different parts of the spectrum
are also labeled. *(Hale Observatories.)*

certain wavelengths of light, peculiar to that gas. The presence of a particular pattern of dark (or bright) lines characteristic of a certain element is evidence of the presence of that element somewhere along the path of the light whose spectrum has been analyzed.

Thus the dark lines (Fraunhofer lines) in the solar spectrum give evidence of certain chemical elements between us and the sun, absorbing those wavelengths of light. It is easy to show that most of the lines must originate from gases in the outer part of the sun itself.

The wavelengths of the lines produced by various elements are determined by laboratory experiment. Most of the thousands of Fraunhofer lines in the sun's spectrum have now been identified with more than 60 of the known chemical elements.

Dark lines are also found in the spectra of stars and in stellar systems. Much can be learned from their spectra, in addition to evidence of the chemical elements present in a star. A detailed study of its spectral lines indicates the temperature, pressure, turbulence, and physical state of the gases in that star; whether or not magnetic and electric fields are present, and the strengths of those fields; how fast the star is approaching or receding from us; and many other data. Information about a star can also be obtained from a study of its continuous spectrum. The study of the spectra of celestial objects is one of the most powerful means at the astronomer's disposal.

(b) The Doppler Effect

In 1842 Christian Doppler (1803–1853) pointed out that if a light source is approaching or receding from the observer, or vice versa, the light waves will be, respectively, crowded closer together or spread out. The principle, known as the *Doppler principle* or *Doppler effect,* is illustrated in Figure 7-11. In (a) the light source is stationary with respect to the observer. As successive wave crests 1, 2, 3, and 4 are emitted, they spread out evenly in all directions, like the ripples from a splash in a pond. They approach the observer at a distance λ behind each other, where λ is the wavelength of the light. On the other hand, if the source is moving with respect to the observer, as in (b), the successive wave crests are emitted with the source at

different positions, S_1, S_2, S_3, and S_4, respectively. Thus, to observer A, the waves seem to follow each other by a distance *less* than λ, whereas to observer C they follow each other by a distance greater than λ. The wavelength of the radiation received by A is shortened; the wavelength of the radiation received by C is lengthened. Because the wavelengths seem to A to follow each other at an accelerated rate, A also observes the light at a higher frequency than if the source were stationary, while C receives the light at a diminished frequency. To observer B, in a direction at right angles to the motion of the source, no effect is observed. The effect is produced only by a motion *toward* or *away* from the observer, a motion called *radial velocity.* Observers between A and B or B and C would observe some shortening or lengthening of the light waves, respectively, for a component of the motion of the source is in their line of sight.

The Doppler effect is also observed in sound. We have all heard the apparent higher than normal pitch of the whistle of an approaching train and the lower than normal pitch of a receding one. It

(a)

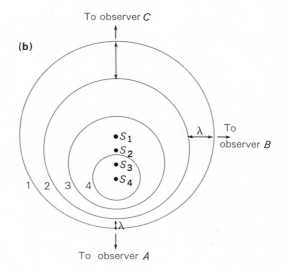

(b)

FIG. 7-11 Doppler effect.

follows from the theory of relativity that for light there is no way of determining whether the source, the observer, or both, is in motion, and the formula for the increase or decrease in the wavelength, the *Doppler shift,* is identical for each case. The wavelengths are shortened if the distance between the source and observer is decreasing, and they are lengthened if the distance is increasing.

The exact formula for the Doppler shift is:

$$\frac{\Delta\lambda}{\lambda} = \frac{\sqrt{1 + v/c}}{\sqrt{1 - v/c}} - 1,$$

where λ is the wavelength emitted by the source, $\Delta\lambda$ is the difference between λ and the wavelength measured by the observer, c is the speed of light, and v is the relative line of sight velocity, called the *radial velocity,* of the observer and source, which is counted as positive if the velocity is one of recession and negative if it is one of approach. If the relative line-of-sight velocity of the source and observer is small compared to the speed of light, however, the formula reduces to the simple form, which is usually used,

$$\frac{\Delta\lambda}{\lambda} = \frac{v}{c}.$$

Solving this last equation for the velocity, we find

$$v = c\frac{\Delta\lambda}{\lambda}.$$

If a star approaches or recedes from us, the wavelengths of light in its continuous spectrum appear shortened or lengthened, as well as those of the dark lines. However, unless its speed is tens of thousands of miles per second, the star does not appear noticeably bluer or redder than normal. The Doppler shift is thus not easily detected in a continuous spectrum (except for very remote galaxies — see Chapter 22) and cannot be measured accurately in such a spectrum. On the other hand, the wavelengths of the absorption lines can be measured accurately, and their Doppler shift is relatively simple to detect. Generally, when the spectrum of a star or other object is photographed at the telescope, sometime during the exposure the light from an iron arc or some other emission-line source is allowed to pass into the same spectrograph, and the spectrum of the arc is then photographed just beside that of the star. The known wavelengths of the bright lines in the spectrum of the arc (or other laboratory source) serve as standards against which the wavelengths of the dark lines in the star's spectrum can be accurately measured. Further illustrations of the Doppler effect are given in later chapters.

7.4 RADIATION LAWS

If a body intercepts electromagnetic radiation, it generally reflects some of it, transmits some, and absorbs the rest. Except for transparent objects the transmitted radiation is negligible, if even measurable. The energy associated with the absorbed radiation heats the body, and would continue to heat it indefinitely if the body did not begin to radiate that energy away. A body (for example, Mars) that is exposed to an approximately constant flow of radiation (sunlight falling on Mars) will eventually reach an equilibrium temperature such that the average rate at which it reradiates energy just equals the rate at which it absorbs it. Therefore, an opaque body (Mars, or some other planet) may be observed both by the radiation it reflects and by the reradiated energy it previously absorbed.

The quality of the reflected radiation depends on the absorbing properties of the body. If, for example, it is exposed to white light and absorbs the same fraction of electromagnetic energy of all wavelengths, it appears, in reflected light, to be white. If it preferentially absorbs short wavelengths, the light it reflects is dominated by long wavelengths and it appears red. Thus the colors of objects depend on their absorbing properties.

Many complicating factors determine the quality of the energy reemitted by the body, but the

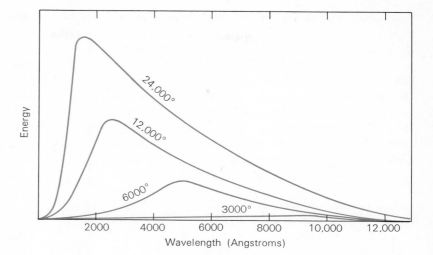

FIG. 7-12   Energy emitted at different wavelengths for black bodies at several temperatures.

most important is its temperature when the energy it reemits is in equilibrium with that which it absorbs. If a body with no internal source of energy were perfectly reflecting, it would absorb and reemit nothing and its temperature would be at absolute zero. Planets typically have temperatures of tens or hundreds of degrees absolute,† and reradiate the solar energy they absorb mostly at infrared and radio wavelengths. At room temperature on earth, almost all objects (including people) are radiating infrared radiation, with a maximum intensity near 0.01 mm wavelength.

(a) Perfect Radiators

Of particular interest is a hypothetical *perfect radiator*, also called a *black body*. This is an idealized body that absorbs all the electromagnetic energy incident on it. Its temperature then depends only on the total radiant energy striking it each second, and the quality of the energy it reradiates can be precisely predicted from the *radiation laws*.

---

†In astronomy, temperature is almost always expressed on the Kelvin or absolute scale — that is, in Centigrade degrees above absolute zero (−273°C); see Appendix 5.

A perfect radiator or black body is black in the sense that it is completely opaque to all wavelengths but of course need not look black. At room temperature the radiated energy is mostly in the invisible infrared and a black body does look black, but a perfect radiator with a temperature of thousands of degrees appears very bright indeed.

Because of the high opacity of the gases that constitute stars, the light from stars resembles, approximately, that from perfect radiators. The resemblance is not complete for various reasons, but the spectral energy distribution of a star (the distribution of its light among various wavelengths) usually can be approximated by the energy emitted by a perfect radiator at a temperature more or less representative of that of the star's photospheric layers. Thus, with fair accuracy, we can apply the laws of black-body radiation to a star.

(b) Planck's Radiation Law

The energy emitted from black bodies had been studied experimentally during the last century. It was only about the beginning of the 20th century, however, that the German physicist Max Planck (1858–1947) found a theoretical interpretation for black-body radiation. He succeeded in this by adopting the hypothesis that light energy is *quantized*, that is, it occurs only in discrete "packets," or *photons*, each of energy (in ergs) hc/λ (Section 7.1e). An equation derived by Planck gives the radiant energy emitted per second at any wavelength from 1 cm² of the surface of a black body of any temperature.† In Figure 7-12, this energy distribution, computed from Planck's formula, is plotted against wavelength for several different temperatures. Three points, in particular, should be noted:

(1) A perfect radiator at any temperature emits *some* radiation at *all* wavelengths, but not in equal amounts.

(2) A hotter black body emits more radiation (per square centimeter) at *all* wavelengths than does a cooler black body.

(3) A hotter black body emits the largest proportion of its energy at shorter wavelengths than a cooler black body does. Hot stars appear blue because most of their energy is at short wavelengths; cool stars appear red because most of their energy is at long wavelengths.

(c) Wien's Law

We notice in Figure 7-12 that there is a particular wavelength, λ_{max}, at which a perfect radiator emits its maximum light. The higher the temperature of the body, the shorter λ_{max}; in fact, λ_{max} is inversely proportional to the temperature:

$$\lambda_{max} = \frac{constant}{T}.$$

If the wavelength is measured in centimeters and the temperature in degrees Kelvin, the constant has the value 0.2897. The foregoing relation is known as *Wien's law*.*

Since stars resemble black bodies, the wavelengths at which they emit their maximum light indicate their temperatures. For the sun, this wavelength is about 4.750×10^{-5} cm, which corresponds to a temperature of 6100°K.

(d) The Stefan-Boltzmann Law

If we sum up the contributions from all parts of the spectrum, we obtain the total energy emitted by a black body over all wavelengths. The total energy emitted per second per square centimeter by a black body at a temperature T is given by the equation known as the *Stefan–Boltzmann law:*

$$E(T) = \sigma T^4.$$

The constant σ, called the *Stefan–Boltzmann constant*, has the value 5.672×10^{-5}, if T is in °K and $E(T)$ is in ergs/cm²/sec. The Stefan–Boltzmann law was derived in the last century from thermodynamics, but it can also be derived from Planck's law.‡ Note that a blue, hot star emits more energy per unit area than does a cool, red star.

7.5 ABSORPTION AND EMISSION OF LIGHT BY ATOMS

The radiation from stars shows that even their outermost, coolest layers have temperatures of thousands of degrees. Stellar temperatures are above the boiling points of all chemical substances. Moreover, except in the atmospheres of the cooler stars, the temperatures are too high for chemical compounds or even for simple molecules to exist.

---

†If T is the temperature in degrees Kelvin, λ the wavelength in centimeters, and k Boltzmann's constant (1.37×10^{-16}), $E(\lambda, T)$, the energy in ergs emitted per unit wavelength interval per second per square centimeter and into unit solid angle, is

$$E(\lambda, T) = \frac{2hc^2}{\lambda^5} \frac{1}{e^{hc/\lambda kT} - 1}.$$

---

*Wien's law can be derived from Planck's formula by finding the wavelength at which the derivative of Planck's formula equals zero.

---

‡By integrating Planck's formula.

Consequently, the gases of stars are in the atomic state. These atoms are responsible for the absorption and emission of light that occurs.

We know that at least some of these atoms can emit a *continuous* spectrum, that is, light over a continuous range of wavelengths, because we see such continuous radiation coming from stars. We know also that at least some of the atoms can absorb radiation (the gas is opaque over large ranges of wavelengths), for there is no wavelength at which we can see into the interiors of stars. Further, we know that the atoms sometimes absorb radiation at certain discrete wavelengths much more strongly than at other wavelengths, for we observe the dark absorption lines in the spectra of the sun and stars, at which wavelengths light escapes from the stars only with greatly reduced intensity. Finally, we know that atoms can also emit light of these same wavelengths, for we observe emission lines in the laboratory spectra of gases and in the spectra of gaseous nebulae (Chapter 19).

We now consider the structure of atoms and how they give rise to the absorption and emission processes described above.

(a) Structure of Atoms

Atoms are the "building blocks" of matter; they are the smallest particles into which a chemical element can be subdivided and still retain its chemical identity. From experiments in the physics laboratory, we have learned that an atom consists of two basic parts: a nucleus and a system of electrons. The nucleus of an atom contains practically all the atom's mass. The nucleus also carries a positive electric charge. The charge on a nucleus is always an integral (or whole) multiple of a certain small unit of charge.† The atoms of the different chemical elements differ from each other in the number of these charge units on their nuclei. The number of charge units on the nucleus of an atom is called

its *atomic number*. An atom of hydrogen (the simplest kind of atom) has an atomic number of 1. The atomic numbers of helium and lithium are 2 and 3, respectively, of oxygen, 8, and of uranium, 92. Atoms of still higher atomic number have been produced in the nuclear physics laboratory, but most of them are unstable and in relatively short times they spontaneously decay into simpler atoms.

The mass of an atomic nucleus is also nearly equal to an integral multiple of a basic unit, called the *atomic mass unit* (amu). The atomic mass unit is about 1.66×10^{-24} gm. The mass of a nucleus, in amu, is called its *atomic weight*. Most hydrogen nuclei have an atomic weight of about 1; a few, however, have atomic weights of 2 or even 3. Atomic nuclei of the same atomic number but different masses are said to compose the different *isotopes* of that element. For example, an isotope of hydrogen in which the atomic nuclei have an atomic weight of 2 is called *deuterium,* or sometimes "heavy hydrogen." Atoms of the most common isotope of helium have a mass of approximately 4, and those of oxygen have a mass of 16.* Most of the heavier atoms have atomic masses 2 to 3 times their atomic numbers.

Nuclei can be decomposed into smaller particles. The most important of these are *protons,* which have a charge of one unit and a mass of approximately one unit (the nucleus of the most common isotope of hydrogen *is* a proton), and *neutrons,* which have a mass of about one unit, but are electrically neutral.

An electron has a mass of $\frac{1}{1835}$ amu and carries a *negative* charge that is numerically equal to the positive charge on a single proton. Under ordinary circumstances, an atom contains just as many electrons as its nucleus has positive units of

---

†The unit of charge is 4.80×10^{-10} electrostatic unit.

*In chemical notation the atomic mass unit is $\frac{1}{16}$ the mean mass of the oxygen atom. In the new physical notation, it is $\frac{1}{12}$ the mass of an atom of the most common isotope of carbon. The nearest integer to the atomic mass of a nucleus is called its *mass number.*

charge, and so is electrically *neutral.* The electrons of an atom are clustered outside the nucleus and revolve about it. The orbital motion of an electron corresponds to a certain amount of energy possessed by the atom, and the larger the orbit, the more the energy. It is possible for an atom to absorb or emit energy (for example, in the form of electromagnetic energy), and in the process one (or more) of its electrons moves, respectively, into a larger or smaller orbit.

(b) Spectrum of Hydrogen

A clue to the structure of atoms came from the study of the spectrum of hydrogen, whose atoms are the simplest in nature. The dark lines of hydrogen that can be observed in the spectra of many stars occur in an orderly spaced series of wavelengths. The bright lines of hydrogen that are observed in the laboratory spectrum of glowing hydrogen are observed in the same series of wavelengths. The Swiss physicist Balmer found a simple formula that gave the wavelengths of this series of lines, now called the *Balmer series* of hydrogen (Figure 7-13).

(c) The Bohr Atom

The Danish physicist Niels Bohr (1885–1962) suggested that the hydrogen spectrum can be explained if the assumption is made that only orbits of certain sizes are possible for the electron in the hydrogen atom. By specifying those permissible sizes for the electron orbits, Bohr was able to compute the values of energy, corresponding to the orbital motion of the electron, that are possible for an individual atom. He assumed that an atom can change from one allowed state of energy to another state of higher energy if its electron moves from a smaller to a larger allowed orbit. Conversely, according to the hypothesis, if the electron moves from a larger to a smaller orbit, the atom changes from a higher to a lower state of energy. One way in which an atom can gain or lose energy is by absorbing or

emitting light. Since light is composed of *photons* whose energies depend on their wavelengths (the energy of a photon is hc/λ), the only wavelengths of light that could be absorbed or emitted are those corresponding to photons possessing energies equal to differences between various allowed energy states of the hydrogen atom. Those energy states are given by a simple formula that Bohr derived from the orderly progression of wavelengths in each series of lines in the hydrogen spectrum. In other words, by assuming that the allowed sizes of the electron orbits and thus the possible energies of the hydrogen atom are *quantized,* Bohr was able to account for the spectrum of hydrogen.

For example, suppose a beam of white light (which consists of photons of all wavelengths) is passed through a gas of atomic hydrogen. A photon of wavelength 6563 Å has the right energy to raise an electron in a hydrogen atom from the second to the third orbit, and can be absorbed by those hydrogen atoms that are in their second to lowest energy states. Since the energy of a photon is inversely proportional to its wavelength, the shorter the wavelength of a photon, the higher its energy. Photons with higher energies corresponding to the other successively shorter wavelengths in the Balmer series, therefore, have the right energies to raise an electron from the second orbit to the fourth, fifth, sixth, and larger orbits, and can also be absorbed. Photons with intermediate wavelengths (or energies) cannot be absorbed. Thus, the hydrogen atoms absorb light only at certain wavelengths, and produce the spectral *lines.* Conversely, hydrogen atoms in which electrons move from larger to smaller orbits emit light — but again only light of those energies or wavelengths that correspond to the energy differences between permissible orbits. The transfer of electrons giving rise to spectral lines is shown in Figure 7-13.

A similar picture can be drawn for kinds of atoms other than hydrogen. However, since they ordinarily have more than one electron each, the

FIG. 7-13 Emission and absorption of light by the hydrogen atom according to the Bohr model.

Balmer series (absorption)

Lyman series (emission)

Balmer series (emission)

Paschen series (emission)

Brackett series (emission)

$n = 1$

2

3

4

5

6

energies of the orbits of their electrons are much more complicated, and the problem of their spectra is much more difficult to handle theoretically.

(d) Energy Levels of Atoms and Excitation

Bohr's model of the hydrogen atom was one of the beginnings of the quantum theory, and was a great step forward in the development of modern physics and in our understanding of the atom. However, we know today that atoms cannot be represented by quite so simple a picture as the Bohr model. Even the concept of discrete orbits of electrons must be abandoned, because according to the modern quantum theory it is impossible at any instant to state the exact position and the exact velocity of an electron in an atom simultaneously. Nevertheless, we still retain the concepts that only certain discrete energies are allowable to an atom. These energies, called *energy levels,* can be thought

of as representing certain mean or average distances of an electron from the atomic nucleus.

Ordinarily, the atom is in the state of lowest possible energy, which, in the Bohr model, would correspond to the electron being in the innermost orbit. However, an atom can absorb energy which raises it to a higher energy level (corresponding, in the Bohr picture, to the movement of an electron to a larger orbit). The atom is then said to be in an *excited state.* Generally, an atom remains excited only for a very brief time; after a short interval, typically a hundred-millionth of a second or so, it drops back down to its lowest energy state, with the simultaneous emission of light, unless it chances to absorb another photon first and go to a still higher state. (In the Bohr model, this corresponds to a jump by the electron back to the innermost orbit.) The atom may return to its lowest state in one jump, or it may make

FIG. 7-14 Energy-level diagram for hydrogen. The shaded region represents energies at which the atom is ionized. (See Section 7.5e for explanation of ionization.)

the transition in steps of two or more jumps, stopping at intermediate levels on the way down. With each jump, it emits a photon of the wavelength that corresponds to the energy difference between the levels at the beginning and end of that jump. An energy-level diagram for a hydrogen atom and several possible *atomic transitions* are shown in Figure 7-14; compare this figure with the Bohr model, shown in Figure 7-13.†

Because atoms that have absorbed light and have thus become excited generally deexcite themselves and emit that light again, we might wonder

————————

†Actually, according to the modern quantum theory, atomic energy levels are not *perfectly* discrete (that is, "sharp"), but are, rather, the *most probable* energies of an atom. The vast majority of atoms will have energies that lie in the immediate neighborhood of one of the allowed levels. At any time, however, a few atoms will possess energies that deviate somewhat from one of those values. Spectral lines caused when atoms jump from one energy level to another, therefore, are not perfectly sharp, but spread over a small, but finite, range of wavelengths. The effect is called *natural line broadening.*

why dark lines are ever produced in stellar spectra. In other words, why doesn't this reemitted light "fill in" the absorption lines? Some of the reemitted light actually *is* received by us and this light does partially fill in the absorption lines, but only to a slight extent. We can observe the reemitted light as emission lines if we can view the absorbing atoms from a direction from which no light with a continuous spectrum is coming — as we do, for example, when we look at gaseous nebulae (Chapter 19). Figure 7-15 illustrates the situation.

We can calculate from theory the allowable energies of the simplest kinds of atoms and thus the wavelengths of the absorption lines that can be produced by those atoms. For the more complicated atoms, however, the wavelengths of the spectral lines are determined empirically by laboratory experiment.

Atoms in a gas are moving at high speeds and are continually colliding with each other. They can be excited and deexcited, therefore, by these colli-

FIG. 7-15   The atoms in the gas cloud produce absorption lines in the continuous spectrum of the white light source when viewed from direction *A,* but produce emission lines (of the light they reemit) when viewed from direction *B.* The spectra are shown as they appear on photographic negatives.

sions as well as by absorbing and emitting light. The mean velocity of atoms in a gas depends upon its temperature (actually, *defines* the temperature), and if we know the temperature of the gas, we are able to calculate what fraction of those atoms, at any given time, will be excited to any given energy level. In the photosphere of the sun, for example, where the temperature is in the neighborhood of 6000°K, only about 1 atom of hydrogen in 100 million is excited to its second energy level by the process of collision. The Balmer lines of the hydrogen spectrum arise when atoms in this second energy level absorb light and rise to higher levels. At any given time, therefore, most atoms of hydrogen in the sun cannot take part in the production of the Balmer lines.

(e) Ionization

We have described how certain discrete amounts of energy can be absorbed by an atom, raising the atom to an excited state, and moving one of its electrons farther from its nucleus. If enough energy is absorbed, the electron can be removed completely from the atom. The atom is then said to be *ionized.* Still greater amounts of energy must then be absorbed by the ionized atom (called an *ion*) to remove a second electron. If enough energy is available (in the form of very short wavelength photons or in the form of a collision with a very fast moving atom) an atom can become *completely* ionized, in which case it loses all of its electrons. A hydrogen atom, having only one electron to lose, can be ionized only once; a helium atom can be ionized twice, and an oxygen atom, eight times. Any energy over and above the energy required to ionize an atom can be absorbed also, and appears as energy of motion (kinetic energy) of the freed electron. The electrons released from atoms that have been ionized move about in the gas as free particles, just as the atoms and ions do.

An atom that has become ionized has lost a negative charge — that carried away by the electron — and thus is left with a net positive charge. It has, therefore, a strong affinity for a free electron, and eventually will capture one and become neutral (or ionized to one less degree) again. During the capture process, the atom emits one or more photons, depending on whether the electron is captured at once to the state corresponding to the lowest energy level of the atom, or whether it stops at one or more intermediate levels on the "way in." Any energy that the electron possessed as kinetic energy before capture can be emitted. Absorption or emission of light over a continuum of wavelengths therefore accompanies the process of ionization or recapture, at least over those wavelengths corresponding to energies higher than the ionization energy of the atom. Atoms of various kinds, being ionized and deionized, and absorbing and emitting light at various wavelengths, account for much of the continuous opacity and continuous spectrum of the sun and stars.

Just as the excitation of an atom can result from a collision between it and another atom, so also can ionization. The rate at which such collisional ionizations occur depends on the atomic velocities and hence on the temperature of the gas. The rate of recombinations of ions and electrons also depends on their relative velocities — that is, on the temperature — but in addition it depends on the density of the gas; the higher the density, the greater the chance for recapture, because the different kinds of particles are crowded closer together. From a knowledge of the temperature and density of a gas, it is actually possible to calculate the fraction of atoms that have been ionized once, ionized twice, and so on. In the photosphere of the sun, for example, we find that most of the hydrogen and helium atoms are neutral, whereas most of the atoms of calcium, as well as many other metals, are once ionized.

The energy levels of an ionized atom are entirely different from those of the same atom when it is neutral. In each degree of ionization, the energy levels of the ion, and thus the wavelengths of the spectral lines it can produce, have their own characteristic values. In the sun, therefore, we find lines of *neutral* hydrogen and helium, but of *ionized* calcium. Ionized hydrogen, of course, having no electron, can produce no absorption lines.

There is an additional mechanism by which ionized atoms can absorb and emit light in the continuous spectrum. When an electron passes near an ion, it is attracted by that ion's positive charge (because opposite electrical charges attract each other). If the electron is not captured, it will pass the ion in a hyperbolic orbit, much like two stars passing in interstellar space. While the electron is passing, however, the ion can absorb or emit a photon which accompanies a corresponding increase or decrease in the kinetic energy of the hyperbolic motion of the electron. The process is called *free-free* absorption or emission (because the electron is "free" of the ion both before and after the encounter), or *bremsstrahlung*. Photons of any wavelength can be absorbed or emitted in bremsstrahlung.

(f) Summary of Emission and Absorption Processes
Atoms are characterized by energy levels which correspond to various distances of their electrons from their nuclei. By absorbing or emitting radiant energy, an atom can move from one to another of these levels, thus raising or lowering its energy. Since only certain discrete energy levels exist for each kind of atom, the absorbed or emitted radiation occurs only at certain energies or wavelengths, producing dark or bright spectral lines.

An atom is said to be *excited* if it is in any but its lowest allowable energy level. It is said to be *ionized* if, by the absorption of energy, it has lost one or more of its electrons. It can be excited

The 40-inch refracting telescope.
(*Yerkes Observatory.*)

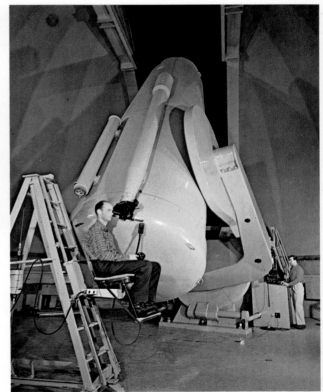

The 48-inch Schmidt telescope.
(*Hale Observatories.*)

The 200-inch telescope dome photographed in moonlight. (Hale Observatories.)

The 120-inch telescope. (Lick Observatory.)

The 200-inch telescope dome. (Hale Observatories.)

Lick Observatory from the east. (Lick Observatory.)

or deexcited by collisions as well as by the absorption and emission of radiation; it can be ionized by collision or by absorbing radiation. Any wavelength of light corresponding to an energy greater than the ionization energy of an atom can be absorbed or emitted in the process of ionization or of deionization when the ion captures an electron.

Ionized atoms (*ions*) can also absorb or emit light at continuous wavelengths, when free electrons pass near them. The relative energy of the ion and passing electron changes by the same amount as the energy absorbed or emitted.

Exercises

1. What kind of electromagnetic energy consists of photons:
(a) Of the highest energy?
(b) Of the lowest energy?
(c) That affect the organs of vision?
(d) To which the earth's atmosphere is opaque?
(e) To which a vacuum is transparent?
2. If a star or other light source were removed to ten times its present distance, how much brighter or fainter would it appear?
3. If the same light source were brought 7 times closer, how much brighter or fainter would it appear?
4. If ocean waves approach the shore at 8 feet per second, and come in at the rate of one wave every fifteen seconds, what is the wavelength?
5. Suppose that a spectral line of some element, normally at 4000 Å, is observed at 3999 Å in the spectrum of a star. How fast is the star approaching us or receding from us?
   Answer: Approaching at about 46 mi/sec.
6. Give some examples in nature:
(a) Of reflection of light;
(b) Of refraction of light;
(c) Of dispersion of light.
7. How could you measure the earth's orbital speed by photographing the spectrum of a particular star at various times throughout the year?
8. What is the approximate temperature of a star whose spectral energy distribution has its maximum intensity at a wavelength of 5794 Å (5.794×10^{-5} cm)?
9. If the sun were suddenly to double in its surface temperature, how would its total output of energy per second change?
10. Most hydrogen atoms in the sun are in their lowest state of energy, so what series of absorption lines would hydrogen produce most strongly in the sun? Do we observe these lines in the solar spectrum? If not, why not?

View looking northwest into Mare Tranquillitatis,
photographed from the Apollo 8 spacecraft in December, 1968. *(NASA.)*

CHAPTER 8

Telescopes

Astronomical instruments have been used since ancient times, but before Galileo they did not employ optical systems. Most of the instruments used by Hipparchus and Ptolemy, and the later, more refined ones of Tycho Brahe, were essentially calibrated "sighting" devices for measuring the directions in the sky of celestial objects. Brahe, for example, designed and built accurate sextants and quadrants, in which sights for observing a star were carried on a long pivoted arm. The end of the arm moved along an arc of a circle with a graduated scale on which the precise angle of the arm and hence of the sights could be read, indicating in turn the angle of the star being observed above the horizon (that is, the star's altitude). Sundials also date from early times.

It is to the astronomical telescope, however, beginning with Galileo (Section 3.2b), that we now turn our attention.

8.1 FORMATION OF AN IMAGE

The most important part of a telescope is the *objective*, the lens or mirror that produces an image. First, therefore, we consider how an image is formed.

(a) Formation of An Image By A Lens

In Figure 8-1 a simple convex lens is shown. Parallel light, say, from a distant star or other light source, is incident upon the lens from the left. A convex lens is a lens thicker in the middle than at the edges. If the curvatures of the surfaces of the lens are correct, light passing through the lens will be refracted in such a way that it converges toward a point. The point where light rays come together is called the *focus* of the lens. At the focus, an *image* of the light source appears. The distance of the focus, or image, behind the lens is called the *focal length* of the lens.

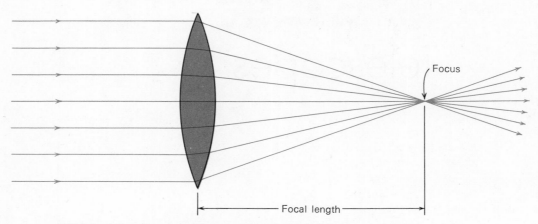

FIG. 8-1 Formation of an image by a simple convex lens.

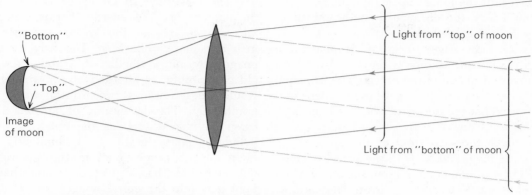

FIG. 8-2   Formation of an image of an extended object.

In Figure 8-2 we see how an image is formed of an extended source, for example, the moon. From each point on the moon, light rays approach the lens along parallel lines. However, from different parts of the moon, the parallel rays of light approach the lens from different directions. The light from each point on the moon strikes all parts of the lens (or *fills* the lens); these rays of light are focused at a point at a distance behind the lens equal to the focal length of the lens. If a screen, such as a white card, is placed at this distance behind the lens, a bright spot of light appears, representing that point on the moon. Light from other points on the moon similarly focuses at other points, producing bright spots on the card in different places. Thus, an entire image of the moon is built up at the focus of the lens. The plane in which the image is formed is called the *focal plane*.

Note that if part of the lens is covered up, or if the middle is cut out, or if ink or mud is splattered over it, as long as part of the lens is still transparent to light, an entire image will be formed. All parts of the lens contribute to each part of the image. Covering up part of the lens cuts down the total amount of light that can strike each portion of the image and thus makes the image fainter, but nevertheless the whole image is formed.

(b) Formation of An Image By A Mirror

Rays of light can also be focused to form an image with a concave mirror — one hollowed out in the middle. Parallel rays of light, as from a star, fall upon the curved surface of the mirror (Figure 8-3), which is coated with silver or aluminum to make it highly reflecting. Each ray of light is reflected according to the law of reflection. If the mirror has the correct concave shape, all the rays are reflected back through the same point, the focus

FIG. 8-3   Formation of an image by a concave mirror.

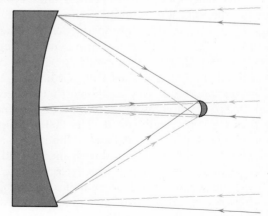

of the mirror. At the focus appears the image of the star. As in a lens, the distance from the mirror to the focus is its focal length.

Rays of light from an extended object are focused by a mirror, exactly as they are by a lens, into an inverted image of the object. The principal difference between image formation by a lens and by a mirror is that the mirror reflects the light back into the general direction from which it came, so that the image forms in front of the mirror. The image can be inspected, as with the lens, by allowing the light to illuminate a screen, such as a white card, held at the focus of the mirror. The card, of course, will block off part of the incoming light, but since, as in the lens, all parts of the mirror contribute to the formation of all parts of the image, the presence of the card will not produce a "hole" in the image but will merely reduce its brightness.

8.2 PROPERTIES OF AN IMAGE

The most important properties of an image are its *scale* or size, its *brightness,* and its *resolution.*

The *scale* of an image is a measure of its size. In all astronomical applications we are dealing with objects whose sizes can be expressed in angular units, and it is generally convenient to express the scale of an image as the linear distance in the image that corresponds to a certain angular distance in the sky. For example, suppose that an image of the moon were produced that is exactly 1 in. across. The moon has an "apparent" or angular size of ½°; that is, it subtends ½° in the sky. The scale of the image is thus 1 in. per half degree, or 2 in. per degree. It can be shown that the scale of an image is directly proportional to the focal length of the lens producing it. The 200-in. mirror of the Hale telescope on Palomar Mountain has a focal length of 660 in. It produces images with a scale of 11.5 in. per degree, which corresponds to 12″.3 per millimeter. Most astronomical telescopes have lenses or mirrors that give image scales in the range 10 to 200″ per millimeter.

The *brightness* of an image is a measure of the amount of light energy that is concentrated into a unit area of the image, say, per square inch or square millimeter. The brightness of an image determines whether it is above the threshold of visibility or, alternatively, how long a time would be required to record the image photographically. The brightness of an image of an extended object, such as the moon, a planet, a nebula, a galaxy, or the faint illumination of the night sky, is proportional to the square of the diameter of the lens (which determines how much light gets into the entire image) and is inversely proportional to the square of the focal length (which determines over how large an area this total light must be spread).

The brightness of the image of a point source, however, is different; stars are effectively point sources, for even through the largest telescopes the stars still appear too small to show any apparent disks. Therefore a lens or mirror of good quality concentrates the starlight into a "point" image regardless of the focal length. For a point-source object such as a star, the amount of light in the image thus depends only on the amount of light gathered by the lens or mirror, and hence is proportional to the square of the aperture.

In summary, the brightness of the image of an extended object is proportional to the square of the aperture-focal length ratio of the lens or mirror, whereas the brightness of the image of a point source is proportional to the square of the aperture.

Resolution refers to the fineness of detail inherently present in the image. Even if the lens or mirror is of perfect optical quality, it cannot produce perfectly sharp and detailed images. Because of the phenomenon of *diffraction,* a point source does not form an image as a true point but as a minute spot of light surrounded by faint, concentric, evenly spaced rings. Therefore, we do not see the geometrical image of a star itself but only a *diffraction pattern* that the telescope lens produces with the star's light. If a star is viewed telescopically under good conditions, the diffraction pattern, consisting of the bright central disk and faint surroundings, is clearly visible.

In the image of an extended source, the diffraction patterns of the various parts of the image, all overlapping each other, wash out the finest details. The ability of an optical system (lens or mirror) to distinguish fine detail in an image it produces, or to produce separate images of two close stars, is called its *resolving power*. In astronomical practice, the resolving power of a lens or mirror is described in terms of the smallest angle between two stars for which separate recognizable images are produced. If the wavelength and aperture are measured in the same units (for example, both in centimeters), the smallest angle (α) in seconds of arc that can be resolved by a lens or mirror of aperture d is given by the equation,

$$\alpha = 2.1 \times 10^5 \times \frac{\lambda}{d} \text{ seconds of arc,}$$

where λ is the wavelength. If λ is chosen as 5.5×10^{-5} cm (5500 Å) — near the middle of the visible spectrum — and if d is measured in inches, the formula becomes

$$\alpha = \frac{4.56}{d} \text{ seconds of arc.}$$

8.3 ABERRATIONS OF LENSES AND MIRRORS

An image produced by any optical system always has imperfections. These are called *aberrations.*

Chromatic or *color aberration,* for example, is a consequence of dispersion. It is the inability of a simple lens to bring light rays of different wavelengths to the same focus. Light of shorter wavelength, being refracted more than that of longer wavelength in passing through the lens, focuses closer to the lens than does the latter.

Spherical surfaces are the most convenient to grind and polish, whether they be the concave or convex surfaces of a lens or mirror. Unfortunately, however, a simple lens or mirror with spherical surfaces suffers from the imperfection called *spherical aberration.* Figure 8-4 illustrates spherical aberration. Light striking nearer the periphery focuses closer to the lens or mirror; light striking near the center focuses farther away.

Both chromatic and spherical aberrations can be reduced greatly by constructing a lens of 2 *elements* — two lenses of different kinds of glass. One lens is concave, the other convex. The combination lens still converges light to a focus, but the two elements cancel out, at least partially, the aberrations introduced by each other.

For a mirror, spherical aberration can be eliminated by grinding and polishing the surface not to a spherical shape, but to a *paraboloid of*

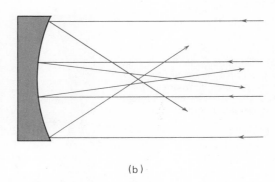

(a)   (b)

FIG. 8-4   Spherical aberration: (a) in lens; (b) in mirror.

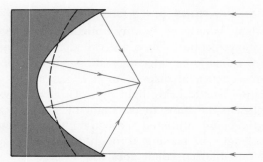

FIG. 8-5 Correction of spherical aberration with a concave parabolic mirror.

revolution, that is, to a surface whose cross section is a parabola (Figure 8-5). A paraboloid has the property that parallel light rays striking all parts of the surface are reflected to the same focus. Similarly, light leaving the focus of a paraboloid is reflected at the surface in a parallel beam. An automobile headlight or searchlight, for example, has a parabolic reflector with the light source at the focus. Mirrors designed for astronomical telescopes generally have parabolic surfaces.

Still other aberrations are *coma, astigmatism, distortion,* and *curvature of field.* Space does not permit a discussion of each here.

8.4 THE SCHMIDT OPTICAL SYSTEM

An ingenious optical system that utilizes both a mirror and a lens was invented by the Estonian optician Bernhard Schmidt of the Hamburg Ob-

servatory. To avoid spherical aberration, parabolic rather than spherical mirrors are used in practice. The principal disadvantage of the parabolic mirror is that it produces good images over only a relatively small field of view, that is, for light that approaches the mirror very nearly on axis.

On the other hand, for a spherical surface, any line reaching the surface through its center of curvature (that is, through the center of the sphere of which the surface is a part) is perpendicular, that is, "square on" to the surface. Thus, all rays of light that pass through an aperture (or opening) located at the center of curvature of a spherical surface must strike the surface "on axis." The Schmidt optical system, utilizing this principle, employs a spherical mirror that is allowed to receive light only through an opening located at its center of curvature (Figure 8-6). Thus there can be no off-axis aberration. The only trouble is that a spherical mirror, suffering as it does from spherical aberration, produces generally poor images for light coming from any direction. Schmidt solved the problem by introducing a thin correcting lens at the aperture at the center of curvature of the mirror. The lens is of the proper shape to correct the spherical aberration introduced by the spherical mirror but does not have to be thick enough to introduce appreciable aberrations of its own. Thus, the Schmidt optical system produces excellent images over a large angular field.

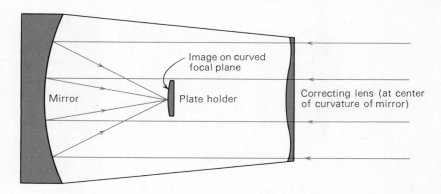

FIG. 8-6 Schmidt optical system. The mirror is generally larger than the lens, so images of unreduced intensity can be produced over an appreciable field of view.

Image on curved focal plane

Mirror

Plate holder

Correcting lens (at center of curvature of mirror)

Today, the Schmidt optical system, and modifications of it invented by Maksutov, Wright, and others, are widely used in science and industry.

8.5 THE COMPLETE TELESCOPE

There are two general kinds of astronomical telescopes in use. They are (1) *refracting* telescopes, which utilize lenses to produce images; and (2) *reflecting* telescopes, which utilize mirrors to produce images. Schmidt telescopes, and modifications thereof, are primarily reflecting telescopes but have refracting correcting lenses.

The refracting telescope is the most familiar. This is the kind of telescope that we can literally "look through." Ordinary binoculars are two refracting telescopes mounted side by side. A lens, generally consisting of two or more elements, is usually mounted at the front end of an enclosed tube. The tube is not really essential — its purpose is merely to block out scattered light — an open framework would suffice as well. In a refracting telescope, the objective, the optical part that produces the principal image, is the lens at the front of the tube. The image is formed at the rear of the tube, where various devices must be available to inspect it, photograph it, or otherwise utilize it.

The reflecting telescope was first conceived by James Gregory in 1663, and the first successful model was built by Newton in 1668. Here a concave mirror (usually a paraboloid) is used as an objective. The mirror is placed at the *bottom* of a tube or open framework. The mirror reflects the light back up the tube to form an image near the front end. Since, in a reflector, the image is formed directly in front of the mirror, it is more inconvenient to get at for observation than the image formed by a refracting telescope. However, this disadvantage is often offset by other advantages inherent in the reflecting telescope.

The principle of reflecting telescopes has been known since Newton's time. Because of the difficulty of precisely producing the reflecting surface,

however, the reflecting telescope did not become an important astronomical tool until the time of William Herschel, a century later.

Schmidt telescopes are used for wide-angle photography in astronomy. The image is formed directly in front of the spherical mirror about halfway from the mirror surface to the correcting lens. Most often the entire system is enclosed in a light-tight tube, and the photographic emulsion, curved to fit the spherical focal surface, is inserted at the focus of the mirror in the center of the tube. The exposure can be started and stopped by uncovering and recovering the correcting lens.

(a) Getting At the Image — Various Focuses

The purpose of a telescope is to produce an image, which must be accessible for inspection, study, or photography. For some types of image analysis, bulky equipment must be placed at the telescope focus. There are various possible arrangements for getting at the focus. Which arrangement is chosen depends on the type of telescope and on the purpose for which the image is to be used.

In a refracting telescope, the image is formed behind the objective lens, at the rear of the telescope tube. Thus even with a large refractor there is no problem in gaining access to the image. All that is required is a suitable ladder or platform to lift the observer to the point from which he can conveniently reach the rear of the tube of the telescope.

With a reflecting telescope the problem of image accessibility arises because a concave mirror produces the image in front of the mirror, in the path of the incoming light. If the image is to be photographed, a small plate or film holder can be suspended at the focus in the middle of the mouth of the telescope tube. The plate holder generally blocks out only a small fraction of the incoming light, so that the brightness of the image is only slightly dimmed. However, if the image is to be inspected visually or with a spectrograph, the observer or the necessary apparatus blocks out so much incoming light that this *prime focus* is often impractical. Thus the light is usually diverted before it comes to a focus so that the image is formed in a more convenient location.

In the *Newtonian*-type reflecting telescope, the problem is solved by a flat mirror mounted diagonally in the middle of the tube so that it intercepts the light just before it reaches the focus. The mirror reflects the light to the side so that it comes to a focus just outside the tube (see Figure 8-7).

Another arrangement for a reflecting telescope is the *Cassegrain* system, in which a small convex mirror rather than a flat mirror is suspended in the telescope tube. The convex mirror intercepts the light before it reaches the prime focus and reflects it back down the tube of the telescope. In many reflecting telescopes (for example, the 200-inch telescope) a small hole is provided in the center of the objective mirror so that the light reflected from the convex mirror can form an image behind the objective. If the objective has no hole (for example, the 100-inch telescope on Mount Wilson), the image must be reflected out to the side of the tube

with a third, flat mirror, mounted diagonally just in front of it. These arrangements are illustrated in Figure 8-7.

In the *Coudé* arrangement, a convex mirror intercepts the light just before the prime focus is reached, and the light is reflected back down the tube until it reaches one of the pivot points about which the telescope tube can be rotated to point to various parts of the sky. There it is intercepted by a flat mirror that reflects the light outside the tube to a stationary observing station. Because the station is not attached to the moving part of the telescope, heavy equipment can be used there. The details of the Coudé focus are illustrated in the diagram of the 200-inch telescope (Figure 8-8).

Most large reflectors in use in astronomical observatories are equipped with auxiliary flat and convex mirrors so that several of the possible focus arrangements can be used.

(b) Telescope Mountings
The telescope tube must be mounted in such a way that it can turn to any direction in the sky. Nearly all astronomical optical telescopes have *equatorial*

FIG. 8-7   Various focus arrangements: (a) prime focus; (b) Newtonian focus; (c) and (d) two types of Cassegrain focus; (e) Coudé focus.

(a)   (b)   (c)   (d)   (e) *mounts.* An equatorial mount (see Figure 8-9) allows the telescope to turn to the north and south about one axis and to the east and west about another. The axis for the east-west motion of the telescope is parallel to the axis of the earth's rotation, and the other axis, about which the telescope can rotate to north or south, is perpendicular to this axis.

Most telescopes are equipped with setting circles that indicate the direction or coordinates in the sky toward which the telescope is pointing. An important advantage of the equatorial mount is that a simple slow motion of the telescope about its axis parallel to the earth's axis — the *polar axis* of the telescope — is sufficient to compensate for the apparent motions of the stars across the sky that result from the earth's rotation. A mechanical or electric mechanism called a *clock drive* is usually employed to drive the telescope automatically about its polar axis, so that it will follow the stars accurately as they move across the sky.

(c) Some Famous Telescopes

The United States, until recently, has had a monopoly on the world's largest telescopes. The "size" of a telescope refers to the diameter of its lens or mirror, or of the correcting lens in the case of a Schmidt. Thus, a 6-inch telescope has a lens or mirror 6 inches in diameter.

The largest existing refractor is the 40-inch refracting telescope at the Yerkes Observatory of the University of Chicago. The Yerkes Observatory is located at Williams Bay, Wisconsin. The second largest refractor is the 36-inch telescope at the Lick Observatory of the University of California, on Mount Hamilton, California. These big refractors were built long ago, before the technique of building large reflectors was perfected.

Some of the largest reflectors are in California. The 200-inch Hale telescope on Palomar Mountain has a focal length of 660 in. It can be used at the prime focus, Cassegrain focus, or Coudé focus. At the prime focus there is a cage 6 ft across in which the observer can ride while he makes his observations. The Hale telescope is one of the few existing telescopes in which the observer is carried "inside" the telescope. The 17-ft Pyrex mirror weighs 14½ tons; the entire instrument has a weight of about 500 tons.

Until the completion of the 200-inch telescope in 1949, the largest reflector was the 100-inch telescope at the Mount Wilson Observatory. (The Mount Wilson and Palomar Observatories are jointly operated by the California Institute of Technology and the Carnegie Institution of Wash-

FIG. 8-8   The 200-inch Hale telescope, sectional view through dome. *(Drawing by Russell W. Porter, Hale Observatories.)*

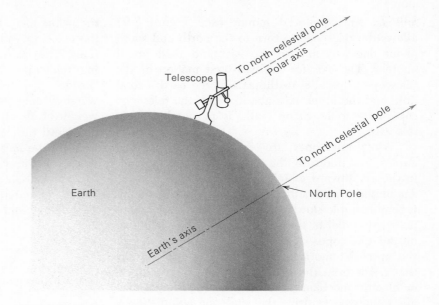

FIG. 8-9   Equatorial mount.

ington, under the name *Hale Observatories;* they have a single director and astronomical staff.)

The Lick Observatory has a 120-inch reflector, of focal length and design features similar to the 200-inch telescope. The University of Texas now operates a new 107-inch reflector at the McDonald Observatory near Fort Davis, Texas. The United States National Observatory at Kitt Peak, near Tucson, Arizona, has an 84-inch reflector, and a telescope of about 158-in. aperture is under construction. A similar 158-in. reflector is planned for the Southern Station of the National Observatory at Cerro Tololo in the Western Andes in Chile. An organization of European nations also plans a large reflector (about 140-in. aperture) in Chile. An additional 150-inch telescope is planned in Australia. The new 98-inch Newton telescope has been installed at the Royal Greenwich Observatory at Herstmonceux, England. East Germany has a 79-in. universal reflector at the Tautenburg Observatory, and a new 79-in. reflector has been installed at Ondrejev, near Prague, Czechoslovakia. The Soviet Union has a

104-in. reflector in use at the Crimea Observatory and a 236-in. reflector nearly completed. The latter, when operational, will be the world's largest telescope of optical quality. There are more than a dozen other reflectors of aperture between 70 and 90 in. scattered over the world and several larger reflectors of nonoptical quality for special purposes.

One of the largest Schmidt telescopes is the 4-foot Schmidt of the Palomar Observatory.† The Palomar Schmidt is ideal for surveying the sky. The first main project assigned to the Schmidt after its completion in 1949 was to produce a photographic atlas of the sky. This sky survey, financed by the National Geographic Society, and called the National Geographic Society–Palomar Observatory Sky Survey, took 7 years to complete. It provides

---

†The universal telescope of the Tautenburg Observatory can be used as a Schmidt by inserting a removable 53-in. correcting lens; it then becomes the world's largest Schmidt.

the most comprehensive coverage of the sky available. Copies of the 1870 photographs that comprise the survey have been distributed to astronomical institutions throughout the world. Large reflectors, like the 200-inch, can probe deeper in space, but only in small regions at a time. It would take the 200-inch telescope at least 10,000 years to complete a survey of the sky comparable to the one completed by the Palomar Schmidt telescope.

8.6 ASTRONOMICAL OBSERVATIONS

The popular view of the astronomer is a man in a cold observatory peering through his telescope all night. Most astronomers do not live at observatories but near the universities or laboratories where they work. A typical astronomer might spend a total of only a few weeks a year observing at the telescope, and the rest of his time measuring or analyzing his data. Some astronomers work at purely theoretical problems and never observe at a telescope. Actually, professional astronomers seldom inspect telescopic images visually except to center the telescope on a desired region of the sky or to make adjustments. On the contrary, the image is generally utilized in one of several other ways.

The most direct way to use the image is to place a screen, for example, a white card, at the focus of the telescope and allow the image to fall on it. This is the most convenient way to view the sun, for example. However, a visual image does not provide a permanent record that can be measured and studied later in the laboratory. Therefore, it is far more common to photograph the image instead of viewing it directly.

(a) Telescopic Photography

To photograph the image, the screen is replaced by a photographic plate or film. The image then forms on a light-sensitive coating which, when developed, provides a permanent record of the image — one that can be measured, studied, enlarged, published, and inspected by many individuals. When used for photography, a telescope becomes nothing more than a large camera; the lens or mirror of the telescope serves as the camera lens.

The most important advantage in using a telescope as a camera is that photographic emulsions can accumulate luminous energy and build up an image during a long exposure. Most astronomical objects of interest are very remote, and hence the light we receive from them is very feeble. However, long time exposures can be made. Astronomical exposures commonly run into hours in length, and occasionally over several successive nights. The longer the exposure, the more faint light gradually accumulates to help build up the photographic image. Objects can be photographed that are about a hundred times too faint to see by just "looking through" a telescope. The layman is often disappointed when he takes his first look through an astronomical telescope, for what he can see visually is nothing compared to the spectacular photographs, such as those reproduced in this book, that are the result of long time exposures.

(b) Visual Inspection with An Eyepiece

Occasionally it is desirable to inspect a telescopic image visually, doing what is commonly called "looking through" a telescope. To best inspect detail in the image, it is customary to view it with a magnifying lens. A common hand magnifier could be used for this purpose. A telescopic *eyepiece* or *ocular* is simply a high-quality magnifying lens that is used to view the image. Figure 8-10 shows how an eyepiece is used in a refracting telescope and in a Newtonian reflector. In practice, eyepieces usually have two or more elements to reduce aberrations.

When an extended celestial object is viewed through a telescope equipped with an eyepiece, it appears enlarged, that is, closer, than when viewed naturally. The factor by which an object appears larger (or nearer) is called the *magnifying power* of the telescope. For example, the moon appears

(a)

(b)

FIG. 8-10   Use of an eyepiece with (a) a refracting telescope; (b) a reflecting telescope.

to subtend an angle of ½° when viewed with the naked eye. If, when viewed through a particular telescope, the moon appears to subtend 10°, the *magnifying power* of the telescope is 20.

Everyone knows that if he has to read fine print he must use a magnifying glass of higher power than when he reads the print in a newspaper. A higher power magnifying glass is one of shorter focal length. Similarly, eyepieces of different focal lengths, used in conjunction with the same telescope objective, produce different image magnifications. It is the purpose of the objective of a telescope to produce an image; it is the purpose of the eyepiece to magnify the image to the point where details in it can be viewed. In principle, any desired magnification can be obtained if an eyepiece of sufficiently short focal length is used. Therefore, it does not make sense to ask an astronomer what "power" his telescope is. The power can be changed at will by using different eyepieces.

At every observatory there is a collection of eyepieces of different focal lengths that can be interchanged for various magnifications.

The exact value of the magnifying power can easily be calculated by dividing the focal length of the objective of the telescope (f_o) by the focal length of the eyepiece (f_e); that is,

$$\text{magnifying power} = \frac{f_o}{f_e}.$$

The highest useful magnification is that with which the finest details resolved in the image can be viewed. Additional magnification reveals no further detail but merely enlarges the blur of the diffraction pattern. Atmospheric disturbances further limit useful magnification, especially in large telescopes.

(c) Spectroscopy

Spectroscopy occupies at least half of the available observing time of most large telescopes. The technique of spectroscopy is made possible by the phenomenon of dispersion of light into its constituent wavelengths. We saw in Chapter 7 that white light is a mixture of all wavelengths and that these can be separated by passing the white light through a prism, producing a spectrum.

A spectrum can also be formed by passing light through, or reflecting it from, a *diffraction grating*. These are of two kinds: A *transmission grating* is a transparent sheet or plate (such as glass) that is ruled with many fine, parallel, closely spaced scratches (typically, thousands per inch), so that light can pass through the material only between the scratches. A *reflection grating* consists of a shiny reflecting surface, say a mirror, upon which are ruled similar fine scratches or grooves, so that light can be reflected only from the places on the surface between the scratches. In either kind of grating, light emerges as if from a series of long parallel finely spaced slits. Because of diffraction, the light fans out in all directions as it passes through these "slits." The waves of light from

different slits interfere with each other in such a way that at any given point beyond the grating, the waves cancel each other out except at one specific wavelength. With an appropriate lens system, successive wavelengths of light can be focused at successive points along a line parallel to the surface of the grating, and in a direction perpendicular to the scratches in it. A spectrum can thus be produced by a grating as well as by a prism. (Space here does not permit a description of the details of how the light waves from a grating interfere with each other to produce the spectrum.)

A *spectrograph* is a device with which the spectrum of a light source can be photographed. Attached to a telescope, the spectrograph can be used to photograph the spectrum of the light from a particular star. The construction of a simple spectrograph is illustrated in Figure 8-11. Light from the source enters the spectrograph through a narrow slit and is then collimated (made into a beam of parallel rays) by a lens. In Figure 8-11

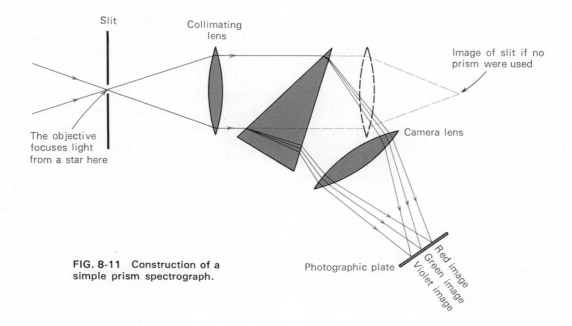

FIG. 8-11 Construction of a simple prism spectrograph.

the collimated light is shown entering a prism, but it can just as well pass through or be reflected from a grating. Different wavelengths of light leave the prism (or grating) in different directions, because of dispersion (or diffraction). A second lens placed behind the prism forms an image of the spectrum on the photographic plate. If the light from the collimating lens had been passed directly into the second lens, rather than through the prism first, the second lens would simply produce an image of the slit through which the light entered the spectrograph (dashed lines in Figure 8-11). However, because of the dispersion introduced by the prism, light of different wavelengths enters the second lens from slightly different directions, and consequently the lens produces a different image of the slit for each different wavelength. The multiplicity of different slit images of different colors lined up at the photographic emulsion is the desired spectrum. If the photographic plate is removed and an eyepiece is used to inspect the spectrum, the whole instrument becomes a *spectroscope*.

All of the parts of a spectrograph — slit, 2 lenses, prism (or grating) and photographic plate — must be attached to a rigid framework so that they do not shift during long exposures. The whole spectrograph may weigh hundreds of pounds and require a lot of space behind the focus.

An *objective prism* is a special plate of glass that is occasionally placed in front of the objective lens of a telescope. This plate is relatively thin; both of its faces are flat but at a slight angle to each other, so that, in effect, it is a thin prism. Starlight entering the telescope must first pass through the objective prism and is slightly dispersed. With the objective prism, if a photograph is taken of a field of stars, the image of each star is not a point but rather a small spectrum. In this way, the spectra of many stars can be photographed at once. The spectra are too small and of too low a dispersion (the colors are not widely spread out) to reveal much detail. Nevertheless,

enough of the major features of the spectra show to enable the stars to be classified roughly according to their spectra.

(d) Photoelectric Photometry

An important datum about a star is the amount of light we receive from it, or its *magnitude*. An accurate device for measuring stellar radiation received at the telescope is the *photomultiplier*, a light-sensitive electron tube placed at the focus of the telescope. Light from a star is gathered by the telescope and is focused on a light-sensitive surface just inside the glass envelope of the tube. Photons striking this *photocathode* dislodge electrons, which are attracted to a positively charged element in the tube, thus creating a feeble current. The current is amplified at a number of successive stages within the tube, and finally in an external amplifier. A measure of the amount of current produced in the tube is a measure of the amount of light striking the photocathode — that is, a measure of the star's brightness.

Other radiation-sensitive devices in which incident photons induce electric currents are the thermocouple, thermopile, and lead sulfide cell. These detectors are used to measure infrared radiation from celestial objects.

(e) Image Tubes

Tremendous improvements in the efficiency of image recording by our present telescopes have come about in recent years through the utilization of photoemissive surfaces, that is, surfaces that emit electrons when exposed to light. One of the devices now in wide use is the *image tube*. The photoemissive surface in the image tube is placed at the focus of the telescope. Starlight, striking the surface, generates a flow of electrons that are focused electronically onto a phosphor screen (such as a television screen). The illuminated screen can be photographed directly, or the electrons can be focused, instead, onto a photographic plate some distance behind the photoemissive surface. With the image tube, a photograph of an astronomical subject, or of its spectrum, can be obtained up to 50 times as fast as by the older photographic procedure.

New and better image-tube devices are constantly being developed, but already we are able to accomplish in a single night at the telescope what hitherto has required weeks. Even so, large telescopes are still neded in astronomy.

The most important property of a telescope is its ability to gather light so that we can detect objects too remote or faint to see with the unaided eye and to collect enough energy to measure and analyze the light from celestial objects. The main reason that we build bigger and bigger telescopes is to obtain more and more light-gathering power. The 200-inch Hale telescope on Palomar Mountain can photograph objects 10 million times fainter than those we can see without optical aid.

8.7 ATMOSPHERIC LIMITATIONS

The earth's atmosphere, so vital to us all, is the biggest headache to the observational astronomer. In at least five ways the air imposes limitations upon the usefulness of telescopes.

(1) The most obvious limitation is that of weather — clouds, wind, rain, and the like. Even on a clear, windless night, however, the atmospheric conditions may be such as to render a telescope virtually useless for many types of work.

(2) Even in visible wavelengths the atmosphere filters out a certain amount of the starlight that strikes it and dims the stars slightly. To wavelengths longer than the near infrared out to the radio wavelength of about 1 cm, the air is almost opaque, and it is opaque also to all wavelengths shorter than 2900 Å. Thus, most of the electromagnetic spectrum remains invisible to ground-based observatories.

(3) In the daytime, the air molecules scatter about so much sunlight that the resulting blue sky hides all celestial objects except the sun, moon, and Venus. At night the sky is darker, but never completely dark. Near cities the air scatters about the glare from city lights, producing an illumination that hides the faintest stars and limits the depths that can be probed by telescopes as well as by the naked eye. Even starlight scattered by the air contributes to the brightness of the night sky.

(4) To make matters worse, the air emits light of its own. Charged particles and x-rays from the sun and outer space, bombarding the upper atmosphere, set the air aglow. When large numbers of these particles are funneled into the atmosphere in the polar regions by the earth's magnetic field, auroras result. In the absence of recognizable auroras, the night air glow still remains. Its brightness does vary from time to time. The faint illumination of the night sky limits the time that a telescopic photograph can be exposed without fogging it completely and thus also limits the faintness of the stars that can be recorded.

(5) When the air is unsteady, star images are blurred. In astronomical jargon, the measure of the atmospheric stability is the *seeing*. When the seeing is good, the stars appear as sharp points and fine detail can be seen on the moon and planets. When the seeing is bad, images of celestial objects are blurred and distorted by the constant twisting and turning of light rays by turbulent air. It is seeing, in fact, that causes stars to "twinkle."

Because stars present too small an angular size for existing telescopes to resolve their disks, their images should be geometrical points or rather tiny diffraction patterns. But in bad seeing, a star may appear as a "mothball" or as a dancing firefly. The star image usually looks "bigger" in bad seeing, but not because the telescope has magnified it, only because the air has distorted it. Use of higher magnification only magnifies these atmospheric disturbances. Only on the rarest of occasions does the image of a star appear smaller than 0.″25 in diameter. Thus seeing limits the resolution attainable with large telescopes.

8.8 RADIO TELESCOPES

In 1931 K. G. Jansky of the Bell Telephone Laboratories was experimenting with antennas for long-range radio communication when he encountered interference in the form of radio radiation coming from an unknown source. He discovered that this radiation came in strongest about 4 minutes earlier on each successive day, and correctly concluded that since the earth's sidereal rotation period is 4 minutes shorter than a solar day, the radiation must be originating from some region of the celestial sphere. Subsequent investigation showed that the source of the radiation was the Milky Way.

In 1936 Grote Reber built the first antenna specifically designed to receive these cosmic radio waves. In 1942 radio radiation was first received from the sun at radar stations in England. After World War II, the technique of making astronomical observations at radio wavelengths developed rapidly, especially in Australia, the Netherlands, England, and more recently the United States.

Radio waves have now been received from many astronomical objects — the sun, moon, some planets, gas clouds in our galaxy, other galaxies, and various other objects. The technique of radio astronomy has become an extremely important tool in observational astronomy.

(a) Detection of Radio Energy from Space

First, it is important to understand that radio waves are not "heard"; they have nothing whatever to do with sound. Although in commercial radio broadcasting, radio waves are modulated or "coded" to carry sound information, the sound itself is not transmitted. The radio waves merely carry the information that a radio receiver must "decode" and convert into sound by means of a loudspeaker or earphones. Sound is a physical vibration of matter; radio waves, like light, are a form of electromagnetic radiation. We can also code visible light to carry sound information, as is done, for example, by the sound track on a movie film.

Many astronomical objects emit all forms of electromagnetic radiation — radio waves as well as light, infrared and ultraviolet radiation, and so on. The radio waves we can receive from space are those that can penetrate through the ionized layers of the earth's atmosphere — those with wavelengths in the range from about 8 mm to about 17 m. The human eye and photographic emulsions are not sensitive to radio waves; we must detect this form of radiation by different means. Radio waves have the property that they induce a current in conductors of electricity. An antenna is such a conductor; it intercepts radio waves which induce a feeble current in it. The current is then amplified in a radio receiver until it is strong enough to measure or record.

(b) Radio Reflecting Telescopes

Radio waves are reflected by conducting surfaces just as light is reflected from an optically shiny surface and according to the same law of reflection. A radio reflecting telescope consists of a parabolic reflector, analogous to a telescope mirror. The reflecting surface can be solid metal, or a fine mesh such as chicken wire. In the professional jargon, the reflecting paraboloid is called a "dish." Radio dishes are usually mounted so that they can be steered to point to any direction in the sky, and gather up radio waves just as an optical reflecting telescope can be directed in any direction to gather up light. The radio waves collected by the dish are reflected to the focus of the paraboloid, where they form a radio image. In an optical telescope, a photographic plate, photomultiplier, spectrograph, or some other device is placed at the focus to utilize the image. In a radio telescope, an antenna or wave guide is placed at the focus of the dish. Radio waves focused on the antenna induce in it a current. This current is conducted to a receiver, not unlike ordinary home radio receivers in principle, where the current is amplified. As the optical astronomer chooses the type of photographic emulsion that is sensitive to the colors of light he wants to detect, the radio astronomer tunes his receiver to amplify the wavelengths he wants to receive from space.

One advantage of astronomical observations at radio wavelengths is that some of the important atmospheric effects discussed in Section 8.7 are not bothersome. In particular, radio observations are not affected by atmospheric seeing (although there is a similar *scintillation* effect due to clouds of ions in interplanetary space). They are less affected by weather and sky brightness, and at some wavelengths can even be made throughout the entire 24-hour day. Manmade radio interference, however, is a serious problem.

The ability of a radio telescope to gather radiation depends on its size. The radio energy received from most astronomical bodies is very small compared to the energy in the optical part of the electromagnetic spectrum. Hence radio dishes are usually built in large sizes; few are under 20 ft across. At first thought, the problem of constructing a large parabolic reflecting surface to a point of sufficient accuracy might seem prohibitive. The 200-in. mirror of the Hale telescope has a surface accurate to about 2 millionths of an inch, about one eighth of the wavelength of visible light. Ideally, radio reflectors should be built to similar accuracy —⅛ wavelength. However, a radio dish designed to receive radio waves of a length of 8 in., for example, need be accurate to only about 1 in.

(c) Radar Astronomy

Radar is the technique of transmitting radio waves to an object and then detecting the radio radiation that the object reflects back to the transmitter. The time required for the radio waves to make the round trip can be measured electronically, and

because they travel with the known speed of light the distance of the object is determined. The value of radar in navigation, whereby surrounding objects can be detected and their presence displayed on a screen, is well known.

In recent years the radar technique has been applied to the investigation of the solar system. Radar observations of the moon and some of the planets have yielded our best knowledge of the distances of those worlds. In addition, as will be discussed in later chapters, radar observations have determined the rotation periods of Venus and Mercury and some information about the nature of the surfaces of those objects.

(d) Famous Radio Telescopes

Among the famous large steerable telescopes (that is, those that can be pointed to various positions in the sky) are the 250-foot dish at Jodrell Bank, England, the 210-foot Parkes dish in Australia, and the 300-foot dish at the National Radio Astronomy Observatory in Greenbank, West Virginia. Several institutions have pairs of large dishes that can be separated to various distances and are used as interferometers — devices that permit an accurate determination of the direction of a radio source. Among these installations is Caltech's Owen's Valley (California) radio observatory, with two movable 90-ft dishes and one 130-ft dish.

Several important radio telescopes are large "bowls" carved out of the ground and lined with reflecting metal meshes. These bowls observe radio sources as they pass overhead, although some directional flexibility is provided by moving the antenna around in the vicinity of the focus of the bowl. One of the best-known installations of this type is the 3300-ft bowl at Arecibo, Puerto Rico.

Still larger installations exist which consist of arrays of antennas spread out over the ground, some more than a mile across. These act as interferometers as well as receivers. There are large arrays in Australia, France, Canada, Italy, the

Netherlands, Germany, the Soviet Union, and the United States.

(e) Long Baseline Interferometers

When two radio telescopes are used as an interferometer, they are both pointed to the same radio source in the sky. The electrical current generated by the radio signals detected at the two telescopes are then brought to one place and mixed together electronically. The way they combine gives information about the direction in the sky from which the original radio source comes. The farther apart the telescopes, the greater is the precision in direction that can be achieved, and the greater is the power of the interferometer to resolve separate sources that are close to the same direction.

It is very critical, however, that the precise differences in time between the instants when the radio waves are received at the two telescopes be known. If the electrical signals are carried over wires of accurately-measured lengths to the electronic equipment, these time differences can be taken care of automatically. On the other hand, there is an obvious limitation to how far the telescopes can be separated and still obtain sufficient control of the timing.

An important advance occurred in 1967, when techniques were developed that allowed tape recordings to be made of the output at each telescope, and simultaneously to record precise time markers on the tape (today atomic time standards are employed). Now telescopes in two different parts of the world can be used to observe the same radio source at the same time and tape record the detected astronomical radio waves along with the time signals. Two two tapes can later be played back, and the recorded data from the two compared electronically to obtain the same information as if the telescopes were wired together directly. By 1972 such *long baseline interferometers* had employed radio telescopes up to 6,600 miles apart — in California and Australia. This intercontinental radio astronomy has achieved an angular resolution in the sky of one-thousandth of a second of arc.

8.9 ROCKET AND SPACE OBSERVATIONS

Since World War II, rockets, satellites, and space probes have provided means of carrying instruments above the earth's atmosphere, and even to the vicinity of the moon, Venus, and Mars, making possible astronomical observations that could never have been made from the earth's surface.

In 1946 the United States Naval Research Laboratory launched a rocket to observe the far ultraviolet radiation from the sun. Since then, many rockets have been launched to make ultraviolet and x-ray photographs of the sun. The photographs were recovered from the rockets after they parachuted back to earth. Spectra have also been obtained of the sun and some other stars at wavelengths in the ultraviolet and in the region of the lower-energy x-rays. High-energy x-rays and gamma rays can be detected (outside the earth's atmosphere) by the ionization they produce in gas chambers. Several sources of x-rays have been found; some are gaseous nebulae, some appear to be associated with particular stars, and some are not yet identified with optically visible objects. In addition, a general background of x-ray radiation of unknown origin has been found.

Many observations have been obtained from earth satellites — both of the earth and of its space environment. In particular, NASA's orbiting solar observatories (OSOs) and orbiting astronomical observatories (OAOs) are designed exclusively for astronomical investigations. The first OSO was launched on March 4, 1962; subsequent ones were launched in 1965 and in 1967.

Probes to the moon, Venus, and Mars have carried instruments to investigate those worlds and the interplanetary space on the way. Television cameras have given us excellent views of the moon and Mars. Other instruments carried on these probes have measured magnetic fields, the intensity of microwave (shortwave radio) radiation, corpuscular radiation (atomic nuclei from the sun), micrometeorite densities in space, and other phenomena.

Exercises

1. If a lens of 120-inch focal length produces an image of the moon 1 in. in diameter, how large an image of the moon would be produced by the 200-inch mirror of the Hale telescope, which has a focal length of 660 in.?

2. If a certain star can barely be photographed in 20 min with a telescope of 10 in. aperture, how long a time would be required to photograph the same star with a telescope of 20 in. aperture? (Ignore photographic complications; assume that the time necessary to photograph a given object with a given telescope is inversely proportional to the amount of light falling on the emulsion.)

3. Most cameras have adjustable iris diaphragms so that their *f*-ratios can be changed at will. Describe how the human eye is an analogy. Consider what happens to the eye in bright sunlight and in very dim light, and why.

4. What kind of telescope would you use to take a color photograph, entirely free of chromatic aberration? Why?

5. What is the theoretical resolving power of the 200-inch telescope? In practice, do you think this high a resolution is ever realized? Why or why not?

6. An eyepiece of ½-in. focal length is used with a telescope whose objective has a focal length of 100 in. What is the magnifying power?

7. Suppose one wants the moon to appear 50° in angular diameter when viewed through a telescope whose objective lens has a focal length of 75 in. What focal length eyepiece must be used?

8. What kind of telescope would you use for each of the following purposes:

(a) To photograph the spectra of many faint stars?

(b) To measure the precise separation of two stars in a double-star system?

(c) To photograph a large field of the Milky Way at once?

(d) To make a photographic survey of the sky?

(e) To record radiation of 21-cm wavelength from a remote gas cloud?

(f) To photograph the sun in x-ray radiation?

(g) To search for ships off the coast of New England?

Apollo 11 astronaut Edwin E. Aldrin, Jr., prepares to deploy laser
beam reflector and seismometer on the surface of the moon.
Note the depth of the footprints in the ejecta as Aldrin walks next
to a small crater off to the left of the picture. *(NASA)*.

CHAPTER 9

The Solar System

The ancient observer, who considered the earth to be central and dominant in the universe, regarded the sun, moon, and planets as luminous orbs that moved about on the celestial sphere through the zodiac.

Our solar system is indeed dominated by one body, but it is the sun, not the earth. Our sun, so important to us, is merely an ordinary, "garden-variety" star. Only careful scrutiny at close range would reveal the tiny planets to an imaginary interstellar visitor. First Jupiter, the largest, would be seen; then Saturn; and perhaps only with the greatest difficulty, the earth and other planets. Almost 99.9 percent of the matter in the system *is* the sun itself; the planets comprise most of what is left — the earth scarcely counts among them. The countless billions of other objects that inhabit the solar system, mostly unknown to the ancients, would probably remain unnoticed by a casual traveler passing through the solar neighborhood.

We turn now to those worlds of the solar system. They will be considered individually in detail in later chapters. Here we take only a brief look at some of the general characteristics of the solar system, and remark on a few of the properties that its constituent worlds have in common.

9.1 INVENTORY OF THE SOLAR SYSTEM

The solar system consists of the sun and a large number of smaller objects gravitationally associated with it. These other objects are the planets, their satellites, the comets, the minor planets or asteroids, the meteoroids, and an interplanetary medium of very sparse gas and microscopic solid particles. The relative prominence of the various kinds of members of the solar system can be seen from Table 9.1, which lists the approximate distribution of mass among the bodies of the solar system. The last four entries in the table are order of magnitude guesses only.

TABLE 9.1 Distribution of Mass in the Solar System

OBJECT	PERCENTAGE OF MASS
Sun	99.86
Planets	0.135
Satellites	0.00004
Comets	0.00003 (?)
Minor planets	0.0000003 (?)
Meteoroids	0.0000003 (?)
Interplanetary medium	<0.0000001 (?)

(a) The Sun

The sun is a typical star — a great sphere of luminous gas. It is composed of the same chemical elements that compose the earth and other objects of the universe, but in the sun (and other stars) these elements are heated to the gaseous state. Tremendous pressure is produced by the great weight of the sun's constituent layers. The high temperature of its interior and the consequent thermonuclear reactions keep the entire sun gaseous. There is no distinct "surface" to the sun; the apparent surface we observe is optical only — the layer in the sun at which its gases become opaque, preventing us from seeing deeper into its interior. The temperature of that region is about 6000°K (11,000°F). Relatively sparse outer gases of the sun extend for millions of miles into space in all directions. The visible part of the sun is 864,000 mi across, which is 109 times the diameter of the earth. Its volume is 1⅓ million times that of the earth. Its mass of 2×10^{33} gm exceeds that of the earth by 333,000 times. The sun's energy output of 4×10^{33} ergs/sec, or about 500 sextillion horsepower, provides all the light and heat for the rest of the solar system. The sun derives this energy from thermonuclear reactions deep in its interior, where temperatures exceed 12 million °K. We shall describe the sun, a typical star, in Chapter 17.

(b) The Planets

Most of the material of the solar system that is not part of the sun itself is concentrated in the planets. The nine known planets include the earth, the five other planets known to the ancients: Mercury, Venus, Mars, Jupiter, and Saturn; and the three discovered since the invention of the telescope: Uranus, Neptune, and Pluto. In contrast to the sun, the planets are small, relatively cool, and (at least mostly) solid. They give off no light of their own but shine only by reflected sunlight.

The masses of the planets, in terms of the mass of the earth, range from 0.05 (Mercury) to 318 (Jupiter). The mass of Jupiter is greater than that of all the other planets combined. In diameter, the planets range from 3000 mi (Mercury) to 86,000 mi (Jupiter). Most, but not all, of the planets are surrounded by gaseous atmospheres. All but three of the planets are known to have natural satellites; Jupiter leads with 12 known moons.

All the planets revolve about the sun in the same direction — from west to east (counterclockwise as viewed from the north). Their mean distances from the sun range from 0.39 AU (36 million mi) for Mercury to 39.46 AU (3670 million mi) for Pluto. Their periods of orbital revolution range from 88 days for Mercury to 248 years for Pluto; the corresponding mean orbital speeds range from about 30 to about 3 mi/sec. The orbits of the planets are all in nearly the same plane, and are also nearly circular.

The planets all rotate as they revolve about the sun. The Jovian planets are rapid rotaters; Jupiter rotates most rapidly, in a period of 9^h50^m.

Mercury rotates 1½ times during its 88-day revolution about the sun. Venus rotates still more slowly, in 243 days, but from east to west, reverse to the rotation of most of the planets (that is, *retrograde*). Some of the planets, especially Jupiter and Saturn, show marked oblateness or flattening because of their rapid rotation.

(c) Satellites

The satellites of the planets are the next most prominent members of the solar system. Only Mercury, Venus, and Pluto do not have known satellites. Jupiter has 12, Saturn has 10, Uranus has 5, Neptune and Mars 2 each, and earth 1. Six of the 32 known natural satellites of the solar system are about as large, or larger, than our moon, although none is so large as the moon in comparison with its primary planet. (The earth-moon system is sometimes referred to as a "double planet.") Only *Titan*, Saturn's largest satellite, is known to have an atmosphere. Most of the satellites revolve about their planets from west to east, and most have orbits that are approximately in the equatorial planes of their primary planets. (The moon is an exception — its orbit is nearly in the ecliptic plane — Chapter 6).

(d) Comets

Comets are loose swarms of small particles that revolve about the sun in orbits of very high eccentricity — that is, very elongated elliptical orbits. Comets spend most of their time in those parts of their orbits that are very far from the sun, where they receive negligible radiant energy. However, as a comet moves in closer to the sun, it warms up and some of the particles vaporize to form a cloud of gas, or *coma*, around the swarm of particles. The particles and surrounding coma make up the *head* of a comet. When a comet is only a few astronomical units from the sun, the pressure of the sun's radiation and solar corpuscular radiation sometimes force particles and gases away from the head to form a *tail*. The heads of comets average from 10,000 to 100,000 mi across, and tails sometimes grow to lengths of many millions of miles. The entire mass of a typical comet, however, is probably less than a billionth — perhaps as little as a trillionth — that of the earth. A comet is, therefore, an extremely flimsy entity.

More than a thousand comets have been observed, and an average of 5 to 10 new ones are discovered telescopically each year. Most newly discovered comets have periods of thousands or even millions of years, and have never before been observed in recorded history. It has been hypothesized that there may be a vast cloud of a hundred billion or so comets surrounding the sun at a distance of trillions of miles, and that occasional perturbations from passing stars start some of those comets moving in toward the sun. Unlike the planets, which move in orbits that are all nearly coplanar, comets approach the sun along orbits that are inclined at all angles to the plane of the earth's orbit.

(e) Minor Planets

The *minor planets*, or *asteroids* as they are often called, are small planets, differing from the nine principal planets primarily only in size. There are probably tens of thousands of them large enough to observe with existing telescopes. Ceres, the largest minor planet, has a diameter of about 500 mi. Only a few hundred minor planets are over 25 mi across; most are only a few miles or less in diameter. None of these objects is large enough to retain an atmosphere.

Like the principal planets, the asteroids move from west to east in elliptical orbits of small eccentricity and lie close to the plane of the earth's orbit. Typical minor planets have mean distances of about 2.5 to 3 AU from the sun and have periods in the range 4 to 6 years; thus, they occupy a region of the solar system that lies between the orbits of Mars and Jupiter. A few of the minor planets, however, have orbits that bring them close to the earth; several minor planets pass closer to the earth than any other known objects except the moon and meteoroids.

(f) Meteoroids

The numbers of small solid objects revolving about the sun that are too small to observe with telescopes are very great indeed, and the number seems to be greater and greater for objects of smaller and smaller size. These tiny astronomical bodies, too small to observe individually as each travels unhindered in its orbit, are called *meteoroids*. Their presence becomes known only when they collide with the earth, and plunging through the earth's atmosphere, heat with friction until they vaporize. The luminous vapors that are produced look like stars moving quickly across the sky and are popularly known as "shooting stars." Such phenomena are correctly called *meteors*. On a typical clear dark night, about half a dozen meteors can be seen per hour from any given place on earth. The total number of meteoroids that collide with the earth's atmosphere during a 24-hour period is estimated at 200 million. Recently, meteoroids have also been recorded as they strike space vehicles.

Rarely, a meteoroid survives its flight through the earth's atmosphere and lands on the ground. It is then called a *meteorite*. A number of such fallen meteorites can be inspected in various museums. The largest known meteorites have masses of about 50 tons. Most are the size of pebbles. Chemical analysis reveals them to be formed of the same chemical elements that exist on earth and elsewhere in the cosmos.

The smallest meteoroids are slowed down by the atmosphere before they have a chance to heat up and vaporize. They are microscopic in size and are known as *micrometeorites*. They eventually settle to the ground as meteoritic dust. Micrometeorites are the most common kind of meteoroids and are the kind most frequently encountered by space rockets.

In fact, the entire region of interplanetary space contains minute, widely spread particles, and very sparse gas. In the neighborhood of the earth, there are only a few ions (of ionized gas) per cubic centimeter. This is a far better vacuum than can be produced in any terrestrial laboratory.

9.2 THE PLANETS

The other planets of the solar system are of special interest, for they, like the earth, are worlds that revolve about the sun and derive their light and warmth from it. The planets are considered individually in Chapter 11. Here we shall summarize how certain information about them can be obtained.

(a) Some Basic Characteristics

The basic characteristics of a planet are its mass, its size, and its distance from the sun. From these data alone we can predict to some extent many of its other characteristics, for example, whether or not it can be expected to have an atmosphere, what kind of atmosphere, what its temperature is likely to be, and many other physical data.

To determine the distance of a planet from the earth, and from the sun, we geometrically survey it by sighting it from different places in the earth's orbit. This procedure, of course, gives the distance to the planet only in terms of the size of the earth's orbit — that is, in astronomical units. The evaluation of the astronomical unit in, say, miles — that is, the determination of the scale of miles of the solar system — is a more difficult problem; it will be discussed in Chapter 13.

The mass of a planet must be determined by measuring the gravitational acceleration that it produces on other objects. Four different methods have been used: (1) observing the acceleration a planet produces upon one of its satellites (see Section 4.4), (2) observing the perturbations a planet produces upon the motions of other planets, (3) observing the perturbations a planet produces upon the motion of a close-approaching minor planet, and (4) observing the effects a planet produces on manmade space probes. The latter technique has been especially important in determining good values for the masses of Venus and Mars.

The diameter of a planet is found from its angular diameter and its distance, exactly as the diameter of the moon is calculated. The procedure was described in greater detail in Section 6.2b.

(b) Rotation of the Planets

All the planets are observed to rotate. At least four techniques have been employed to determine planetary rotation rates.

The most direct method to observe a planet's rotation is to watch permanent surface features move across its disk. If a particular surface feature is observed for a large number of rotations, a very accurate value for the time required for one rotation is obtained. In this way, for example, the rotation period of Mars has been well determined.

If surface features cannot be seen, or are indistinct, the rotation period of a rapidly rotating planet can be found from the Doppler effect in the lines in the spectrum of sunlight reflected from the planet. Since one side of the planet is turning toward us and the other side away, light reflected from the approaching limb is Doppler-shifted to slightly shorter wavelengths (relative to light reflected from the center of the planet's disk) and light from the opposite limb is shifted to longer wavelengths. Suppose the planet's image is focused across the slit of a spectrograph at the telescope; if the apparatus is oriented so that the slit runs along the equator of the planet's image, all the lines in the spectrum will appear tilted (the top of the spectrum being light from one limb and the bottom being light from the other). A measure of the amount of tilt reveals the rotation rate. The rotation periods of Uranus and Neptune are found in this way.

If the planet is too small or distant to show an appreciable disk through the telescope, its rotation may be detected from variations in its brightness. Pluto, for example, seems to have surface irregularities so that different sides reflect light with different intensity. Thus as the planet rotates, it varies in brightness (as seen from the earth) with a period equal to the rotation period.

The most recent technique is that of *radar*. Radio waves beamed to a planet from earth are reflected back but are Doppler-shifted according to the relative line of sight velocity of the earth and planet. The waves reflected from the limb of the planet turning toward us are shifted to shorter wavelengths than those that are reflected from the center of the planet's disk, and radio waves reflected from the receding limb are shifted to longer wavelengths. Thus, although the transmitted beam may consist of waves of a very narrow wavelength range, those reflected back will comprise a somewhat larger range of wavelengths. Measurement of this range of wavelengths (or *bandwidth*) gives the planet's rate of rotation. In this way, the rotation rates of Mercury and Venus have been found.

(c) Atmospheres

All the planets except Mercury and probably Pluto are surrounded by appreciable gaseous atmospheres. A planetary atmosphere manifests itself in several ways. In most cases, opaque clouds in the atmosphere reflect light brilliantly. Among the planets possessing atmospheres, only Mars has an atmosphere thin enough so that we can see through it to examine features on the surface of the planet. Even the Martian atmosphere, however, blurs the markings on the planet, especially when viewed in blue or violet light of the shorter wavelengths, because gas molecules scatter short wavelengths more effectively than long ones. The atmosphere of Venus scatters considerable sunlight around the terminator into its night side, producing an easily observable twilight zone.

Spectrographic analysis of sunlight reflected from a planet can reveal what some of the gases are that compose its atmosphere. By carefully comparing the spectrum of direct sunlight and the spectrum of sunlight reflected from a planet, the astronomer can ascertain which of the multitude of dark lines in the latter spectrum must actually originate in that planet's atmosphere. Proper identification of the lines in the spectrum of a planet reveals the identity of the gases in its atmosphere that produced them.

Not all gases in the atmosphere of a planet can be detected spectrographically. Any oxygen, water vapor, carbon dioxide, methane, or ammonia that is abundant in the planet's atmosphere produces lines that are easy to detect. On the other hand, at the prevailing temperatures of planets, the gases hydrogen, helium, and nitrogen do not produce conspicuous lines in the observable spectrum — that is, at wavelengths at which radiation can penetrate the earth's atmosphere.

Which gases are present in a planet's atmosphere must depend, at least in part, on how that planet was formed and on its subsequent history. Most of the gases in the earth's atmosphere, for example, must have escaped (that is, *outgassed*) from the earth's crustal rocks, and have changed subsequently due to photosynthesis and other chemical activity, whereas the constituents of Jupiter's atmosphere may reflect the original composition of the material from which Jupiter was formed.

On the other hand, the kind of atmosphere a planet has also depends on its ability to hold the various gases. The molecules of a gas are always in rapid motion; if their speeds exceed the velocity of escape of a planet, (Section 4.6a), that kind of gas can gradually "evaporate" from the planet into space.

(d) Retention of Atmospheres

All gases are composed of molecules, which are

in rapid motion, darting to and fro and constantly colliding with each other. The mean speed of molecules of a given kind depends on the temperature (in fact, *defines kinetic temperature*), being greater the greater the temperature, and also on the mass of one molecule, being less the greater the molecular mass. Whether a planet can retain a given gas in its atmosphere depends on whether the molecules are moving fast enough to escape the planet (that is, whether their speeds exceed the planet's escape velocity).

The temperature of a planet depends on the amount of solar radiation it absorbs. Just outside the earth's atmosphere, 1.37×10^6 ergs/sec of solar radiation are incident on each square centimeter. This is called the *solar constant*. The earth's distance from the sun is 1 AU. Because of the inverse-square law, a planet r AU from the sun would receive $1.37 \times 10^6/r^2$ ergs/sec/cm². Of this, some is reflected and some absorbed. The fraction of incident energy reflected is called the *albedo*. The albedo (or reflecting power) of a planet can be determined by comparing its observed brightness with the calculated brightness it would have if it were perfectly reflecting. Thus we can estimate the mean temperature a planet should have. The temperature can also be determined from the intensity of the infrared radiation emitted by the planet if we assume that it radiates like a black body (Section 7.4a).

The velocity of escape from a planet is the same as the speed at which an artificial satellite would have to be launched from its surface in order for it to have a parabolic orbit about the planet (Section 4.6a). It can be shown that the escape velocity is $\sqrt{2\,GM/R}$.

To know whether a planet can retain a particular gas in its atmosphere, therefore, we must find the mean speed that molecules of that gas would have at the temperature expected for the planet, and compare that speed with the escape velocity. Now at any given time, some molecules

are moving slower and some faster than the mean molecular speed. Some may have been given good "jolts" in recent collisions and be moving far faster than average. Even they, however, cannot escape unless they are high in the atmosphere — in the "escape layer," from which, if they are moving in the right direction, collisions with other molecules will not deflect them back toward the ground.

Calculations show that if the mean molecular speed is as much as one third the velocity of escape, the planet will lose half of that gas from its atmosphere in a few weeks. If the mean molecular speed is as much as one fifth the escape velocity, that gas will disperse into space in a few hundred million years. To be conservative, we shall suppose that for a planet to hold a particular gas in its atmosphere for several billion years, its velocity of escape must be at least six times the mean molecular speed for that gas.

Analysis shows that Mercury, Ceres, and the moon would not be expected to retain any of the common gases, even if they ever did possess atmospheres. Uranus, Neptune, and Pluto would be able to retain any gas, except that all but hydrogen and helium would probably be in the liquid or solid state because of the low temperatures there. Pluto's low albedo (Chapter 11) suggests that it has no atmosphere at all, whereas the spectra of Uranus and Neptune show strong absorption lines of methane; apparently parts of the atmospheres of these planets are warm enough to allow some methane to sublimate to the gaseous state.

Mars could not hold hydrogen and helium but would be expected to retain the heavier gases. The giant planets, Jupiter and Saturn, could retain any gas. Hydrogen (although it is difficult to detect spectroscopically) and probably also helium are abundant in the atmospheres of all the Jovian planets. One would not expect to find atmospheric oxygen on these planets, for the free hydrogen present would combine chemically with oxygen to form water, which in turn would freeze out of the atmospheres. Hydrogen is observed on those planets, both in molecular form and in combination with carbon in the form of methane (CH_4). Ammonia (NH_3) is also present in the atmosphere of Jupiter.

Hydrogen could not be retained for billions of years in the atmospheres of Venus or Earth. Nor could appreciable helium be retained at the high temperatures expected for the escape layers. Moreover, substantial amounts of hydrogen and helium may never have been present in the gaseous state on the earth and Venus. If, as is widely hypothesized, the planets coalesced from a preplanetary material, that material may have been swept clear of gas before the planets formed. In other words, the fact that a planet may now be able to retain a particular gas in its atmosphere does not necessarily mean that that gas is present.

(e) Interiors of Planets

The study of the internal structure of planets is a very difficult subject. Yet, some clues enable us to draw a few conclusions about the interiors of the planets. A knowledge of the mass and radius of a planet are sufficient for calculation of its mean density. Knowing the density, we can say something about the material that composes the planet. Clues to the distribution of mass in a planet can be gathered from the degree to which the planet flattens under the influence of its rotation, and from the perturbative effects that its flattening has on its satellites and on the planet's rate of precession. With reference to the earth, additional valuable information is obtained from the transmission and reflection of seismic waves at various levels in the interior (Chapter 10).

As far as their internal structures are concerned, the planets fall naturally into two groups: the terrestrial planets, Mercury, Venus, Earth, Mars, and Pluto (sometimes the moon is also included); and the Jovian planets, Jupiter, Saturn, Uranus, and Neptune. The terrestrial planets are relatively small and dense and are believed to be composed mostly of rocky and metallic material.

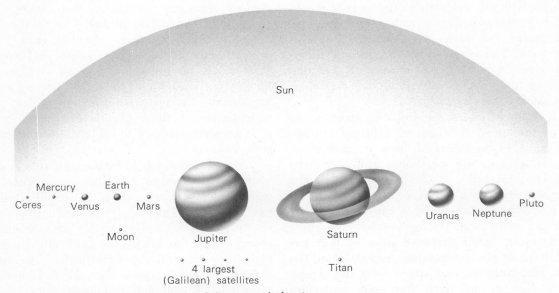

FIG. 9-1   Relative sizes of the sun and planets.

The Jovian planets are relatively large and of low mean density (indeed, Saturn has a mean density less than that of water). The chemical composition of the Jovian planets is probably more nearly typical of the general cosmic abundances of the various chemical elements — for example, the relative abundances found in the sun. Hydrogen and helium, the most abundant elements in the universe, probably dominate in the Jovian planets. Jupiter and Saturn, at least, are almost certainly composed mostly of hydrogen compressed to a solid state.

(f) Summary

The various planets of the solar system differ greatly, and in many ways. These differences can be largely understood, however, in terms of the respective distances of the planets from the sun and their respective sizes and masses. These three basic parameters are important to an understanding of Chapters 10 and 11, which deal with the planets individually.

In Appendices 9 and 10 are tabulated various data pertaining to the planets. However, many precise figures are often more confusing than helpful to the nonscientific reader. Therefore, in Table 9.2 we have summarized, very roughly, a few of the most important data concerning the planets. It is easier to remember that Jupiter is about 5 AU from the sun than to remember the figure 778,730,000 km. The relative sizes of the planets and sun are shown in Figure 9-1.

TABLE 9.2   Some Approximate Planetary Data

	MERCURY	VENUS	EARTH	MARS	JUPITER	SATURN	URANUS	NEPTUNE	PLUTO
Distance from sun (AU)	$\frac{2}{5}$	$\frac{3}{4}$	1	$1\frac{1}{2}$	5	10	20	30	40
Mass in terms of earth's mass	$\frac{1}{20}$	$\frac{4}{5}$	1	$\frac{1}{9}$	318	95	15	17	$\frac{1}{10}(?)$
Radius in terms of earth's radius	$\frac{2}{5}$	1	1	$\frac{1}{2}$	11	9	4	4	$\frac{1}{2}(?)$
Rotation rate	59 days	243 days retrograde	24 hours	$24\frac{1}{2}$ hours	10 hours	10 hours	11 hours	16 hours	6 days
Velocity of escape (mi/sec)	$2\frac{1}{2}$	$6\frac{1}{2}$	7	3	37	22	13	14	(?)
Approx effective temperature† (°K)	450	235	240	220	100	75	50	40	40
Observed gases in atmosphere	—	CO_2, H_2O, HCl, HF, $O_2(?)$	N_2, O_2, A, CO_2, H_2O, etc.	CO_2, H_2O	H_2, CH_4, NH_3	H_2, CH_4	H_2, CH_4	H_2, CH_4	—
Number of known satellites	0	0	1	2	12	10	5	2	0

†Temperatures on the sunlit sides of planets, or at their surfaces, may be much higher. The sunlit side of Mercury, for example, has a temperature of over 600°K, and the surface temperature of Venus is also near 600°K (Chapter 11).

Exercises

1. From the data given in the chapter, show that Kepler's third law holds for Pluto.

2. With what types of objects in the solar system might it be possible for the earth to collide?

3. It has been suggested that the minor planets are the debris of a planet that broke up. If this hypothesis (which is *not* generally accepted) were correct, in what respect would that planet have had to be very unusual, compared with the other planets?

4. Mercury and Mars have rather similar velocities of escape, yet Mars has an atmosphere and Mercury does not. Can you supply the explanation? Do so.

5. Explain how observations of the infrared emission from a planet can lead to a determination of its temperature.

6. Which planet in the solar system has the highest surface gravity?

7. Would you expect the larger satellites of Jupiter and Saturn to more nearly resemble the Jovian or terrestrial planets in their internal structures? Why?

8. Uranus has its axis of rotation lying nearly in its orbital plane. Describe what seasons must be like on Uranus.

9. Arrange the following in order of increasing (a) diameter; (b) mass:

Halley's comet   Mercury
A typical micrometeorite   A typical meteoroid
Jupiter   Ceres
Sun   A typical minor planet
Moon   Neptune

Interior of the observer's cage at the prime
focus of the 200-inch telescope.
(Hale Observatories.)

The 200-inch telescope. *(Hale Observatories.)*

The earth from 22,300 miles, photographed with the Applications Technology Satellite in synchronous orbit on November 10, 1967. *(NASA.)*

The approach to the Apollo 11 landing site, photographed from an altitude of 63 miles. The landing zone is in the center of the picture, near the top. *(NASA.)*

Lunar terrain near lunar module. Apollo 11 mission, July 20, 1969. *(NASA.)*

India and Ceylon, looking north with the Bay of Bengal on the right.
Gemini XI photograph taken from an altitude of 410 nautical miles. *(NASA.)*

CHAPTER 10

The Double Planet– Earth and Moon

Having described the general properties of the solar system, we turn our attention to its individual members. Except for the sun, which we shall discuss in Chapter 17, the planets are the most important solar system objects. Most of the planets have satellites, and several of the satellites are larger than the moon. Of all satellites, however, the moon is the largest in comparison to its primary; the earth-moon system has, in fact, often been called the "double planet." Moreover, the earth and moon are the objects we have explored most directly, so it is fitting that we should begin our investigation of the solar system with them.

10.1 GROSS PROPERTIES OF THE EARTH AND MOON

In previous chapters it has been explained how the mass of the earth (Section 3.6c) and of the moon (Section 6.3) are found. The mass of the earth is 5.98×10^{24} kg (about 6.6×10^{21} English tons); the moon's mass, smaller by a factor of 81.3, is 7.35×10^{22} kg. The equatorial diameter of the earth is 12756 km (7927 mi). The moon's diameter is 3476 km (2160 mi), or 0.273 times that of the earth. From their diameters we find the volumes of these bodies. The volume of the earth is 1.08×10^{27} cm³. The mean, or average, density of the earth is the ratio of its mass (5.98×10^{27} gm) to its volume (see Section 3.4a); that is,

$$\text{density of earth} = \frac{5.98 \times 10^{27}}{1.08 \times 10^{27}} = 5.5 \text{ gm/cm}^3.$$

The earth thus outweighs its volume in water by 5½ times. The mean density of the moon, less than

the earth's, is only 3.34 gm/cm³.

In section 3.6 we found that the acceleration of gravity at the surface of a spherical body of mass M and radius R (or the weight of a unit mass on the surface of that body) is given by the formula

$$a = G \frac{M}{R^2},$$

where G is the gravitational constant. At the earth's surface the acceleration of gravity is 980 cm/sec². Since the moon's mass is 0.0123 of the earth's and its radius is 0.273 of the earth's radius, we find that the acceleration of gravity at the surface of the moon is in proportion to that at the surface of the earth in the amount

$$\frac{a(\text{moon})}{a(\text{earth})} = \frac{0.0123}{(0.273)^2} = 0.165,$$

or about one sixth. On the moon, therefore, objects weigh about one sixth what they do on the surface of the earth, and if they are dropped at the surface of the moon, they accelerate to the ground only one sixth as fast as on earth.

The velocity of escape from a planet or satellite is $\sqrt{2GM/R}$ (Chapter 9). The escape velocity from the earth is 11.19 km/sec (about 7 mi/sec), but from the moon is only 2.38 km/sec (about 1.5 mi/sec). For comparison, the sun's escape velocity is 618 km/sec.

10.2 ATMOSPHERES OF THE EARTH AND MOON

From a knowledge of the escape velocities from the earth and moon, we would expect the earth to be able to retain all but the lightest gases in

FIG. 10-1   The nearly full earth photographed by Lunar Orbiter V. *(NASA.)*

into the extremely sparse gases of interplanetary space. *Auroras,* discharge in the upper atmosphere caused when charged particles from the sun bombard the upper air, have been observed to 600 mi, and analysis of drag on earth satellites reveals evidence for some atmosphere at heights of 1000 to 1300 mi.

The chemical composition of the earth's atmosphere is determined by quantitative chemical analysis. At the earth's surface, the constituent gases are found to be 78 percent nitrogen, 21 percent oxygen, and 1 percent argon, with traces of water (in the gaseous form), carbon dioxide, and other gases. At lower elevations, variable amounts of dust particles and water droplets are also found suspended in the air. The gases nitrogen and argon are relatively inert chemically. It is the oxygen, of course, that sustains animal life on earth.

its atmosphere and the moon to have no atmosphere at all. Such is, in fact, the case.

We, on earth, live at the bottom of the ocean of air that envelops our planet. The atmosphere, weighing down upon the surface of the earth under the force of gravitation, exerts a pressure which at sea level amounts to nearly 15 lb/in.^{2}. If the weight of the air over 1 in.^{2} is 15 lb, the total weight of the atmosphere may be found by multiplying this figure by the surface area of the earth in square inches. We find, thus, that the total weight of the atmosphere is about 6×10^{15} tons, so the atmosphere has a mass of about a millionth that of the earth.

There probably is no real upper limit to the earth's atmosphere. At higher and higher altitudes the air thins out more and more until it disappears

FIG. 10-2   The full moon. *(Hale Observatories.)*

(a) The Lack of an Atmosphere and Water on the Moon

The earth's atmosphere is believed to have escaped (outgassed) from its crustal rocks, largely through volcanism. Water vapor, the major gas to escape, has, of course, condensed to form the oceans. Carbon dioxide was next most important, and has mostly combined chemically with surface rocks. Third in abundance was nitrogen, the major constituent of the atmosphere today. The oxygen now present is believed to have been produced from carbon dioxide by photosynthesis during the past few hundred million years.

In the absence of air, the moon can have no water on its surface. It is well known that water boils more easily (at a lower temperature) at high altitudes in the mountains than at sea level. If all air pressure could be removed above water, it would boil away. On the moon, therefore, if there ever were any liquid water, it would have evaporated and then, as a gas, dispersed into space with the rest of the moon's atmosphere.

In the absence of air and water, there can be no weather on the moon. There are no clouds, winds, rain, snow, or even smog, which accounts for many of the differences between the moon and earth. Weather on the earth and the running of water over its surface have been major sources of erosion that have washed and worn away entire mountain ranges, and even the faces of continents many times over during geologic time. On the moon, where air and water cannot contribute to erosion, features are more nearly permanent. There are formations of all ages on the moon, standing side by side; features that were formed probably billions of years ago are often still intact, standing with those formed in the recent past. We find the history of the moon written upon its face.

10.3 TEMPERATURES ON THE EARTH AND MOON

Because of the moon's lack of atmosphere its tem-

perature range between day and night is rather extreme. The earth's atmosphere serves to reflect and disperse much of the sun's radiation during the day and to blanket heat in near the surface at night.

The moon rotates with respect to the sun in about 29½ days. Therefore, for any one place on the moon the sun is shining for about 2 weeks at a time. Without any atmosphere to absorb or reflect some of the sun's radiation, the surface temperature rises to a high value, as more than 90 percent of the incident energy from the sun is absorbed. Eventually this energy is reradiated into space, mostly as infrared radiation, and some at radio wavelengths as well.

The rate at which a body like the moon radiates electromagnetic energy at various wavelengths depends on its surface temperature. The infrared radiation from the moon can be measured with a heat or infrared-sensitive device, such as a lead sulfide cell or thermopile, placed at the focus of a telescope, and radio telescopes can measure the radio energy it emits. Thus, earth-based observations had determined the lunar temperatures long before they were measured directly by the *Surveyor* lunar probes and *Apollo* astronauts. The moon's temperature ranges from just above the boiling point of water, where the sun is shining, down to about $100°K$ ($-173°C$) on its dark side.

10.4 THE MAGNETIC FIELDS OF THE EARTH AND MOON

The earth has a magnetic field similar to that produced by a bar magnet. Nearly everyone is familiar with the way iron filings align themselves along the lines of force that extend between the north and south poles of a magnet. The magnetic poles of the earth are located at about latitude 78° north and south, or about 825 miles from its geographical poles; the north magnetic pole is in Northeast

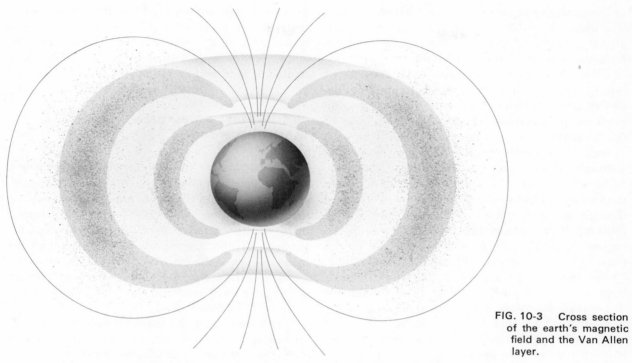

FIG. 10-3   Cross section of the earth's magnetic field and the Van Allen layer.

Canada. Between the magnetic poles of the earth stretch lines of force along which compass needles align. The positions of the magnetic poles themselves, and the orientations of the lines of magnetic force, gradually shift around. During periods of solar activity (Chapter 17) rapid short-period fluctuations in the earth's magnetic field also occur. The over-all strength of the field is fairly weak. Its origin is believed to be in the earth's core. Fluid motions in the electrically-conducting core are thought to cause it to act like a dynamo. The rotation of the earth causes the magnetic field to be aligned approximately with the rotation axis. The energy source of this dynamo is still uncertain.

(a) The Van Allen Layer

An important discovery made by the early artificial satellites was the presence of a zone of highly energetic charged particles high above the earth.

The inner part of this radiation zone was detected by instruments carried aboard the Army satellite Explorer I, launched on January 31, 1958. More information about the nature and extent of the belt of radiation was obtained with instruments on Explorer III, Explorer IV, and Pioneer I. In December 1958, the Army space probe Pioneer III revealed an additional region of radiation high above the one first discovered. The radiation-detecting experiments on Explorer I were under the direction of James A. Van Allen, physicist at the State University of Iowa; the radiation zone is called, in his honor, the Van Allen layer.

Figure 10-3, showing a cross section of the magnetic field surrounding the earth, displays the approximate location and extent of the most intense part of the Van Allen layer. It surrounds the earth, rather like a broad doughnut. The inner intense region is centered about 2000 mi above the surface of the earth and has a thickness of 3000 mi or more. The outer region of the Van Allen layer is about 10,000 to 12,000 mi from the earth's surface and has a thickness of from 4000 to 6000 mi.

The Van Allen radiation zone consists of rapidly moving charged particles trapped in the earth's magnetic field. Most of the fastest moving particles are electrons, although the inner part of the Van Allen layer contains high energy protons as well. The particles in the inner region are moving with speeds far greater than those in the outer one. Some are moving with energies as great as those possessed by electrons that have been accelerated through an electric field of billions of volts. The radiation strengths, heights, and thicknesses of the regions in the Van Allen layer, especially of the outer one, vary considerably from time to time. These regions are not really sharply defined; trapped charged particles must exist throughout the earth's magnetic field. The inner and outer parts of the Van Allen belt may be thought of as regions where the density of the moving particles is much greater than elsewhere.

The origin of the trapped particles is not known with certainty. It seems likely, however, that the outer region consists mostly of particles captured from the "wind" of corpuscular radiation from the sun (see Chapter 17). It has been suggested that the protons and electrons in the inner region may be produced in interactions between molecules of the earth's atmosphere and cosmic rays.

(b) Lunar Magnetism

Magnetometers carried to the moon by Apollo astronauts, and those on earlier lunar probes and orbiters, have not detected a permanent general lunar magnetic field. There are some very weak local fields, indicating slight magnetism in some of the surface rocks. On the other hand, to be magnetized at all those rocks must have been exposed to a somewhat stronger field in the past. Possibly the rocks were magnetized in the earth's field at a time in the past when the moon was nearer the earth. In any event, there is no evidence for a molten or fluid core in the moon.

10.5 INTERNAL STRUCTURES OF THE EARTH AND MOON

(a) Clues Obtained from the Earth's Mean Density

Our knowledge of the mean density of the earth tells us something about its interior. By sampling surface rocks, we have found that the average density of the outer crust of the earth is about 2.7 gm/cm^3. On the other hand, we have seen that the mean density of the earth as a whole is 5.5 gm/cm^3. Because of the low surface density but high average density, we must conclude that the interior of the earth is very dense indeed.

The weights of the various layers of the earth bearing down upon its interior causes the pressure to increase inward. Calculations indicate that the pressure at the earth's center must be close to 50 million $lb/in.^2$. When subjected to this great pressure, matter is highly compressed and heated. We can expect the central regions of the earth, then, to be hot and dense.

(b) Seismic Studies

Stresses that build up gradually in the crust of the earth are often released by slippages along fissures or *faults*. The energy released in these sudden movements in the earth's crust results in earthquakes. Vibrations are sent out and travel to all parts of the earth. Some types of the vibrations travel along the surface; others pass directly through the interior.

The study of these waves that originate in earthquakes is *seismology*. The seismic vibrations are picked up and recorded by delicate instruments stationed around the earth's surface. An analysis of the time required for seismic waves to reach these various stations provides information about the earth's interior.

The transmission of the waves through most of the earth's interior indicates that the major part of the earth is extremely rigid — more so than steel. This major part of the earth is called its *mantle*. The density in the mantle increases downward from about 3½ to 5½ g/cm³. It is believed to be composed mostly of basic silicate rocks.

At the upper boundary of the mantle is the *crust*, a shell extending on the average about 20 mi inward from the earth's surface under the continents. The crust consists of surface rocks such as granite and basalt, with overlying oceans and sedimentary rocks. Evidence from seismic studies suggests that the crust extends somewhat deeper under the continents than under the ocean floors, where it may be only a few miles thick.

At the inner boundary of the mantle, about 1800 mi below the surface, is the *core* of the earth. The outer part of the core acts like a liquid, for it does not transmit certain kinds of seismic waves. The innermost part of the core is extremely dense and hot (in the thousands of degrees).

(c) Age of the Earth

The age of the earth can be estimated from the degree of radioactive decay of certain elements in the crust. Atoms of uranium and thorium disintegrate spontaneously but very gradually through various elements, including radium, and end as atoms of lead. The exact rate of this radioactive decay from one element to another has been measured in the laboratory. The decay is found to occur at a very constant and regular rate.

Studies of the relative proportions of lead, uranium, and radium in mineral deposits containing these elements indicate how long the disintegration process has been at work and give the ages of these surface rocks, from which we derive the age of the earth itself. The results of different investigations differ somewhat, but all confirm that the age of the earth is in the billions of years. The oldest rocks on the surface have ages of 3.5 billion years. The age of the earth itself is estimated at about 4.5 billion years, on the basis of more subtle studies of the ratios of those isotopes of lead that do not originate from radioactive decay.

(d) Interior of the Moon

Analysis of Apollo seismic and magnetic studies of the moon gives us information about its interior. It is found that half way to the moon's center the temperature is about 1000°C, confirming our expectations, based on the relatively small weight of the moon's overlying layers, that its center is not molten. Analysis of perturbations on lunar satellites also show that the moon is less centrally condensed than earth. However, surface irregularities in the moon's mass distribution, called *mascons*, have been detected. These relatively dense subsurface regions are possibly submerged material that collided with the moon in the past.

10.6 THE SURFACE OF THE MOON

The most conspicuous of the moon's surface features are easily visible to the unaided eye. In medieval Europe these markings were generally regarded as colorations rather than actual irregularities in the surface of the moon. Galileo's telescope observations, however, revealed mountains, craters, valleys, and what appeared to be seas. The idea developed that the moon might be a world, perhaps not unlike our own. Many other lunar observers followed in Galileo's lead. In 1647 John Hevel of Danzig (1611–1687) published his comprehensive treatise on the moon, *Selenographia*. In a series of carefully prepared plates, Hevel identified many of

FIG. 10-4   The western hemisphere of the moon. *(Lick Observatory.)*

SOUTH

MARE AUSTRALE

Rhoita Valley

Stevinus

MARE SMYTHII

WEST

MARE MARGINIS

MARE HUMBOLDTIANUM

Moretus
Cuvier
Faraday
Steinheil
Stöfler
Janssen
Fabricius
Maurolycus
Aliacensis
Metius
Apianus
Werner
Rheita
Piccolomini
Mayro
Fracastorius
Catherina
Langrenus
MARE NECTARIS
Cyrillus
Abulf
Albategnius
Pyrenees Mts.
Madler
Theophilus
Hipparchus
Langrenus
Torricelli
Horrocks
Messier
Gutenberg
Sabine
Delambre
Godin
Triesnecker
MARE
FOECUNDITATIS
Pickering
Maskelyne
Ritter
Agrippa
Hyginus Cleft
Boscovitch
Ariadaeus Cleft
Taruntius
MARE TRANQUILLITATIS
Julius Caesar
Manilius
Plinius
Proclus
MARE
CRISIUM
Bessel
Macrobius
Linne
Le Monnier
Autolycus
MARE SERENETATIS
Aristillus
Posidonius
Caucasus Mts.
Cassini
Eudoxus
Piton
Martis
Alpine Valley
Atlas
Aristoteles
Hercules
MARE FRIGORIS
Endymion
Gärtner

NORTH

FIG. 10-5   The eastern hemisphere of the moon. (Lick Observatory.)

SOUTH

Clavius

Maginus

Longomontanus

Tycho

Schickard

Deslandres

Walter · Pitatus

Aliacensis

Werner

Purbach

MARE HUMORUM

Thebit

Straight Wall

Arzachel

Bullialdus

Gassendi

Alphonsus

Letronne

Albategnius

Ptolemaeus

Grimaldi

Hipparchus

Herschel

EAST

Flammarion

Mösting

Riccioli

MARE NUBIUM

OCEANUS

SINUS
MEDII

Copernicus · Kepler

Apennine

Eratosthenes

Caucasus Mts.

PROCELLARUM

Pytheas

Caucasus Mts.

MARE

Lambert

Aristarchus

Palochoris

Autolycus

Herodotus

Archimedes

Aristillus

IMBRIUM

Cassini

Piton

Jura Mts.

Pico

Alps Mts.

Plato

MARE FRIGORIS

John Herschel

Fontinelle

Pythagoras

Philolaus

NORTH

the moon's features and gave them names, in many cases in honor of similar features on the earth. The names he gave the mountain ranges on the moon survive today.

The early lunar observers regarded the moon as having continents and oceans and as being a possible abode of life. We know today, however, that the resemblance of lunar features to terrestrial ones is superficial. The moon's lack of air and water makes most of its features unlike anything we know on earth.

(a) Modern Observations

Many excellent photographs of the moon have been obtained with earth-based telescopes, but today most of the lunar photographs obtained from earth have been rendered obsolete by those obtained by the Apollo astronauts, and earlier by the spectacular Ranger and Surveyor lunar probes and the lunar Orbiters, whose pictures were transmitted back to earth by television. Soviet lunar probes have also produced high-quality photographs, and in fact a Soviet space vehicle was the first to send to earth photographs of the moon's far side (in October, 1959). Moreover, the Soviet Luna 9 made the first soft landing on the moon on January 31, 1966, and transmitted close-up photographs to earth for 3 days. The American Ranger, Surveyor, and Orbiter missions, however, were far more fruitful.

In the Ranger series in 1964 and 1965, Rangers 7, 8, and 9 made successful "hard" landings on the moon; that is, they crashed directly onto the moon as planned. On the way, however, a great many views of the moon were televised directly back to earth, with resolution on the lunar surface as high as 1 ft. The best observations from earth telescopes gave a resolution of about ½ mi on the moon.

The first successful unmanned American soft landing on the moon was by Surveyor I, launched on May 30, 1966. Surveyor I alone sent over 11,000 photographs back to earth. In the next year and a half, Surveyors III, V, VI, and VII made successful landings. Each had television cameras which showed both the lunar landscape and close-up views of experiments done on the lunar "soil" by equipment carried on the probe. In addition to obtaining photographs, the Surveyors analyzed the chemical nature of lunar surface material and studied its physical and magnetic properties. Observations were even made of part of the solar corona far from the sun's surface (Chapter 17).

The five lunar Orbiters were all highly successful. The first was launched on August 10, 1966, and the last on August 1, 1967. Each was sent near the moon and then put into a lunar orbit by firing a retrorocket; a final maneuver changed its orbit to make its closest approach to the moon (perilune) conveniently near the surface (usually about 28 mi). Each Orbiter carried a 260-ft roll of 70-mm film, on which were made simultaneous exposures with a wide angle camera and with a second camera with a telephoto lens. The film was processed automatically and then scanned with a television camera, and the picture information was transmitted to earth. The entire "front" side of the moon (the side turned toward us) was photographed with a resolution averaging 10 times better than that of the best photographs obtained from earth. About 99.5 percent of the far side of the moon was photographed with better resolution than that of the best earth photographs of the near side. In addition, the Orbiters delineated several possible landing sites for astronauts in the Apollo project, photographed the earth, made measures of micrometeorites and the moon's magnetic field (not detected), and measured charged particles in the vicinity of the moon.

(b) The Lunar Seas

The largest of the lunar features are the so-called seas, still called *maria* (Latin for "seas"). It is they that form the features of the "man in the moon." The maria are of course dry land. They are great plains, with relatively smooth, flat floors that appear darker than the surrounding regions.

The largest of the 14 lunar "seas" on the moon's visible hemisphere is *Mare Imbrium* (the "Sea of Showers"), about 700 mi across. The other maria have equally fanciful names, such as Mare Nubium ("Sea of Clouds"), Mare Nectaris ("Sea of Nectar"), Mare Tranquilitatis ("Tranquil Sea"), Mare Serenitatis ("Serene Sea"), and so on. Most of the maria are roughly circular in shape, although many of them are interconnected or overlap slightly, and

FIG. 10-6   Apollo 16 photographs of landing site in the region of the lunar crater, Descartes. *(NASA).*

FIG. 10-7   Part of the moon's far side, photographed by Lunar Orbiter V. *(NASA.)*

all have irregularities and baylike inlets, such as Sinus Iridum ("Bay of Rainbows") on the north "shore" of Mare Imbrium.

Even inspection of good photographs shows that the maria do not have perfectly smooth floors, as they appear to have at first glance. They are speckled with thousands of tiny craters resembling potholes ranging in size down to a few inches or less across, and inside some maria there are large craters. Some of the mare floors show wavelike ripples when the sunlight strikes them at a glancing angle. In a few places cliffs are found, suggesting slippages of the moon's crust along faults — fissures or lines of weakness in the crust. The best example is the "Straight Wall" in Mare Nubium, which is a cliff 600 ft high and 80 mi long.

Apollo and Surveyor landings in several different maria found them to be remarkably similar to each other. The mare floors are covered with a layer of material that analysis shows to be volcanic in origin, although severely eroded, probably by meteoritic bombardment. The material consists mostly of fine grains with a variety of sizes, typically about $\frac{1}{1000}$ in. in diameter. However, mixed in are numerous aggregates or "clots" of fine grains and some hard rocks. The material is, on the average, very porous and is compressible, but, of course, it is able to support easily the weight of an astronaut. Its density is between 0.7 and 1.2 g/cm^3.

(c) The Lunar Craters

Even from earth there can be observed on the moon some 30,000 craters, circular depressions ranging in size from less than 1 mile to over 100 miles across. The largest of the craters are *Clavius* and *Grimaldi,* both nearly 150 miles in diameter. Following the custom started by John Riccioli in 1651, craters are generally named after famous scientists and philosophers. Other famous craters are Tycho, Copernicus, Kepler, Aristarchus, and Plato. Craters are sometimes still named after outstanding selenolo-

gists (astronomers who study the moon) today.

Although craters can be seen on the moon at almost any time, even with binoculars, the best time to view them is when the moon is near the first or last quarter phase. Then the terminator, dividing day and night on the moon, runs about down the middle of the apparent disk of the moon in the sky. The sun strikes craters near the terminator at a glancing angle and they cast long shadows, which makes them stand out in bold relief.

In general, craters are found over those regions of the moon that are not covered by the seas. There are comparatively few craters in the maria themselves, which suggests that the maria may be more recent features on the moon than most of the craters. The existence of crater tops barely visible in some of the mare floors supports this view. Most of the craters are found in the rugged regions of the moon. There they occur in all sizes and probably all ages as well. Frequently they overlap, one piled on top of another. An especially rugged region of overlapping craters is found in the vicinity of Tycho, near the moon's south pole.

The largest craters are often called *walled plains.* Examples are Clavius, Plato, and Ptolemaeus. A walled plain often appears to be a sunken region with little or no outside wall, sloping up to the crater rim. Some of the walled plains do not have circular walls, but are irregular in shape.

Most of the craters of moderate size are quite circular and have outside walls. In a few cases the crater floors are higher than the surrounding landscape, but in the majority of cases they are lower. Their inside walls are almost always steeper than the outside walls and rise to heights of as much as 10,000 ft above the crater floors. Many of these craters have mountain peaks in their centers. They somewhat resemble craters left by bomb explosions and may have had a violent origin. An excellent example is the crater Copernicus.

The most numerous craters are the smallest ones, those ranging in size down to craters less

FIG. 10-8   A region of many craters, including Tycho; photographed by Lunar Orbiter V. *(NASA.)*

than a foot across photographed by astronauts. Often craters, especially the smaller ones, occur in clumps or clusters or in long lines or rows of craters. In some cases lines of craters seem to follow shallow cliffs or what appear to be faults.

(d) The Lunar Mountains

There are several mountain ranges on the moon. Most of them bear the names of terrestrial ranges — the *Alps, Apennines, Carpathians,* and so on. The similarity of mountains on the moon and the earth ends, however, with their names. Because of the absence of water, the lunar mountain ranges are devoid of the drainage features so characteristic of our own mountain ranges, and the lack of weather

erosion on the moon results in a different local appearance of mountains there.

The heights of lunar mountains above the surrounding plains can be determined by measuring the length of the shadows they cast. Pains were taken, in fact, that the photographs made with lunar Orbiters would be at such angles that the features on the moon would have shadows — enabling the determination of the heights of those features.

The highest lunar peaks range up to elevations of 25,000 ft or more, comparable to the highest peaks in the Himalayas on the earth. It is not certain, however, that any mountain on the moon is quite as high as Mount Everest. (Of course, on

FIG. 10-9   The crater Copernicus, viewed from the south; photographed by Lunar Orbiter II. (NASA.)

earth mountain heights are measured from sea level rather than from the elevations of the surrounding plains.)

(e) Other Lunar Features

There are many other lunar features besides the seas, craters, and mountains. The *Alpine Valley*, for example, is a deep straight gorge cutting through the Alps Mountains. Also on the moon are many crevasses or clefts, half a mile or so across, called *rilles*.

Especially interesting are the *rays* — bright streaks that seem to radiate out from certain of the craters that appear to be of explosive origin, notably Tycho and Copernicus. The rays have widths of 5 or 10 mi or more and in many cases extend for distances of hundreds of miles. Some of the rays from Tycho seem to extend completely around the visible hemisphere of the moon. They follow nearly great circles on the moon's surface, and sometimes are *tangent* to the craters from which they originate rather than coming from the exact centers of the craters. They cast no shadows and are seen best near full moon, when the sun shines down on them most directly. At full moon the ray system around Tycho gives the moon almost the appearance of a peeled orange. Ranger photographs show some of the rays to be associated with lines of craterlets.

(f) Composition of the Lunar Surface

The chemical composition of the lunar surface material was first analyzed by alpha-particle scattering experiments carried on by the soft landers, Surveyors V, VI, and VII. More recently, of course, chemical analysis has been carried out directly on lunar rocks and soil samples returned to earth by the Apollo astronauts.

The results from widely separated places on the moon are strikingly similar. The most abundant element is oxygen, accounting for about 58 percent of the atoms present. Most of the oxygen atoms are

FIG. 10-10   The Alpine Valley, photographed by Lunar Orbiter V. *(NASA.)*

FIG. 10-11   Apollo 16 photograph of South Ray Crater from the slope of Stone Mountain. *(NASA.)*

chemically united with silicon, next most abundant, and accounting for about 20 percent of the atoms. Also abundant are aluminum, calcium, iron, magnesium, and other terrestrial elements. The chemical composition of the lunar material, and also the mineral structure, is similar to that of volcanic basalts on earth. The most common lunar minerals (which are also common in terrestrial volcanic rocks) are plagioclase and pyroxene. Some of the lunar minerals, however, are unknown on earth, so while the moon has had a similar chemical history to earth's, it is not identical. In particular, the moon can never have been part of the earth itself—an old theory long ruled out by dynamical arguments.

FIG. 10-12   Rocky interior of crater near Stone Mountain, photographed by Apollo 16 astronauts. *(NASA.)*

FIG. 10-13   Lunar scene from the Apollo 16 Lunar Roving Vehicle. *(NASA.)*

(g) Origin of the Lunar Features

With the Apollo missions, as well as the unmanned lunar probes, we have learned a great deal about the history of the moon's surface. Still, we have a great deal more to learn before we can begin to understand the origin of most of the moon's features. This should not surprise us, for we do not even thoroughly understand all of the features of the earth, which has been studied, so far, enormously more comprehensively than has the moon.

We have learned that much of the lunar surface, especially in the maria, is covered with material of volcanic origin. Thus, whereas we believe that the moon has never had a molten core, there has at least been some melting of its crustal rocks. The necessary heat may have been produced by the decay of radioactive elements in the outer layers of the moon, or it may have been produced by the energy of impact of meteorites with the lunar surface, or both. In any event, the solid lava that filled the mare floors has been broken down to dust by gradual erosion caused by temperature changes, cosmic rays, and meteoritic bombardment.

The origin of the large lunar craters is now believed to be due to meteoritic impacts. Craters 50 miles in diameter can be produced by the infall of a body with a mass of about 8×10^{12} tons (or about 10 mi in diameter). Although most experts lean to the meteoritic hypothesis for the origin of the large craters, some or many of the smaller ones, those less than 10 mi in diameter, may have had a volcanic origin.

If meteoritic collisions with the moon have formed so many craters there, one might wonder why we do not find such features on the earth. There may indeed have been many thousands of large craters of meteoritic origin on the earth in the past, but they would have been washed away long since by the erosion of wind and rain; only those formed in the very recent past would still be intact. A few meteorite craters are known on earth; one of the largest of suspected meteoritic origin is the Reiss Kessel crater, which is 15 miles across. The best known is the Barringer (or Great Meteor) Crater, near Winslow, Arizona, which is nearly 1 mile across (Chapter 12).

The origin of the rays is probably associated with the formation of the craters from which they radiate. It is possible that they may have been formed from secondary particles blown out over the moon's surface at the time of formation of the craters.

Exercises

1. Describe the appearance of the earth and moon in the sky as seen by a hypothetical observer on (a) Venus; (b) Mars. Include a discussion of the phases the earth and moon would go through.

2. To the observer on Mars, approximately how many times as bright as the moon would the earth appear?

3. Would the earth appear resolved as a disk when viewed with the naked eye from Venus, or would it appear as a point of light?

4. Imagine a baseball game on the moon. Describe such phenomena as:

(a) How far a ball might be hit;

(b) The distance between the bases for an optimum game;

(c) The likelihood of rain checks;

(d) The necessity of calling the game because of darkness;

(e) The difficulty of bawling out the Umpire (consider how sound travels on the moon).

5. If the moon did happen to have any atmosphere, what kind of gases would you expect to find in that atmosphere? Why?

6. Describe the ways in which lunar mountains would appear different to a tourist on the moon from the way terrestrial mountains appear to earthbound tourists.

7. If you were an astronomer-astronaut on an Apollo mission to the moon, what are some of the experiments you would like to perform on the moon in order to increase our knowledge of that world?

CHAPTER 11

The Other Planets

11.1 MERCURY

Mercury is one of the brightest objects in the sky. At its brightest it is inferior in brilliance only to the sun, the moon, the planets Venus, Mars, and Jupiter, and the star Sirius. Yet, most people — including even Copernicus, it is said — have never seen Mercury. The planet's elusiveness is due to its proximity to the sun. Its orbit is only about one third the size of the earth's; it can never appear further from the sun than about 28°. It is visible to the unaided eye, for a period of only about 1 week, at times when it is near eastern elongation and can appear above the western horizon just after sunset and also when it is near western elongation, and rises in the east shortly before sunrise. Because the synodic period of Mercury is 116 days, the intervals of its visibility as an "evening star" after sunset, and as a "morning star" before sunrise, occur about three times a year. However, Mercury sets so soon after the sun (or rises so shortly before the sun), that only rarely can it appear above the horizon when the sky is completely dark; generally, one must look for it in twilight.

Mercury was well known to the ancients of many lands. The earliest observers, however, did not recognize it as the same object when it appeared as an evening star and as a morning star. The early Greeks, for example, called it Mercury when it was visible in the evening and Apollo when it was seen in the morning twilight. The corresponding names given it by the Egyptians were Horus and Set, and the Hindus called it Raulineya and Buddha.

(a) Mercury's Orbit

Mercury is the nearest to the sun of the nine major planets and, in accordance with Kepler's third law, has the shortest period of revolution about the sun — 88 of our days; it is appropriately named for the fleet-footed winged messenger god. Its mean orbital speed is nearly 30 mi/sec.

The semimajor axis of the orbit of Mercury, that is, the planet's median distance from the sun, is 36 million miles, or 0.39 AU. However, because Mercury's orbit has the high eccentricity of 0.206, its actual distance from the sun varies from 28.6 million miles at perihelion to 43.4 million miles at aphelion. Pluto is the only major planet with a more eccentric orbit. Furthermore, the 7° inclination of the orbit of Mercury to the plane of the ecliptic is also greater than that of the orbit of any other planet except Pluto.

(b) General Properties and Structure

The mass of Mercury, which has no known satellite, is determined from the perturbations its gravitational influence produces on other bodies, such as occasional comets or minor planets that pass near it. The minor planet Icarus, for example, passed within about 10 million miles of Mercury in April 1968. Mercury's mass can also be found from its effects on the Venus space probes, for example, Mariners II and V. The best estimate today of the planet's mass is about $\frac{1}{18}$ that of the earth. Mercury is the least massive planet in the solar system.

Mercury is also the smallest of the planets, having a diameter of only about 3030 mi, or less than half that of the earth. Its mean density is about 5½ times that of water — about the same as the mean density of the earth.

(c) Telescopic Appearances

Mercury, like Venus, presents different portions of its illuminated hemisphere to us as it revolves about the sun and hence goes through phases similar to those of the moon (see Section 3.2b). The alternate crescent and gibbous shape of Mercury is its only conspicuous telescopic characteristic.

Observations of Mercury are very difficult because of its nearness to the sun. At night it can be seen only close to the horizon, and then for but a short while. Its light must traverse a long path through the earth's atmosphere, which not only dims the planet, but subjects the rays to disturbances in the air ("seeing") which are especially troublesome for all observations of objects near the horizon. Consequently, nearly all telescopic observations of the planet are carried out in daylight, when it is high in the sky. Unfortunately, daytime observational conditions are seldom good. Nevertheless, experienced observers have been able to detect permanent surface markings on Mercury. They appear simply as darker areas, something like the maria on the moon, but less conspicuous.

(d) Rotation

Visual studies of Mercury's indistinct surface markings seemed to indicate that the planet kept one face to the sun, and for many years it was widely believed that Mercury's rotation period equaled its period of revolution about the sun of 88 days.

Radar observations of Mercury in the mid 1960s, however, showed conclusively that Mercury does rotate with respect to the sun (Section 9.2b). Its sidereal period of rotation (that is, with respect to the distant stars) is about 59 days. G. Colombo first pointed out that this is very nearly two thirds of the planet's period of revolution, and subsequently Goldreich and Peale showed that there are theoretical reasons for expecting that Mercury can rotate stably with a period *exactly* two thirds that of its revolution — 58.56 days.

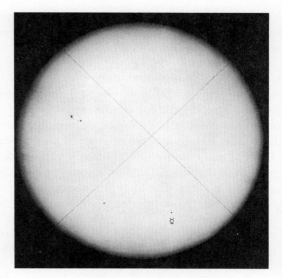

FIG. 11-1 Photograph of Mercury in transit across the disk of the sun. *(Royal Observatory, Greenwich, England.)*

(e) Temperature

The temperature of the middle of the daylight hemisphere of Mercury has been measured at the Mount Wilson Observatory with a thermopile. Infrared radiation focused onto the thermopile by the telescope generates a feeble electric current whose strength is proportional to the amount of infrared energy radiated by the planet (radiation from the sun that is merely reflected to the earth by the planet is very weak at those wavelengths). From the thermopile current reading, the temperature can be calculated. The lighted side of Mercury has a temperature of about 610°K (640°F), hot enough to melt tin and lead. The dark side of the planet is too cold to radiate a measurable amount of radiation in the infrared. A recent observation places an upper limit of 150°K on the night-side temperature — that is, it is colder than − 123°C. Since all sides of Mercury receive sunlight during its revolution (the period from noon to noon is 176 days), we conclude that its surface material cannot hold heat well; heat is not conducted well into depths beneath its surface. Like the moon, the surface material of Mercury has low conductivity.

(f) The Lack of a Mercurian Atmosphere

Because of the high temperature of the illuminated side of Mercury and the low velocity of escape, we would not expect it to retain an appreciable atmosphere. The lack of a Mercurian atmosphere seems to be confirmed by its low albedo; it reflects only 6 percent of the light incident upon it — less than any other planet. This low reflecting power is characteristic of a solid surface, like the moon's; an atmosphere would reflect light far more efficiently. Moreover, there is no visible twilight zone on Mercury. An atmosphere would scatter light a little distance around the planet into the night side. No trace of an atmosphere is revealed in the spectrum of sunlight reflected from the planet.

11.2 VENUS

Venus, named for the goddess of love and beauty, is sometimes called the "earth's sister," for it is most like the earth in mass and size of all the planets. It approaches the earth more closely than any other planet — at its nearest it is only 25 million miles away.

Venus is a beautiful object in the night sky; its brilliance is exceeded only by that of the sun and moon, save for very rare comets and bright meteors. At night, it can cast a shadow; at its brightest it can be seen easily in broad daylight if one knows exactly where to look for it.

Like Mercury, Venus is an inferior planet (nearer the sun than the earth); consequently, it appears to swing back and forth in the sky, during its synodic period, from one side of the sun to the other. Like Mercury, therefore, it appears sometimes as an "evening star" and sometimes as a "morning star." The early Greeks, thinking it to be two objects, called it *Phosphorus* and *Hesperus*

when it was seen in the morning and in the evening, respectively. Pythagoras, in the sixth century B.C., is credited with being the first to recognize that Phosphorus and Hesperus were one and the same planet.

Because Venus is farther than Mercury from the sun, it reaches much greater eastern and western elongations, and can appear as far as 47° from the sun. It can only be seen in the west in the early evening or in the east before sunrise; however, it is visible for a longer time than Mercury, after sunset or before sunrise. Indeed, it is often so conspicuous in the evening sky that people surprised at its brilliance have called observatories to inquire whether it is the star of Bethlehem.

(a) Venus's Orbit, Mass, and Size

The median distance of Venus from the sun is 67,270,000 mi. Its orbit is the most nearly circular of any of the planetary orbits, having an eccentricity of only 0.007. Its distance from the sun, consequently, varies by only about 1 million miles.

To complete its revolution about the sun in 225 days, Venus moves with a mean orbital speed of about 22 mi/sec. Its synodic period — the interval between successive conjunctions or oppositions — is 584 days. The inclination of its orbit to the ecliptic is 3°24′.

The mass of Venus, which, like Mercury, has no satellite, must be calculated from the perturbations it produces upon the motions of the earth and other bodies. The best value, found from the Mariner V perturbations, is 0.82 times the mass of the earth. Venus has a diameter of 7526 mi, only about 400 mi less than the earth's. Its mean density is 5.2 times that of water.

(b) Telescopic Appearance

Venus, like Mercury, goes through phases. Galileo was the first to report the phases of Venus, a discovery that had great importance in disproving the validity of the Ptolemaic cosmology (Section 3.2b).

Venus is not observable at the full phase, because then it is at superior conjunction — on the other side of the sun from the earth — and is too nearly in line with the sun. Even if it could be seen, however, Venus would not show its greatest brilliance because at that time it is also at its greatest distance from the earth and subtends an angle of only 10″. Venus has its largest angular size of 64″ when it is nearest the earth, but then it is at the new phase (at inferior conjunction) and, again, is unobservable. Venus is at its greatest brilliance when it is a crescent, with an elongation of about 39°; this occurs about 36 days before and after inferior conjunction.

The surface of Venus is not visible because it is shrouded by dense clouds (or at least an opaque atmosphere). These clouds reflect sunlight very efficiently; measures of the albedo of the planet indicate that somewhere between one half and three quarters of the incident sunlight is reflected, a circumstance that contributes greatly to its brightness. Some rather indistinct dark markings can be observed on Venus. They show up best when the planet is photographed in violet light. They appear as broad streaks across the disk of the planet and may exhibit slow changes. They doubtless originate in the clouds or upper atmosphere.

(c) Rotation

The rotation period of Venus, like that of Mercury, has been determined by radar observations. The first radar observations of the planet's rotation were made in the early 1960s. Surprisingly, they showed Venus to rotate from east to west — in the reverse direction from the rotation of most other planets — in a period of about 250 days. Subsequently, R. Carpenter, at the Jet Propulsion Laboratory's Goldstone Radar Tracking Station in Southern California, identified permanent features on the planet that show up in the radar (reflected radio) signals. These are regions on Venus that have high reflectivity at radio wavelengths. Today, rather detailed radar maps have been made of the otherwise invisible surface of the planet, not only at the Jet Propulsion Laboratory but elsewhere as well. The rate at which the features appear to cross the disk of Venus gives us the most accurate determination of its rotation period — retrograde (east to west) in 243 days.

(d) Temperature

Venus, being closer than the earth to the sun, receives about twice as much solar energy per unit area and should, therefore, have a warmer climate than ours. Measures of the radiation from the planet, made at the Mount Wilson and Palomar Observatories, indicate temperatures, for both sunlit and dark hemispheres, in the range 230 to 240°K (−27 to −45°F). These temperatures are based on measures of infrared radiation and refer to the upper atmospheric layers of Venus, from which that radiation originates. But a heated body emits all wavelengths of electromagnetic radiation, radio waves as well as infrared rays. The strength of the radio radiation from Venus at wavelengths greater than 3.2 cm has also been measured and is found to correspond to that expected from a body at a temperature of about 700°K (nearly 900° F). Since radio waves would penetrate the Venusian cloud layer, they presumably originate from a region at or near the planet's surface. The high surface temperature of Venus was also indicated by the United States Mariner Venus probes in 1962 and 1967, and by the Soviet Venera 4 and 7 probes in 1967 and 1970. The Venera probes actually parachuted instruments into the atmosphere. Venera 4 is believed not to have reached the ground when it stopped transmitting data, but Venera 7 recorded a temperature of about 750°K at the surface.

(e) Atmosphere

The first gas to be identified spectrographically in the atmosphere of Venus was carbon dioxide, in 1932. More recently, traces of hydrogen fluoride and hydrogen chloride have been detected.

Aug. 20   Aug. 21   Aug. 23

FIG. 11-2   Ultraviolet photographs of Venus in 1962. *(Lick Observatory.)*

FIG. 11-3   Mars, photographed from a distance of
293,200 mi. with the television system carried by
the Jet Propulsion Laboratory's Mariner 7 space vehicle
that flew by the planet in August 1969. The television
observations of Mars from Mariners 4, 6, and 7
were all under the direction of R. B. Leighton of the
California Institute of Technology. *(NASA)*

FIG. 11-4   Phobos, Mars' larger satellite,
in a computer-enhanced photograph taken by Mars
probe Mariner 9 and televised to earth on November 30,
1971. The sun's illumination is from the left, and
the right (night) portion of the satellite is hidden
in shadow. Note the many craters produced by
meteorite impacts. *(NASA.)*

It is difficult to detect water vapor on a planet by means of spectrographic observations from the surface of the earth, because water in the earth's atmosphere absorbs light and contaminates the planetary spectrum. In 1959 the spectrum of Venus was photographed from a balloon at an altitude high above most of the terrestrial water vapor; an analysis of that spectrum shows evidence for the presence of water in the atmosphere of the planet. Subsequent observations by telescopes on earth confirmed a small concentration of water vapor.

A partial chemical analysis was performed on the Venusian atmosphere by Venera 4, the Soviet probe launched on June 11, 1967, which passed Venus on October 18 of that year and dropped, by parachute, an instrument package to the surface of the planet. The Venera 4 data are reported to show that the atmosphere of Venus is 90 to 95 percent carbon dioxide. Small amounts of oxygen and water vapor were also reported, but these data — especially on oxygen — are open to serious question. The data obtained from all of the Mariner and Venera Venus probes, as well as theoretical calculations based on earth-based observations of the planet's upper atmosphere, indicate that the surface pressure of the Venusian atmosphere is very high — about 90 or 100 times sea-level pressure on the earth.

The American Mariner V space probe, launched to Venus about the same time as the Soviet Venera 4 and passing about 2500 miles from the planet's surface on October 19, 1967, also carried instruments to measure the conditions in the atmosphere of Venus and in its immediate space environment. Although Mariner V made no chemical analysis of the Venusian atmosphere, it did transmit radio waves through it to the earth, and the attenuation of those signals provided information on the density and some other properties of the atmosphere of Venus. Other instruments carried by Mariner V attempted to measure any magnetic field around Venus, the density of hydrogen in its outermost atmosphere, and the flow of solar cor-puscular radiation near the planet. No magnetic field was detected, indicating that if there is any field at all it is at most a few thousandths as intense as the field around the earth.

The nature of the Venusian "clouds" is still a mystery. It is possible that the clouds are water clouds, perhaps contaminated by impurities; some experts consider it *probable* that they are water clouds. On the other hand, there are also other kinds of droplets or crystals that could display the characteristics of the Venusian cloud layer.

What the surface of Venus is like is an even bigger guess. Because of the high surface temperature and because of the low abundance of water vapor in the atmosphere of Venus, most investigators regard the surface as being hot and dry. Some even envision vast dust storms blowing across the planet.

11.3 MARS

Mars is sometimes called the "newspaper planet"; it is the most favorably situated of the planets for observation from the earth and has excited more interest and comment than any other. Only on Mars and Mercury can we see the solid surface, and Mercury is too near the sun to observe easily. Venus comes a little closer to the earth, but at its closest it is at the new phase when it cannot be studied. Mars, on the other hand, is sometimes favorably situated for observation all night long.

(a) Orbit of Mars

The median distance of Mars from the sun is 141,690,000 mi, but its orbit is somewhat eccentric (eccentricity of 0.093) and its heliocentric distance varies by 26 million mi. The sidereal period of revolution is 687 days. The orbit is inclined to the ecliptic plane by 1°51′.

At intervals of the synodic period of Mars, 780 days, the earth passes between it and the sun. Then Mars, at opposition, is above the horizon all night. It is also at its nearest to the earth and is most

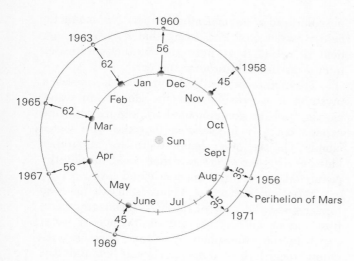

FIG. 11-5   Orbits of the earth and Mars, with the positions of the planets shown for several oppositions.

favorably disposed for observation. All oppositions of Mars are not equally favorable, however; at such times its distance from earth can be anywhere from 35 to 63 million mi, depending on where the planet is in its orbit when the earth passes it. This circumstance is due to the eccentricity of Mars' orbit and, to a lesser extent, that of the earth's orbit (see Figure 11-5). The most favorable oppositions are those that occur when Mars is near perihelion — its closest approach to the sun and hence to the earth's orbit. These are the oppositions that take place during our late summer or early fall. Such optimum opportunities to investigate Mars arise about once or twice every 15 to 17 years.

(b) Satellites

Centuries ago, Kepler, hearing of Galileo's discovery of four satellites of Jupiter, speculated that Mars should have two moons. There was, of course, no scientific justification for this speculation. Again, in 1726, Jonathan Swift in his satire *Gulliver's Travels* described Gulliver's visit to the land of

Laputa, where he found scientists engaged in many interesting investigations. In one of them, Gulliver reported, Laputian astronomers had discovered

". . . [two] satellites, which revolve about Mars, whereof the innermost is distant from the centre of the primary planet exactly three of the diameters, and the outermost five; the former revolves in the space of ten hours, and the latter in twenty one and a half; so that the squares of their periodical times are very near in the same proportion with the cubes of their distance from the centre of Mars, which evidently shows them to be governed by the same law of gravitation, that influences the other heavenly bodies."

It is an interesting coincidence that in 1877, 150 years later, Asaph Hall of the Naval Observatory actually discovered two small satellites of Mars that closely resemble those described by Swift. Those moons are named *Phobos* and *Deimos,* meaning "fear" and "panic" — appropriate companions of the god of war, Mars. Phobos is 5800 mi from the center of Mars, and revolves about it in 7^h39^m; Deimos has a distance of 14,600 mi and a period of 30^h18^m. The "month" of Phobos is less than the rotation period of Mars; consequently, Phobos would appear to an observer on Mars to rise in the west.

Both satellites are too small to present disks that can be measured from earth. They have been photographed, however, by Mariner 9 (Figure 11-4). They are slightly elongated and are pitted by meteoritic impacts. The greatest diameters of Phobos and Deimos are about 15 and 8 miles, respectively.

(c) Physical Properties

The mass of Mars can be found conveniently and accurately from the distances and periods of its satellites applied to Newton's formulation of Kepler's third law (Section 4.3a) and also from Mars' perturbations on interplanetary probes; it is only 0.107 times as massive as the earth. The diameter of the planet is about 4200 mi, which gives it a mean density of approximately four times that of water and a surface gravity of 38 percent of the earth's. A 200-lb man would weigh about 75 lb on Mars.

(d) Telescopic Appearance

Until Mariner IV, launched November 28, 1964, flew by Mars 228 days later and televised to earth photographs obtained at an altitude of 6000 mi, our only views of the Martian surface were through ground-based telescopes. The layman, having heard so much about Mars, is usually disappointed when he first sees it through a telescope. At its nearest, Mars is 150 times the moon's distance; a feature 1 mi wide on the moon can be resolved about as well as a feature 150 mi wide on Mars. At its nearest, the planet subtends only about 25″.

Since Mars is farther than the earth from the sun, it does not go through crescent phases. When it is near quadrature, however, it does appear distinctly gibbous. Except for its occasional gibbous shape, Mars usually resembles a shimmering orange ball — the shimmering is caused, of course, by the earth's atmosphere. Most of the Martian surface appears yellowish orange or red. Often white areas, the *polar caps,* can be seen at one or both of the planets' poles. Under good conditions large dark areas can be seen as well. They have a grayish color and somewhat resemble oceans; they were once believed to be bodies of water and so were named *maria,* like the "seas" on the moon.

Under the best observing conditions, a large amount of detail on Mars is telescopically visible. The finest features, however, are seen best visually.

FIG. 11-6   The region of Nix Olympica, a gigantic volcanic mountain on Mars. Photographed by Mariner 9. (*NASA.*)

during fleeting instants of steadiness of the earth's atmosphere, and are blurred during even brief time exposures needed for photographs. Consequently, the interpretation of Martian surface markings as seen from earth has always been controversial. The so-called "canals" provide an important example.

These straight-appearing streaks were first observed in 1877 by the Italian astronomer Schiaparelli, who called them *canali* ("channels"). Many other observers thought they saw straight lines on Mars, notable among them the American astronomer Percival Lowell, who charted more than 400 "canals," about 50 of which he thought were double, and many of which intersected at what he called "oases." Lowell believed these markings to be too nearly geometrically perfect to be natural, and he envisioned them to be actual canals built by intelligent beings to carry the melting

waters from the polar caps across the desert to irrigate Martian crops.

The eye is easily deceived into interpreting poorly-observed complex features as simpler ones. The Mariner Mars probes have shown that Mars does, indeed, abound in detailed surface features, but there are no canals.

(e) Rotation and Seasons

The permanence of the surface features of Mars enables us to determine its rotation period with great accuracy; its sidereal day is $24^h37^m23^s$, very near the rotation period of the earth. This high precision is not obtained by watching Mars for a single rotation, but by noting how many turns it makes in a long period of time. Good observations of Mars date back for more than 200 years, during which period tens of thousands of Martian days have passed. The value accepted today is accurate

FIG. 11-7   Nix Olympica region. (*Left*) Summit crater of the volcanic mountain; (*Right*) Detail of white rectangular region of left photograph, showing lava flow. Mariner 9 photograph. (*NASA*.)

to within a few hundredths of a second.

The equator of Mars is inclined to its orbital plane by about 25° — very close to the 23½° angle between the earth's orbit and its equator. Thus each of Mars' poles is alternately tipped toward and away from the sun, and the planet goes through seasons, much like those on earth. Because of the longer Martian year, however, seasons there each last about six of our months.

Certain of the Martian seasonal changes can be observed telescopically. Most conspicuous are the changes in the polar caps. The southern polar cap, in winter, reaches a maximum diameter of about 3700 mi, and extends halfway to the planet's equator. In the spring, the cap shrinks, and often disappears entirely in summer. The north polar cap, in a region where the range of seasonal temperature is less extreme, does not reach so large a size, and never quite disappears. Results of experiments carried on the Mariner probes indicate that the white caps consist mostly of frozen carbon dioxide (dry ice), which sublimes directly into the Martian atmosphere as warmer weather reaches the polar regions. There may also be a small amount of frozen water.

Telescopic observers of the planet have also reported the maria to change with the seasons, not much in size but in intensity of color. Some have even reported the maria to change in color, turning blue or even green in spring and summer. At one time these alleged color changes were believed to be evidence for vegetation of Mars, but Mariner observations rule out the possibility of large areas of plant life. The changes in the maria most likely result from the shifting of dust or other matter with seasonal winds.

(f) Martian Temperatures

The temperatures determined for Mars are based on measures of infrared radiation from its surface, obtained from earth-based observatories as well as by detectors carried by Mariners 6, 7, and 9.

The maximum equatorial temperatures are slightly over 300°K (about 80°F). The maria tend to be from 10 to 30°F warmer than the surrounding regions. The sunrise and sunset temperatures at the equator, although more uncertain, may be around −4° and 40°F, respectively. At night, the temperature drops to nearly −100°F.

The temperature at the south polar cap was measured by Mariner 7, and was found to range down to −250°F, which is near the frost point of carbon dioxide. The temperature at the north polar cap is believed to be less extreme, but it probably does not get above −100°F.

(g) The Martian Atmosphere

The existence of a Martian atmosphere is easily demonstrated by the way it scatters sunlight. Measures of the sunlight reflected from Mars reveal the presence of a Martian atmosphere; measures by all four Mariners showed that the atmospheric pressure at the surface of the planet is everywhere less than 10 millibars; this is somewhat under 1 percent of the sea-level pressure of the earth's atmosphere. The relatively low albedo of Mars (0.15) confirms the fact that its atmosphere is not dense. It is dense enough, however, to produce a distinct twilight zone. Also, as in the earth's atmosphere, light of shorter wavelengths is scattered more than that of longer wavelengths. The surface markings on Mars, therefore, show up much better on photographs made in red light than in blue.

The first gas to be identified spectrographically in the Martian atmosphere was carbon dioxide, which is estimated to comprise at least 80 percent of the gas present. Spectrograms obtained at Mt. Wilson in 1963 revealed evidence for a very low concentration of water vapor on Mars. Oxygen is absent except for a minute trace in the atomic form, which is believed to result from the photodissociation of water vapor and carbon dioxide. Traces of atomic hydrogen and carbon are also present.

Even telescopic observations from the earth have revealed clouds in the Martian atmosphere. The Mars probes, especially Mariner 9, confirm the existence of clouds at high altitudes. Clouds about 45 km above the planet's surface are believed to consist of crystals of carbon dioxide, while lower ones, with altitudes of 15 to 30 km, are probably water ice. The Martian winds sometimes also raise vast dust clouds. In late 1971, almost the entire planet was shrouded in a great dust storm that rendered the atmosphere opaque to visible light at altitudes below 11 km.

(h) The Martian Surface

The first successful space exploration of Mars was by Mariner 4, which flew past the planet in 1965 and televised about 20 pictures back to earth. The most striking discovery was that Mars has very many craters superficially resembling those on the moon. In 1969 Mariners 6 and 7 flew by Mars and transmitted about 200 high-quality pictures to earth. Even more spectacular was the success of Mariner 9, which went into orbit about the planet in late 1971, and in the succeeding months televised many thousands of photographs. The entire planet has now been observed within a range of 2000 to 4000 km, and over parts of its surface, features as small as 100 m across have been resolved. In contrast, the smallest features that can be resolved with earth-based telescopes are about 30 mi across.

Much of Mars is covered with craters, which range in size up to hundreds of miles across. Their appearance reminds us of the craters on the moon, except that the Martian craters show greater evidence of erosion, which is expected, because of winds and possibly other weather effects. Indeed, there is a much smaller percentage of small craters on Mars than on the moon; evidently most of the smaller craters have been eroded away entirely or perhaps buried in dust. Many craters appear to have floors filled with dust or lava, and surprisingly, there are large areas of the planet that are very smooth, almost devoid of craters altogether. The latter may be regions where prevailing winds have deposited deep layers of dust or other material.

Most of the Martian craters are presumed to have been formed by meteoritic impact. Some, however, are almost certainly volcanic. Most spectacular is the Nix Olympica region (Figure 11-6) which contains a giant mountain more than 300 miles across at the base. Its summit crater or caldera is at least 40 miles across. Elevations on Mars can be estimated from the surface atmospheric pressures that were measured with Mariner 9's ultraviolet spectrometer. An altitude of 6 km was indicated just part way up the Nix Olympica slope. The entire system is larger than the Hawaii volcanic pile, the largest on Earth.

Lava-like slopes are seen outside the Nix Olympica mountain (Fig. 11-7). Other regions of Mars are also very rugged. In some, terrace-like features are observed. There are huge canyons; one extends for thousands of kilometers in an east-west direction across the planet, and is 6 km deep. There are also meandering channels running for hundreds of kilometers. Some have branching tributaries, suggesting drainage patterns. Most experts, however, suspect erosion by lava and wind rather than by water, which in any case would usually be frozen.

Of all the other planets, Mars is the one that has always been the best candidate for supporting some form of life. Percival Lowell was one of the most ardent supporters of the hypothesis of life on Mars. He built the Lowell Observatory in Arizona primarily for the purpose of studying Mars. The supposed canal system on the planet and the changing colors of the maria seemed to give credence to the idea. Moreover, at one time there seemd to be spectroscopic evidence for the hydrocarbon radical — basic to organic molecules — in the reflected sunlight from the Martian maria. Subsequent observations, however, discredited the evidence for the hydrocarbon radical, and we have

Apollo 11 astronaut Edwin Aldrin deploying the seismometer. The laser reflector is to his right and further in the background. In the center background is the lunar module. *(NASA.)*

Mars, photographed with the 200-inch telescope. *(Hale Observatories.)*

Jupiter, photographed with the 200-inch telescope. *(Hale Observatories).*

Saturn, photographed with the 200-inch telescope.
(Hale Observatories.)

Comet Humason (1961a), photographed with
the 48-inch Schmidt telescope.
(Hale Observatories.)

Comet Ikeya-Seki, photographed by
J. B. Irwin in Chile, October, 1965.

Section through Neenach, California stoney meteorite, showing
flecks of nickel-iron metal. *(From the collection of Ronald Oriti, Griffith
Observatory; photograph by Ivan Dryer.)*

Slice of Kamkas iron meteorite which has been polished and then etched
with dilute nitric acid to show the criss-cross Widmanstätten figures. *(From the
collection of Ronald Oriti, Griffith Observatory; photograph by Ivan Dryer.*

already seen how Mariner observations have evaporated the canals on Mars. Furthermore, sophisticated observations of the surface of the planet by the Mars probes have found no evidence for growing vegetation on the planet's surface. Nor has any evidence been found for roads or cities or other artifacts of a civilization.

All of this, however, is neither evidence for or against life on Mars. To be sure, Mars has but a trace of water above the surface, and has temperatures and an atmosphere far too extreme for the higher forms of life known on earth. But whereas there is no evidence for life, we cannot rule it out. The question of life on Mars must await further exploration for a final answer.

11.4 JUPITER

Jupiter is well named for the leader of the gods. Next to the sun, it is the largest and most massive object in the solar system.

(a) Orbit of Jupiter

The median distance of Jupiter from the sun is 484 million mi, 5.2 times that of the earth. Its orbit has an eccentricity of 0.048, so its distance from the sun varies by 47 million mi. Jupiter's mean orbital speed of 8.1 mi/sec carries it once around its orbit in 11.96 years. Its orbital plane is inclined $1°18'$ to the ecliptic.

(b) Satellites

Jupiter has twelve known satellites, the largest number of any planet in the solar system. The four largest moons were discovered by Galileo; they are so bright that they would be visible to the unaided eye on a clear dark night were it not for their proximity to the bright planet. Since the four *Galilean* satellites present observable disks, their angular sizes can be measured directly with a filar micrometer, and their linear sizes calculated from the known distance of Jupiter. Their diameters can also be determined from a knowledge of their orbital velocities and from the time it takes them to enter or leave the umbra of Jupiter's shadow. The two largest are about 3000 mi across, $1\frac{1}{2}$ times the size of our moon; the other two are about the same size as the moon. Their distances from Jupiter's center range from about $\frac{1}{4}$ million to more than 1 million mi; their corresponding periods run from $1^d18\frac{1}{2}^h$ to $16^d16\frac{1}{2}^h$. Their discovery provided strong support for the Copernican theory (Section 3.2b).

Subsequently, eight additional satellites have been discovered, the latest in 1951. They are all much smaller than the Galilean moons, and their sizes must be estimated from their brightnesses and assumed albedos. Their probable diameters range from about 15 to 150 mi. They are known by the numbers 5 through 12, assigned to them in order of their discovery. The most distant is about 15 million mi from Jupiter, and has a period of more than 2 years.

The five inner satellites of Jupiter (including the four Galilean) have nearly circular orbits that lie near the planet's equatorial plane. The outer seven, however, have rather eccentric orbits, some of which have large inclinations to Jupiter's equator. The four most distant satellites revolve from east to west, contrary to the motions of most of the objects in the solar system. None of the moons of Jupiter has an observable atmosphere.

(c) Physical Properties

The mass of Jupiter has been calculated both from the periods of its satellites and the sizes of their orbits and from the perturbations it produces on the orbits of passing minor planets. It is found to be 318 times as massive as the earth, a value which is very close to $\frac{1}{1000}$ the mass of the sun. Jupiter is also the largest planet, having a diameter of 87,000 mi, or about 11 times the earth's. If it were as near as Mars, it would be the brightest planet in the sky, often visible in daylight.

Jupiter's surface gravity is higher than the earth's; a body at its surface would weigh 2.64 times what it would weigh on earth. Jupiter has the highest velocity of escape of any planet, 37 mi/sec; it can easily retain all kinds of gases in its atmosphere. Its mean density, however, is much less than the earth's — only 1.33 times the density of water.

Because of the tremendous gravitational attraction of Jupiter for its constituent parts, it would most certainly be compressed to a far greater mean density unless it were composed almost entirely of hydrogen and helium — the lightest and most common elements in the universe. Even these elements are probably compressed to the solid state — like ice — throughout practically all of Jupiter's interior.

(d) Telescopic Appearance

Because Jupiter is so far from the sun, it can appear only very slightly gibbous, and usually presents a nearly full disk to the telescopic viewer. Its relatively high albedo (0.51) makes it quite bright, despite the small amount of sunlight that reaches it. Its four brightest satellites are easy to see, even with binoculars; the shadows of some of the satellites occasionally can be observed crossing the planet's disk.

The disk of the planet itself is crossed with alternate light and dark cloud bands parallel to its equator. Details in their cloud bands are numerous, and exhibit gradual changes, showing that we are looking at atmospheric phenomena, not the solid surface of the planet. The light bands are not pure white, but show various colored regions: yellow, red, and sometimes blue.

One of the most striking surface features of Jupiter is the Great Red Spot that was seen telescopically at least as far back as 1831. It may have been the "spot" on Jupiter described by Cassini in 1660 and used by him to determine the planet's rotation period. It changes in size and shape slightly and also in the intensity of its color. It has been as large as 30,000 mi across. It has been described as a floating "island" of liquid or solid particles; however, there is no suggestion as to what the particles may be, and there are strong hydrodynamic arguments against the hypothesis. A modern theory for the Red Spot is that it is the top of a *Taylor column,* a vertical flow of gases over a large topographical feature on the solid surface of the planet.

(e) Rotation and Flattening

Distinct details in the cloud patterns on Jupiter allow us to determine its rotation rate; it is the most rapidly spinning of all the planets. However, it does not rotate quite as a solid body, or at least the clouds that we observe do not. For most of the planet, the rotation rate is fairly constant, averaging one rotation every 9^h55^m. A broad band about 10,000 to 15,000 mi wide along the equator, however, moves about 200 mi/hr faster than the surrounding regions, and completes a rotation in 9^h50^m. The rotation of the planet can be measured spectroscopically as well as visually. Jupiter's equator is inclined at only 3° to its orbit plane, so the planet has no appreciable seasons.

Jupiter's rapid rotation has caused it to become noticeably oblate, just as the earth is slightly flattened because of its rotation. When observed visually through a telescope, Jupiter's flattening is readily apparent; its polar diameter is less than its equatorial diameter by 1 part in 15. The polar and equatorial diameters of the earth differ by only 1 part in 298.

Radio radiation is observed to come from regions high above Jupiter's surface, indicating the interaction of ions with a strong magnetic field around the planet, a field made possible by its rapid rotation. There are also ionospheric disturbances that give rise to sharp bursts of radio energy. Some of the bursts seem to be correlated with the revolution of Jupiter's satellite, Io.

(f) Atmosphere

The temperature of Jupiter, calculated from the intensity of radiation that comes, probably, from its visible clouds, is about 130°K, or −220°F, just about what we would expect for a planet of Jupiter's albedo and distance from the sun. Because of the high velocity of escape from Jupiter, we would expect it to retain even the lightest gases; hydrogen and helium, therefore, should be abundant constituents of its atmosphere. Nitrogen, neon, and other inert gases also may be fairly abundant. The gases on Jupiter that are observed most easily by spectrographic means are, however, methane and ammonia; these gases are abundant but still comprise only a fraction of 1 percent of the total mass of Jupiter's atmosphere. Hydrogen, when cold, is difficult to observe spectrographically because its principal spectral lines lie in the far ultraviolet, to which the earth's atmosphere is opaque. However, *molecular* hydrogen (the H_2 molecule) does produce a series of absorption lines in the infrared.

A recent spectrographic analysis of the gases in Jupiter's atmosphere shows that the amount of hydrogen above each square inch in the opaque layer of Jupiter's clouds is from 10 to 20 times as great as the entire amount of atmosphere over each square inch of the earth's surface at sea level. Jupiter's clouds are already high in its atmosphere, so the total amount of atmospheric hydrogen there is immense. Methane has an abundance of about $\frac{1}{1000}$ that of the hydrogen, and ammonia is only $\frac{1}{20}$ as plentiful as methane. The ratio of hydrogen to carbon (the latter in combination with hydrogen in the methane) in Jupiter's atmosphere is found to lie between 1200 and 1800. The higher figure is close to the corresponding ratio in the sun.

Theoretical studies indicate that the total atmosphere cannot extend downward more than about 1 percent of the planet's radius. Within 100 or 200 mi below the visible clouds, the pressures are high enough to liquefy hydrogen. The atmosphere is pictured as gradually becoming thicker and "slushier" with depth, until it finally is condensed to an "ocean" of liquid hydrogen and other elements. The ocean probably gradually hardens to the solid planet in about another 150 or so miles down.

11.5 SATURN

Saturn is the solar system's second largest planet. Its unique *ring* system makes it one of the most impressive of telescopic objects. It is named for the Titan god of seed sowing, the father of Jupiter. It is the most remote of the planets known before the invention of the telescope.

(a) Orbit and Satellites

The orbit of Saturn has an eccentricity of 0.056, about the same as that of Jupiter. The planet's distance from the sun varies from about 840 to 935 million mi and has a median value of 886,100,000 mi. The orbit is inclined 2½° to the ecliptic plane. At a mean orbital speed of 6 mi/sec, Saturn completes one sidereal revolution in 29½ years.

In addition to its ring system, Saturn has 10 known satellites. Several of them are easily visible with small telescopes. The largest, Titan, is larger than our moon and is the only satellite in the solar system that is known definitely to possess an atmosphere; Kuiper identified methane in its spectrum in 1944. The second largest of Saturn's satellites is about 1000 mi in diameter. The most distant satellite is Phoebe, with a distance and period of about 8 million miles and 550 days, respectively. The last satellite to be discovered, and the innermost one, is Janus, found by Dollfus in France. A small gap in Saturn's ring system led Dollfus to suspect the existence of the satellite very close to the outer rim of the ring; the gravitational influence of the satellite would produce the gap. The new object was found during a successful search in December 1966, when the rings were turned edge on to the earth and thus

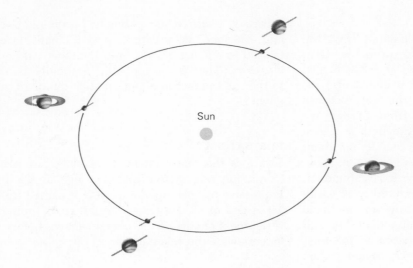

Sun

FIG. 11-8 Orientation of the rings as seen from the sun with Saturn at different places in its orbit.

appeared very faint. Janus is less than 100,000 mi from the center of Saturn and revolves about it in 18 hours.

(b) The Ring System

Galileo first saw the rings of Saturn, but he was unable to discern them clearly with his crude telescopes. It was not until half a century later, in 1655, that Huygens described their true form. The rings have approximately the appearance of the brim of a straw hat surrounding the planet in its equatorial plane. There are three concentric portions of the rings. The brightest and broadest is the central, or *bright* ring. Surrounding this is the *outer* ring with an outside diameter of 171,000 mi. The faintest is the inner or *crape* ring, whose inside diameter of about 88,000 mi allows only a 8000-mi gap between it and the ball of the planet itself.

The rings are not a solid sheet. Background stars can be seen shining through them. Moreover, the Doppler shift in the spectrum of sunlight reflected from them shows the inner portions of the rings to be revolving about the planet in a shorter period than the outer portions, in accord-

ance with Kepler's third law (the periods would be the same if the rings were a solid sheet). The intensity of the reflected light from the rings, and its spectrum, show that they cannot be gaseous. There is no question that the rings are composed of many billions of minute solid particles, probably gravel-sized or smaller. Sunlight reflected from the countless myriads of particles gives the illusion of solid rings.

The rings lie in the equatorial plane of Saturn, whose orientation remains constant during the planet's revolution about the sun. This plane is inclined at about 28° to the ecliptic, so that during part of Saturn's orbital revolution we see one face of the rings, and when the planet is on the opposite side of the sun, we see the other face. At intermediate points, the rings may appear edge on to our line of sight; in that condition they virtually disappear (Figure 11-8), which shows that they are very thin. The thickness of the rings is too small to measure; observations show it not to exceed 10 or 15 mi. There are also theoretical reasons to believe that the rings might be extremely thin — perhaps as little as a few feet deep.

(c) Mass, Size, and Structure

Saturn has a mass equal to 95 earth masses, and the mean diameter of the planet proper is just under 72,000 mi. These figures give it an average density (in terms of water) of only 0.71 — the lowest of any planet in the solar system. In fact, Saturn would be light enough to float, if an ocean existed large enough to launch it. Its surface gravity is only 17 percent greater than the earth's; nevertheless, it has a velocity of escape of 22 mi/sec, great enough for Saturn to hold the lightest gases. Calculations show that Saturn must be at least 60 percent hydrogen.

(d) Telescopic Appearance and Rotation

Except for its rings, Saturn looks very much like Jupiter. It, too, has parallel alternately dark and light cloud bands, although the details, color, and irregularities in them are much less distinct than in Jupiter's clouds. Small light spots are occasionally seen on the planet.

The rotation period of the planet, as determined both from the Doppler shift in its spectrum and from the apparent motions of the spots on its disk, is just over 10 hours at the equator. Like Jupiter, however, Saturn rotates more slowly at latitudes away from the equatorial regions. The mean rotation period for most of the planet is near 10^h38^m. Because of its rapid rotation, Saturn is the most oblate of all the planets; its equatorial diameter is about 10 percent greater than that through its poles.

(e) Atmosphere

Measures of the radiant heat from Saturn indicate a temperature of 120 to 130°K, or near −230°F. This is some 30 to 40°F higher than might be expected for a planet of Saturn's albedo and distance from the sun. It therefore may be appropriate for a depth somewhat below the visible cloud layer. The only gases detected spectrographically are hydrogen and methane. Evi-

dently, because of the somewhat lower temperature of Saturn (compared to Jupiter), the ammonia has "snowed" out of the atmosphere. The structure of Saturn's atmosphere is thought to be similar to Jupiter's. The gaseous region probably merges into a hydrogen "ocean" a few hundred miles below the cloud layers. The ocean may extend a little deeper than in Jupiter, but probably, at a depth of not more than 200 or 300 mi., it solidifies into the icelike interior of the planet.

11.6 URANUS

Uranus is a planet that must surely have been seen by the ancients and yet was unknown to them. It was discovered on March 13, 1781, by the German-English astronomer, William Herschel, who was making a routine telescopic survey of the sky in the constellation of Gemini. Herschel noted that through his telescope the planet did not appear as a stellar point but seemed to present a small disk. He believed it to be a comet and followed its motion for some weeks. Several months later, a preliminary solution for its orbit was computed and it was found to be a nearly circular one, lying beyond that of Saturn; the object was unquestionably a new planet.

Uranus can be seen by the unaided eye on a dark clear night, but is near enough to the limit of visibility so that it is indistinguishable from a very faint star. It is so inconspicuous that its motion escaped notice until after its telescopic discovery. However, it turned out that Uranus had been plotted as a star on charts of the sky on at least 20 previous occasions since the year 1690. These earlier observations were later of use in the determination of how perturbations were altering the planet's orbit.

Herschel proposed to name the newly discovered planet *Georgium Sidus,* in honor of George III, England's reigning king. Others suggested the name Herschel; the name finally adopted, in keeping with the tradition of naming planets for gods of Greek mythology, was Uranus, father of the Titans, and grandfather of Jupiter.

(a) Orbit and Satellites

The orbit of Uranus lies more nearly in the plane of the ecliptic than that of any of the other planets, its inclination being only 46′. Its eccentricity is 0.047, and it varies by a little more than 80 million mi from its median distance of 1783 million mi from the sun. Being so far from the sun, its orbital speed is low (4½ mi/sec), and its period is long (84 years).

Uranus has five known satellites; the last to be discovered, and the faintest, was found by Kuiper in 1948. Herschel himself found the two brightest moons of the planet. None of the satellites is probably much over 1000 mi in diameter. Their distances range from 81,000 to 364,000 mi from the center of Uranus.

(b) Gross Characteristics

The mass of Uranus is found from the motions of its satellites and from its perturbative effects upon the motion of Saturn; it is 14½ times the mass of the earth. It presents a small angular diameter, just under 4″, because of its great distance from the sun. Its linear diameter, therefore, is difficult to measure accurately, but it is about 30,000 mi. With these values for its mass and size, its mean density is calculated to be 1½ times that of water, its surface gravity about the same as on the earth, and its velocity of escape about 20 km/sec. Micrometric measures and also studies of the perturbations on its satellites both show Uranus to be somewhat oblate.

The spectrum of Uranus reveals a strong concentration of methane, and also a series of infrared absorption lines attributed to molecular hydrogen. The identification of hydrogen was achieved in 1952 by Kuiper and Herzberg. Uranus

FIG. 11-9   Uranus and its satellites, photographed with the 120-inch telescope *(Lick Observatory.)*

is the first planet in which hydrogen was identified spectrographically. Ammonia does not appear in the spectrum, probably because it is frozen out of the atmosphere; the temperatures to be expected on Uranus are below −300°F.

(c) Appearance and Rotation

Uranus appears as a greenish disk when seen through the telescope. The green color is probably due to its atmospheric methane. A few observers report faint markings, but these are too indefinite to indicate the rotation. The rotation period, therefore, must be obtained from the Doppler shift in the spectrum of light from different parts of its disk. The accepted value, somewhat uncertain, is 10h45m.

A unique feature about Uranus is that its axis of rotation lies almost in the plane of its orbit; during some parts of its revolution, it is so oriented that we look almost directly at one or the other of its poles. The actual inclination of its equatorial plane to that of its orbit is 82°. Its direction of rotation is the same as that of the revolution of its satellites, and both are in the *reverse* direction from the rotation of the other planets. Its direction of orbital revolution, however, is normal — that is, from west to east.

11.7 NEPTUNE

Whereas the discovery of Uranus was quite unexpected, Neptune was found as the result of mathematical prediction. The discoveries of the two planets could hardly have been made under more different circumstances, yet in other respects Uranus and Neptune are more alike than any two other worlds in the solar system.

(a) Discovery

By 1790, an orbit had been calculated for Uranus (first by Delambre) on the basis of observations of its motion in the decades following its discovery. Even after allowance was made for the perturbative effects of Jupiter and Saturn, however, it was found that Uranus did not move on an orbit that fitted exactly the earlier observations of it made since 1690, before it was known as a planet (see Section 11.6). By 1840, the discrepancy between the positions observed for Uranus and those predicted from its computed orbit amounted to about 2′ — an angle barely discernible to the unaided eye but still much larger than the probable errors in the orbital calculations. In other words, Uranus did not seem to move on an orbit that would have been predicted from Newtonian theory.

In 1843 John Couch Adams, a young Englishman who had just completed his work at Cambridge, began an analysis of the irregularities in the motion of Uranus to see whether they could be produced by the perturbative action of an unknown planet. His calculation indicated the existence of a planet more distant than Uranus from the sun. This unknown planet was pulling on Uranus, accelerating it slightly ahead in its orbit until 1822, when Uranus passed between the planet and the sun. After 1822, the motion of Uranus was being retarded slightly. In October 1845 Adams sent his results to Sir George Airy, the Astronomer Royal, informing him where in the sky he should look to find the new planet. We know today that Adams' predicted position for the unknown body was correct to within 2°. Airy, however, apparently had little faith in the ability of so young and unknown a mathematician, and posed a simple problem to him as a test. When Adams did not bother to respond, Airy let the whole matter drop.

Meanwhile, Leverrier, a French mathematician, unaware of Adams or his work, attacked the same problem, and published its solution in June 1846. Airy, noting that Leverrier's predicted position for the unknown planet agreed to within 1° with Adams', posed the same test problem to the Frenchman and received a prompt and correct reply.

Airy then suggested to Challis, director of the Cambridge Observatory, that he begin a search for the new object. The Cambridge astronomer, having no up-to-date star charts of the region of the sky in *Aquarius* where the planet was predicted to be, proceeded by plotting all the faint stars he could observe with his telescope in that location. It was Challis' plan to repeat such plots at intervals of several days, in the hope that the planet would reveal its presence and distinguish itself from a star by its motion. Unfortunately, he was negligent in examining his observations; although he had actually seen the planet, he did not recognize it.

About 1 month later, Leverrier suggested to Galle, an astronomer at the Berlin Observatory, that he look for the planet. Galle received Leverrier's letter on September 23, 1846, and, possessing new charts of the *Aquarius* region, he found and identified the planet that very night. It was only 52′ from Leverrier's predicted position.

The discovery of the eighth planet, now known as Neptune (named for the god of the sea), was a major triumph for gravitational theory and ranks as one of the great scientific achieve-

FIG. 11-10   Neptune and a satellite, photographed with the 120-inch telescope. *(Lick Observatory.)*

ments. The honor for the discovery is quite properly shared by the two mathematicians, Adams and Leverrier.

Although a great triumph, the discovery of Neptune was not a complete surprise to astronomers, who had long suspected the existence of the planet. On September 10, 1846, Sir John Herschel, son of the discoverer of Uranus, remarked: "We see it as Columbus saw America from the shores of Spain. Its movements have been felt trembling along the far-reaching line of our analysis with a certainty hardly inferior to ocular demonstration."

(b) Orbit, Satellites, and Rotation

Neptune is almost exactly 30 AU from the sun. Its orbit is very nearly circular, having an eccentricity of only 0.009; its distance from the sun varies by only 24 million mi on each side of its median value of 2797 million mi. Traveling with an orbital speed of only 3⅓ mi/sec, it requires 165 years to complete one revolution around the sun. It will not have completed its first revolution since its discovery until 2011.

Its axial rotation period, determined from the Doppler shift in its spectrum, is 15.8 hours; the value is uncertain by 1 hour or so. Its equatorial plane is inclined about 29° to that of its orbit. Its oblateness is small, the equatorial diameter exceeding the polar diameter by only about 1 part in 40.

Neptune has two satellites. *Triton,* the larger, is somewhat greater in diameter than our moon. Its distance from the center of Neptune is only about 220,000 mi, and its motion is backward — that is, east to west. Neptune's other satellite, discovered by Kuiper in 1949, is very much smaller, and has the most eccentric orbit of any satellite in the solar system; its distance from Neptune varies from 1 to 6 million mi.

(c) Other Properties

Neptune's mass is 17.2 times that of the earth. Since it subtends an angle of only about 2″, its diameter is difficult to measure, but it is approximately 30,000 mi. Its mean density is somewhat more than twice that of water — higher than the density of Uranus. Its internal structure is believed to resemble that of Uranus.

Neptune appears telescopically as a small greenish disk. There are no conspicuous markings. Its high albedo of 0.5 to 0.6, however, suggests that light is reflected from a gaseous atmosphere. Both methane and hydrogen have been detected spectrographically, as in Uranus. Helium is probably also present but cannot be observed in the spectrum. Ammonia appears to be absent in the gaseous state, as we would expect, since the temperature of Neptune must be near −350°F.

11.8 PLUTO

After the perturbative effects of Neptune upon Uranus had been taken into account, the discrep-

FIG. 11-11   Two photographs of Pluto, showing its motion among the stars in a 24-hour period; photographed with the 200-inch telescope. *(Hale Observatories.)*

ancies between the observed and predicted positions of Uranus were reduced to only $\frac{1}{60}$ of what they had been. Today those remaining discrepancies are realized to be smaller than the errors of the original observations and thus are almost certainly not real. Yet, several investigators made attempts to account for these remaining deviations by a gravitational influence of a ninth planet, beyond the orbit of Neptune. Among them were Gaillot, W. H. Pickering, and Percival Lowell. It was Lowell's solution to the problem that led to the discovery of Pluto.

(a) Discovery

By the beginning of the twentieth century, when Lowell made his calculations, Neptune had moved such a short distance in its orbit that a knowledge of the perturbations upon it was not yet available. Therefore Lowell, as did the others, based his calculations entirely on the minute remaining irregularities of the motion of Uranus. His computations indicated two places where a perturbing planet could be, the more likely of the two being in the constellation of *Gemini*. Lowell predicted a mass for the planet intermediate between that of the earth and that of Neptune (his calculations gave about 6.6 earth masses), and also an angular diameter in excess of 1″. Lowell searched for the

unknown planet at his Arizona Observatory from 1906 until his death in 1916, without success. Subsequently, Lowell's brother donated to the observatory a 13-in. photographic telescope that could record a 12° by 14° area of the sky on a single photograph. The new camera went into operation in 1929, and the search was continued for the ninth planet.

In February 1930, Clyde Tombaugh, comparing photographs made on January 23 and 29 of that year, found an object whose motion appeared to be about right for a planet far beyond the orbit of Neptune. It was within 6° of Lowell's predicted position for the unknown planet; subsequent investigation of the object showed its orbit to have elements very similar to those Lowell had calculated. Announcement of the discovery was made on March 13, 1930. The new planet was named for Pluto, the god of the underworld. (Appropriately, the first two letters of Pluto are the initials of Percival Lowell.)

(b) Orbit of Pluto

Pluto's orbit has the highest inclination to the ecliptic (17°9′) of any planet and also the largest eccentricity (0.249). Its median distance from the sun is 39.52 AU, or 3,675,000,000 mi, but its aphelion distance is over 4½ billion mi, and its perihelion distance under 2.8 billion mi. Part of its orbit is closer to the sun than the orbit of Neptune. There is no danger of collision between the two planets, however; because of its high inclination, Pluto's orbit clears that of Neptune by 240,000,000 mi. Moreover, according to calculations by Cohen and Hubbard, dynamical effects between Pluto and Neptune, due to their gravitational interactions, will never allow them to approach closer to each other than 18 AU's. Pluto completes its orbital revolution in a period of 248.4 years, poking along at a mean velocity of only 2.9 mi/sec.

(c) Nature of Pluto

The diameter of Pluto was measured by Kuiper at the 200 inch telescope; he found the angular diameter to be about 1/4″, which at Pluto's distance corresponds to a linear diameter of about 6000 km (3700 mi). An opportunity for a check was provided in the Spring of 1965, when Pluto was expected to occult a star. However, careful observations of the star from several observatories in the United States revealed no trace of the star's light being dimmed. Excellent measures of the positions of the star and Pluto were made at Flagstaff. They showed that the center of Pluto must have passed only 1/8″ from the star, as seen from the McDonald Observatory in Texas, but actual observations made at McDonald indicated that Pluto did *not* cover the star at any time. Thus the radius of Pluto must be smaller than 1/8″, and its corresponding linear diameter after allowances for all suspected errors are made, cannot be more than 6400 km — about the size found by Kuiper. A lower limit to the size of the planet is found by assuming that it is perfectly reflecting — that is, has an albedo of 1.0. Even then, it would have to be at least 2000 km in diameter to account for its observed brightness. Thus Pluto's diameter lies between 2000 and 6400 km, and is probably closer to the larger value, because its albedo is believed to be small.

A recent determination of the mass of Pluto, based on its perturbations on Neptune, gives 0.18 times that of the earth. This value, rather uncertain, and the upper limit to Pluto's diameter corresponds to a mean density for the planet of 1.4 times the earth.

Kuiper and Harris reported that they thought Pluto might be variable in its brightness. Walker and Hardie, therefore, carefully observed the planet with a photoelectric photometer attached to a telescope and found that it does, in fact, vary in light with a period of 6.3867 days, with an uncertainty of about 0.0003 day. They interpret the variation as a rotation of Pluto, alternately turning to our view hemispheres of greater and lesser albedo. Evidently, the surface is not uniform.

Pluto is several thousand times too faint to see with the unaided eye. Even under the best conditions a telescope of aperture 6 in. or more is required to see it, and even then it appears only as a star. No gases are revealed in its spectrum, but at its expected temperature of only about 40°K (nearly 400°F below zero) all common gases except neon, hydrogen, and helium would be frozen, and of these all but neon would probably escape the planet, and neon is not a gas that one would expect to find there in appreciable abundance. In any case, none of the gases that would remain unfrozen would be easy to observe spectrographically. Probably Pluto has little or no atmosphere.

(d) The Problem of Pluto's Discovery

Pluto's discovery was based on alleged residuals in the positions of Uranus after account was taken of the perturbations that could have been produced on Uranus' motion by Neptune and other known planets. These residuals, only about 4″ to 5″, were in observations of the positions of Uranus made in the early eighteenth century before it had been recognized as a planet. They were, in fact, smaller than the errors expected in observations made that long ago, and have now been generally discredited by the experts as having any significance. In view of this, one may well wonder whether there can have been any real validity in the calculations of the predicted orbit of Pluto.

Even if the residuals in the positions of Uranus *were* real, to produce them with perturbations by Pluto, Pluto would need a mass substantially greater than that of the earth (note that Lowell predicted a mass of 6.6 earth masses). Recent determinations of Pluto's mass and size, however, show that its mass is too small to have produced perturbations in the amount required, and that the prediction of Pluto's position, although mathematically correct, was based on invalid data; Pluto's discovery was accidental!

The fact that Pluto *was* discovered, with so nearly the orbital elements that Lowell predicted, is one of the startling coincidences in the history of astronomy. Be that as it may, it was Lowell's faith and enthusiasm that led to our knowledge of Pluto, and he has justly earned the honor for its discovery.

(e) The Solar System from Pluto

From Pluto, the solar system must appear a bleak and empty place. The sun would appear as a bright star, although it would still provide Pluto with 250 times as much light as we on earth receive from our full moon. The earth as seen from Pluto would scarcely ever be at more than 1° angular separation from the sun and would be slightly too faint to be seen with the unaided eye. Even Jupiter could only be 7° from the sun and would appear only as a medium-bright star.

11.9 ARE THERE UNKNOWN PLANETS

The possibility of undiscovered planets exists. However, a careful search by Tombaugh after his discovery of Pluto failed to reveal any other trans-Neptunian planet. He should have picked up any object as large as Neptune within a distance of 270 AU.

In the spring of 1972, Joseph Brady, at Lawrence Livermore Laboratory, suggested that an undiscovered planet (dubbed Planet X) might be able to produce enough perturbations on Halley's comet to account for some of the irregularities in its recent times of perihelion passage. The planet would have to have a mass about equal to that of Jupiter, and an orbit of 60 AU radius inclined at about 60° to the ecliptic plane. Most experts, however, regard it as highly unlikely that so major a planet can have so inclined an orbit, and suspect that the behavior of Halley's comet can be explained in other ways.

11.10 LIFE ON OTHER WORLDS

Among the worlds in our own solar system, only the planets Venus, Earth, and Mars possess conditions under which we can even remotely imagine forms of living organisms to exist. Of these three, only earth offers conditions compatible with the existence of higher forms of plant or animal life as we know them. We cannot rule out the possibility of life processes unknown to us, but neither have we evidence for their existence. The best we can say is that we have no knowledge of any life on other planets of the solar system.

It is often pointed out that the sun is a typical star, and that there must be many other suns with planetary systems similar to our own. To be sure, among the countless billions of stars known to exist in the universe, we have no reason to expect that there are not many with planets, and that some of these planets might have conditions approximating those on earth. On the other hand, we do not know how life began on our own planet and have no objective means of assessing the probability of life beginning elsewhere, much less of evaluating the likelihood of the critical steps of evolution which may lead to intelligent beings. Life may abound in the universe, or it may be unique to the earth; on this topic we can at this time do no more than speculate.

Exercises

1. Give several reasons why Mercury would be a particularly unpleasant place to live.

2. Explain with a diagram why Mercury is above the horizon after sunset when it is at greatest eastern elongation.

3. From the data given in the chapter, calculate what your weight would be on each of the planets. For Pluto, you must make some assumptions; what are they?

4. Venus requires only 144 days to move from eastern elongation to western elongation, but 440 days to move from western to eastern elongation. Use a diagram as a help, and explain why.

5. From what planet would an astronomer have the best opportunity to observe Venus?

6. At its nearest, Venus comes within about 25 million miles of earth. How far away is it at its farthest?

7. What problems would be encountered by settlers trying to live on Venus?

8. When might it be possible to observe both Mercury and Venus, at the same time, when they are at inferior conjunction? Why would this be an extraordinarily rare event?

9. Use the data in Appendix 11, and show that the satellites of Mars obey Kepler's third law.

10. Which satellite would have the greatest period, one a million miles from the center of Mars or one a million miles from the center of Saturn (assume both have circular orbits)? Why?

11. If you were a scientist working on the planning of a space probe designed to land on Mars, what are some experiments you would like the probe to perform?

12. The satellite *Io* of Jupiter is 262,000 mi from the center of the planet, and has a nearly perfectly circular orbit and a period of $1^d18^h28^m$. It requires 3.6 min from the time it starts to enter Jupiter's shadow until it is completely eclipsed. What is its diameter?

　　Answer: About 2,300 mi.

13. If the orbit of Saturn were to lie in the plane of the ecliptic, how often would the planet's rings be turned edge on to the earth?

14. If Pluto had a mean density of 5 gm/cm³, what would be the largest value its mass could have, in terms of the earth's mass?

15. Compare and contrast the discoveries of Uranus, Neptune, Pluto, and "Planet X."

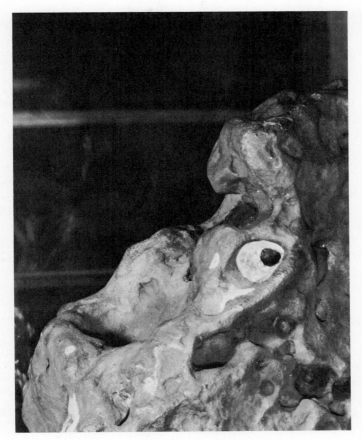

CHAPTER 12

Minor Members
of the Solar System

The planets are the most important objects revolving about the sun, but they are only 9 in number. We turn now to the countless billions of minor members of the solar system.

12.1 THE MINOR PLANETS

Of the many thousands of minor planets, the vast majority appear telescopically like stars — that is, as small points of light. The term *asteroid* (meaning "starlike") is therefore often applied to these tiny worlds. Another synonym is *planetoid*. The term *minor planet,* preferred by many, is used in this text.

The orbits of most of the minor planets, including that of Ceres, the largest and first to be discovered, lie between the orbits of Mars and Jupiter. From the time of Kepler it was recognized that this region of the solar system constituted a considerable gap in the spacing of the planetary orbits; consequently, when Ceres was found there in 1801 it occasioned no great surprise. Moreover, the median distance of Ceres from the sun fitted well into Bode's law, a sequence of numbers that gives the approximate distances of the planets from the sun in astronomical units.

(a) "Bode's Law"

The term "Bode's law" is a complete misnomer, for it was not discovered by Bode and is not a law. A law in science is a statement of an invariable order or relation of phenomena. Bode's law can be described more accurately as a scheme for remembering the distances of the planets from the sun. It was discovered by Titius of Wittenberg in 1766,

and was brought into prominence by Bode, director of the Berlin Observatory, in 1772.

Titius' progression can be obtained easily by writing down the numbers: 0, 3, 6, 12, . . . , each succeeding number in the sequence (after the first two) being obtained by doubling the preceding one. If now 4 is added to each of the numbers, and in each case the sum is divided by 10, the numbers obtained give the approximate distances from the sun (in astronomical units) of the planets known in 1766. The progression is illustrated in Table 12.1.

TABLE 12.1 Bode's Law

TITIUS' PROGRESSION	PLANET	PLANET'S ACTUAL DISTANCE (AU)
(0 + 4)/10 = 0.4	Mercury	0.387
(3 + 4)/10 = 0.7	Venus	0.723
(6 + 4)/10 = 1.0	Earth	1.000
(12 + 4)/10 = 1.6	Mars	1.524
(24 + 4)/10 = 2.8		
(48 + 4)/10 = 5.2	Jupiter	5.203
(96 + 4)/10 = 10.0	Saturn	9.539
(192 + 4)/10 = 19.6	Uranus	19.18
(384 + 4)/10 = 38.8	Neptune	30.6
(768 + 4)/10 = 77.2	Pluto	39.4

When Uranus was discovered in 1781, it was found to fit fairly well into Bode's law. There is no major planet at a distance corresponding to the 2.8 position, and a search for the "missing planet" was actually organized. The discovery of the minor planets was independent of this search, but they fell nicely into the gap at 2.8 AU; the orbit of Ceres has a semimajor axis of 2.767 AU. Bode's law, however, breaks down entirely for Neptune and Pluto.

(b) Discovery of the Minor Planets

The Sicilian astronomer Giuseppe Piazzi (1746–1826) was engaged in mapping a region of the sky in Taurus, when on January 1, 1801 (the first night of the nineteenth century) he observed an uncharted "star." During the next two nights, he noted that the new object had shifted its position slightly. At first, he suspected that it was a comet, but on January 14 it changed its direction of motion from westward to eastward, characteristic of a superior planet completing its retrograde motion when near opposition (Section 2.3d). Piazzi continued to observe the object until February 11, after which he was forced to interrupt his work because of illness.

In mid-January Piazzi wrote of his discovery to Bode, at the Berlin Observatory. Unfortunately, his letter did not reach Bode until March 20, when the object was too nearly in the direction of the sun to be observed. Bode suspected that Piazzi's object might be the "missing planet," but since it had only been observed for the brief span of 6 weeks, astronomers were not able to calculate an orbit for it.

The discovery of the new planet was rescued by the brilliant young German mathematician, Karl Friedrich Gauss (1777–1855), who had devised new mathematical techniques that could be applied to orbit calculations. By November, Gauss had successfully calculated the orbit of the object from Piazzi's short series of observations. The object was found very near Gauss' predicted position in the constellation of Virgo by von Zach, on New Year's Eve, the last night of the year of its first discovery. At Piazzi's request, the new planet was named for *Ceres,* the protecting goddess of Sicily.

Ceres was widely assumed to be the missing planet predicted by "Bode's law." It came as a complete surprise, therefore, when in March 1802 Heinrich Olbers discovered a second moving starlike object — the minor planet to be named *Pallas.*

It was a natural speculation that if there were room for *two* minor planets, there could be room for others as well, and a search for such objects began in earnest. The discovery of *Juno* followed in 1804 and of *Vesta* in 1807. It was 1845 before Karl Hencke discovered the fifth minor planet after 15 years of search. Subsequently, new ones were found with increasing frequency, until by 1890 more than 300 were known.

In 1891 Max Wolf, of Heidelberg, introduced the technique of astronomical photography as a means of searching for minor planets. The angular motion of a minor planet is large enough (especially if it is near opposition) so that during a long time exposure its image will form a trail on the emulsion. The object appears on the photograph, therefore, as a short dash rather than a starlike point image. *Brucia,* the three-hundred twenty-third minor planet to be discovered, was the first to be found photographically.

FIG. 12-1   Time exposure showing trails left by two minor planets (marked by arrows). *(Yerkes Observatory.)*

Today, discoveries of minor planets are usually accidental; they most often occur when the tiny objects leave their trails on photographs that are taken for other purposes. The total number of minor planets bright enough to leave trails on telescopic photographs is estimated, from sample surveys, to be from 50,000 to 100,000.

(c) Orbits of the Minor Planets

A standard reference in which the orbital elements and ephemerides of minor planets are published is the Soviet *Minor Planet Ephemeris*. The 1969 edition gives data for 1735 minor planets.

The minor planets all revolve about the sun in the same direction as the principal planets (from west to east), and most of them have orbits that lie nearly in the plane of the earth's orbit; the mean inclination of their orbits to the plane of the ecliptic is 9°5. About two dozen, however, have inclinations greater than 25°; the orbit of Betulia is the most inclined (52°) to the ecliptic.

The orbits of most of the minor planets have semimajor axes that lie in the range 2.3 to 3.3 AU. The sidereal periods of most of them are in the range 3.5 to 6 years. Icarus has the smallest orbit — with a semimajor axis equal to 1.0777 AU. It also has the orbit of highest eccentricity among the known minor planets (0.83), and its perihelion distance from the sun is only 17 million mi. It is the only known object in the solar system other than comets and meteorites that passes within the orbit of Mercury. Hildago is the minor planet with the largest known orbit (its semimajor axis = 5.79 AU), and second largest eccentricity (e = 0.66). At aphelion it is farther from the sun than the orbit of Jupiter. The mean value of the eccentricities of the minor planet orbits is 0.15, not much greater than the average for the orbits of the planets.

Three minor planets are known to have passed within less than 3 million mi of the earth. They are Apollo (1932), Adonis (1936), and Hermes (1937). None of the three, however, was observed long enough to allow accumulation of sufficient data to compute its orbit, and each has been lost. The two minor planets that most closely approach the earth and that are "in captivity," that is, whose orbits are known and whose returns can be predicted, are Icarus, which on June 14, 1968, passed within about 4 million mi of earth, and Geographos, discovered on the National Geographic Society–Palomar Observatory Sky Survey and accordingly named for "the geographer." In 1969 Geographos passed within about 6 million mi of earth, but in the most favorable approaches it can pass as close as Icarus did in 1968.

(d) Physical Nature of the Minor Planets

On those occasions when the minor planet Vesta is at opposition while it is also at perihelion, it is faintly visible to the unaided eye. With this single exception, the minor planets are visible only telescopically. The four brightest, and presumably largest, show measurable disks. Even these, however, can be measured only with considerable difficulty; the diameters that have been found for them are therefore only approximate. Those values are given in Table 12.2, but the diameters are not as precisely known as the figures in the table would indicate.

TABLE 12.2 Sizes of Minor Planets

OBJECT	DIAMETER	
Ceres	785 km	488 mi
Pallas	489	304
Vesta	399	248
Juno	190	118

The albedo, or light-reflecting power, can be calculated for each of the minor planets of known diameter, because once we know its size, we can calculate how much sunlight falls upon it. From measurements of its brightness, we can then deduce how much light it reflects into space. By comparing the two quantities, we find the reflecting power. The albedos of the four minor planets of known size are found to average about 0.1; that is, they reflect about 10 percent of the light incident upon them. If we assume that the other minor planets have similar albedos, we can calculate the sizes they must have to account for their observed brightness.

It is thus found that there are probably fewer than a dozen minor planets with diameters greater than 100 mi. There may be a few hundred that are more than 25 mi across. Most observable minor planets must have diameters not much greater than 1 mi or so.

If it is assumed that the minor planets have mean densities comparable to that of the crust of the earth, or to that of the moon, their masses can be estimated from their volumes. The mass of Ceres, the largest, is probably only about $\frac{1}{8000}$ that of the earth. The velocity of escape from its surface must be only about $\frac{1}{3}$ mi/sec. Even Ceres, therefore, could not retain an atmosphere. From the smaller minor planets, a good pitcher could easily pitch a baseball into space. These small ones cannot even have enough gravity to pull themselves into a spherical shape. The combined mass of all minor planets is not known; a guess is about $\frac{1}{20}$ that of the moon.

Many minor planets have elongated shapes or irregularities such that in different orientations they reflect light in different amounts. We can sometimes detect rotation periods of these objects from the periodic variations in brightness they display.

(e) Naming Minor Planets

By modern custom, after a newly found minor planet has had its orbit calculated, and has been observed again after having completed at least one circuit of the sun since its first discovery, it is given a name and a number. The number is a running index that indicates the order of discovery among the minor planets. The discoverer is customarily given the honor of supplying the name. The full designation of the minor planet contains both number and name, with the number encircled and preceding the name, thus: ① Ceres, ② Pallas, ④③③ Eros, ①⑤⑥⑥ Icarus, and so on.

Originally, the names were chosen from those of gods in Greek and Roman mythology. These, however, were soon used up. After exhausting the names of heroines from Wagnerian operas, discoverers chose names of wives, friends, flowers, cities, colleges, pets, and even favorite desserts. One is named for a computer (NORC). The thousandth minor planet to be discovered was named Piazzia, and number 1001 is Gaussia. Other minor planets bear the names Washingtonia, Hooveria, and Rockefellia.

Today, feminine names are generally given to "run of the mill" minor planets and masculine names to those of unusual interest. For example, Geographos has a masculine name because it can approach so near to the earth.

(f) Origin of the Minor Planets

Most authorities regard it as probable that the minor planets were formed from the same material that formed the principal planets, and at about the same time. It was once speculated that the minor planets, and perhaps the meteorites as well, may have originated from the breakup of a larger body, presumably a planet that once revolved about the sun at a distance of 2.8 AU. It must be remembered, however, that the combined mass of all the minor planets may well be less than a thousandth of the mass of the earth, itself a relatively small planet.

On the other hand, there is evidence that many of the minor planets may have originated from the breakup of several somewhat larger bodies. Studies by the Japanese astronomer K. Hirayama in 1917, and by D. Brouwer in 1950 have revealed several "families" or groups of minor planets of similar orbital characteristics. It is hypothesized that each family may have resulted from a breakup of a larger body, or by the collision of two bodies. Slight differences in the initial velocities given the fragments of a collision can result in the relatively small differences now observed among the orbits of the different minor planets in a given family. Collisions among a relatively few larger objects would be extremely rare. If they once occurred, however, subsequent collisions would be more likely among the larger number of smaller bodies that resulted. If this collision and fragmentation hypothesis for the origin of the minor planets should be shown to be correct, it would be inferred that the fragmentation process is still going on.

12.2 COMETS

Comets have been observed from the earliest times; accounts of spectacular comets are found in the histories of virtually all ancient civilizations. Yet, until comparatively recently, comets were not generally regarded as celestial objects.

A typical comet that is bright enough to be conspicuous to the unaided eye has the appearance of a rather faint, diffuse spot of light, somewhat smaller than the full moon and many times less brilliant. There may be a very faint nebulous tail, extending for a length of several degrees away from the main body of the comet. Like the moon and planets, comets slowly shift their positions in the sky from night to night, remaining visible for periods that range from a few days to a few months. Unlike the planets, however, most comets appear at unpredictable times. In medieval Europe, comets were usually regarded as poisonous vapors in the earth's atmosphere and as bad omens. More fear and superstition have been attached to comets than to any other astronomical objects.

(a) Early Investigations

When Newton applied his law of gravitation to the motions of the planets, he wondered whether comets might similarly be gravitationally accelerated by the sun. If so, their orbits should be conic sections. If comets, like planets, had nearly circular orbits, they should be visible at regular and frequent intervals. On the other hand, if a comet moved in an elongated elliptical orbit of large size, it would necessarily be visible during only that relatively brief period of time during which it passed near the sun (perihelion); over most of its orbit the comet would be so far from the sun as to be invisible. Furthermore, the periods of comets, moving in such orbits, would be very great. A comet that had been seen for a short time, and then was seen again many tens or hundreds of years later when it next appeared near perihelion, would naturally be mistaken for a new object. One end of an ellipse, of very great length and of eccentricity near unity, is nearly indistinguishable from a parabola; thus the motion of a comet moving on a parabolic orbit would closely resemble that of a comet whose orbit is a very long ellipse. Newton concluded that comets are gravitationally attracted about the sun in orbits that are either very long ellipses or parabolas.

Edmond Halley greatly extended Newton's studies of the motions of comets. In 1705 he published calculations relating to 24 cometary orbits. In particular, he noted that the elements of the orbits of the bright comets of 1531, 1607, and 1682 were so similar that the three could well be the same comet, returning to perihelion at intervals of about 75 or 76 years. If so, he predicted that the object should return about 1758. The comet was sighted by an amateur astronomer, George Palitzsch, on Christmas night, 1758. The comet has been named *Halley's comet,* in honor of the man who first recognized it to be a permanent member of the solar system. Subsequent investigation has shown that Halley's comet has been observed and recorded on every passage near the sun at intervals of from 74 to 79 years since 240

FIG. 12-2   Halley's comet. *(Yerkes Observatory.)*

B.C. The period varies somewhat because of perturbations upon its orbit produced by the Jovian planets. It last appeared in 1910, and is due again about 1986.

(b) Discovery and Designation

Observational records exist for about a thousand comets. Today, new comets are discovered at an average rate of about a dozen per year. Some of these are discovered accidentally on astronomical photographs taken for other purposes. Many are discovered by amateur astronomers.

Most of the new comets found each year never become conspicuous, and are visible only on photographs made with large telescopes. Every few years, however, a comet may appear that is bright enough to be seen with the unaided eye. About

two or three times each century, there have appeared spectacular comets that reached naked-eye visibility even in daylight. A recent "daylight comet" was Comet Ikeya-Seki, whch appeared briefly in 1965.

A newly found comet is named for its discoverer or discoverers (there are often more than one). It is also given a temporary designation consisting of the year of its discovery followed by a lowercase letter indicating its order of discovery in that year. For example, Comet 1969a is the first comet discovered in 1969, 1969b is the second, and so on. Later, when the orbits of all recently observed comets have been calculated, each of the comets is given a permanent designation consisting of the year in which it passed nearest the sun, followed by a Roman numeral designating its order among the comets that passed perihelion during that particular year. For example, Comet 1969 I would be the first comet to pass perihelion in 1969; 1969 II, the second, and so on.

(c) Orbits of Comets

The orbit of a comet, like that of a planet, can be determined from three or more fairly well spaced observations of its position in the sky among the stars. Most comet orbits are indistinguishable from parabolas. Some are definitely elliptical; a very few comets have been observed whose orbits appear to be hyperbolas.

If comets were from beyond the solar system, however, their orbits should nearly all be hyperbolic. An elliptical orbit is possible only for a body permanently revolving about the sun, not for one coming from outside the solar system. Even a parabolic orbit would be possible only for a body that, before being attracted toward the sun, was moving through space in almost exactly the same direction and at the same speed as the sun, that is, which was *at rest with respect to the sun.* Such motion would be extremely improbable for an interstellar visitor.

Those few comets that do appear to have moved on hyperbolic orbits have been investigated, and it is found that in every case the gravitational influences of the planets had changed the orbit from a nearly parabolic one to a hyperbola. Although a comet, through planetary perturbations, may thus escape from the solar system, there is no case where a comet is known to have *approached* the sun on a hyperbolic orbit.

It is believed, therefore, that comets are members of the solar system, and that they travel on elliptical orbits, most of which are extremely long and have eccentricity near unity. A comet can be observed from the earth only when it traverses that end of its orbit where it passes close to the sun. If the orbit has an eccentricity greater than about 0.99, that part of it which can be observed is indistinguishable from a parabola. Thus "parabolic" orbits observed for comets are probably only the ends of very long ellipses that carry comets tens of thousands of astronomical units from the sun to the "outposts" of the solar system. Such comets have periods that probably range into millions of years and thus may have been visible only once during man's existence on the earth.

A minority of comets have orbits of low enough eccentricities that their semimajor axes can be determined from the relatively small portions of the orbits that are actually observed. These comets, whose periods can be determined, are called *periodic comets*. Technically, all observed comets could be regarded as periodic, inasmuch as they are all presumed to revolve about the sun in elliptical orbits. The diffuse appearance of a comet, however, makes it difficult to measure its position with great precision; consequently, only those comets whose orbits differ rather substantially from parabolas can be definitely classed as periodic. The comets of longest definitely established periods are Pons-Brooks (71 years), Halley (76 years), and Rigollet (151 years). The comet of the shortest known period is Comet Encke (3.3 years).

Comets are such flimsy objects that they often suffer severe damage as the result of tidal forces produced by the sun when they pass perihelion. It is estimated that most periodic comets are completely disintegrated after a hundred or fewer perihelion passages.

Comet Biela, for example, with a period of 7 years, was discovered in 1772. During its approach to the sun in 1846, it was observed to break into two separate comets. Both comets returned on about the same orbit in 1852, but neither has been seen since. However, spectacular meteor showers (see Section 12.3) were observed on November 27, 1872, and on November 27, 1885, on which occasions the earth passed through the orbit of Comet Biela and encountered swarms of particles traveling in the path of the disintegrated comet. The orbit of the swarm has subsequently been altered by Jupiter; consequently, spectacular displays of Bielid meteors have not been observed in the present century.

Because of their high mortality rates, periodic comets cannot have been periodic for long but must originally have been comets of very long periods having nearly parabolic orbits. Within the recent past (perhaps the last few thousand years) their orbits have been drastically altered to their present relatively small size by perturbations produced by the planets, especially Jupiter. Those occasional comets that are highly spectacular, and hence cannot have suffered appreciable disintegration, almost invariably have nearly parabolic orbits, and also they have not been seen before in recorded history.

(d) Physical Nature of Comets

No two comets are alike, but they all have one characteristic in common — the *coma*, which appears as a small, round, diffuse, nebulous glow. As a comet approaches the sun, the coma usually grows in brightness, but sometimes it becomes so diffuse that it disappears. Often, but not always, there is a small bright *nucleus* in the center of the coma. Together, the coma and the nucleus (if any) comprise the *head* of the comet. Many comets, as they approach the sun, develop tails of luminous material that extend for millions of miles away from the head.

The coma of a typical comet is very large, of the order of a hundred thousand miles across, although individual comets differ enormously from one another in size. The coma of a comet is composed of extremely tenuous material; stars can easily be seen shining through it. Only a few miles of the terrestrial atmosphere substantially dims starlight; yet stars shone in full brillance through more than 40,000 miles of the head of Halley's comet in 1910.

Comets are seldom observed when they are much farther away from the sun than the planets Jupiter or Saturn. At that distance, they are usually very faint, and their spectra resemble that of the sun, a fact which shows them to be composed of solid particles, reflecting sunlight.

When a comet is within about 3 AU of the sun, volatile materials begin to vaporize from the swarm of particles, thus producing the coma. Then the spectrum changes. Superimposed upon the spectrum of reflected sunlight, there are bright emission lines (or series of many closely spaced lines called *bands*) due to the chemical combinations C_2, OH, CN, NH, and NH_2. These chemical *radicals* are probably dissociated from the molecules CH_4 (methane), NH_3 (ammonia), and H_2O (water) and shine by the process of *fluorescence;* that is, they have been set aglow by ultraviolet radiation from the sun.

Some comets, at perihelion, pass within a few hundred thousand miles of the solar surface. Such a comet was Ikeya-Seki, which passed perihelion in October 1965. In such proximity to the sun, comets can become as hot as 4500°K. If a comet approaches very close to the sun, its spectrum changes again. Bright emission lines produced by atoms of various metals appear: sodium, iron, silicon, magnesium, and others.

Most of the solid particles, which comprise a comet's nucleus, must be in the form of microscopic dust grains. It is not known how large the largest may be. The evidence suggests, however, that they are probably not over a few miles in diameter. At least two comets, the great "daylight" comet of 1882 and Halley's comet in 1910, passed directly between the earth and the sun. In each case efforts were made to observe the nucleus silhouetted against the disk of the sun, but no trace whatever of any opaque body was visible. The largest particles in the nucleus must have been less than 50 mi across.

Many comets develop tails as they approach the sun. Some, however, especially certain short-period comets, have either lost the ability to form a tail or never possessed it. In the sixteenth century, the astronomers, Fracastor and Apian described many comets, and remarked that their tails generally point *away* from the sun. This observation was later confirmed by Kepler. Newton attempted to account for the tails of comets as being produced by a repulsive force of sunlight driving particles away from the comets' heads, an idea close to our modern view.

The repulsive forces from the sun that produce comet tails are of two kinds: the pressure of radiation pushing on small dust particles and some atoms and molecules, and the force of corpuscular radiation comprising the *solar wind* — a continuous rain of atomic nuclei ejected from the sun into space. The solar wind, acting on charged atoms (ions) in the comet, is the greater force, and produces the straightest comet tails.

Tails generally grow in size as comets near the sun; some comet tails have reached lengths of more than 100 million miles. As the tail material recedes farther from the sun, its *angular velocity* about the

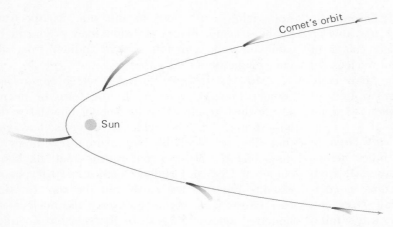

FIG. 12-3 Shape of a typical comet tail as the comet passes perihelion.

sun decreases, and the material lags behind the comet. In general, therefore, comet tails lie in the plane of the comet's orbit, point more or less away from the sun, but curve somewhat backward, away from the direction of the comet's motion (see Figure 12.3). Many comets have two or more tails showing different curvatures; presumably these tails consist of different kinds of particles, accelerated in different amounts by the repulsive forces on them.

Many experts estimate that a typical comet may have only a trillionth (10^{-12}) the mass of the earth. Thus, despite the large sizes of comets, they are mostly "empty space." A comet has been described as the nearest thing to nothing that anything can be and still be something.

(e) Source of Supply of Comets
The most popular hypothesis for the origin of comets is probably that of the Dutch astronomer, Jan H. Oort. He suggests that there might be a vast "cloud" of comets revolving about the sun at distances of from 50,000 to 150,000 AU. Their orbits, ordinarily, cannot bring them close to the sun, because a comet that comes within 2 AU of the sun can survive, at most, only a few million years before suffering disintegration or being lost to the solar system. If the comets we now observe

had come so close as this to the sun since the formation of the solar system, they would long ago have been destroyed by solar tidal forces, or, alternatively, would have been deflected by planetary perturbations away from the solar system.

On the other hand, the outer fringe of the cloud of comets cannot extend beyond about 60 percent of the distance of the nearest star (Alpha Centauri) because beyond that point their orbits would be unstable. Even so, perturbations produced by those stars that pass, every few million years, within a few light-years of the sun must alter some of the primeval cometary orbits. Some comets might be lost to the solar system as the result of such stellar perturbations, but others, retarded in their motion by the star's attraction, would fall in toward the sun. Thus, Oort accounts for the steady influx of comets into the central part of the solar system.

Those comets that are so deflected toward the sun move subsequently on orbits of very high eccentricity, until, after one or a few perihelion passages, further perturbations by the planets either convert them into "periodic" comets, or deflect them into interstellar space on hyperbolic orbits. In the former case, they are eventually decomposed by tidal forces; in the latter, they are permanently lost to the solar system.

FIG. 12-4   Comet Seki-Lines photographed from Frazier Mountain, California, August 9, 1962. *(Alan McClure, Los Angeles.)*

Although the supply of comets in the "cloud" must be gradually diminishing, the remaining number of comets may still be very large. Oort estimates that there are at least 100 billion, and that their combined mass is between $\frac{1}{100}$ and $\frac{1}{10}$ that of the earth.

12.3 METEOROIDS, METEORITES, AND METEORS

Although the layman often confuses comets and meteors, these two phenomena could hardly be more different. Comets, when seen from the earth, are swarms of particles and glowing gases that revolve about the sun. They can be seen when they are many millions of miles away from the earth, and may be visible in the sky for weeks, or even months, slowly shifting their positions from day to day. They rise and set with the stars, and during a single night appear motionless to the casual glance. Meteors, on the other hand, appear only when small solid particles enter the earth's atmosphere from interplanetary space. Since they are moving at speeds of many miles per second, they vaporize as a result of the high friction they encounter with the air. The spot of light caused by the luminous vapors formed in such an encounter appears to move rapidly across the sky, and it fades out within a few seconds. Meteors are commonly called "shooting stars."

On rare occasions, an exceptionally large particle may survive its flight through the earth's atmosphere and land on the ground. Such an occurrence is called a *meteorite fall,* and if the particle is later recovered, it is known as a fallen *meteorite.* Frequently, the term "meteorite" is reserved for the fallen particle, and the particle, when still in space, is called a *meteor.* In this book, however, we shall use the terminology recommended by the International Astronomical Union in 1961. The particle, when it is in space, is called a *meteoroid;* the luminous phenomenon caused when the particle vaporizes in the earth's atmosphere is a

FIG. 12-5 Photograph of the trail of a bright meteor that happened to cross the field of view of the telescope while photographing the Andromeda galaxy. *(Photograph by F. Klepesta.)*

meteor; and if it survives and lands on the ground, the particle is a *meteorite.*

(a) Phenomenon of a Meteor

On a typical dark moonless night an alert observer can see half a dozen or more meteors per hour. However, to be visible, a meteor must be within 100 to 150 mi of the observer; over the entire earth, the total number of meteors bright enough to be visible must total about 25 million per day. Faint meteors are far more numerous than bright ones; the number of meteors that are potentially visible with binoculars or through telescopes, and that range down to a hundred times fainter than naked-eye visibility, must number near 8 billion per day.

More meteors can generally be seen in the hours after midnight than in the hours before.

Moreover, there are certain times when meteors can be seen with much greater than average frequency — up to 60 or more per hour. These unusual meteor displays are called *showers.* Meteor showers occur when the earth collides with swarms of many meteoroids moving together in space. A distinction is made between shower meteors and nonshower, or *sporadic,* meteors.

Occasionally an exceptionally bright meteor is seen, attracts much attention, and is reported by many observers. These bright meteors are called *fireballs,* or *bolides.* It is estimated that a few thousand fireballs appear every day over the entire earth, most of them over oceans or uninhabited regions. A few fireballs are visible in broad daylight; some have even reached the brilliance of the full moon. Sometimes they break up in midair with explosions that are audible from the ground.

(b) Orbits of Meteoroids

The paths of many fainter, more common meteors have been determined accurately from routine photographic surveys. Recently, meteors have been detected by radio and radar as well as visually and photographically. It is found that meteoroids produce meteors at an average height of about 60 mi (95 km). The highest meteors form at heights of 80 mi. Nearly all meteoroids completely disintegrate and their meteors disappear, by the time they reach altitudes of 50 mi. A few meteoroids doubtless skim out of the atmosphere, returning to space before they are completely burned up.

The vast majority of meteoroids have speeds between 8 and 45 mi/sec. The faster ones tend to be brighter and are more likely to leave trains, because of their greater energy upon entry into the earth's atmosphere. On the other hand, the faster-moving particles "burn up" at higher altitudes and are less likely to survive their flights and reach the ground.

The orbits of those meteoroids producing fireballs and the brighter meteors are found to lie close to the plane of the ecliptic and to have small eccentricities, and the revolution of the particles is *direct* — that is, from west to east, as is the revolution of the planets. These meteoroids are thought to have orbits similar to those of the minor planets.

For orbits of the vast majority of typical meteoroids, on the other hand, there seems to be no preference for the ecliptic plane. Like the comets, they approach the earth from all directions. Also, like the comets, most of them are found to be moving at speeds very near the velocity of escape from the solar system; that is, they travel on near-parabolic heliocentric orbits.

Meteoroids are believed to be permanent members of the solar system. The argument is exactly the same as that for the solar-system membership of comets. If meteoroids were chance intruders from interstellar space, they would have hyperbolic orbits. Only a very few meteoroid orbits have been observed to be slightly hyperbolic, and these orbits may have been altered from parabolic ones as a result of perturbations produced by other planets. Like the orbits of comets, the apparent parabolic orbits that we observe for meteoroids must actually be the ends of very long ellipses. At most, less than 1 percent of all observed meteoroids can be interstellar visitors. Most meteoroids are believed to be the debris of disintegrated comets. Some of those producing the brighter meteors, however, may have had an origin in common with the minor planets, which have similar orbits.

FIG. 12-6   A meteor shower radiant (a) and the apparent divergence or convergence of parallel lines (b).

(a)

(b)

(c) Meteor Showers

Several times each year meteors can be seen with much greater than average frequency. These *meteor showers* occur when the earth encounters swarms of particles moving together through space. Many such swarms of particles that the earth intercepts at regular intervals are known. Showers produced by them are predictable. On rare occasions the earth unexpectedly encounters swarms of particles that produce spectacular meteor displays.

Unlike sporadic meteors, which seem to come from any direction, meteors belonging to a shower all seem to radiate or diverge away from a single point on the celestial sphere; that point is called the *radiant* of the shower. Recurrent showers are named for the constellation within which the radiant lies or for a bright star near the radiant.

The phenomenon that shower meteors seem to diverge away from a common point is easily explained. The meteoroids producing a meteor shower are members of a swarm; they are all traveling together in closely spaced parallel orbits about the sun. When the earth passes through such a swarm, it is struck by many meteoroids, all approaching it from the same direction. As we, on the ground, look toward the direction from which the particles are coming, they all seem to diverge from it. Similarly, if we look along railway tracks, those tracks, although parallel to each other, seem to diverge away from a point in the distance (see Figure 12-6).

On about August 11 of each year, the earth passes through a swarm of particles that approach from the direction of Perseus. In 1866 it was observed that the particles producing this *Perseid shower* travel in an orbit that is almost identical to that of Comet 1862 III. It was then realized that those meteoroids encountered each August are debris from the comet that has spread out along the comet's orbit.

Subsequently, it has been found that the elements of the orbits of many other meteoroid swarms are similar to those of the orbits of known comets. Not all meteor showers have yet been identified with individual comets, but it is presumed that all showers have had a cometary origin. These swarms of debris, provided by the gradual disintegration of comets, give further evidence of the flimsy nature of comets.

Sometimes, if a meteoroid swarm is old enough, its particles are strewn more or less uniformly along their entire orbit [Figure 12-7(a)]; an example is the swarm of particles that produces the Perseid shower. Consequently, the earth meets about the same number of particles each time it passes through the orbit of the swarm, and the Perseid shower is of about equal intensity each year. During the shower, if the night is clear and moonless, up to 60 Perseid meteors can be counted per hour. More often, a swarm will be bunched up [Figure 12-7(b)], and we will experience a spectacular display only on those occasions when the earth passes through the orbit of the swarm at just the right time to complete a rendezvous with the meteoroids as they reach that same point in their revolution about the sun. A good example is provided by the Leonid meteors in the last century, when the earth met the debris of Comet 1866 I in 1833 and again in 1866 (after an interval of 33 years — the period of the comet). Those were among the most spectacular showers ever recorded. As many as 200,000 meteors could be seen from one place within a span of a few hours. The last good Leonid shower was on November 17, 1966, where in some southwestern states, up to 140 meteors could be observed per second. Even in such dense swarms the individual particles of the swarm are separated by distances of 20 mi or more; in most meteoroid swarms, the particles are more than 100 mi apart.

The characteristics of some of the more famous meteor showers are summarized in Table 12.3. Other spectacular meteor showers can occur, however, at almost any time, just as some bright

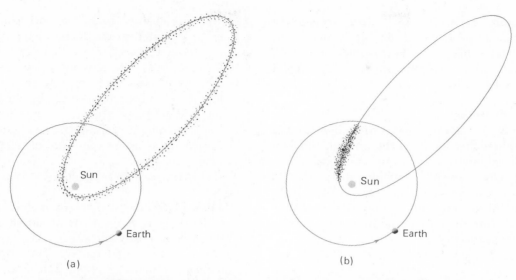

FIG. 12-7   Meteoroids in a swarm may be strewn more or less uniformly along their orbit (a), or bunched up (b).

comets appear unexpectedly. There are also several daytime showers, observable only with radio techniques.

(d) Masses of Meteoroids

The masses of meteoroids are estimated from the amount of energy (in the form of light, for example) emitted in their meteors. It is found that the mass of a meteoroid producing a typical bright meteor is about ¼ gm (about $\frac{1}{100}$ ounce). A few meteoroids, however, have masses greater than a few grams. Many visible meteors are produced by particles with masses of only a few milligrams. Most meteoroids, therefore, are much smaller than pebbles.

TABLE 12.3 Characteristics of Some Meteor Showers

SHOWER	DATE OF MAXIMUM DISPLAY	VELOCITY (MI/SEC)	ASSOCIATED COMET	PERIOD OF COMET (YR)
Quadrantid	Jan 3	27	—	7.0
Lyrid	Apr 21	30	1861 I	415
Eta Aquarid	May 4	37	Halley	76
Delta Aquarid	Jul 30	27	—	3.6
Perseid	Aug 11	38	1862 III	105
Draconid	Oct 9	15	Giacobini-Zinner	6.6
Orionid	Oct 20	41	Halley	76
Taurid	Oct 31	19	Encke	3.3
Andromedid	Nov 14	10	Biela	6.6
Leonid	Nov 16	45	1866 I	33
Geminid	Dec 13	23	—	1.6

The rate at which the earth accumulates meteoritic material is not accurately known. However, the total mass of meteor-producing meteoroids that enters the atmosphere each day is probably a few tons.

(e) Fallen Meteorites

Occasionally, a meteoroid survives its flight through the atmosphere and lands on the ground; this happens with extreme rarity in any one locality, but over the entire earth probably more than a thousand meteorites fall each year. Fallen meteorites have been found for centuries; it was 1800, however, before their association with "shooting stars" was appreciated.

Today, fallen meteorites are found in two ways:

(1) Sometimes bright fireballs are observed to penetrate the atmosphere to very low altitudes. A search of the area beneath the point where the fireball was observed to burn out may reveal one or more remnants of the meteoroid. *Observed falls,* in other words, may lead to discoveries of fallen meteorites.

FIG. 12-8   Full-scale model of the Goose Lake Meteorite. The original weighs 2573 lb. The white card in the photograph is 9 inches long. *(Griffith Observatory.)*

(2) Unusual-looking "rocks" are occasionally discovered that turn out to be meteoritic. These are termed "finds." Now that the public has become "meteorite conscious," many suspected meteorites are sent to experts each year. The late F. C. Leonard, a specialist in the field, referred to these objects as "meteorites" and "meteorwrongs." Genuine meteorites are turned up at an average rate of about 25 per year.

Meteorites can be classified in three general groups†:

(1) *Irons (siderites)* — alloys of metals. From 85 to 95 percent of the mass of siderites is iron; the rest is mostly nickel.

(2) *Stony irons (siderolites)* — relatively rare meteorites that are about half iron and half "stony" silicates. The stony materials generally decompose more rapidly than the iron, leaving the latter in separated fragments.

(3) *Stones (aerolites)* — composed mostly of silicates and other stony materials. Aerolites contain about 10 to 15 percent iron and nickel in metallic flakes.

Stones are probably the most common kind of fallen meteorites. Ninety-three percent of those meteorites obtained from observed falls are stones. However, the aerolites gradually break up under erosive action. Moreover, they resemble terrestrial rocks, and generally only experts can ascertain their meteoritic nature from the iron pieces embedded in them. Consequently, about two thirds

---

†*Tektites,* somewhat rounded glassy bodies of indefinite shape, have been found in Indonesia, Australia, and elsewhere; some investigators suspect them to be of extraterrestrial origin, but their identity as true meteorites has never been established. They show evidence of having been molten at high pressure, and having undergone rapid cooling. They contain small spheres of iron and nickel, and schreibersite, a mineral known only in meteorites. The latter is believed to be produced by melting of the surface layer of crater-producing meteorites.

of the total number of recovered meteorites are irons. (Irons are seldom found, however, in those civilized regions of the world where for centuries man has made use of iron for tools.) When irons are cut, polished, and etched, about 80 percent show a characteristic crystalline structure (*Widmanstätten figures*) that makes their identification as meteorites certain.

So far as is known, all recovered meteorites are remnants of *sporadic* meteors. There has never been an observed fall from a *shower* meteor. The spectra of shower meteors, however, indicate that they contain some of the same elements as stones.

The chemical composition of many meteorites has been determined by laboratory analysis. It is found that meteorites contain the same common chemical elements that we find in the crust of the earth. When allowance is made for the fact that stones are at least 4 to 9 times more common than irons, the following abundances (by weight) of elements are derived: iron, 30 to 40 percent; oxygen, 30 percent; silicon, 15 percent; magnesium, 12 percent; sulfur and nickel, 2 percent each; calcium, 1 percent; and small amounts of carbon, sodium, aluminum, lead, chlorine, potassium, titanium, chromium, manganese, cobalt, copper, and other elements.

A skin-deep layer on the surface of an iron meteorite may show evidence that the surface has been molten. Meteoroids pass through the air too quickly, however, for their interiors to heat up. By the time a meteoroid is slowed by air friction to 2 mi/sec, the molten surface of the iron hardens; when its speed has slowed to 1 mi/sec, it stops glowing altogether. Popular opinion to the contrary, meteorites are not "red hot" when they strike the ground; meteorites that have been picked up just after they have fallen are generally cool enough to handle. The surfaces of the stones may also become molten during their atmospheric passage and turn glassy. Both stones and irons are sometimes ablated to conical shapes.

Studies of the relative concentrations of various elements that undergo natural radioactivity indicate how long the radioactive decay process has been going on, and hence the age of the meteorite. The derived age, of course, applies only to the age of the solid particle in its present mineral state. The age of the oldest meteorites, like that of the oldest moon rocks, is 4.6 billion years — comparable to the age of the earth.

The largest meteorite ever found on the earth is *Hoba West,* near Grootfontein, South West Africa. It has a volume of about 9 cubic yards and an estimated mass of more than 50 tons. The largest meteorite on display in a museum has a mass of 34 tons; it was found by Peary in Greenland in 1897 and is now at the American Museum of Natural History in New York. The largest fallen meteorite found in a single piece in the United States was discovered in a forest near Willamette, Oregon, in 1902; it has a mass of 14 tons. (The discoverer spent about 3 months hauling the meteorite to his own property, where he put it on display for an admission price. Among those interested in the new exhibit were the attorneys of the Oregon Iron and Steel Company, owner of the land on which the meteorite was found. After litigation, it was decided that the company was the rightful owner of the object.)

(f) Meteorite Falls

A total of 1800 meteorite falls are listed in the *Hey Catalogue of Meteorites;* of these, 785 are falls, and 1015 are finds.

The two most spectacular meteorite falls in recent times took place in Siberia. On June 30, 1908, a brilliant fireball occurred that was seen in broad daylight. Remnants of its meteoroid landed in a forest near the Tunguska River in central Siberia. The impact produced shock waves that were registered on seismographs in distant Europe, and also set up an air-compression wave that was reported detected in Europe. Trees were seared of their branches and were felled radially from the impact over an area more than 20 mi in radius. About 1500 reindeer were killed, and a man standing on the porch of his home 50 mi away was knocked unconscious. No pieces of the original particle have been recovered, but its original mass before it entered the earth's atmosphere is estimated to have been about 10^5 tons.

A second spectacular Siberian fall occurred on February 12, 1947, near Vladivostok. The approaching fireball was described as "bright as the sun." The impact produced 106 craters ranging in size up to 30 yards across and 10 yards deep. Trees were felled radially around each of the large craters. The entire region covers nearly 2 square miles. More than 5 tons of iron meteorite fragments have been recovered in the area.

It is estimated that in each century several meteorites strike the earth with enough force to produce craters more than 10 yards across. A large meteorite produces a crater larger than itself, for when it strikes the ground its kinetic energy is dissipated with explosive violence.

The first crater to be discovered, and the most famous definitely known to be meteoritic, is the Barringer Meteorite Crater, near Winslow, Arizona. The crater was shown to be meteoritic as a result of research instigated by the Barringers (owners of the land) early in the twentieth century. The crater is 4200 ft across and 600 ft deep, and its rim rises 150 ft above the level of the surrounding ground. In the area over 30 tons of iron meteorite fragments have been found, some buried near the crater and many more scattered at distances up to 4 miles around. Sizable chunks of iron can still be found shallowly buried in the vicinity of the crater. The age of the crater is not accurately known. Estimates run between 5000 and 75,000 years; it cannot be much older and have survived erosion.

A larger crater is the New Quebec Crater (formerly Chubb Crater) in Quebec. It was discovered in 1950 on aerial photographs of the

FIG. 12-9 The Barringer meteorite crater in Arizona. *(American Meteorite Museum.)*

region. It is similar in appearance to the Barringer Crater but is about twice as large, being more than 2 mi in diameter. It is currently filled with water and forms a lake in solid granite. No trace of meteoritic fragments has been found in the area, and it is not absolutely certain that the crater is meteoritic; however, the absence of volcanic activity in the region suggests that a meteoritic origin of the crater is likely.

Older "fossil" craters must exist. The largest suspected fossil meteorite crater is in South Africa and is about 20 mi across. In size it is reminiscent of the lunar craters. If it is meteoritic, it has lost much of its original appearance. We do not yet know enough about meteorite craters to identify such fossils with certainty.

(g) Micrometeorites

Many very tiny meteorites strike the earth each day. Those that are only a few microns in diameter (1 micron is 10^{-4} cm, or about 4×10^{-5} in.) will be slowed in the air before they have a chance to heat up; they will eventually settle to the ground. These particles, too small to make meteors, are called *micrometeorites*. The total daily accretion of

FIG. 12-10   Region of the lunar crater Goclenius, photographed by the crew of Apollo 8 in December, 1968. *(NASA.)*

micrometeorites by the earth is estimated at 50 to 100 tons.

(h) Organic Compounds in Meteorites

Some fallen meteorites, the most famous found near Alès, France, are found to contain organic compounds. Organic material suggests the possibility of fossils of living organisms. However, the chance that these meteorites have been contaminated by terrestrial organisms has always been hard to rule out. Special interest, therefore, was generated by a meteorite that fell near Murchison, Australia on September 28, 1969, which did have organic compounds that were unquestionably present in the meteoroid before it struck the earth. Critical tests, however, showed that those compounds are definitely not of biological origin. At present, there is no generally-accepted evidence favoring any extra-terrestrial fossils of living organisms.

12.4 THE INTERPLANETARY MATERIAL

We see, then, that the space between the planets contains a vast number of objects, ranging down to the extremely numerous micrometeorites. These particles comprise a distribution of *interplanetary dust*. Micrometeorites are a few microns or more in diameter. Particles less than 1 micron in size are "blown" out of the solar system by radiation pressure from the sun, just as small particles are blown out of comets in the form of comet tails. Particles the size of micrometeorites are not blown away but revolve about the sun like tiny planets, which, indeed, they are.

In addition to micrometeorites, or interplanetary dust, space vehicles have found evidence for interplanetary gas. Large numbers of charged atoms (ions) have been encountered. It is now well established that these ions have been expelled from the sun, and this outflow of gas is known as the *solar wind*. In the neighborhood of the earth,

FIG. 12-11 Zodiacal light behind telescope dome. The time exposure also shows star trails. *(Yerkes Observatory.)*

the ion density is from 1 to 10 particles per cubic centimeter. Further evidence of the solar wind is provided by its effect on comets in producing the straight tails.

There is also additional evidence for the interplanetary dust (micrometeoritic material): the *zodiacal light* and the *gegenschein*.

(a) Zodiacal Light

The zodiacal light is a faint glow of light along the zodiac (or ecliptic). It is brightest along those

Stoney-iron meteorite, Glorieta Mountains, N.M. The specimen has been polished and etched to show the metallic structure. *(From the collection of Ronald Oriti, Griffith Observatory; photograph by Ivan Dryer.)*

Albin, Wyoming stoney meteorite. *(From the collection of Ronald Oriti, Griffith Observatory; photograph by Ivan Dryer.)*

Polished slice of Albin, Wyoming stoney meteorite. This type of meteorite consists of nickel-iron metal with inclusions of the green mineral olivine. *(From the collection of Ronald Oriti, Griffith Observatory; photograph by Ivan Dryer.)*

A thin section of the Lucerne Valley stoney meteorite, polished to a thickness of 0.001 inch. This photograph was made in polarized light transmitted by the section. Note the rounded grains, or *chondrules*, found only in stoney meteorites. *(From the collection of Ronald Oriti, Griffith Observatory; thin section cut by John DeGrosse, UCLA.)*

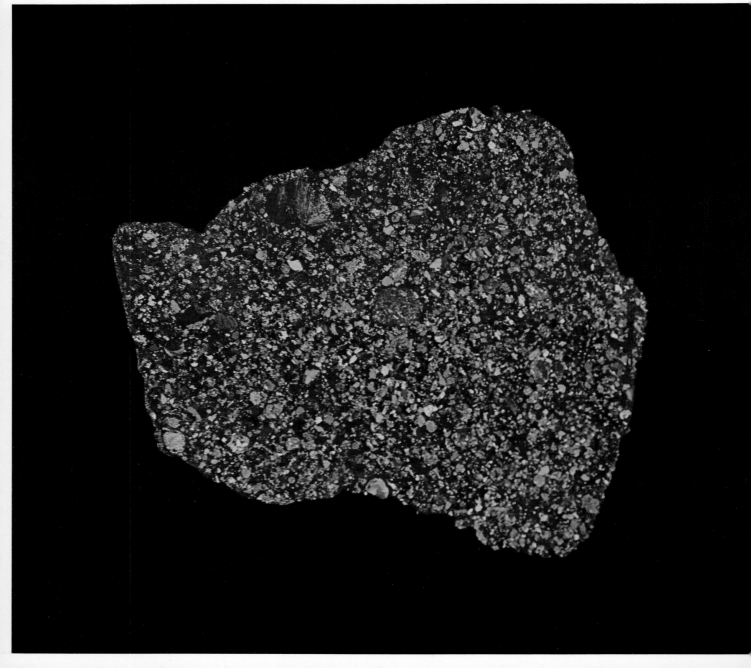

parts of the ecliptic nearest the sun and is best seen in the west in the few hours after sunset or in the east in the hours before sunrise. Under the most favorable circumstances, the zodiacal light rivals the Milky Way in brilliance. It is sometimes called the "false dawn" because of its visibility in the morning hours before twilight actually begins. The zodiacal light has the same spectrum as the sun, which shows it to be reflected sunlight. The present interpretation of the zodiacal light, then, is that the interplanetary dust is concentrated most heavily in the plane of the ecliptic. The dust reflects enough sunlight to produce the faint glow along the zodiac.

(b) The Gegenschein

The *gegenschein,* German for "counterglow," is a faint glow of light, centered on the ecliptic, which is exactly opposite the sun in the sky. Its angular size is from 8° to 10° by 5° to 7°. It is much more difficult to see than the zodiacal light, but it can be measured photoelectrically, and it has also been photographed. Like the zodiacal light, the gegenschein appears to be reflected sunlight.

Several explanations have been advanced for the gegenschein. It is generally thought to be sunlight reflected from solid particles revolving about the sun in orbits that lie in the vicinity of those of the minor planets.

Exercises

1. What would be the semimajor axis of the orbit of a minor planet with a period exactly one-third that of Jupiter? Do you think that such minor planets might exist?

    Answer: About 2.5 AU

2. Two minor planets are observed, both of which have the same distance from the earth and from the sun. One is 49 times as bright as the other. The brighter has a measured diameter of 56 miles. What is the diameter of the fainter, if both have the same albedo?

3. Suppose you weigh 200 lb on the earth. What would you weigh on the surface of a minor planet with a diameter 1/20 that of the earth and a mass of 1/10,000 that of the earth? Can you name a minor planet that fits this description?

4. Suppose a comet had an orbit of semimajor axis 20,000 AU. What would be its period? If it could survive 100 perihelion passages, how long could it exist as a comet in this orbit?

5. Why do you suppose that most comets that are very bright are most conspicuous in the hours after sunset or before sunrise?

6. Contrast comets to planets, typical satellites, meteoroids, and the sun, as regards their:

(a) sizes

(b) masses

(c) densities

7. When can the tail of a comet lie in the direction toward which the comet is moving?

8. Can you think of a way in which we could estimate at least a lower limit to the mass of a meteoroid from the light emitted in the meteor it produces?

9. The earth intercepts the orbit of Halley's comet twice, and we observe two meteor showers each year associated with this comet. Show, by a diagram, how this is possible.

10. What do you suppose determines the angular size in the sky of the radiant of a meteor shower? Why?

11. Do meteoroids of cometary or minor planet origin strike the earth's atmosphere with the greatest speed, on the average? Why?

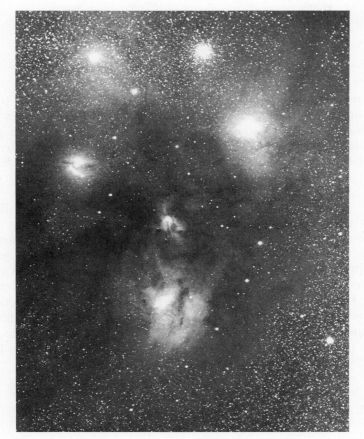

CHAPTER 13

Surveying Space–Distances and Motions

Nearly all measurements of astronomical distances, as well as of distances on the earth, depend directly or indirectly on the principle of triangulation. Some of the concepts of triangulation were known to the Egyptians; the art was developed by the Greeks.

13.1 TRIANGULATION

Six qualities describe the dimensions of a triangle: the lengths of the three sides and the values of the three angles. It is a well-known theorem in elementary geometry that any three of these quantities in succession around the perimeter of the triangle (for example, two sides and an included angle or two angles and an included side) determine the triangle uniquely. In surveying applications, the parts of the triangle that are measured are two angles and the included side—the *base line*.

In astronomy we frequently have to measure distances that are very large in proportion to the length of the available base line, and the triangle to be solved is thus long and "skinny." Suppose (Figure 13-1) that it is desired to measure the distance to the moon or a minor planet, located at O. Two observation stations, A and B, are set up on the earth. The base line AB is known, since the size of the earth and the geographical locations of A and B are known. From A, O appears in direction AS, in line with a distant point, such as some very remote star. Because of the great distance of that star, observers at A and B would look along parallel lines to see it; thus from B the same star is in direction BS'. Observer B sees O, on the other hand, in direction BT, at an angle p away from

direction BS'. This angle, p, is the same as the angle at O between the lines AO and BO (because the line BO intercepts the parallel lines AS and BS'). Note that p is the difference in the directions of O as seen from A and B; this is the apparent displacement (or parallax) of an object as seen from two different points, (Section 2.2d). The procedure for finding the distance from the base line to O is to measure the parallax, P, and then calculate how distant O must be in order for the base line AB to subtend that angle.

Because observers at A and B both see O projected against remote stars on the celestial sphere, each can measure O's direction among those stars with great accuracy; the difference between those directions, the parallax, can be found, therefore, with considerable precision. The parallax, however,

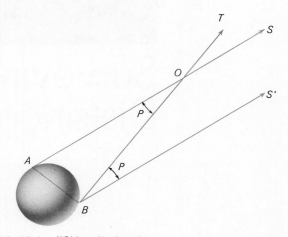

FIG. 13-1   "Skinny" triangle.

can be determined only as accurately as the direction O can be measured with respect to the background of stars on the celestial sphere. Under the most favorable circumstances parallaxes can be measured to within a few thousandths of a second of arc. The more distant the object, the smaller its parallax and the greater the uncertainty of its value in comparison to its size — that is, the greater its *percentage error*, and the greater the uncertainty in the distance to the object that is derived from its parallax. It is advantageous, therefore, that the base line be made as long as possible, so that the parallax will be as large as possible in comparison to the error of its measurement.

13.2 RELATIVE DISTANCES IN THE SOLAR SYSTEM

The planets and other bodies that revolve about the sun in the solar system are so far away that at any time each one is seen in almost the same direction by all terrestrial observers. The parallaxes of these bodies are, therefore, very small and are difficult to measure accurately; a larger base line than the diameter of the earth is required. As the earth moves about the sun, however, it carries us across a base line that can be as large as the diameter of the earth's orbit — nearly 200 million mi. Thus, we can determine accurate distances to the other members of the solar system by observing them at times when the earth is in two different places in its orbit. We must, of course, take into account the motion of the object during the interval between the sightings of it. There are various mathematical techniques for unscrambling the effects of the combined motions of the body and the earth (which will not be gone into here).

The measures described above — measures within the solar system — are not obtained directly in miles or kilometers; they are rather found in terms of the *astronomical unit*, the semimajor axis of the earth's orbit. It is a different problem to determine the length of the astronomical unit in miles or kilometers. The foregoing procedure for surveying the distances to the planets provides us with an accurate map of the solar system, but does not give us the absolute scale of the map. On a map of the United States that is correct proportionally, the relative distances between cities, for example, can be determined simply by measuring distances on the drawing itself. Distances so obtained cannot be converted to miles or kilometers, however, until the scale of the map is established — that is, one inch equals so many miles, kilometers, or other units of distance. The scale can be determined if the actual distance in miles or kilometers between two places (perhaps New York and Philadelphia) is known. Similarly, to find the scale of the solar system — that is, to evaluate the astronomical unit — we must find the distance to some object that revolves about the sun, both in astronomical units and in miles or other units.

(a) Radar Measures

We have already seen that the distance to the moon is accurately found from the time required for radar waves to make the round trip to the moon and back. In February 1958 the Lincoln Laboratories of the Massachusetts Institute of Technology made the first attempt at radar contact with the planet Venus. Since then, Venus has been observed more or less routinely at various radar stations, especially at Lincoln Laboratories and at the Jet Propulsion Laboratory's Goldstone Tracking Station. The radar waves are beamed to Venus, and the return echo is observed. The distance in miles traversed by the beam is found from the observed travel time of the signal and the accurately known velocity of light. Comparison of this distance with the known distance of Venus in astronomical units determines the length of the unit. The best determination of the astronomical unit obtained from radar observations of Venus gives the value 149,597,893 km (or about 92,960,000 mi) with an uncertainty of only a few kilometers.

13.3 SURVEYING DISTANCES TO STARS

The measurement of the parallax of a star enables us to calculate its distance in terms of the earth-sun distance — that is, in astronomical units. However, the stars are so remote that their parallaxes are imperceptible to the unaided eye, even when the stars are observed from both ends of a diameter of the earth's orbit — a base line of 2 AU. When it was conceded that the earth is in orbital motion, therefore, it was realized that the stars are thousands of times more distant than the planets.

With telescopes, the parallaxes of the nearest stars actually can be measured by observing their apparent displacements against the background of more distant stars as the earth traverses its orbit. Even the nearest star, however, shows a total displacement of only about 1″.5; small wonder that Tycho Brahe was unable to observe the stellar parallaxes and thus concluded that the earth was not in motion. The first successful measurements of stellar parallaxes took place in about the year 1838, when Friedrich Bessel (Germany), Thomas Henderson (Cape of Good Hope), and Friedrich Struve (Russia) detected the parallaxes of the stars 61 Cygni, Alpha Centauri, and Vega, respectively.

(a) The Parallactic Ellipse

As the earth moves about its orbit, the place from which we observe the stars is continually changing. Consequently, the positions of the comparatively near stars, projected against the more remote ones, are also continually changing. If a star is in the direction of the ecliptic, it seems merely to shift back and forth in a straight line as the earth passes from one side of the sun to the other. A star that is at the pole of the ecliptic (90° from the ecliptic) seems to move about in a small circle against the background of more distant stars, as we view it from different positions in our nearly circular orbit. A star whose direction is intermediate between the ecliptic and the ecliptic pole seems to shift its position along a small elliptical path during the year. The eccentricity of the ellipse ranges from that of the earth's orbit (nearly a circle) for a star at the ecliptic pole to unity (a straight line) for a star on the ecliptic (see Figure 13-2). This small ellipse is called the *parallactic ellipse*.

(b) Stellar Parallaxes and Stellar Distances

The angular semimajor axis of the parallactic ellipse is called the *stellar parallax* of the star. Since the major axis of the ellipse is the maximum apparent angular deflection of the star as viewed from opposite ends of a diameter of the earth's orbit, the stellar parallax, the semimajor axis of this ellipse, is the angle, at the star's distance, subtended by 1 AU perpendicular to the line of sight.

Until now we have implied that the star whose parallax is observed is motionless with respect to the sun. Actually, the stars are all moving, at many kilometers per second. The effects of the relative motion of the star and the sun can be separated from the effects of the earth's motion (the star's parallax) by observing the star's direction at the same time of the year several years apart. Any observed change in its direction then must be due to its own motion relative to the sun. The change indicates the corrections that must be applied to observed total changes in the direction of the star to obtain its true parallax.

(c) Units of Stellar Distance

If a line segment of length D subtends an angle of p seconds of arc as seen from a distant object, the distance r of that object is given by the formula (compare with the formula derived in Section 6.2):

$$r = 206{,}265\,\frac{D}{p}.$$

Since the parallax of a star is the angle, in seconds, subtended by 1 AU at the star's distance, the distance of a star, in astronomical units, is $206{,}265/p$.

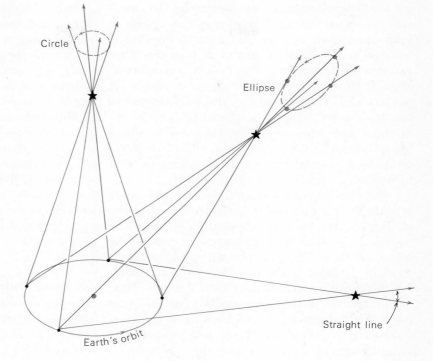

Circle

Ellipse

Earth's orbit

Straight line

FIG. 13-2   Parallactic ellipse.

The length 206,265 AU is defined as a *parsec* (abbreviated pc). One parsec is, therefore, the distance of a star whose *parallax* is 1 *sec*ond of arc. The distance of any star, in parsecs, is thus given by

$$r = \frac{1}{p},$$

where p is the parallax of the star. A star with a parallax of ½″, for example, has a distance of 2 pc; one with a parallax of $\frac{1}{10}$″ has a distance of 10 pc. Parallaxes of stars are usually measured in seconds, so the parsec is a convenient unit of distance — a star's distance in parsecs is simply the reciprocal of its parallax. One parsec is 1.92×10^{13} mi, or 3.08×10^{13} km.

Since 1 pc is 206,265 AU, the sun's distance is $\frac{1}{206,265}$ pc, and its stellar parallax is 206,265 sec.

[This should not be confused with the *solar parallax*, which is defined as the angle subtended by the earth's radius at 1 AU; the latter has the value of approximately 8″.8.]

Another unit of stellar distance is the *light-year*, which is the distance traversed by light in 1 year at the rate of 186,000 mi/sec. One light-year (abbreviated LY) is 5.88×10^{12} mi or 9.46×10^{12} km; 1 pc contains 3.26 LY.

(d) The Nearest Stars

The nearest stellar neighbors to the sun are three stars that make up a multiple system. To the naked eye the system appears as a single bright star, Alpha Centauri, which is only 30° from the south celestial pole and hence is not visible from the United States. Alpha Centauri itself is a double star — two stars revolving about each other that are too close together to be separated by the naked eye. Nearby is the third member of the system, a faint star known as *Proxima Centauri.* Proxima is believed to be slightly closer to us than the other two stars of the system. All three have a parallax of about $0\rlap{.}''76$, and a distance of $\frac{1}{0.76}$, or about 1.3 pc (4.3 LY). The nearest star visible to the naked eye from most parts of the United States is the brightest appearing of all the stars, *Sirius.* Sirius has a distance of 2.6 pc, or about 8 LY. It is interesting to note that light reaches us from the sun in 8 minutes and from Sirius in 8 years.

Parallaxes have been measured for thousands of stars. Only for about 700 stars, however, are the parallaxes large enough (about $0\rlap{.}''05$ or more) to be measured with a precision of 10 percent or better. Of those 700 or so stars within about 20 pc, most are invisible to the unaided eye and actually are intrinsically less luminous than the sun. Most of the stars visible to the unaided eye, on the other hand, have distances of hundreds or even thousands of parsecs and are visible not because they are relatively close, but because they are intrinsically very luminous. The nearer stars are described more fully in Chapter 16.

The vast majority of all known stars are too distant for their parallaxes to be measured, and we must resort to other methods to determine their distances. Most of these methods are either statistical or indirect; they are discussed in later chapters. Considerable information on stellar distances, however, can be obtained from a statistical study of stellar motions, which we now consider.

13.4 MOTIONS OF STARS

The ancients distinguished between the "wandering stars" (planets) and the "fixed stars," which seemed to maintain permanent patterns with one another in the sky. The stars are, indeed, so nearly fixed on the celestial sphere that the apparent groupings they seem to form — the constellations

— look today much as they did when they were first named, more than 2000 years ago. Yet the stars are moving with respect to the sun, most of them with speeds of many kilometers per second. Their motions are not apparent to the unaided eye in the course of a single human lifetime, but if an ancient observer who knew the sky well — Hipparchus, for example — could return to life today he would find that several of the stars had noticeably changed their positions relative to the others. After some 50,000 years or so, terrestrial observers will find the handle of the Big Dipper unmistakably "bent" more than it is now. Apparent changes in the positions of the nearer stars can be measured with telescopes after an interval of only a few years.

(a) Proper Motion

The *proper motion* of a star is the rate at which its direction in the sky changes, and is usually expressed in seconds of arc per year. It is almost always an angle that is too small to measure with much precision in a single year; in an interval of 20 to 50 years, on the other hand, many stars change their directions by easily detectable amounts. The modern procedure for determining proper motions is to compare the positions of the star images on two different photographs of the same region of the sky taken at least two decades apart. Most of the star images on such photographs do not appear to have changed their positions measurably; these are, statistically, the more distant stars that are relatively "fixed," even over the time interval separating the two photographs. With respect to these "background" stars, the motions of a few comparatively nearby stars can be observed. Modern practice is to measure proper motions of stars with respect to remote galaxies, which can show no measurable proper motions themselves.

The star of largest known proper motion is *Barnard's star,* which changes its direction by

FIG. 13-3   Two photographs of Barnard's Star, showing its motion over a period of 22 years. *(Yerkes Observatory.)*

10″.25 each year. This large proper motion is partially due to the star's relatively high velocity with respect to the sun but is mostly the result of its relative proximity. Barnard's star is the nearest known star beyond the triple system containing Alpha Centauri; its distance is only 1.8 pc, but it emits less than ¹⁄₁₀₀₀ as much light as the sun and is nearly a hundred times too faint to see with the unaided eye. There are about 330 other stars with proper motions as great as 1″.0. The mean proper motion for all naked-eye stars is less than 0″.1; nevertheless, the proper motions of most stars are larger than their stellar parallaxes.

The compete specification of the proper motion of a star includes not only its angular rate of motion, but also its *direction* of motion in the sky.

(b) Radial Velocity

The *radial* velocity (or line-of-sight velocity) of a star is the speed with which it approaches or recedes from the sun. This can be determined from the Doppler shift of the lines in its spectrum (Section 7.3b). Unlike the proper motion, which is observable only for the comparatively nearby stars, the radial velocity can be measured for any star that is bright enough for its spectrum to be photographed. The radial velocity of a star, of course, is only that component of its actual velocity that

is projected along the line of sight — that is, that carries the star toward or away from the sun. Radial velocity is usually expressed in kilometers per second, and is counted as *positive* if the star is moving *away* from the sun, and *negative* if the star is moving *toward* the sun. Since motion of either the star or the observer (or both) produces a Doppler shift in the spectral lines, a knowledge of the radial velocity alone does not enable us to decide whether it is the star or the sun that is "doing the moving." What we really measure, therefore, is the speed with which the distance between the star and sun is increasing or decreasing — that is, the star's radial velocity *with respect to the sun*.

Actually, the radial velocity of a star with respect to the sun is not obtained directly from the measured shift of its spectral lines, because we must observe the star from the earth, whose rotational and orbital motions contribute to the Doppler shift. Since the direction of the star is known, however, as are the speed and direction in which the moving earth carries the telescope at the time of observation, it is merely a problem in geometry (albeit a slightly complicated one) to correct the observed radial velocity to the value that would have been found if the star had been observed from the sun.

FIG. 13-4   Calculation of tangential velocity.

(c) Tangential Velocity

Radial velocity is a motion of a star along the line of sight, while proper motion is produced by the star's motion *across*, or at right angles to, the line of sight. Whereas the radial velocity is known in kilometers per second, the proper motion of a star does not, by itself, give the star's actual *speed* at right angles to the line of sight. The latter is called the *tangential* or *transverse* velocity. To find the tangential velocity of a star, we must know both its proper motion and *distance*. A star with a proper motion of 1″.0, for example, might have a relatively low tangential velocity and be nearby, or a high tangential velocity and be far away.

The relation between tangential velocity and proper motion is illustrated in Figure 13-4. As seen from the sun S, a star A, at distance r, is in direction SA. During 1 year it moves from A to B and then appears in direction SD, at an angle μ (the proper motion) from SA. The star's radial motion is AC — it has moved a distance AC farther away during the year; its *tangential* motion is AD — it has moved that distance across the line of sight. The motions shown in the figure are grossly exaggerated; actual stars do not move enough to change their distances by an appreciable percentage, even in 100 years.

The tangential motion AD can be approximated very accurately by a small arc of a circle of radius r centered on the sun. The arc AD is the same fraction of the circumference of the circle, $2\pi r$, as the proper motion is of 360°. The proper motion is expressed in seconds of arc, so we have

$$\frac{\mu}{1,296,000} = \frac{AD}{2\pi r}$$

(there are 1,296,000″ in 360°). If we solve for AD and note that AD is the product of the star's tangential velocity in kilometers per second, and the number of seconds in a year (3.16×10^7), and express r in parsecs, we obtain

$$T = 4.74 \, \mu r \text{ km/sec} = 4.74 \, \frac{\mu}{p} \text{ km/sec,}$$

where p is the parallax. The above equations show how the tangential velocity of a star is related to its proper motion and distance (or parallax).

(d) Space Velocity

The *space velocity* of a star is its total velocity, in miles or kilometers per second, with respect to the sun. The radial velocity is the distance the star

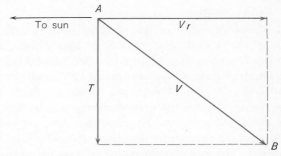

FIG. 13-5   Space velocity; the star moves from A to B in one second.

moves toward or away from the sun in 1 second; the tangential velocity is the distance it moves at right angles to the line of sight in 1 second. The space velocity, therefore, being the total distance the star moves in 1 second, is simply the hypotenuse of the right triangle whose sides are the radial and tangential velocities (Figure 13-5). It is found immediately by the theorem of Pythagoras,

$$V^2 = V_r^2 + T^2,$$

where V and V_r are the space and radial velocities, respectively.

13.5 THE SOLAR MOTION

As we know, the sun is a member of the Galaxy, a system of a hundred billion or so stars. The Galaxy is flat, like a pancake, and is rotating. The sun, partaking of this general rotation of the Galaxy, moves with a speed of about 250 km/sec or 150 mi/sec to complete its orbit about the galactic center in a period of about 200 million years. At first thought it might seem that the galactic center is the natural reference point with which to refer the stellar motions. However, our observations of the proper motions and radial velocities of the stars that surround the sun in space, all in our own local "neighborhood" of the Galaxy, do not directly give us the motions of these stars about the galactic center. The reason is that the stars' orbits around the galactic center and their orbital velocities are both nearly the same as those of the sun. The motions we observe are merely small "residual" or *differential* motions of these stars with respect to the sun. These small residual motions arise because our neighboring stars' orbits about the galactic center are not absolutely identical to our own. We are overtaking and passing some stars, while others are passing us; the slightly different eccentricities and inclinations of our respective orbits bring us closer to some stars and carry us farther from others. We can study these residual motions without knowing anything about the actual motions of stars around the center of the Galaxy. Our situation is analogous to that of a man driving an automobile on a busy highway. All the cars around him are going the same direction and at roughly the same speed, but some are changing lanes and others are passing each other. More or less like the highway traffic, the residual motions of the stars around us seem to be helter skelter.

(a) The Solar Motion and Solar Apex

We can, however, deduce the motion of the sun with respect to its neighboring stars by analyzing the proper motions and radial velocities of the stars around us. The easiest way to understand how the sun's motion can be found is to consider the effect it has on the apparent motions of the other stars.

First, consider the radial velocities of stars with respect to the sun. It should be clear that in a direction that is at right angles to the direction in which the sun is actually moving, the solar motion cannot affect the observed radial velocities of the stars. In those directions, indeed, we find as many stars approaching us as receding from us — their radial velocities *do* average to zero. On the other hand, if we look in the direction *toward* which the sun is moving we find that most of the stars are approaching us, because, of course, we are moving forward to meet them. The only stars in that direction that have radial velocities of recession are those that are moving in the same direction we are going, but at a faster rate, so that they are pulling away from us. The observed radial velocities of all the stars in the direction toward which the sun is moving do not average to zero, but to -20 km/sec, (about 12 mi/sec), showing that we are moving toward them at about 20 km/sec. Similarly, stars in the opposite direction have an average radial velocity of $+20$ km/sec, because we are pulling away from them at that speed.

Now consider the proper motions of stars. Part of a star's proper motion, in general, is due to its own motion and part is due to the sun's motion. However, the sun's motion can contribute nothing to a star's proper motion if the star happens to lie in the direction toward which the sun is moving. Therefore, if we look at stars that lie in a path along the direction of the solar motion, as many of their proper motions should be in one direction as any other; the average of the motions of many stars in those directions, therefore, should be zero. On the other hand, the maximum effect on the proper motions of stars should occur in directions that are at right angles to the direction of the solar motion. If the stars were at rest, they would all show a backward drift due to our forward motion. As it is, the stars have motions of their own, but only those moving in the same direction we are, but at a faster rate than we, appear to have "forward" proper motions — the rest, by far the majority, *do* appear to drift backward.

William Herschel was the first to attempt to detect the direction of the solar motion from the proper motions of stars. In 1783 he analyzed the proper motions of 14 stars and deduced that the sun was moving in a direction toward the constellation Hercules — a nearly correct result.

Modern analysis of the proper motions and radial velocities of the stars around the sun has shown that the sun is moving approximately toward the direction now occupied by the bright star Vega in the constellation of Lyra at a speed of 20 km/sec (about 12 mi/sec or 4.2 AU/year), with an uncertainty of about 0.5 km/sec. The direction in the sky toward which the sun is moving is called the *apex* of solar motion,† and the opposite direction, away from which the sun is moving, is called the *antapex.*

The velocity of a star, corrected for the sun's motion, is called its *peculiar velocity.* The *space velocity* of a star, being its motion with respect to the sun, is made up of both the star's peculiar velocity and a component due to reflection of the solar motion. Since the solar motion is known, the peculiar velocity can be calculated for a star of known space velocity.

(b) Distances from Stellar Motions

The proper motions of stars can be expected to be largest, statistically, for the nearest stars. If, for example, a star is only a few parsecs away, its proper motion will almost certainly be observable after a few years (but see Exercise 11). The proper motion of a very distant star, on the other hand, may only be detectable after a long time, and then only if the star has a very great space velocity. Searches for nearby stars, therefore, are usually conducted by searching for stars of large proper motion. Conversely, remote stars can be identified by their lack of observable proper motions; the latter are needed to serve as standards against which we can measure the parallaxes of the nearby stars to determine their distances. The proper motion of an individual star does not uniquely indicate its distance. However, proper motions and distances of stars are inversely correlated, and investigations of such motions do give information about stellar distances. Proper motions yield only average distances to certain classes of stars, but these statistical methods of obtaining stellar distances are very important and have been our only methods of obtaining distances to some kinds of stars.

(c) Moving Clusters

The stars that belong to a group or cluster, moving more or less together through space, all have about the same space velocity. If such a cluster is close enough to the sun, however, the proper motions of its member stars may not all be exactly parallel. If, for example, the cluster is approaching us, its stars appear to radiate away from a distant point

†The equatorial coordinates of the *solar apex* are (1950) $\alpha = 18^h4^m \pm 7^m$, $\delta = +30° \pm 1°$.

FIG. 13-6   Proper motions of the stars in a moving cluster. Note that here we see the cluster projected on the plane of the sky. Although the stars are moving parallel to one another in space, their proper motions appear to converge toward a point infinitely far away. The direction from the earth to that convergent point is also parallel to the space motions of the cluster stars.

in the direction from which the cluster is coming. Conversely, if the cluster recedes from us, the stars appear to converge toward the direction in which the cluster is moving. In either case, the proper motions of the member stars indicate the direction in which the cluster is traveling. The situation for a hypothetical cluster is illustrated in Figure 13-6, which shows the apparent convergence of the proper motions of its stars toward a distant "convergent point," and in Figure 13-7, which shows the actual space velocities of the stars. In these diagrams the cluster is viewed from two different directions.

In studying a moving cluster for which the direction of convergence or divergence can be observed, we have the great advantage (see Fig. 13-7) of knowing the angle A between that direction of motion and the line of sight from the sun to the cluster. Now consider any star in the cluster, such as star S. Its space velocity, radial velocity, and tangential velocity are SP, SQ, and QP, respectively. Angle SQP is a right angle, and angle A and the star's radial velocity are both observed. Thus the tangential velocity QP can be calculated by solving the triangle.†

Then the parallax (or distance) of the star S, and thus that of the cluster, can be found from the observed proper motion of the star and from the formula relating tangential velocity and proper motion:

$$QP = 4.74\mu r = 4.74\,\frac{\mu}{p}\ \text{km/sec.}$$

There are several groups and clusters that are near enough for us to observe the convergence or divergence of the proper motions of their member stars. The three best-known examples are the *Hyades* in Taurus, the *Ursa Major group* (which contains most of the naked-eye stars in the Big Dipper), and the *Scorpio-Centaurus group*. These

FIG. 13-7   Space velocities of the stars in a moving cluster.

---
†$QP = SQ\tan A.$

three clusters contain, in all, more than 500 stars. This is almost as great a number of stars as the number whose parallaxes can be directly observed to an accuracy of 10 percent or better. Thus, this method of using moving clusters to obtain stellar distances is very important and fruitful. It has contributed greatly to our knowledge of fundamental data about stars.

Exercises

1. A radar astronomer claims that he beamed radio waves to Jupiter, and received an echo exactly 48 minutes later. Do you believe him? Why?

2. Saturn is approximately 10 AU from the sun. In the course of one year, how large a parallax can we observe of Saturn from different parts of the earth's orbit?

3. At greatest eastern elongation, an inferior planet is 30° from the sun. If the planet and earth have circular orbits, what is the radius of the planet's orbit in AU's?

4. Give the distances to stars having the following parallaxes:

    a) $p = 0\rlap{.}{''}1$ c) $p = 0\rlap{.}{''}005$

    b) $p = 0\rlap{.}{''}05$ d) $p = 0\rlap{.}{''}001$

5. Give parallaxes of stars having the following distances:

    a) 10 pc   c) 412530 AU

    b) 32.6 LY   d) 1 AU

6. The value of the astronomical unit can be obtained from radar observations of a planet, such as Venus, not only from the time required for the radio waves to travel the round-trip to Venus and back, but also from the difference between the frequencies of the waves beamed to Venus and those received waves that are reflected back to earth. Explain how this is possible.

7. A spectral line that has a laboratory wavelength of exactly 6000 angstroms is observed, in the spectrum of a star, at 6002 angstroms. What is the radial velocity of the star? Is the star approaching or receding from us?

8. A star is observed on the ecliptic 90° from the sun in the sky. It has an observed radial velocity (as seen from the earth) of −10 km/sec. Six months later the star is observed again and it has a radial velocity of +50 km/sec.

a) What is the orbital speed of the earth about the sun?

b) What is the radial velocity of the star?

9. Following are data on 5 stars:

STAR	DISTANCE	PARALLAX	PROPER MOTION	RADIAL VELOCITY
1	1000pc		$0\rlap{.}{''}05$	−80 km/sec
2		$0\rlap{.}{''}0001$	$0\rlap{.}{''}1$	+75 km/sec
3	40pc		$0\rlap{.}{''}25$	+20 km/sec
4		$0\rlap{.}{''}1$	$0\rlap{.}{''}01$	− 7 km/sec
5	100pc		$1\rlap{.}{''}0$	−18 km/sec

a) Which star has the largest space motion?

b) Which has the smallest space motion?

c) Which is most distant?

d) Which is nearest?

e) Which approaches the fastest?

10. A star has a parallax of 0".474, a proper motion of 3".0, and a radial velocity of 40 km/sec. What is its space motion?

11. Under what conditions would the proper motion of a star be zero if:

a) It were in the direction of the solar apex?

b) It were *not* in the direction of the solar apex or antapex?

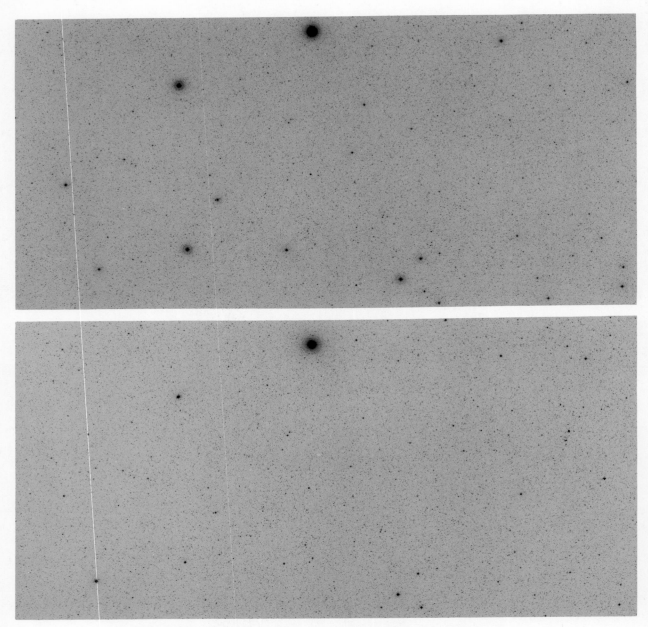

Two photographs of the same region of the sky (in Coma Berenices).
The upper is in blue light, and the lower in red; note how the colors of the stars are apparent from comparison of the two. These are negative prints, the stars showing as black dots on a white sky. *(National Geographic Society-Palomar Observatory Sky Survey.)*

CHAPTER 14

Analyzing Starlight

The most casual glance at the sky shows that stars differ from one another in apparent brightness. They differ not only because of actual differences in their output of luminous energy, but also because they are at widely varying distances. Measures of the amounts of light energy, or *luminous flux,* received from stars are among the most important and fundamental observational data of astronomy; they are used in estimating both the distances and the actual output of energy of stars.

14.1 STELLAR MAGNITUDES

In the second century B.C., Hipparchus compiled a catalogue of about a thousand stars (Section 2.3c). He classified these stars into six categories of brightness, which are now called *magnitudes*. The brightest-appearing stars were placed by him in the *first magnitude;* the faintest naked-eye stars were of the *sixth magnitude*. The other stars were assigned intermediate magnitudes. This system of stellar magnitudes, which began in ancient Greece, has survived to the present time, with the improvement that today magnitudes are based on precise measurements of apparent or total luminosity rather than arbitrary and uncertain eye estimates of star brightnesses.

(a) Measuring Starlight — Photometry

That branch of observational astronomy which deals with the measurement of the intensity of starlight is called *photometry*. In the latter part of the eighteenth century, William Herschel devised a simple and direct, although only approximate, method of stellar photometry. The principle of his method depends on the fact that the light-gather-ing power of a telescope is proportional to the area of its aperture (lens or mirror). The aperture size of a telescope can be varied at will by the use of circular diaphragms or an iris diaphragm, like the one on a camera. As an example, consider two stars that appear of different brightness in the field of view of a telescope. Suppose, now, that the dimmer of the stars becomes just barely visible when the telescope is closed down to an 8-in. diameter, while the brighter one becomes barely visible after the aperture is stopped down to 4 in. The two stars, when made just barely visible, appear equally bright. However, since only one fourth as much light enters the telescope with the 4-in. diaphragm as with the 8-in. one, we must, in fact, actually be receiving four times as much light from the brighter star as from the fainter. By such reasoning, Herschel estimated the relative amounts of luminous flux received from different stars. He determined, for example, that the average first-magnitude star delivers to the earth somewhat more than 100 times as much light as a star which is just barely visible on a dark night — that is, a star of approximately sixth magnitude.

A widely used modern method of comparing the amounts of light received from stars is that of *photographic photometry,* in which the degrees of blackness and the sizes of star images on photographic negatives are measured with special instruments. Actually, even large telescopes do not resolve the stars into disks. In principle, stars make point-like images. The photographic images of bright stars, however, are larger than those of faint ones because of the *turbidity* of the photographic emulsions — the fact that during the time exposure the light from the brighter star images is scattered about within the emulsion, exposing a larger area than that occupied by the true image. Bright star images, in other words, are very much overexposed in long time exposures. Atmospheric turbulence (seeing) also enlarges star images.

The most modern and precise method of stellar photometry, *photoelectric photometry,* has come into widespread use since World War II. The light from a star, coming to a focus at the focal plane of the telescope, is allowed to pass

FIG. 14-1   Effect of telescope aperture and exposure time. *(Top left)* 14 in., 1 min. Stars to 12th magnitude. *(Top right)* 60 in., 1 min. Stars of 15th magnitude. *(Lower left)* 60 in., 27 min. Stars to 18th magnitude. *(Lower right)* 60 in., 4 hr. Stars to 20th magnitude. *(Hale Observatories.)*

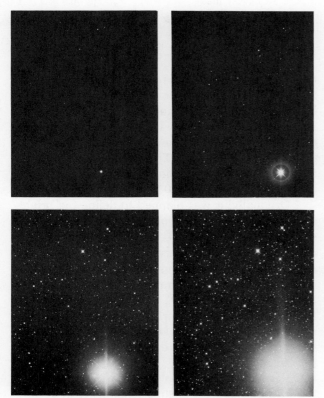

through a small hole in a metal plate and thence onto the photosensitive surface of a photomultiplier (Section 8.6d). The electric current generated in the photomultiplier is amplified and recorded and provides an accurate measure of the light passing through the hole. Because even the darkest night sky is not completely dark, however, not all the light striking the photomultiplier is due to the star, but some is provided by the light of the night sky, which is, of course, also gathered by the telescope and passed through the hole. Consequently, the hole is next moved to one side of the star image, so that only light from the sky passes through and is recorded; the difference between the first and second readings is a measure of the star's light entering the telescope.

(b) The Magnitude Scale

In 1856, after early methods of stellar photometry had been developed, Norman R. Pogson proposed the quantitative scale of stellar magnitudes that is now generally adopted. He noted, as did Herschel, that we receive about 100 times as much light from a star of the first magnitude as from one of the sixth, and that, therefore, a difference of five magnitudes corresponds to a ratio in luminous flux of 100:1. It is a widely accepted assumption in the physiology of sense perception that what appear to be equal intervals of brightness are really equal *ratios* of luminous energy. Pogson suggested, therefore, that the ratio of light flux corresponding to a step of one magnitude be the fifth root of 100, which is about 2.512. Thus, a fifth-magnitude star gives us 2.512 times as much light as one of sixth magnitude, and a fourth-magnitude star, 2.512 times as much light as a fifth, or 2.512×2.512 times as much as a sixth-magnitude star. From stars of third, second, and first magnitude, we receive 2.512^3, 2.512^4, and 2.512^5 ($= 100$) times as much light as from a sixth-magnitude star. By assigning a magnitude of 1.0 to the bright stars Aldebaran and Altair, Pogson's new scale gave magnitudes that agreed roughly with those in current use at the time.

TABLE 14.1 Magnitude Differences and Light Ratios

DIFFERENCE IN MAGNITUDE	RATIO OF LIGHT
0.0	1:1
0.5	1.6:1
0.75	2:1
1.0	2.5:1
1.5	4:1
2.0	6.3:1
2.5	10:1
3.0	16:1
4.0	40:1
5.0	100:1
6.0	251:1
10.0	10,000:1
15.0	1,000,000:1
20.0	100,000,000:1
25.0	10,000,000,000:1

Table 14.1 gives the approximate ratios of light flux corresponding to several selected magnitude differences. Note that a given ratio of light flux, whether between bright or faint stars, always corresponds to the same magnitude interval. Further, note that the numerically *smaller* magnitudes are associated with the *brighter* stars; a numerically *large* magnitude, therefore, refers to a faint star.†

With optical aid, stars can be seen that are beyond the reach of the naked eye. The *limiting magnitude* of a telescope is the magnitude of the faintest stars that can be seen with that telescope under ideal conditions. A 6-in. telescope, for example, has a limiting magnitude of about 13. The *photographic limiting magnitude* of a telescope is the magnitude of the faintest stars that can be photographed with it. The photographic limiting magnitude of the 200-inch telescope on Palomar Mountain is about 23.5.

---

†If m_1 and m_2 are the magnitudes corresponding to stars from which we receive light flux in the amounts l_1 and l_2, the difference between m_1 and m_2 is defined by

$$m_1 - m_2 = 2.5 \log \frac{l_2}{l_1}.$$

The so-called first-magnitude stars are not all of the same apparent brightness. The brightest-appearing star, Sirius, sends us about 10 times as much light as the average star of first magnitude, and so has a magnitude of $1.0 - 2.5$, or of about -1.5. Several of the planets appear even brighter; Venus, at its brightest, is of magnitude -4. The sun has a magnitude of -26.5. Some magnitude data are given in Table 14.2. It is of interest to note

TABLE 14.2 Some Magnitude Data

OBJECT	MAGNITUDE
Sun	-26.5
Full moon	-12.5
Venus (at brightest)	-4
Jupiter; Mars (at brightest)	-2
Sirius	-1.4
Aldebaran: Altair	1.0
Naked-eye limit	6.5
Binocular limit	10
6-in. telescope limit	13
200-inch (visual) limit	20
200-inch photographic limit	23.5

that the brightness of the sun and Sirius differ by 25 magnitudes — a factor of 10 billion (10^{10}) in light energy, or flux — and that we also receive 10 billion times as much light from Sirius as from the faintest stars that can be photographed with the 200-inch telescope. The entire range of light flux represented in Table 14.2 covers a ratio of about 100 quintillion (10^{20}) to 1.

Originally, it was planned to define the brightest stars as having first magnitude. The brightest stars, on the other hand, are not always conveniently disposed for observation. At one time, by agreement among astronomers, it was decided to define as standards several stars in the vicinity of the north celestial pole, which is always visible from northern observatories. This group of stars was called the *North Polar Sequence*. The magnitudes of these stars were intended to serve as a permanent basis of the magnitude system. The actual magnitude values were assigned so that the brightest-appearing stars in the sky, in accordance with previous convention, would still be reasonably near the first magnitude.

Since the development of photoelectric photometry, however, it became clear that the North Polar Sequence, which had been based largely upon photographic photometry, was inadequate for calibrating the precise measures that are obtained today. Consequently, the old standards have been superseded by accurate photoelectric measures of a large number of new standard stars distributed over the sky.

14.2 THE "REAL" BRIGHTNESSES OF STARS

Even if all stars were identical, and if interstellar space were entirely free of opaque matter, stars would not all appear to have the same brightness, because they are at different distances from us, and the light that we receive from a star is *inversely proportional to the square of its distance* (Section 7.1a). The apparent brightnesses of stars therefore do not provide a basis for comparing the amounts of light that they actually emit into space. To make such a comparison we would first have to calculate how much light we would receive from each star if all stars were at the same distance from us. Fortunately, if we know a star's distance we can make such a calculation.

(a) Absolute Magnitudes

The sun gives us billions of times as much light as any of the other stars, but on the other hand, it is hundreds of thousands of times closer to us than any other star is. To compare the intrinsic luminous outputs of the sun and of other stars, we first have to determine what magnitude the sun would have if it were at a specified distance, typical of the distances of the other stars. Suppose we choose 10 pc as a more or less representative distance of the nearest stars. Since 1 parsec is about 200,000 AU, the sun would be 2,000,000 times as distant as it is now if it were removed to a distance of

10 pc; consequently, it would deliver to us $\frac{1}{2,000,000}^2$ or $1/(4 \times 10^{12})$ of the light it now sends. A factor of 4×10^{12} corresponds to about $31\frac{1}{2}$ magnitudes (which can be verified by raising 2.512 to the 31.5 power). The sun, therefore, if removed to a distance of 10 pc, would appear fainter by some $31\frac{1}{2}$ magnitudes than its present magnitude of -26.5; that is, the sun would then appear as a fifth-magnitude star.

Similarly, we can use the inverse-square law of light to calculate how luminous all other stars of known distance would appear if they were 10 pc away. Suppose, for example, that a tenth-magnitude star has a distance of 100 pc. If it were only 10 pc away, it would be only one tenth as far away, and hence 100 times as bright — a difference of five magnitudes. At 10 pc, therefore, it too would appear as a fifth-magnitude star.

We now define the *absolute magnitude* of a star as the magnitude that star would have if it were at the standard distance of 10 pc (about 32.6 LY). The absolute magnitude of the sun is about $+5$. Most stars have absolute magnitudes that lie in the range 0 to $+15$. The extreme range of absolute magnitudes observed for normal stars is -9 to $+20$, a range of a factor of nearly 10^{12} in intrinsic light output. The absolute magnitudes of stars are measures of how bright they "really" are; they provide a basis for comparing the actual amount of light emitted by stars.

(b) Stellar Distances from Magnitude Measures

The absolute magnitude of a star is the magnitude that the star *would* have if it were at a distance of 10 pc, and therefore is a measure of the actual rate of emission of visible light energy by the star, which, of course, is independent of the star's actual distance. On the other hand, the magnitude of a star (sometimes called the *apparent magnitude* to avoid confusion with *absolute magnitude*) is a measure of how bright the star *appears* to be, and this depends on both the star's actual rate of light output and its distance. The *difference* between the star's apparent magnitude, symbolized *m*, and absolute magnitude, symbolized *M,* can, however, be calculated from the inverse-square law of light and from a knowledge of how much greater or less than 10 pc the star's distance actually is. The difference $m - M$ therefore depends only on the distance of the star and is called its *distance modulus.* The distance modulus of a star, then, is a measure of its distance, from which the actual distance, in parsecs, can be calculated.† For example, suppose the distance modulus of a star is 10 magnitudes. Ten magnitudes (see Table 14.1) corresponds to a ratio of 10,000:1 in light. Thus, we actually receive from the star $\frac{1}{10,000}$ of the light that we would receive if it were 10 pc away; it must, therefore, be 100 times as distant as 10 pc, or at a distance of 1000 pc. In succeeding sections we shall see that the absolute magnitude of a star can often be inferred from its spectrum or from some other observable characteristic. Since the apparent magnitude of a star can be observed, a knowledge of its absolute magnitude is equivalent to a knowledge of its distance.

Most of us make use of this same principle, subconsciously, in everyday life. Every experienced motorist, for example, has an intuitive notion of

the actual brightness of a stop light. If, while driving down the highway at night, he sees a stop light, he judges its distance from its apparent faintness. In other words, the difference between the light's *apparent* and *real* brightness indicates its distance. The computation of a star's distance from its distance modulus is analogous.

14.3 COLORS OF STARS

Until now, this discussion has ignored the fact that stars have different colors and that all colors do not produce an equal response in the human eye. The apparent brightness of a star can depend to some extent, therefore, upon its color.

(a) Color Response of the Eye and Other Detecting Devices

Every device for detecting light has a particular color or spectral sensitivity. The human eye, for example, is most sensitive to green and yellow light; it has a lower sensitivity to the shorter wavelengths of blue and violet light and to the longer wavelengths of orange and red light. It does not respond at all to ultraviolet or to infrared radiation. The eye, in fact, responds roughly to the same kind of light that the sun emits most intensely; this coincidence is probably not accidental — the eye may have evolved to respond to the kind of light most available on earth.

Another detecting device is the photographic plate (or film). The early photographic emulsions, before the development of yellow- and red-sensitive and panchromatic emulsions, were sensitive only to violet and blue light and did not respond to light of wavelengths longer than about 5000 Å (in the blue-green). The basic photographic emulsion is still sensitive to violet and blue; dyes must be added to the basic emulsion to make it sensitive to longer wavelengths.

Suppose, now, that the total amount of light energy entering a telescope from each of two stars

---

†Let $l(r)$ be the observed light of a star at its actual distance, r, and $l(10)$ the amount of light we would receive from it if it were a distance of 10 pc. From the definition of magnitudes, we have

$$m - M = 2.5 \log \frac{l(10)}{l(r)},$$

and from the inverse-square law of light,

$$\frac{l(10)}{l(r)} = \left(\frac{r}{10}\right)^2.$$

Combining the above equations, we obtain

$$m - M = 5 \log \frac{r}{10}.$$

The quantity, $5 \log (r/10)$, then, is the distance modulus.

is exactly the same if light of all wavelengths is considered, but that one star emits most of its light in the blue spectral region and the other in the yellow spectral region. If these stars are observed visually (that is, by looking at them through the telescope) the yellow one will appear brighter, that is, will have a numerically smaller magnitude, because the eye is less sensitive to most of the light emitted by the blue star. If the stars are photographed on a blue-sensitive photographic plate, however, the blue star will produce the more conspicuous image; measures of the photographic images will show the blue star appearing brighter and having the smaller magnitude. Consequently, when a magnitude system is defined, it is necessary also to specify how the magnitudes are to be measured — that is, what detecting device is to be used.

Magnitudes, whether apparent or absolute, that are based on stellar brightness as they are observed with the human eye are called *visual magnitudes* or *absolute visual magnitudes* and are symbolized m_v and M_v, respectively. Magnitudes based on measures of star images on standard violet- and blue-sensitive photographic emulsions are called *photographic magnitudes* and *absolute photographic magnitudes* and are symbolized m_{pg} and M_{pg}. When green- and yellow-sensitive photographic emulsions were perfected, it was found that they could be used in conjunction with a suitable color filter to approximate closely the spectral response of the human eye; stellar photographs so obtained provide *photovisual magnitudes* (m_{pv} and M_{pv}). Today photographic plates or photomultipliers can be used in conjunction with many kinds of color filters to produce a great variety of different magnitude systems — for example, red magnitudes, ultraviolet magnitudes, infrared magnitudes, and so on. A certain few spectral bands have become more or less standard, however, and are now widely used to define magnitudes.

(b) Color Indices

Until about 1955, the most commonly used magnitudes were photographic magnitudes and photovisual magnitudes. The *difference* between the photographic and photovisual magnitudes of a star is called its *color index, CI:*

$$CI = m_{pg} - m_{pv}.$$

Since the inverse-square law of light applies equally to all wavelengths, the color index of a star would not change if the star's distance were changed. The color index of a star, therefore, can be defined in terms of either its apparent or absolute photographic and photovisual magnitudes:

$$M_{pg} - M_{pv} = m_{pg} - m_{pv} = CI.$$

The system of photographic and photovisual magnitudes and color index is called the international system. (Actually, space is not completely transparent; microscopic particles of "cosmic dust" not only dim the light of stars in some parts of the Galaxy, but also *redden* their light. Chapter 19 includes a discussion of the effect of this interstellar absorption of starlight and how it can be allowed for. In this chapter, however, space is assumed to be perfectly transparent.)

A very blue star will appear brighter on a blue-sensitive photograph than on a yellow-sensitive one; its photographic magnitude, therefore, will be algebraically *less* than its photovisual magnitude, and its color index will be *negative*. A yellow or red star, on the other hand, will have a brighter (smaller) photovisual magnitude, and a *positive* color index. Color indices, therefore, provide measures of the *colors* of stars. Colors, in turn, indicate the temperatures of stars. Photographic and photovisual magnitudes are adjusted to be the same, so as to give a color index of zero to a star with a temperature of about 10,000°K (a spectral-type A star — see Section 14.4). Color indices of stars range from −0.6 for the bluest to more than 2.0 for the reddest.

FIG. 14-2   Two photographs of the same region of the Milky Way taken with the
48-inch Schmidt telescope; *(left)* in blue light, *(right)* in red light. Note
the difference with which the stars and nebulae show up in the different colors.
(National Geographic Society-Palomar Observatory Sky Survey.)

If a star field is photographed on separate blue-sensitive and yellow-sensitive plates, the photographic and photovisual magnitudes and color indices so obtained comprise a powerful tool for determining the temperatures of all the stars appearing on the photographs. Since the star images on these two black and white photographs can be accurately measured, the color-index method is a more precise means of obtaining colors and temperatures of stars than the method of color photography would be.

Since the development of the more precise methods of photoelectric photometry, it has become customary to measure the light of stars in three or more spectral regions instead of only two, and thereby to increase the amount of information obtained about their light. In the early 1950s, H. L. Johnson and W. W. Morgan defined the U,B,V magnitude system, which is now rather generally adopted. The U,B,V system utilizes measures in the ultraviolet (U), blue (B), and green-yellow or visual (V) spectral regions. It provides two independent color indices, $B-V$ and $U-B$. There are many other more or less standardized magnitude systems.

The maximum information is obtained in photoelectric spectral scanning. In this technique, a photomultiplier replaces the photographic plate in a spectrograph. The photomultiplier scans the spectrum (usually by slowly rotating the grating), and the output of the tube is recorded, yielding a point by point record of the distribution of energy according to wavelength in the star.

(c) Bolometric Magnitudes and Luminosities

All the magnitude systems so far discussed refer only to certain spectral regions. A magnitude system based on *all* the electromagnetic energy reaching the earth from the stars would seem to be more fundamental. Magnitudes so based are called *bolometric magnitudes,* m_{bol}, and the bolometric magnitudes that stars would have at a distance of 10 pc are *absolute bolometric magnitudes,* M_{bol}. Unfortunately, bolometric magnitudes cannot be observed directly because some wavelengths of electromagnetic energy do not penetrate the earth's atmosphere. It is true that most of the radiation from stars such as the sun does reach the earth's surface, and for such stars bolometric magnitudes can be determined approximately. A large part of the energy from stars that are substantially hotter or cooler than the sun, however, lies in the far ultraviolet or in the infrared, and is either blocked by the earth's atmosphere or cannot be observed easily; for those stars, bolometric magnitudes can only be estimated or calculated from theoretical considerations. Recently, observations from rockets and artificial satellites have begun to provide measures in spectral regions to which the earth's atmosphere is opaque, and have been of much value in improving our knowledge of bolometric magnitudes of stars — especially of those stars of extreme temperatures.

The absolute bolometric magnitude of a star is a measure of the rate of its entire output of radiant energy. The rate at which a star pours radiant energy into space, usually expressed in ergs per second, is called its *luminosity*. Absolute bolometric magnitudes of stars bear the same relation to their luminosities as their visual magnitudes do to the amount of light we receive from them.†

(d) The Solar Luminosity

The first step in determining the luminosity of the sun is to measure the rate at which its radiation falls upon the earth. There are various devices for measuring this quantity. In one method, the sun's radiation is allowed to fall upon the blackened surface of some metal (for example, platinum or silver), which absorbs this radiant energy and rises in temperature. The rate of the temperature rise can be measured by the increased resistance that the metal offers to an electric current. A simpler device measures the rate at which the temperature of water rises as the sun's radiation is absorbed by it. The measures obtained of the sun's radiation must, in any case, be corrected for the attenuation of the sunlight passing through the earth's atmo-

sphere. (Measures of stellar magnitudes must similarly be corrected.)

It is found that a surface of 1 cm² just outside the atmosphere and oriented perpendicular to the direction of the sun receives from the sun 1.36×10^6 ergs/sec, or 1.95 cal/min. This value is known as the *solar constant*.

The total energy that leaves the sun during an interval of 1 second diverges outward, away from the sun, in all directions. Since one astronomical unit is 1.49×10^{13} cm, the area of the spherical surface over which the solar radiation has spread by the time it reaches the earth's distance from the sun (about 8 minutes later) is 2.8×10^{27} cm². The solar constant of 1.36×10^6 ergs/sec/cm² is the energy that crosses just one of those square centimeters. The total energy that leaves the sun in 1 second — its luminosity — is thus $1.36 \times 10^6 \times 2.8 \times 10^{27} = 3.8 \times 10^{33}$ ergs/sec. The corresponding absolute bolometric magnitude of the sun is +4.6.

The sun's luminosity is equivalent to 5×10^{23} horsepower. A dramatic illustration of the magnitude of that amount of energy is obtained by imagining a bridge of ice 2 mi wide and 1 mi thick and extending over the nearly 100-million-mile span from the earth to the sun. If all the sun's radiation could be directed along the bridge, it would be enough to melt the entire column of ice in 1 second.

14.4 SPECTRA OF STARS

About 1665 Newton showed (perhaps not for the first time) that white sunlight is really a composite of all colors of the rainbow (Section 7.2), and that the various colors, or wavelengths, of light could be separated by passing the light through a glass prism. William Wollaston (Section 7.3) first observed dark lines in the solar spectrum, and Joseph Fraunhofer catalogued about 600 such dark lines. As early as 1823, Fraunhofer observed that

---

†If two stars of absolute bolometric magnitudes, $M_{bol}(1)$ and $M_{bol}(2)$, have luminosities L_1 and L_2,

$$M_{bol}(1) - M_{bol}(2) = 2.5 \log \frac{L_2}{L_1}.$$

stars, like the sun, also have spectra that are characterized by dark lines crossing a continuous band of color. Sir William Huggins, in 1864, first identified some of the lines in stellar spectra with those of known terrestrial elements.

When the spectra of different stars were observed, it was found that they differed greatly among themselves. In 1863 the Jesuit astronomer Angelo Secchi classified stars into four groups according to the general arrangement of the dark lines in their spectra. Secchi's scheme subsequently was modified and augmented, until today we recognize seven such principal *spectral classes.*

(a) The Spectral Sequence

As we have seen (Sections 7.3 and 7.5), each dark line in a stellar spectrum is due to the presence of a particular chemical element in the atmosphere of the star observed. It might seem, therefore, that stellar spectra differ from each other because of differences in the chemical makeup of the stars. Actually, the differences in stellar spectra are due mostly to the widely differing temperatures in the outer layers of the various stars. Hydrogen, for example, is by far the most abundant element in all stars, except probably in those at an advanced stage of evolution (Chapter 20). In the atmospheres of the very hottest stars, however, hydrogen is completely ionized, and can, therefore, produce no absorption lines. In the atmospheres of the coolest stars hydrogen is neutral and can produce absorption lines, but in these stars, since practically all the hydrogen atoms are in the lowest energy state (unexcited), they can absorb only those photons that can lift them from that first energy level to higher ones. The photons so absorbed produce the *Lyman series* of absorption lines, which lies in the unobservable ultraviolet part of the spectrum. In a stellar atmosphere with a temperature of about 10,000°K, many hydrogen atoms are not ionized; nevertheless, an appreciable number of them are excited to the second energy level, from which they can absorb additional photons and rise to still higher levels of excitation. These photons correspond to the wavelengths of the *Balmer series,* which is in the part of the spectrum that is readily observable. Absorption lines due to hydrogen, therefore, are strongest in the spectra of stars whose atmospheres have temperatures near 10,000°K, and they are less conspicuous in the spectra of both hotter and cooler stars, even though hydrogen is, roughly, equally abundant in all the stars. Similarly, every other chemical element, in each of its possible stages of ionization, has a characteristic temperature at which it is most effective in producing absorption lines in the observable part of the spectrum.

Once we have ascertained how the temperature of a star can determine the physical states of the gases in its outer layers, and thus their ability to produce absorption lines, we need only to observe what patterns of absorption lines are present in the spectrum of a star to learn its temperature. We can therefore arrange the seven classes of stellar spectra in a continuous sequence in order of decreasing temperature. In the hottest stars (temperatures over 25,000°K) only lines of ionized helium and highly ionized atoms of other elements are conspicuous. Hydrogen lines are strongest in stars with atmospheric temperatures of about 10,000°K. Ionized metals provide the most conspicuous lines in stars with temperatures from 6000° to 8000°K. Lines of neutral metals are the strongest in somewhat cooler stars. In the coolest stars (below 4000°K), bands of some molecules are very strong. The most important among the molecular bands are those due to titanium oxide, a tenacious chemical compound which can exist at the temperatures of the cooler stars. The sequence of spectral types is summarized in Table 14.3 and Figure 14-3.

The spectral classes of stars listed in Table 14.3 can be subdivided into tenths; thus a star of

FIG. 14-3   Relative intensities of different absorption lines in stars at various places in the spectral sequence.

TABLE 14.3 Spectral Sequence

SPECTRAL CLASS	COLOR	APPROXIMATE TEMPERATURE (°K)	PRINCIPAL FEATURES	STELLAR EXAMPLES
O	Blue	>25,000	Relatively few absorption lines in observable spectrum. Lines of ionized helium, doubly ionized nitrogen, triply ionized silicon, and other lines of highly ionized atoms. Hydrogen lines appear only weakly.	10 Lacertae
B	Blue	11,000–25,000	Lines of neutral helium, singly and doubly ionized silicon, singly ionized oxygen and magnesium. Hydrogen lines more pronounced than in O-type stars.	Rigel Spica
A	Blue	7,500–11,000	Strong lines of hydrogen. Also lines of singly ionized magnesium, silicon, iron, titanium, calcium, and others. Lines of some neutral metals show weakly.	Sirius Vega
F	Blue to white	6,000–7,500	Hydrogen lines are weaker than in A-type stars but are still conspicuous. Lines of singly ionized calcium, iron, and chromium, and also lines of neutral iron and chromium are present, as are lines of other neutral metals.	Canopus Procyon
G	White to yellow	5,000–6,000	Lines of ionized calcium are the most conspicuous spectral features. Many lines of ionized and neutral metals are present. Hydrogen lines are weaker even than in F-type stars. Bands of CH, the hydrocarbon radical, are strong.	Sun Capella
K	Orange to red	3,500–5,000	Lines of neutral metals predominate. The CH bands are still present.	Arcturus Aldebaran
M	Red	<3,500	Strong lines of neutral metals and molecular bands of titanium oxide dominate.	Betelgeuse Antares

FIG. 14-4   Spectra of several stars of representative spectral classes. *(UCLA Observatory.)*

spectral class A5 is midway in the range of A-type stars — that is, halfway between stars of type A0 and F0. The sun is of spectral class G2 — two tenths of the way from class G0 to K0.

The famous American astronomer Henry Norris Russell proposed a scheme by which every student can remember the order of classes in the spectral sequence: The class letters are the first letters in the words, "Oh, Be A Fine Girl, Kiss Me!"

(b) Spectrum Analysis

At least half of the observing time of many large telescopes is assigned to the study of stellar spectra. Analyses of stellar spectra yield an enormous amount of information about the stars. Unfortunately, spectrum analysis is a very complicated subject. Here we shall only summarize how spectra are observed and what some of the more important data are that can be obtained.

The design of a spectroscope or spectrograph is described in Section 8.6e. Stellar spectra were first observed by means of a spectroscope placed at the focus of a telescope. With a magnifying glass (eyepiece) the observer viewed the spectrum of the light from a star gathered by the telescope and passed through the apparatus. Today, however, stellar spectra are usually observed photographically. With a long time exposure we can photograph a spectrum that is far too faint to be seen visually; moreover, a photograph is a permanent record that can be measured and reproduced at will.

In practice, the spectra of the star and a laboratory source are both photographed on the same negative. The laboratory source is usually iron vaporized in an electric arc; the light emitted from the glowing iron vapor is passed through portions of the slit of the spectrograph adjacent to the portion through which the starlight is passed. The spectrum of the iron arc consists of many bright emission lines whose wavelengths have been measured accurately in the laboratory. They serve as a convenient standard with which the absorption (or emission) lines in the star's spectrum can be compared, so that the wavelengths of the stellar lines can be determined. In this way, the chemical elements that give rise to the lines can be identified, and the Doppler shifts of those lines (due to the radial velocity of the star) can be measured.

We now summarize some of the more important kinds of data that can be obtained from spectrum analysis:

TEMPERATURE

The kind of temperature measured by an ordinary thermometer is called *kinetic temperature;* it is a measure of the kinetic energies of the atoms or molecules. Since we cannot place a thermometer in the photosphere of a star, we cannot measure its kinetic temperature directly, but must infer its temperature from the quality of the radiation that we receive from the star. Because that radiation emerges from a variety of depths in the outer layers of the star, a region called the stellar *photosphere,* it can not be expected to indicate a unique temperature that corresponds to any one layer. The temperature that we derive is a representative temperature, corresponding to some representative depth in the photosphere. What that depth is depends on the manner in which the temperature is measured. If the star's continuous spectrum, or a portion of it, is fitted to the spectral energy distribution of a black body (perfect radiator), for example, the temperature of that black body is called the star's *color temperature.* A temperature that is estimated from the extent to which lines of ionized atoms appear in the spectrum is called an *ionization temperature.* A temperature that is estimated from the extent to which lines of certain atoms in certain states of excitation are present is called an *excitation temperature.* The temperature of a perfect radiator that would emit the same total amount of radiant energy per unit area as the star does is the *effective*

temperature of the star; the effective temperature cannot be determined directly from the spectrum alone, but is found, rather, from the luminosity and radius of a star if they are known (see Chapter 15). If a star were a perfect radiator, and were in complete thermodynamic equilibrium, all these temperatures would agree. As it is, the different kinds of temperatures usually do agree with each other fairly well — a circumstance that results from the fact that the observable radiation from most stars emerges from a relatively small range of depths in their photospheres.

PRESSURE

The degree of ionization of atoms depends on the gas density (and thus the pressure) as well as the temperature. Thus comparison of the intensities of lines of an element in various states of ionization gives some indication of the pressure of the gases in a star's photosphere, as well as of the temperature.

THE PRESENCE OF PARTICULAR ELEMENTS

Dark lines of a majority of the known chemical elements have now been identified in the spectra of the sun and stars. Of course, the lines of *all* elements are not observable in the spectrum of a star, nor are lines of any one element visible in the spectra of *all* stars. As we have seen, because of variations among the stars in temperature and pressure of the photosphere, only certain of the prevailing kinds of atoms are able to produce absorption lines in any one star. The absence of the lines of a particular element, therefore, does not necessarily imply that that element is not present. Only if the physical conditions in the photosphere of a particular star are such that lines of an element *should* be visible, were the element present in reasonable abundance in that star, can we conclude that the absence of observable spectral lines implies low abundance of the element. On the other hand, spectral lines of an element,

FIG. 14-5 A portion of the solar spectrum and a comparison spectrum of
iron photographed with the same spectrograph. Notice that many of the dark lines
in the solar spectrum are matched by the bright lines in the comparison
spectrum, showing that iron is present in the sun. *(Lick Observatory.)*

in the neutral state or in one of its ionized states, certainly imply the presence of that element in the star.

RELATIVE ABUNDANCES OF THE ELEMENTS

Once due allowance has been made for the prevailing conditions of temperature and pressure in a star's photosphere, analyses of the strengths of absorption lines in its spectrum can yield information regarding the relative abundances of the various chemical elements whose lines appear. It is found that the relative abundances of the different chemical elements in the sun and in most stars (as well as in most other regions of space that have been investigated) are approximately the same. Hydrogen seems to comprise from 50 to 80 percent of the mass of most stars. Hydrogen and helium together comprise from 96 to 99 percent of the mass; in some stars they comprise more than 99.9 percent. Among the 4 percent or less of "heavy elements," neon, oxygen, nitrogen, carbon, magnesium, argon, silicon, sulfur, iron, and chlorine are among the most abundant. Generally, but not invariably, the elements of lower atomic weight are more abuandant than those of higher atomic weight.

RADIAL VELOCITY

The radial or "line-of-sight" velocity of a star can be determined from the Doppler shift of the lines

in its spectrum (Section 7.3b). Huggins made the first radial-velocity determination of a star in 1868. He observed the Doppler shift in one of the hydrogen lines in the spectrum of Sirius and found that the star was receding from the solar system.

ROTATION

If a star is rotating, unless its axis of rotation happens to be directed exactly toward the sun, one of its limbs approaches us and the other recedes from us, relative to the star as a whole. For the sun or a planet, we can observe the light from one limb or the other and measure directly the Doppler shifts that arise from the rotation. A star, however, appears as a point of light, and we are obliged to analyze the light from its entire disk at once. Nevertheless, if the star is rotating, part of the light from it, including the spectral lines, is shifted to shorter wavelengths and part is shifted to longer wavelengths. Each spectral line of the star is a composite of spectral lines originating from different parts of the star's disk, all of which are moving at different speeds with respect to us. The effect produced by a rapidly rotating star is that all its spectral lines are broadened in a characteristic manner. Fortunately this broadening is so characteristic that it usually can be distinguished from broadening produced by other sources. The amount of rotational broadening of the spectral lines, if observable, can be measured, and a lower

FIG. 14-6   Rotational broadening of spectral lines: *(top)* spectrum
of WZ Ophiuchi, which does not rotate rapidly and whose lines are relatively
sharp; *(bottom)* spectrum of UV Leonis, a rapidly rotating star
whose lines are very broadened and have a washed out appearance.
These spectrograms are negative prints.

limit to the rate of rotation of the star can be calculated.

TURBULENCE

If large masses of a star's photospheric layers have vertical turbulent motion, this motion, like rotation, causes a broadening of the spectral lines. Anomalous strengths of the lines in the spectra of some stars have been attributed to such turbulence in their photospheres.

MAGNETIC FIELDS

The Zeeman effect splits or broadens the spectral lines produced by gases in a strong magnetic field. The actual splitting of the spectral lines, for example, can be observed in light coming from the magnetic fields associated with spots on the sun (Chapter 17). However, in the spectra of those stars that possess appreciable magnetic fields, the lines are usually only broadened, because the separate components of a line are not resolved. Zeeman broadening can still be differentiated, however, because the light that is not absorbed in the components of the split lines is polarized. The

amount by which a spectral line is broadened by a magnetic field can be detected, therefore, by photographing the star's spectrum through a polarizing filter. A few stars have been found to have very strong magnetic fields. In some cases, these field strengths are variable. One possible explanation is that the field on such a star is localized, and rotation of the star alternately carries the face with the magnetic field into and out of our view.

SHELLS AND EJECTED GASES

In the spectra of some stars, absorption lines are observed that do not appear to originate in the photospheres. Sometimes such lines can be associated with shells or rings of material ejected by the star. If enough material is ejected, and if the star has radiation of high enough energy to excite or ionize the gas, the latter produces emission lines instead of absorption lines, superposed on the stellar spectrum. Examples are Wolf-Rayet stars, and early type emission-line stars. Stars that are ejecting matter are further discussed in Chapter 18.

Exercises

1. Star A is barely visible through a 100-in. telescope, and star B is barely visible through a 10-in. telescope of the same design and at the same site. What is the approximate difference in magnitudes of the two stars?

2. Find the absolute magnitudes of the following stars:

(a) Apparent magnitude $m = 6$; distance $r = 10$ parsecs.

(b) $m = 15$; $r = 100$ pc.

(c) $m = 10$; $r = 5$ pc.

(d) $m = 5$; $r = 100,000$ pc.

3. Find distances for the following stars:

(a) $m = 7.5$; $M = 7.5$

(b) $m = 8$; $M = -12$.

(c) $m = 8.5$; $M = 3.5$.

(d) $m = 20$; $M = 10$.

4. Draw a diagram showing how two stars of equal bolometric magnitude, but one of which is blue and the other red, would appear on photographs sensitive to:

(a) blue light;

(b) yellow light.

5. What is the distance, in parsecs, of the following star:

$m_{pv} = 8.9$; $M_{pg} = 4.7$; $CI = +0.8$.

6. Saturn is about 10 AU from the sun. Approximately what is the sun's apparent magnitude as seen from Saturn?

7. Arrange the stars of the following spectral types in order of *increasing* color index: (a) K; (b) B; (c) M; (d) O; (e) A; (f) G.

8. What are the probable approximate spectral types of stars described as follows:

(a) Balmer lines of hydrogen are very strong; some lines of ionized metals are present.

(b) Strongest lines are those of ionized helium.

(c) H and K lines of ionized calcium are the strongest in the spectrum; hydrogen lines show with only moderate strength; lines of neutral and ionized metals are present.

(d) Strongest features are molecular bands of titanium oxide.

9. Most stars have masses in the range 0.1 to 10 times that of the sun. Stellar radii, however, range from 0.01 to 1000 times the sun's radius. Find the surface gravity, relative to the sun's, of each of two stars, both of one solar mass, but one with a radius of 0.01 the sun's and the other with a radius of 1000 times the sun's.

10. Explain why we can only find a lower limit to the rotation rate of a star from the rotational broadening of its spectral lines.

CHAPTER 15

Stellar Masses and Sizes–Double Stars

Chapters 13 and 14 have dealt with the methods by which we determine the distances, motions, luminosities, temperatures, and atmospheric characteristics of stars. Now we shall consider the methods by which their masses and radii are determined. The observational problems involved in "weighing" and "measuring" the stars directly are very difficult to overcome, and we have accurate measures for only a handful of stars. This handful, however, makes it possible for us to devise and calibrate indirect methods that we can use to estimate, at least, the masses and radii of many other stars. Most of the data on the masses and sizes of stars come from analysis of *binary* or *double stars:* in this chapter, therefore, we shall be concerned especially with a discussion of binary stars.

15.1 DETERMINATION OF THE SUN'S MASS

The masses of stars must be inferred from their gravitational influences on other objects. We can measure the force exerted by the sun, for example, on the earth and other planets; and the mass of the sun can be determined more reliably than that of any other star.

The most direct way to calculate the mass of the sun is from the acceleration of the earth. For the sake of illustration, let us assume that the orbit of the earth is circular. Then the acceleration required to keep the earth in its orbit is the centripetal acceleration (Section 3.5):

$$a = \frac{v^2}{R},$$

where v is the earth's orbital speed and R is the radius of its orbit. This acceleration must be provided by the gravitational attraction of the sun on a unit mass at the earth's distance; that is,

$$a = \frac{v^2}{R} = \frac{GM_s}{R^2},$$

where G is the universal gravitational constant and M_s is the desired mass of the sun. Solving the above equation for the sun's mass, we obtain

$$M_s = \frac{v^2 R}{G}.$$

Both v and R are known from observation and G is determined from laboratory measurements (Section 3.6c). In metric units, $v = 3 \times 10^6$ cm/sec, $R = 1.49 \times 10^{13}$ cm, and $G = 6.67 \times 10^{-8}$. Substituting these values into the above formula, we find for the mass of the sun,

$$M_s = 2 \times 10^{33} \text{ gm} = 2 \times 10^{27} \text{ metric tons.}$$

Because the earth's orbit is nearly circular, the value thus found for the mass of the sun is very nearly the correct one.

15.2 BINARY STARS

It would be helpful if we could observe directly planets (if any exist) in orbital revolution about stars other than the sun, but unfortunately we are unable to do so. We do observe, however, many cases of double stars — systems of two stars that revolve about each other under the influence of their mutual gravitational attraction. Frequently,

FIG. 15-1   Three photographs, covering a period of about 12 years, which show the mutual revolution of the components of the double star Kruger 60. *(Yerkes Observatory.)*

we even find systems of three or more stars that belong to the same dynamical system.

(a) Discovery of Binary Stars

In 1650, less than half a century after Galileo turned a telescope to the sky, the Italian astronomer Jean Baptiste Riccioli observed that the star Mizar, in the middle of the handle of the Big Dipper, appeared through his telescope as *two* stars; Mizar was the first *double star* to be discovered. In the century and a half that followed, many other closely separated pairs of stars were discovered telescopically.

Usually, one star of a pair is brighter than the other. William Herschel assumed that the fainter star was the more distant, and that the brighter one was a foreground object. It occurred to him that it might be possible to measure the parallax (Section 13.3) of the nearer star with respect to the more distant one, which was so nearly in line with it. Accordingly, he began a search for such pairs, and between 1782 and 1821 he published three catalogues, listing more than 800 double stars. Actually, only rarely does a double star consist of one nearby and one distant star; the vast majority of these systems found by Herschel are *physical pairs* of stars, *revolving about each other.* This had

been suggested as early as 1767 by John Michell, who pointed out the extreme improbability of there being more than a very few close pairs of stars that are really only *optical doubles,* that is, lined up in projection.

One famous double star is Castor, in Gemini. The telescope reveals Castor to be two stars separated by an angle of about 5″. By 1804 Herschel had noted that the fainter component of Castor had changed, slightly, its direction from the brighter component. Here, finally, was observational evidence that one star was moving about another; it was the first evidence that gravitational influences exist outside the solar system. Herschel had failed in his program to facilitate parallax determinations but had found something of far greater interest. As he put it, he was like Saul, who had gone out to seek his father's asses and had found a kingdom. His son, John Herschel, continued the search for double stars, and prepared a catalogue (published posthumously) of more than 10,000 systems of two, three, or more stars.

If the gravitational forces between stars are like those in the solar system, the orbit of one star about the other must be an *ellipse.* The first to show that such is the case was Felix Savary, who in 1827 showed that the relative orbit of the two stars in the double system ξ *Ursae Majoris* is an ellipse, the stars completing one mutual revolution in a period of 60 years.

Another class of double stars was discovered by E. C. Pickering, at Harvard, in 1889. He found that the lines in the spectrum of the brighter component of Mizar (the first double star to be discovered) are usually *double,* but that the spacing of the components of the lines varies periodically, and at times the lines even become single. He correctly deduced that the brighter component of Mizar (Mizar A) itself, is really *two* stars that revolve about each other in a period of 104 days. When one star is approaching us, relative to the center of mass of the two, the other star is receding from us; the radial velocities of the two stars, and therefore the Doppler shifts of their spectral lines, are different, so that when the composite spectrum of the two stars is observed, each line appears double. When the two stars are both moving across our line of sight, however, they both have the same *radial* velocity (that of the center of mass of the pair), and hence the spectral lines of the two stars coalesce.

Stars like Mizar A, which appear as single stars when photographed or observed visually through the telescope, but which the spectroscope shows really to be double stars, are called *spectroscopic binaries;* systems that can be observed visually as double stars are called *visual binaries.*

In 1908 Frost found that the fainter component of Mizar, Mizar B, is also a spectroscopic binary.

Almost immediately following Pickering's discovery of the duplicity of Mizar A, Vogel discovered that the star Algol, in Perseus, is a spectroscopic binary. The spectral lines of Algol were not observed to be double, because the fainter star of the pair gives off too little light compared to the brighter for its lines to be conspicuous in the composite spectrum. Nevertheless, the periodic shifting back and forth of the lines of the brighter star gave evidence that it was revolving about an unseen companion; the lines of both components need not be visible in order for a star to be recognized as a spectroscopic binary.

The proof that Algol is a double star is significant for another reason. In 1669 Montonari had noted that the star varied in brightness; in 1783 John Goodricke established the nature of the variation. Normally, Algol is a second-magnitude star, but at intervals of $2^d20^h49^m$ it fades to one third of its regular brightness; after a few hours, it brightens to normal again. Goodricke suggested that the variations might be due to large dark spots on the star, turned to our view periodically by its rotation, or that the star might be eclipsed regularly by an invisible companion. Vogel's discovery that Algol is a spectroscopic binary verified the latter hypothesis. The plane in which the stars revolve is turned nearly edgewise to our line of

FIG. 15-2   Two spectra of the spectroscopic binary κ Arietis. When the components are moving at right angles to the line of sight *(bottom),* the lines are single. When one star is approaching us and the other receding *(top),* the spectral lines of the two stars are separated by the Doppler effect. *(Lick Observatory.)*

sight, and each star is eclipsed once by the other during every revolution. The eclipse of the fainter star is not very noticeable because the part of it that is covered contributes little to the total light of the system; the second eclipse can, however, be observed. A binary such as Algol, in which the orbit is nearly edge on to the earth so that the stars eclipse each other, is called an *eclipsing binary.*

Binary stars are now known to be very common; they may be the rule — not the exception. In the stellar neighborhood of the sun, somewhere between one half and two thirds of all stars are members of binary or multiple star systems.

We may summarize the types of binary stars before proceeding to explain how they are used to determine stellar masses.

OPTICAL BINARIES

An optical binary consists of two stars, in nearly the same line of sight, of which one is far more distant than the other; they are not true binary stars, and they are relatively rare.

VISUAL BINARIES

A visual binary is a gravitationally associated pair of stars; the members are either so near the sun, or so widely separated from each other (usually, both), that they can be observed visually (in a telescope) as two stars. Typical separations for the two stars in a visual binary system are hundreds of astronomical units; thus the orbital speeds of the stars are usually quite small and their orbital motion may not be apparent over a few decades of observation. Nevertheless, two closely separated stars are generally assumed to comprise a visual binary system if there is no reason to doubt that they are at the same distance from us and if they have the same proper motion and radial velocity, indicating that they are moving together through space. The *Index Catalogue of Double Stars* (1963) lists 64,247 visual binaries.

ASTROMETRIC BINARIES

Sometimes one member of what would otherwise be a visual binary system is too faint to be observed; its presence may be detected, however, by the "wavy" motion of its companion, revolving about the center of mass of the two stars as they move through space. In 1844 Bessel discovered that the bright star Sirius displays such a sinusoidal motion with a period of 50 years. Sirius remained an *astrometric binary* until 1862, when Alvan G. Clark found its faint companion — a member of the class of stars known as *white dwarfs* (discussed in Chapter 16).

SPECTROSCOPIC BINARIES

When the binary nature of a star is known only from the variations of its radial velocity (or of both radial velocities if the spectral lines of both stars are visible), it is said to be a spectroscopic binary. Over 700 systems have been analyzed.

SPECTRUM BINARIES

If the orbit of what would otherwise be a spectroscopic binary is turned nearly "face on" to us (that is, perpendicular to our line of sight), or if the masses of the member stars are so low that they have very small orbital velocities, we can see no radial-velocity variations. It may still be obvious, however, that there are two stars if the composite spectrum contains lines that are characteristic of both hot and cool stars and which would not be expected to occur in the spectrum of a single star. Such systems are called spectrum binaries.

ECLIPSING BINARIES

If the orbit of a binary system is turned nearly edge on to us, so that the stars eclipse each other, it is called an eclipsing binary.

The different kinds of binaries are not mutually exclusive. An eclipsing binary, for example, may *also* be a spectroscopic binary, if it is bright enough that its spectrum can be photo-

FIG. 15-3   Apparent relative orbit of a hypothetical visual binary system.

graphed, and if its radial-velocity variations have been observed.

(b) Mass Determinations of Visual Binary Stars

The two stars of a binary pair revolve mutually about their common center of mass (or barycenter), which in turn moves in a straight line among the neighboring stars. Each star, therefore, describes a wavy path around the course followed by the barycenter. With careful observations it is possible to determine these individual motions of the member stars in a visual binary system. It is far more convenient, however, simply to observe the motion of one star (by convention the fainter), about the other (the brighter). The observed motion shows the *apparent relative orbit*. The periods of mutual revolution for visual binaries range from a few years to thousands of years, but only for those systems with periods less than a few hundred years can the apparent relative orbits be determined with much precision; even then, a long series of observations covering a number of

decades is usually necessary. A typical apparent relative orbit, assembled from many observations, is shown in Figure 15-3.

The true orbit of the binary system will not, in general, happen to lie exactly in the plane of the sky (that is, perpendicular to the line of sight). Consequently, the apparent relative orbit is merely the *projection* of the *true relative orbit* into the plane of the sky. Now it is easy to show that when an ellipse in one plane is projected onto another plane (that is, is viewed obliquely), the projected curve is also an ellipse, although one of different eccentricity. However, the foci of the original ellipse do *not* project into the foci of the projected ellipse. Therefore, the brighter star, although it is located at one focus of the *true* relative orbit, is not at a focus of the *apparent* relative orbit. This circumstance makes it possible to determine the inclination of the true orbit to the plane of the sky. The problem is simply one of finding the angle at which the ellipse of the true relative orbit

FIG. 15-4   Relation between a true and apparent relative orbit.

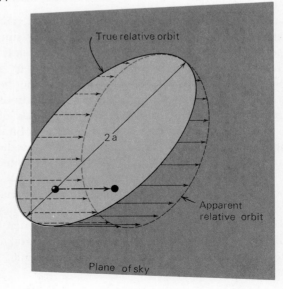

must be projected in order to account for the amount of displacement of the brighter star from the focus of the apparent relative orbit (see Figure 15-4). There are several techniques for solving this geometry problem, so that the shape and angular size of the true orbit can be found. The period of mutual revolution, of course, is observed directly if the system has completed one revolution during the interval it has been under observation; otherwise, the period must be calculated from the rate at which the fainter star moves in the relative orbit.

If the semimajor axis of the true relative orbit (that is, the semimajor axis the orbit would appear to have if it were seen face on) has an angular length, in seconds of arc, of a'', and if the system is at a distance of r pc, the semimajor axis, in astronomical units, is $r \times a''$ (see Exercise 2). The sum of the masses of the two stars, in solar units, is given by Kepler's third law:

$$m_1 + m_2 = \frac{(r \cdot a'')^3}{P^2}.$$

To find what share of the total mass belongs to each star, it is necessary to investigate the individual motions of the stars with respect to the center of mass of the system. The distance of each star from the barycenter is inversely proportional to its own mass.

As an example, consider the visual binary, Sirius A and Sirius B. The semimajor axis of the true relative orbit is about $7\frac{1}{2}''$ and the distance from the sun to Sirius is 2.67 pc. The period of the binary system is 50 years. The sum of the masses of the stars, then, is

$$m_1 + m_2 = \frac{(2.67 \times 7.5)^3}{(50)^2} = 3.2 \text{ solar masses.}$$

Sirius B, the fainter component, is about twice as far from the barycenter as Sirius A, and so has only about half the mass of Sirius A. The masses of Sirius A and Sirius B, therefore, are about 2 and 1 solar masses, respectively.

(c) Barnard's Star—An Interesting Astrometric Binary

Often, interesting information can be learned from the study of an astrometric binary. A very interesting example of a probable astrometric binary is Barnard's star — the star of largest known proper motion (Section 13.4a). Careful observations of the motion of Barnard's star over 51 years were analyzed in 1968 by P. van de Kamp of the Sproul Observatory. Van de Kamp's study shows that the star has a very slight wave in its proper motion with a period of 25 years, which suggests that it revolves about the barycenter of it and an unseen companion with this period. Barnard's star, however, has a very small mean distance from the barycenter — only 0".028, which at its distance of 1.83 pc corresponds to 0.051 AU. Barnard's star is an M5 main-sequence star; such stars have masses of only about 0.15 solar mass. If we denote the masses of Barnard's star and its companion, in solar units, by m_B and m_c, respectively, their mean distances from the barycenter by a_B and a_c AU, and their period of mutual revolution by P years, we have the following two equations that apply to them:

$$\frac{m_B}{m_c} = \frac{a_c}{a_B}$$

and

$$m_B + m_c = \frac{(a_B + a_c)^3}{P^2}.$$

If we now substitute the known quantities $P = 25$, $m_B = 0.15$, and $a_B = 0.051$, and solve the two equations for the two unknowns m_c and a_c, we find the surprising result

$$a_c = 4.5 \text{ AU}$$

and

$$m_c = 0.0017 \text{ solar mass;}$$

that is, the companion of Barnard's star has a

mass only 80 percent greater than that of Jupiter, and must therefore be a planet.

This is the only object of planetary mass known beyond the solar system. It is, of course, far too faint to observe directly with any telescope (its apparent magnitude is estimated to be about 30). It was discovered only because Barnard's star is so near the sun and because it has a small enough mass itself that its own motion about the barycenter could be detected. A planet of appreciably smaller mass, let alone anything as small as the earth, would never have been discovered.

(d) Spectroscopic Binaries

If the two stars of a binary system have a small linear separation, that is, if their relative orbit is small, there is little chance that they will be resolved as a visual binary pair. On the other hand, they have a shorter period, and their orbital velocities are relatively high, as compared with the stars of a visual binary system; unless the plane of orbital revolution is almost face on to our line

of sight, there is a good chance that we will be able to observe radial-velocity variations of the stars due to their orbital motions. In other words, they comprise a spectroscopic binary system.

Most spectroscopic binaries have periods in the range from a few days to a few months; the mean separations of their member stars are usually less than 1 AU. If the two stars of a spectroscopic binary are not too different in luminosity, the spectrum of the system will display the lines of both stars, each set of lines oscillating back and forth in the period of mutual revolution. More often, lines of only one star are observed. A graph showing the radial velocity of a member of a binary star system plotted against time is called a *radial-velocity curve* or, simply, a *velocity curve*.

In general, the orbit of each star about the barycenter is an ellipse, and the radial-velocity curve is skewed (Figure 15-5). Both stars of the system, of course, have the same period, for they stay on opposite sides of the barycenter as they revolve about it. Also, the orbits of both stars have

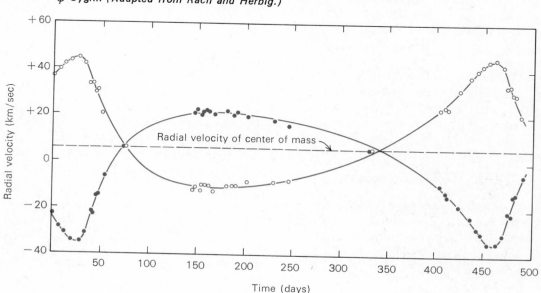

FIG. 15-5   Radial-velocity curves for the spectroscopic binary system φ Cygni. *(Adapted from Rach and Herbig.)*

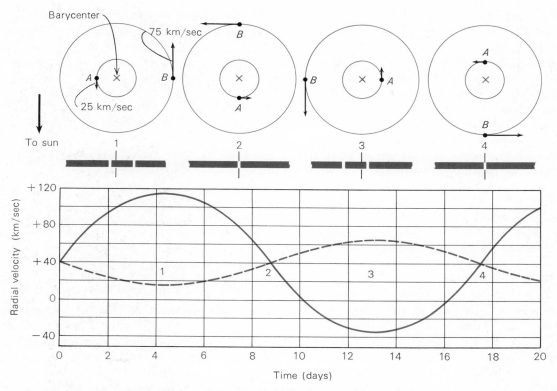

FIG. 15-6   Hypothetical spectroscopic binary system with circular orbits.

the same eccentricity but differ in size by a ratio that is in inverse proportion to their masses (since the masses of the stars are in inverse proportion to their distances from the barycenter — Section 4.1). Consequently, if the spectral lines of both stars are visible, so that radial-velocity curves can be plotted for each, those velocity curves are mirror images of each other, differing only in scale (or in *amplitude*); this is true for the binary system whose velocity curves are shown in Figure 15-5. From the shape of each radial velocity curve, it is possible to determine the eccentricity of the orbits of the stars, as well as the orientation of the major axis of the orbit in the orbital plane.

Here we shall consider, as an example of the analysis of a spectroscopic binary, the simple case in which the orbits of the stars are circular, and in which the spectral lines of both stars are observed. Figure 15-6 shows the two stars in their orbits, the Doppler shifts of a hypothetical spectral line, and the two radial-velocity curves. A complete cycle of variation of radial velocity — the period of the system — is seen to be 17.5 days. At position 1, star A has its maximum possible component of velocity toward the sun, and star B has its maximum possible component of velocity away from the sun. The conditions are reversed at position 3. At positions 2 and 4, both stars are moving across our line of sight and neither has a component of velocity in our line of sight due to orbital motion; both have the same radial velocity as that of the center of mass of the system, 40 km/sec. The radial velocity of star B ranges from 115 to −35 km/sec, a range of 150 km/sec. The maximum difference between the radial velocities of star B and that of the barycenter, then, is 75 km/sec. The corresponding value for star A is only 25 km/sec. Because both stars have the same period and since star A moves only one third as fast as star B, with respect to the barycenter, star A must have only one third as far to go to get around its orbit; its orbit must be one third the size of that of star B, and its mass, therefore, must be three times as great.

Stars A and B are moving in opposite directions with respect to their center of mass; thus the maximum radial velocity (as observed from the solar system) of star B with respect to star A must be the sum of the radial velocities of the two stars with respect to the barycenter, or 100 km/sec. If the orbital plane of the system were in our line of sight, this would be the relative orbital velocity of one star with respect to the other. As it is, however, the orbital plane is tilted at some unknown angle to our line of sight, and the 100 km/sec is only the maximum *radial component* of the relative velocity; the actual relative velocity will be greater by an unknown factor. Now, the distance around the relative orbit — its circumference — is the relative orbital velocity multiplied by the period — the time one star takes to get around the other. The distance between the stars, a, is the circumference of the orbit divided by 2π; that is,

$$a = \frac{V \times P}{2\pi},$$

where V is the relative velocity. If we substitute the observed lower limit to the relative velocity — in our example, 100 km/sec — for V in the above equation, we will obtain a lower limit to the separation of the stars. If this lower limit to a is applied to Kepler's third law, we obtain a lower limit to the sum of the masses of the stars:

$$m_1 + m_2 = \frac{a^3}{P^2}.$$

The calculation for the numerical example

given in Figure 15-6 may be illustrated as follows: A velocity of 100 km/sec is equivalent to about 20 AU/year, and a period of 17.5 days is about 0.048 year; thus a lower limit to a is given by

$$a = \frac{20 \times 0.048}{2\pi} = 0.153 \text{ AU}.$$

A lower limit to the sum of the masses, then, is

$$m_1 + m_2 = \frac{0.153^3}{0.048^2} = 1.6 \text{ solar masses.}$$

Since star A is three times as massive as star B, it has 75 percent of the total mass; lower limits to the individual masses of the stars are therefore

$$m_A \geqq 1.2 \text{ solar masses}$$

and

$$m_B \geqq 0.4 \text{ solar mass,}$$

where the symbol "$\geqq$" means "greater than or equal to."

The analysis is more difficult in the general case of elliptical, rather than circular, orbits, but it can, nevertheless, be carried out. The results are similar except that a is then the semimajor axis of the relative orbit instead of the constant separation between the stars that exists if the orbits are circular. In either case, individual masses of the stars are not found, only lower limits to their masses.† If the spectral lines of only one star are visible, we do not even find lower limits to the masses of the individual stars but only a relation between their masses, known as the *mass function*.* A recent (1967) catalogue lists orbital elements for 737 spectroscopic binary systems.

---

†What is actually found for each star is $m \sin^3 i$, where i is the inclination angle of the plane of orbital revolution to the plane of the sky.

*The *mass function* is

$$\frac{m_2^3 \sin^3 i}{(m_1 + m_2)^2},$$

where m_2 is the mass of the star whose spectrum is not observed.

The analyses of spectroscopic binaries, therefore, yield only lower limits to stellar masses. If large numbers of spectroscopic binary systems are investigated, and if it is assumed that their planes of orbital revolution are oriented at random in space, we can apply statistical corrections to the lower limits to find the *average* masses of the stars in the pairs that are studied. To find the masses of *individual* binary systems, however, we must have some way of determining the angle at which the orbital plane of the system is inclined to our line of sight (or to the plane of the sky); we can find this angle only when the orbit is almost edge on, so that the spectroscopic binary is also an eclipsing binary.

(e) Eclipsing Binaries

To simplify our discussion of eclipsing binaries we shall assume that the two stars in such a system revolve about each other in circular orbits. If the orbits were highly eccentric, some of the remarks

in this section would have to be qualified. Keeping in mind, therefore, that in practice far more sophisticated mathematical techniques are employed in the analysis of eclipsing binaries — techniques that take account of the eccentricities of the orbits — we pretend that the orbits are all circular. Actually the assumption is not so bad. Most stars in binary systems that are seen to eclipse each other are relatively close to each other. Their mutual proximity enhances the effects of their perturbations on each other's motions, and these perturbations tend to produce circular, or nearly circular, orbits.

During the period of revolution of an eclipsing binary, there are two times when the light from the system diminishes — once when the smaller star passes behind the larger one and is eclipsed, and once when the smaller star passes in front of the larger one and eclipses part of it. If the smaller star goes completely behind the larger one, that eclipse is *total* and the other eclipse half a period later is *annular* (see Figure 15-7). If the smaller

FIG. 15-7   Light curve of hypothetical eclipsing binary star with total eclipses.

FIG. 15-8   Light curve of hypothetical eclipsing binary star with partial eclipses.

star is never completely hidden behind the larger star, both eclipses are partial (Figure 15-8). Each interval during an eclipse when the light from the system is farthest below normal is called a *minimum*. Both minima will not, in general, be equally low in light. The same area of each star is covered during the time it is eclipsed; for example, if the eclipses are total or annular, an area equal to the total cross-sectional area of the smaller star is eclipsed in each minimum. The relative amount of light drop at each minimum, however, depends on the relative surface brightnesses of the two stars, and hence on their temperatures. *Primary minimum* occurs when the hotter star is eclipsed (whether it is a total, an annular, or a partial eclipse), and *secondary minimum* occurs when the cooler star is eclipsed. A graph of the light from an eclipsing binary system, plotted against time through a complete period, is called a *light curve*.

The most important data that are obtained from the analysis of the light curve of an eclipsing binary system are the sizes of the stars, relative to their separation, and the inclination of the orbit to our line of sight. To illustrate how the sizes of the stars are related to the light curve, we may consider a hypothetical eclipsing binary in which the stars are very different in size, and in which the orbit is exactly edge on, so that the eclipses are *central* (Figure 15-9). When the small star is at point *a* (*first contact*), and is just beginning to pass behind the large star, the light curve begins to drop. At point *b* (*second contact*), it has gone entirely behind the large star and the total phase of the eclipse begins. At *c* (*third contact*) it begins to emerge, and when it has reached *d* (*last contact*) the eclipse is over. During the time interval between first and second contact (or between third and last contacts) the small star has moved a distance equal to its own diameter. That time interval is in the same ratio to the period of the system as the diameter of the small star is to the circumference of the relative orbit. During

FIG. 15-9   Contacts in the light
curve of a hypothetical eclipsing
binary with central eclipses.

gives, respectively, the diameters of the small and large stars.

In actuality the orbits are not, generally, exactly edge on, and the eclipses are not central. To see how the inclination affects the light curve, consider the system shown in Figure 15-10. Here the smaller star does not pass behind the larger one in a direction perpendicular to its limb, but instead passes obliquely, and thus takes longer to be eclipsed. Moreover, it does not pass behind a full diameter of the disk of the large star but along a shorter chord, so that the entire eclipse, especially the total phase, is of shorter duration than when the eclipses are central. The effects of the sizes of the stars and of the inclination of the orbit are interdependent. However, it is a relatively

the time interval from first to third contacts (or from second to last contacts) the small star has moved a distance equal to the diameter of the large star; that time interval is to the period as the diameter of the large star is to the circumference of the relative orbit. We see, therefore, that the light curve alone gives the sizes of the stars in terms of the size of their orbit. If the lines of both stars are visible in the composite spectrum of the binary, both radial velocity curves can be found. Then the size of the relative orbit can be found, and we can determine the actual (linear) radii of the stars. In other words, the velocity of the small star with respect to the large one is known, and, when multiplied by the time intervals from first to second contacts and from first to third contacts,

FIG. 15-10   Contacts in the light
curve of a hypothetical eclipsing
binary with noncentral eclipses.

simple geometry problem, at least in principle, to separate these effects, and from the depths of the minima and the exact instants of the various contacts to calculate both the inclination of the orbit and the sizes of the stars, relative to their separation.

The foregoing discussion applies only to eclipses that are total and annular. If they are partial, the analysis is far more difficult, although it can still be accomplished. There are also various other complications which we have had to ignore, but a complete analysis must take them into account.

To summarize: From the analysis of the light curve of an eclipsing binary we can find the inclination of the orbit and the sizes of the stars, relative to their separation. If, in addition, we can measure the Doppler shifts of the spectral lines of both stars during their period of revolution, we can obtain their velocity curves. The analysis of the velocity curves, as described in the preceding section, leads to a determination of lower limits to the masses of the stars. The knowledge of the inclination of the orbit now allows us to convert these minimum values for the masses to actual masses for the individual stars. We can also convert the lower limit to the separation of the stars (or the semimajor axis of their relative orbit) to the actual value when the inclination is known; since the sizes of the stars relative to this separation are found from the light curve, we find their actual diameters, or radii. Finally, from the relative depths of the primary and secondary minima, we can calculate the relative surface brightnesses of the stars and hence their effective temperatures (the surface brightness of a star is proportional to the fourth power of its effective temperature). Among the thousands of known eclipsing binaries, however, there are only a few dozen that are so favorably disposed for observation that all the necessary data can be obtained. Only for these few binaries have complete analyses led to fairly reliable values of the masses and radii of their member stars.

15.3 THE MASS-LUMINOSITY RELATION

Studies of binary stars have provided a fairly accurate knowledge of the masses of a few dozen individual stars. When the masses and luminosities of those stars for which both of these quantities are well determined are compared, it is found that, in general, the more massive stars are also the more luminous. This relation, known as the *mass-luminosity relation*, is shown graphically in Figure 15-11. Each point represents a star of known mass and luminosity; its horizontal position (abscissa) indicates its mass, given in units of the sun's mass, and its vertical position (ordinate) indicates its luminosity, in units of the sun's luminosity.

It is seen that most stars fall along a narrow sequence running from the lower left (low mass, low luminosity) corner of the diagram to the upper right (high mass, high luminosity) corner. The relation between the mass and luminosity of a star is not accidental or mysterious but results from the fundamental laws that govern the internal structures of stars. We return to the matter in Chapter 20. It is estimated that about 90 percent of all stars obey the mass-luminosity relation. These, as we shall see later, are the so-called *main-sequence stars* (Chapter 16). Stars that are still contracting after recent formation from interstellar matter (Chapter 20), giants, and white dwarfs (Chapter 16) are the most numerous among those stars to which the relation does *not* necessarily apply. We can usually identify those nonconforming stars by their spectral characteristics; for most other stars the mass-luminosity relation provides a useful means of estimating the masses of stars of known luminosity that do not happen to be members of visual or eclipsing binary systems.

In particular, it should be noted how very much greater the range of stellar luminosities is than the range of stellar masses. Luminosities of stars are roughly proportional to their masses raised to the 3.5 power. Most stars have masses between $\frac{1}{10}$ and 50 times that of the sun; according to the mass-luminosity relation, however, the correspond-

FIG. 15-11   Mass-luminosity relation. The three points at the bottom represent white dwarf stars, which do not conform to the relation.

ing luminosities of stars at either end of the range are respectively less than 0.01 and about 10^6 solar luminosities. The intrinsically faintest known star has a luminosity of 10^{-6} that of the sun; its mass is probably greater than 0.01 of the sun's. If two stars differ in mass by a factor of 2, their luminosities would then be expected to differ by a factor of 10.

15.4 DIAMETERS OF STARS

The eclipsing binary systems, as we have seen, provide one means of determining the diameters of stars, but of course only of those stars that happen to be members of eclipsing systems for which the necessary analysis can be carried out. It would be convenient if the angular sizes could be measured directly for many stars of known distances; then their linear diameters could be calculated just as they are for the moon or planets. The sun is unfortunately the only star whose angular

size can be resolved optically and whose diameter can be calculated simply. There are a few other stars, however, whose angular sizes are only slightly beyond the limit of resolution of the largest telescopes and which can be measured with a device known as the *stellar interferometer.*

It is beyond the scope of this book to explain the stellar interferometer, invented by the American physicist, A. E. Michelson. It can be employed to increase, effectively, the resolving power of a telescope. In the interferometer light reflected from a mirror at each end of a long beam to a telescope is brought to a focus by the telescope as an interference pattern. Analysis of the pattern can lead to a determination of the angular size of the light source. In 1920 Michelson and Pease mounted an interferometer with a 20-ft beam on the front of the 100-inch telescope at Mount Wilson. With the apparatus, they were able to measure the angular sizes of seven giant stars that are large enough and near enough so that their disks, although too small to photograph or to see directly, are still just large enough to measure with the interferometer. The data for these stars are given in Table 15.1.

TABLE 15.1 Stars Measured with the Stellar Interferometer

STAR	ANGULAR DIAMETER	DISTANCE (IN PC)	LINEAR DIAMETER (IN TERMS OF SUN'S)
Betelgeuse	0″.034†	150	500
(α Orionis)	0.042		750
Aldebaran			
(α Tauri)	0.020	21	45
Arcturus			
(α Bootis)	0.020	11	23
Antares			
(α Scorpii)	0.040	150	640
Scheat			
(β Pegasi)	0.021	50	110
Ras Algethi			
(α Herculis)	0.030	150	500
Mira			
(ο Ceti)	0.056	70	420

†Variable in size.

An electronic analogue of the stellar interferometer has been applied more recently to the measurement of star diameters. Separate optical telescopes, placed up to hundreds of feet apart, and equipped with photomultiplier tubes, are used to observe the same star simultaneously. The electric currents generated in the two tubes are brought together to a single amplifier. The electrical impulses carry information that can be made to interfere in a way somewhat analogous to the production of fringes in the optical interferometer. The radio astronomers Brown and Twiss first used the device to measure the angular diameter of the star Sirius.

A recent Brown-Twiss type of interferometer, assembled in Australia, uses two telescopes that can be separated up to 200 m. Each uses a mosaic mirror 6.5 m in diameter, which in turn consists of 251 separate small mirrors, all mounted to reflect the light from a star to one point. These special-purpose mirrors are not of high optical quality and could not be used for ordinary telescopic observations. They are adequate, however,

to focus starlight on the light-sensitive surface of a photo-multiplier. This large intensity interferometer can achieve a resolution down to 5×10^{-4} seconds of arc and by late in 1969 it had been used to measure the angular diameters of 15 stars. These range from Sirius, whose angular diameter is 0″.00585, to Epsilon Orionis, with an angular diameter of only 0″.00070.

The vast majority of stars, however, are too remote for even sophisticated interferometers to be able to measure their angular sizes directly.

(a) Stellar Radii from Radiation Laws

For most stars we must use an indirect method by which we can calculate their radii from theory. The theory involved is the Stefan-Boltzmann law (Section 7.4d); we calculate the radius of a spherical perfect radiator that has the same luminosity and effective temperature that a star does.

The luminosity of a star can be obtained by the procedure discussed in Chapter 14, and the temperature of a star can be obtained in various ways, as from its color or its spectrum. Now the energy emitted per unit area of a star (given by Stefan's law), multiplied by its entire surface area, gives the star's total output of radiant energy — that is, its luminosity. Since the surface area of a sphere of radius R is $4\pi R^2$, the luminosity of a star is given by

$$L = 4\pi R^2 \times \sigma T^4.$$

The above equation can be solved for the radius of the star.

(b) Summary of Stellar Diameters

The few dozen good geometrical determinations of stellar radii come from (1) direct measure of the sun's angular diameter, (2) measures of the angular diameters of seven nearby giant stars with the stellar interferometer, (3) measures of the angular sizes of 15 other stars with an electronic adaptation of the stellar interferometer (the method may well have been applied to more stars by the time this is read), and (4) analyses of the light curves and radial-velocity curves of eclipsing binary systems. All other determinations of stellar radii make use of the radiation laws; the validity of this indirect method is verified by noting that it gives the correct radii for those stars whose sizes can also be measured by geometrical means.

Exercises

1. Describe how Kepler's third law (as formulated by Newton) can be used to calculate the sun's mass. What are the most convenient units? If you can, do the calculation and check the figure given in the text for the solar mass.

2. Show that the linear distance corresponding to the angular separation a'', in seconds of arc, of two objects at a distance of r pc, is $r \times a''$ AU.

3. Many eclipsing binaries are observed that are *not* observed as spectroscopic binaries. Can you offer an explanation?

4. Why do you suppose most visual binaries have long periods, and most spectroscopic binaries shorter periods?

5. Explain why the two stars in a binary system must have orbits of identical eccentricity.

6. Although the periods of known eclipsing binaries range from 4^h39^m to 27 years, the average of their periods is less than the average period of all known spectroscopic binaries. Can you give the explanation for this?

7. Suppose both components of a binary star were of the same mass as the sun. Find their period of mutual revolution if the semimajor axis of their relative orbit is:

(a) 2 AU?

(b) 8 AU?

8. Find the combined mass of the two stars in a binary system whose period of mutual revolution is 2 years and for which the semimajor axis of the relative orbit is 2 AU.

9. A hypothetical spectroscopic-eclipsing binary star is observed. The period of the system is three years. The maximum radial velocities, with respect to the center of mass of the system, are as follows:

Star A: $\frac{4}{3}\pi$ AU/yr

Star B: $\frac{2}{3}\pi$ AU/yr

(a) What is the ratio of the masses of the stars?

(b) Find the mass of each star (in solar units). Assume the eclipses are central.

10. Most eclipsing binaries are stars of high luminosity. Why do you suppose this to be the case?

11. Use the mass luminosity relation to estimate the mass of a star with (a) a luminosity 100 times the sun's; (b) a luminosity 10,000 times the sun's.

12. Give the radii, in units of the solar radius, of stars of the following luminosities (L) and temperatures (T), where L and T are in terms of the solar values:

(a) $L = 100$; $T = 2$.

(b) $L = 10,000$; $T = \frac{1}{2}$.

(c) $L = 1/100$; $T = 1$.

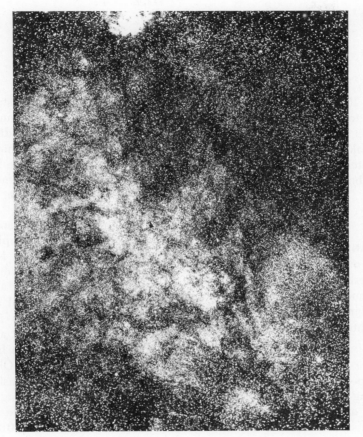

CHAPTER 16

The Stars in General

Chapters 13 through 15 describe the methods by which we are able to obtain basic information about individual stars — their distances, masses, luminosities, radii, colors, temperatures, motions, and so on. This chapter summarizes the data that have been gathered about the normal stars and the stellar population.

16.1 THE NEAREST AND THE BRIGHTEST STARS

Let us consider first our most conspicuous stellar neighbors, the brightest-appearing stars in the sky. Appendix 13 lists some of the properties of the 20 brightest stars. Many of these are double- or triple-star systems; data are given, in such cases, for each component.

(a) The Brightest Stars

The most striking thing about the brightest-appearing stars is that they are bright not because they are nearby, but because they are actually of high intrinsic luminosity. Of the 20 brightest stars listed in Appendix 13, only 6 are within 10 pc of the sun. Even among the approximately 3000 stars with apparent magnitudes less than 6.0, only about 60 are within 10 pc. Most naked-eye stars are tens or even hundreds of parsecs away and are many times more luminous than the sun. Figure 16-1 is a histogram showing the distribution among various absolute visual magnitudes of the 30 brightest-appearing stars (the absolute visual magnitude of the sun is 4.7).

From an inspection of Appendix 13 or Figure 16-1 we might gain the impression that the sun is far below average among stars in luminosity. This is not so. Stars of high luminosity are rare — so rare that the chance of finding one within a small volume of space, say, within 10 pc of the sun, is very slight. Why, then, are the most common, intrinsically faint stars not among the most common naked-eye stars, while rare, highly luminous stars are?

The question is best answered with the help of some numerical examples. The sun, whose absolute visual magnitude is +4.7, would appear as a very faint star to the naked eye if it were 10 pc away. Stars much less luminous than the sun would not be visible at all at that distance. Stars with absolute magnitudes in the range +10 to +15 are very common, but a star of absolute magnitude +10 would have to be within 1.6 pc to be visible to the naked eye. Only Alpha Centauri is closer than this. The intrinsically faintest star observed has an absolute magnitude of about +20. For this star to be visible to the naked eye, it would have to be within 0.016 pc, or 3200 AU. The star could not even be photographed with the 200-inch telescope if it were more distant than 50 pc. It is clear, then, that the vast majority of nearby stars, those less luminous than the sun, do not send enough light across the span of interstellar distances to be seen without optical aid.

In contrast, consider the highly luminous stars. Stars with absolute magnitudes of 0 have luminosities of about 100 times that of the sun. They are far less common than stars less luminous than the sun, but they are visible to the naked eye even out to a distance of 160 pc. A star with an absolute magnitude of −5 (10,000 times the sun's luminosity)

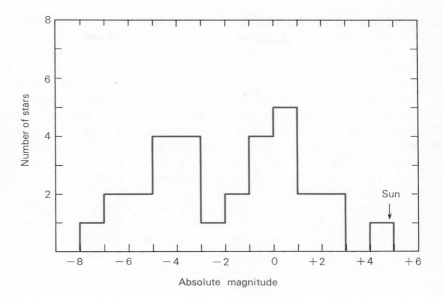

FIG. 16-1   Distribution among absolute magnitudes of the 30 brightest appearing stars.

can be seen without a telescope to a distance of 1600 pc (if there is no dimming of light by interstellar dust — see Chapter 19). Such stars are very rare, and we would not expect to find one within a distance of only 10 pc; the volume of space included within a distance of 1600 pc, however, is about 4 million times that included within a distance of only 10 pc. Hence many stars of high luminosity are visible to the unaided eye.

(b) The Nearest Stars

Evidently, the brightest naked-eye stars do not provide a very representative sample of the stellar population. Let us turn then to the nearest known stars. Appendix 12 lists data for the 37 known stars within 5 pc of the sun (some are double or multiple systems). The table must not be taken as the "final word," for additional nearby stars are discovered from time to time, and the measurements of distances, luminosities, and so on, for known nearby stars are being continually refined.

On the other hand, the table does indicate some general characteristics of the sun's nearest stellar neighbors.

First, the table shows that only three of the 37 stars are among the 20 brightest stars: Sirius, Alpha Centauri, and Procyon. This fact is further confirmation that the nearest stars are not the brightest-appearing stars. The nearby stars also tend to have large proper motions, as would be expected (Section 13.4). In fact, the large proper motions of many of these stars led to the discovery that they are located nearby. Another interesting observation is that 14 of the 37 stars are really binary- or multiple-star systems; the table contains, actually, 54 rather than 37 stars. Thirty-one of these 54 stars are members of systems containing more than one star.

The most important datum concerning the nearest stars is that most of them are intrinsically faint. Only eight of the nearest stars are visible to the unaided eye. Only three are as intrinsically luminous as the sun; 26 have absolute magnitudes fainter than +10. If the stars in our immediate stellar neighborhood are representative of the stellar population in general, we must conclude that the most numerous stars are those of low luminosity.

We can estimate a lower limit to the mean density of stars in space in the solar neighborhood. There are at least 54 stars within 5 pc (counting the members of binary- and multiple-star systems). A sphere of radius 5 pc has a volume of about 520 pc^3. Since this volume of space contains at least 54 stars, the density of stars in space is at least one star for every 10 pc^3; the actual stellar density, of course, can be greater than this figure if there are undiscovered stars within 5 pc. The mean separation between stars is the cube root of 10, or about 2.1 pc. If the matter contained in stars could be spread out evenly over space, and if a typical star has a mass 0.4 times that of the sun, the mean density of matter in the solar neighborhood would be about 3×10^{-24} gm/cm^3.

(c) The Luminosity Function
Once the numbers of stars of various intrinsic luminosities have been found, the relative numbers of stars in successive intervals of absolute magnitude within any given volume of space can be established. This constitutes what is called the *luminosity function*. Figure 16-2 shows the luminosity function for stars in the solar neighborhood, as it has been determined by W. Luyten. Compare Figure 16-2 with Figure 16-1.

The sun, we see, is more luminous than the vast majority of stars. Most of the stellar mass is contributed by stars that are fainter than the sun. On the other hand, the relatively few stars of higher luminosity than the sun compensate for their small numbers by their high rate of energy output. It takes only 10 stars of absolute magnitude 0 to outshine 1000 stars fainter than the sun, and only one star of absolute magnitude −5 to outshine 10,000 stars fainter than the sun. Most of the starlight from our part of space, it turns out, comes from the relatively few stars that are more luminous than the sun.

16.2 THE HERTZSPRUNG-RUSSELL DIAGRAM
In 1911 the Danish astronomer E. Hertzsprung

FIG. 16-2 Luminosity function of stars in the solar neighborhood.

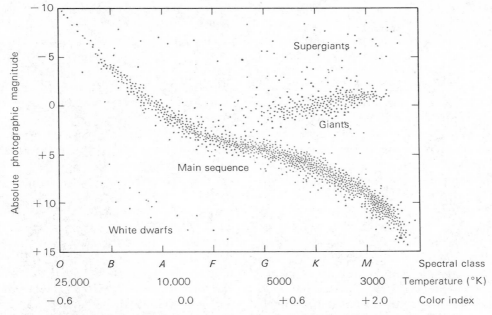

FIG. 16-3   Hertzsprung-Russell diagram for stars of known distance.

compared the colors and luminosities of stars within several clusters by plotting their magnitudes against their colors. In 1913 the American astronomer Henry Norris Russell undertook a similar investigation of stars in the solar neighborhood by plotting the absolute magnitudes of stars of known distance against their spectral classes. These investigations, by Hertzsprung and by Russell, led to an extremely important discovery concerning the relation between the luminosities and surface temperatures of stars. The discovery is exhibited graphically on a diagram named in honor of the two astronomers — the *Hertzsprung-Russell,* or *H-R, diagram.*

(a) Features of the H-R Diagram

Two easily derived characteristics of stars of known distances are their absolute magnitudes (or luminosities) and their surface temperatures. The abso-

lute magnitudes can be found from the known distances and the observed apparent magnitudes. The surface temperature of a star is indicated either by its color or its spectral class. Before the development of yellow- and red-sensitive photographic emulsions and, of course, photoelectric techniques, spectral classes of stars were usually used to indicate their temperatures. Now that stellar colors can be measured with precision, the color index is more often employed, even though the use of spectral classes is still of great value.

If the absolute photographic magnitudes of stars are plotted against their temperatures (or spectral classes, or color indices), an H-R diagram like that of Figure 16-3 is obtained. The most significant feature of the H-R diagram is that the stars are not distributed over it at random, exhibiting all combinations of absolute magnitude and temperature, but rather cluster into certain parts of the diagram. The majority of stars are aligned along a narrow sequence running from the upper left (hot, highly luminous) part of the diagram to the lower right (cool, less luminous) part. This band of points is called the *main sequence*. A substantial number of stars, however, lie above the main sequence on the H-R diagram, in the upper right (cool, high luminosity) region. These are called *giants*. At the top part of the diagram are stars of even higher luminosity, called *supergiants*. Finally, there are stars in the lower left (hot, low luminosity) corner known as *white dwarfs*. To say that a star lies "on," or "off," the main sequence does not refer to its position in space, but only to the point that represents its luminosity and temperature on the H-R diagram.

An H-R diagram, such as Figure 16-3, that is plotted for stars of known distance does not show the relative proportions of various kinds of stars, because only the nearest of the intrinsically faint stars can be observed. To be truly representative of the stellar population, an H-R diagram should be plotted for all stars within a certain distance. Unfortunately, our knowledge is reasonably complete only for stars within a few parsecs of the sun, among which there are no giants or supergiants. It is estimated that about 90 percent of the stars in our part of space are main-sequence stars and about 10 percent are white dwarfs. Only 1 percent or less are giants or supergiants.

Chapter 20 deals with the theoretical interpretation of the distribution of stars on the H-R diagram.

(b) Method of Spectroscopic Parallaxes

One of the most important applications of the H-R diagram is in the determination of stellar distances. Suppose, for example, that a star is known to be a spectral class G2 star on the main sequence. Its absolute magnitude could then be read off the H-R diagram at once; it would be about +5. From this absolute magnitude and the star's apparent magnitude, its distance can be calculated (Section 14.2b).

In general, however, the spectral class alone is not enough to fix, unambiguously, the absolute magnitude of a star. The G2 star described in the last paragraph could have been, for example, a main-sequence star of absolute magnitude +5, a giant of absolute magnitude 0, or a supergiant of still higher luminosity. We recall however, that pressure differences in the atmospheres of stars of different sizes result in slightly different degrees of ionization for a given temperature. It will be seen in the next subsection that giant stars are larger than main-sequence stars of the same spectral class and that supergiants are larger still. In 1913 Adams and Kohlschütter, at the Mount Wilson Observatory, first observed the slight differences in the degrees to which different elements are ionized in stars of the same spectral class but different luminosities (and therefore different sizes). It is now possible to classify a star by its spectrum, not only according to its temperature (spectral class) but also according to whether it is a main-sequence star, a giant, or a supergiant.

Knowing its spectral class, and whether it is a main-sequence star, a giant, or a supergiant, we can locate a star's position on the H-R diagram uniquely. Its absolute magnitude, therefore, is also known, and its distance can be calculated. Distances determined in this way, from the spectral and luminosity classes, are said to be obtained from the *method of spectroscopic parallaxes*.

There are some stars that do not fit into the standard classification scheme (see, especially, Chapter 18). The method of spectroscopic parallaxes does not apply to them.

(c) Extremes of Stellar Luminosities, Radii, and Densities

Let us now investigate the extremes in size, luminosity, and density found for stars. The most massive stars are the most luminous ones, at least for main-sequence stars. These stars have absolute magnitudes of -6 to -8. A few stars are known that have absolute bolometric magnitudes of -10; they are a million times as luminous as the sun. These super-luminous stars, most of which are at the upper left on the H-R diagram, are very hot spectral-type O and B stars, and are very blue. These are the stars that would be the most conspicuous at very great distances in space.

Consider now the stars at the upper right corner of the H-R diagram, both giants and super-giants. Some have surface temperatures less than half that of the sun, so each unit area of the surface of such a star must emit only $\frac{1}{16}$ or less as much light as the sun does (Section 15.4a). Yet they are at least a few hundred times as luminous as the sun (if they are giants) or some thousands of times as luminous (if they are supergiants). We see, then, that the cooler stars of high luminosity are called giants or supergiants because they are truly large stars.

Let us consider an illustration that is rather extreme but uses actually known stars. Suppose a red, cool supergiant has a surface temperature of 3000°K and an absolute bolometric magnitude of -5. This star has 10,000 times the sun's luminosity but only half its surface temperature. Since each unit area of the star emits only $\frac{1}{16}$ as much light as a unit area of the sun, its total surface area must be 160,000 times the sun's. Its radius, therefore, is 400 times the sun's radius. If the sun and its planets could be placed in the center of such a star, the star's surface would lie beyond the orbit of Mars.

In the past several years special telescopes with infrared detectors have been built — especially by R. Leighton and his associates at the Mount Wilson and Palomar Observatories. These telescopes have been used to search for very cool stars that emit most of their energy in the invisible infrared. Many such stars have been discovered.

These extreme red giants, or *infrared giants*, can have sizes even more extreme than the supergiant described in the last paragraph. It is difficult, however, to determine distances to them, so their absolute magnitudes are not well known.

Red giant stars have extremely low mean densities. The volume of the star described in the example in the paragraph before last would be 64 million times the volume of the sun. The masses of such giant stars, however, are probably at most only 50 solar masses, and very likely much less. Plaskett's Star, a spectral-type O star, has a mass of at least 50 solar masses but is one of the most massive stars known. If we assume that the supergiant star with 64 million times the sun's volume has only 10 times its mass, we find that it has just over 1 ten-millionth the sun's mean density, or only about 2 ten-millionths the density of water; the outer parts of such a star would probably constitute an excellent laboratory vacuum.

In contrast, the very common red, cool stars of low luminosity at the lower end of the main sequence are much smaller and more compact than the sun. As an example, consider such a "red dwarf," the star Ross 614B, which has a surface temperature of 2700°K and an absolute bolometric magnitude of about $+13$ ($\frac{1}{2300}$ of the sun's luminosity). Each unit area of this star emits only $\frac{1}{20}$ as much light as a unit area of the sun, but to have only $\frac{1}{2300}$ the sun's luminosity, the star must have only about $\frac{1}{115}$ the sun's surface area, or $\frac{1}{11}$ its radius. A star with such a low luminosity also has a low mass (Ross 614B has a mass of about $\frac{1}{12}$ that of the sun), but still would have a mean density about 100 times that of the sun's. Its density must be higher, in fact, than that of any known solid found on the surface of the earth.

The faint red main-sequence stars are not the stars of the most extreme densities, however. The white dwarfs, at the lower left corner of the H-R diagram, have the highest densities known to exist in nature.

(d) The White Dwarfs

The first white dwarf stars to be discovered were the companions to the stars 40 Eridani, Sirius, and Van Maanen's Star. Sirius, the brightest appearing star in the sky, is the most conspicuous star in the constellation of Canis Major (the big dog). It is a rather interesting coincidence that Procyon, the brightest star in the constellation of Canis Minor (the little dog) also has a white dwarf companion. Both Sirius and Procyon are visual binaries within 5 pc of the sun. One other star within 5 pc, 40 Eridani, is a multiple-star system that contains a white dwarf.

The white-dwarf companion of 40 Eridani, 40 Eridani B, is a good example of a typical white dwarf. Its absolute magnitude is 10.7 and its temperature is about 12,000°K; it has, therefore, 2.1 times the sun's surface temperature and $1/200$ its luminosity. Its surface area is only $1/(200 \times 2.1^4)$, or $1/3900$, of the sun's, which gives it a radius of 0.016 and a volume of 4.1×10^{-6} the sun's. Its mass, however, is 0.43 times that of the sun, so its mean density is $0.43/(4.1 \times 10^{-6})$, or about 100,000, times the mean density of the sun, and nearly 200,000 times the density of water.

One of the tests that Einstein suggested for the general theory of relativity is that light waves should show a slight shift to longer (redder) wavelengths in the presence of a strong gravitational field, (Section 3.7a). Light leaving the surface of the sun should suffer such a red shift, but the effect is so small compared to other effects that it cannot be observed with certainty. At the surfaces of white dwarfs, on the other hand, there exist very strong gravitational fields. The theory of relativity predicts that the redshift of light leaving the white-dwarf companion of 40 Eridani should be equivalent to that produced by a radial velocity of 17 km/sec. Now the actual radial velocity of 40 Eridani B is known from the velocities of its non-white-dwarf companions. In 1954, D. M. Popper observed a redshift for this star that cor-responds to a velocity 21 km/sec greater than that which should be produced by its known radial velocity. Popper's observed redshift agrees within the uncertainty of the observations with that predicted by the theory of relativity. Since then, gravitational redshifts have been observed for other white dwarf stars as well.

Since white dwarfs are intrinsically faint stars, they must be relatively nearby to be observed. Today they are usually discovered by searching for faint stars of large proper motion. Some hundreds of white-dwarfs have now been found, largely due to the efforts of the astronomer W. J. Luyten of the University of Minnesota.

The theory of white dwarfs predicts that the masses of all white dwarfs should range from 0.1 to 1.2 solar masses, and their corresponding radii should range from 4 times to less than half that of the earth (the more massive white dwarfs being the smaller). The predicted mean densities of these stars range from about 50,000 to over 1,000,000 times that of water. At such densities, matter cannot exist in its usual state. Although it is still gaseous, its atoms are completely stripped of their electrons, and the latter are obliged to move according to certain restrictive laws. The matter in white dwarfs is said to be *degenerate*. The white dwarfs are believed to be a final state of stellar evolution (see Chapter 20).

16.3 THE DISTRIBUTION OF THE STARS IN SPACE

In the immediate neighborhood of the sun, the stars seem to be distributed more or less at random (except for their tendency to form small clusters). The larger the volume of space we survey, the more stars we find, and if allowance is made for the fact that the faintest stars become invisible at larger distances, it is found that the number of stars we can count is roughly proportional to the cube of the distance to which we look. Eventually, however, the stars do thin out

FIG. 16-4   The Great Star Cloud in Sagittarius.
(Yerkes Observatory.)

more rapidly in some directions than in others. The way they thin out is a clue to the nature of the stellar system to which the sun belongs. The idea that the sun is a part of a large system of stars was suggested as early as 1750 by Thomas Wright in his *Theory of the Universe.* Immanuel Kant, the great German philosopher, suggested the same hypothesis 5 years later. It was the German-English astronomer William Herschel, however, who first demonstrated the nature of the stellar system.

(a) Herschel's Star Gauging

Herschel sampled the distribution of stars about the sky by a procedure he called *star gauging.* He observed that in some directions he could count more stars through his telescope than in other directions. In 1785 he published the results

of gauges or counts of stars that he was able to observe in 683 selected regions scattered over the sky. While in some of these fields he could see only a single star, in others he was able to count nearly 600. Herschel reasoned that in those directions in which he saw the greatest numbers of faint stars, the stars extended the farthest, and in other directions they thinned out at relatively short distances. As a result of his star gauging, Herschel arrived at the conclusion (only partially correct, as we shall see) that the sun is, indeed, inside a great sidereal system, and that the system is disk-shaped, roughly like a grindstone, with the sun near the center.

(b) The Phenomenon of the Milky Way

All of us who have looked at the sky on a moonless night away from the glare of city lights are aware of the Milky Way, a faint, luminous band of light that completely encircles the sky. Galileo solved the first mystery of the Milky Way when he turned his telescope on it and saw that it really consists of myriads of faint stars. Herschel's grindstone hypothesis solved the second mystery by explaining why the Milky Way should appear as a band all the way around the sky.

It must be recalled that we view our sidereal system from the inside. Figure 16-5 shows a portion of the "grindstone," viewed edge-on. The sun's position is at *O.* If we look from *O* toward either face of the wheel, that is, in directions *a* or *b,* we

FIG. 16-5   Explanation of the Milky Way.

FIG. 16-6   NGC 5897, open star cluster in Libra.
Photographed with the 200-inch telescope.
(Hale Observatories.)

see only those stars that lie between us and the nearest boundary of the stellar system. In these directions in the sky, therefore, we see only scattered stars. On the other hand, if we look edge-on through the wheel, say in directions *c* or *d*, we encounter so many stars along our line of sight that we get the illusion of a continuous band of light. Since the greatest dimensions of the grindstone extend in all directions along its flat plane, the band of light extends completely around the sky. This band of light is the Milky Way; it is simply the light from the many distant stars that appear lined up in projection when we look edge-on through our own flattened stellar system.

(c) The Galaxy

We call our stellar system the *Galaxy,* or sometimes, the *Milky Way Galaxy.* In our modern view,

the Galaxy is more complicated than Herschel's image of the grindstone. It is a vast, wheel-shaped system of some 100 billion stars, with a diameter that probably exceeds 30,000 pc (100,000 LY). The flattened shape of the Galaxy is a consequence of its rotation. The sun, about two-thirds of the way from the center out to the rim of the wheel, moves at a speed of about 250 km/sec (about 150 mi/sec), to complete its orbital revolution about the galactic center in some 200 million years. (The rotation of the Galaxy is not quite like that of a wheel, for it does not rotate as a solid body — see Chapter 19.)

At the center of the Galaxy is a huge hub or *nucleus* of stars, where the stars are somewhat closer together than they are in the solar neighbor-

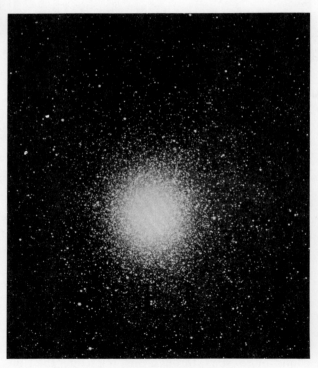

FIG. 16-7   Globular star cluster, Omega Centauri.
(Harvard Observatory.)

hood (although still light-months or light-years apart). Extending outward from the nucleus, and winding through the disk of the Galaxy, like the spirals of light in a gigantic pinwheel, are the *spiral arms*. The spiral arms consist of vast clouds of gas and cosmic dust — the so-called interstellar medium. Associated with these gas and dust clouds are many young stars, a few of which are very hot and luminous. It is in the interstellar gas and dust clouds of the spiral arms that star formation is believed to be still taking place. The sun is thought to be located in or near a spiral arm.

In addition to individual stars and clouds of interstellar matter, the Galaxy contains many *star clusters* — groups of stars having a common origin and probably a common age. The most common star clusters, numbering in the thousands, are the *open* or *galactic clusters*. Typically, an open cluster consists of a few hundred stars, loosely held to-

gether by their mutual gravitation, and moving together through space. The open clusters are located in the main disk of the Galaxy and are usually in or near spiral arms. Besides the open clusters, there are over a hunderd *globular clusters* — beautiful, spherically symmetrical clusters, each containing hundreds of thousands of member stars. Most of the globular clusters are scattered in a roughly spherical distribution about the main wheel of the Galaxy, grouped around it rather like bees around a flower. They form a more or less spherical *halo* or *corona* surrounding the main body of the *Galaxy*.

Our Galaxy is not alone in space. Today we know that far beyond its borders there are countless millions of other galaxies, extending as far as we can see in all directions in space. Our Galaxy, interstellar matter, star clusters, and other galaxies are the subjects of some of the later chapters.

Exercises

1. From the data in Appendix 12 plot the luminosity function (in the form of a histogram) for the stars nearer than 5 pc. How does your plot compare with Luyten's function shown in the text? How does it compare with the histogram given in the text for the brightest stars? Discuss the differences.

2. Now from the data in Appendix 12 plot the nearest stars on a Hertzsprung-Russell diagram. How does your plot differ from the Hertzsprung-Russell diagram shown in the text? Discuss the reasons.

3. Now plot the brightest stars, given in Appendix 13, on the same Hertzsprung-Russell diagram you made for Exercise 2; use different symbols to represent the brightest stars, so you can distinguish them from the nearest ones. Are there any points that you have already plotted (in Exercise 2)? Now how does your diagram compare with the one in the text? Discuss any differences. Why is this diagram not now representative of all stars in a given volume of space?

4. Following are data on 5 stars:

STAR	m	SPECTRAL CLASS	
1	8	A5	white dwarf
2	20	O9	main sequence
3	5	M5	main sequence
4	10	M3	supergiant
5	10	M2	giant

(a) Which is most distant?
(b) Which is nearest?
(c) Which is hottest?
(d) Which is coolest?
(e) Which is most dense?
(f) Which is least dense?

5. Calculate what your own weight would be at the surface of a typical red supergiant and at the surface of a typical white dwarf.

6. Suppose the Milky Way were a band of light extending only half way around the sky (that is, a semicircle). What then would you conclude about the sun's location in the Galaxy?

CHAPTER 17

A Typical Star–
The Sun

The sun, a typical garden-variety star, is the subject of the present chapter. We shall, however, postpone a consideration of the solar interior to Chapter 20.

17.1 OUTER LAYERS OF THE SUN

The only parts of the sun that can be observed directly are its outer layers, collectively known as the sun's *atmosphere*. The solar atmosphere is not stratified into physically distinct layers with sharp boundaries. Yet there are three general regions, each having substantially different properties, even though there is a gradual transition from one region to the next. These are the *photosphere,* the *chromosphere,* and the *corona.*

(a) The Solar Photosphere

What we see when we look at the sun is the solar photosphere. The photosphere is not a discrete surface but covers the range of depths from which the solar radiation escapes. As one looks toward the limb of the sun, his line of sight enters the photosphere at a grazing angle, and the depth below the outer surface of the photosphere to which he can see is even less than at the center of the sun's disk. The light from the limb of the sun, therefore, comes from higher and cooler regions of the photosphere. Analysis of this *limb darkening* can be used to determine the variation of temperature with depth in the photosphere. From the data obtained from observations of limb darkening, using our knowledge of the physics of gases, and the way in which atoms absorb and emit light, we can calculate how the temperature and density

vary through the photosphere. It is found that within a depth of about 160 mi the pressure and density increase by a factor of 10, while the temperature climbs from 4500 to 6800°K. The corresponding range of temperature on the Fahrenheit scale is 7600 to 11,700°. At a typical point in the photosphere, the pressure is only a few hundredths of sea-level pressure on the earth, and the density is about 1 ten-thousandth of the earth's atmospheric density at sea level.

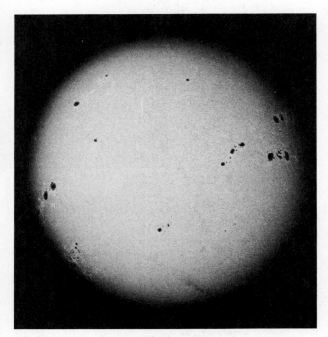

FIG. 17-1   The sun, photographed under excellent conditions, showing a large number of sunspots; September 15, 1957. *(Hale Observatories.)*

Stellar spectra of Pleiades
photographed with an objective prism.
(Warner and Swasey Observatory.)

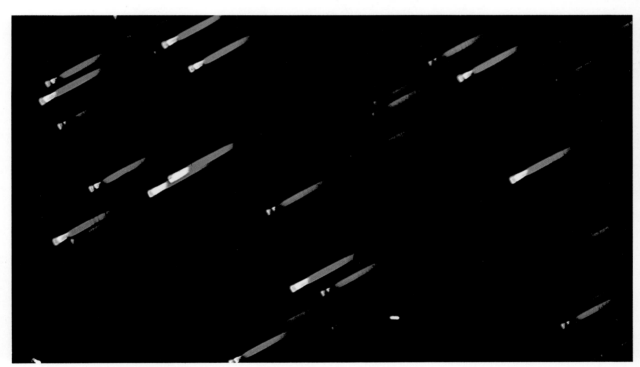

Stellar spectra of Hyades photographed with an objective prism.
(Warner and Swasey Observatory.)

The Sun in the red light of the first
Balmer line of hydrogen. *(Carl Zeiss.)*

Six photographs, over a 40-minute interval, of an
active region of the sun showing a spot group
and a filament. Each is a composite of two black
and white photographs made through filters that
pass light of wavelength slightly less than and
greater than that of the first Balmer line of
hydrogen. The former is printed through a green
filter and the latter through a red one. When the
light from the two photographs is mixed, objects
not moving in the line of sight appear yellow.
Because of the Doppler shift, regions approaching
us (rising in the solar chromosphere) appear
red in the above prints, and those receding
(falling) appear green. Thus we can follow
the motions of the filament by the color
changes. *(Lockheed Solar Observatory.)*

More than 60 of the elements known on the earth have now been identified in the solar spectrum. Those that have not been identified in the sun either do not produce lines in the observable spectrum or are so rare on the earth that they cannot be expected to produce lines of observable strength on the sun unless, proportionately, they are far more abundant there. Most of the elements found in the sun are in the atomic form, but more than 18 types of molecules have been identified. Most of the molecular spectra are observed only in the light from the cooler regions of the sun, such as the sunspots.

The relative abundances of the chemical elements in the sun are similar to the relative abundances found for other stars. About 60 to 80 percent of the sun (by weight) is hydrogen; 96 to 99 percent is hydrogen and helium. The remaining few percent is made up of the other chemical elements, in approximately the amounts that are described in Section 14.4b.

(b) The Chromosphere

There is a change in the physical state of the gases just above the photosphere. The photosphere ends at about the place where the opacity has dropped to a value too low to result in appreciable absorption of light. Gases extend far beyond the photosphere, but they are transparent to most radiation. The region of the sun's atmosphere that lies immediately above the photosphere is the chromosphere.

Until this century the chromosphere could be observed only when the photosphere was occulted by the moon during a total solar eclipse. In the seventeenth century several observers described what appeared to them as a narrow red "streak" or "fringe" around one limb of the moon during a brief instant after the sun's photosphere had been covered. Not until the careful observations of the solar eclipses of 1842, 1851, and 1860, however, was much attention paid to the chromosphere.

Some of these observations were made by photographic methods and established beyond doubt the existence of the chromosphere. In 1868 the spectrum of the chromosphere was first observed; it was found to be made up of bright lines, which showed that the chromosphere consists of hot gases that are emitting light in the form of emission lines. These bright lines are difficult to observe against the bright light of the photosphere but appear in the spectrum of the light from the extreme limb of the sun just after the moon has eclipsed the photosphere. They disappear within a few seconds, when the moon has covered the chromosphere as well. Because of the brief instant during which the chromospheric spectrum can be photographed during an eclipse, its spectrum; when so observed, is called the *flash spectrum*. The element *helium* (from *helios,* the Greek word for "sun") was discovered in the chromospheric spectrum before its discovery on earth in 1895.

Today it is possible to photograph both the chromosphere and its spectrum outside of eclipse. One instrument used for this purpose is the *coronagraph,* a telescope in which a black disk at the focal plane occults the photosphere, producing an artificial eclipse. The chromosphere can also be photographed in limited narrow spectral regions through monochromatic filters.

The chromosphere is usually considered to be a zone of the solar atmosphere about 1000 to 2000 mi thick, but in its upper region it breaks up into a forest of jets (called *spicules*), and other irregular motions, so the position of the upper boundary of the chromosphere is somewhat arbitrary. Its reddish color (from whence comes its name) arises from the fact that one of the strongest emission lines in the visible part of its spectrum is the bright red line due to hydrogen (the Hα line — first line in the Balmer series — Section 7.5b). The density of the chromospheric gases decreases upward above the photosphere, but spectrographic studies show that the temperature *increases* through the

FIG. 17-2   The solar corona, photographed during total eclipses. Note the difference in the shape of the corona during sunspot minimum *(left)* and sunspot maximum *(right)*. *(Yerkes Observatory.)*

chromosphere, from 4500°K at the photosphere to 100,000°K or so at the upper chromospheric levels. The processes that heat the corona (see below) evidently also heat the upper chromosphere.

(c) The Corona

The chromosphere merges into the outermost part of the sun's atmosphere, the corona. Like the chromosphere, the corona was first observed only during total eclipses, but unlike the chromosphere, the corona has been known for many centuries; it is referred to by Plutarch and was discussed in some detail by Kepler. Many of the early investigators regarded the corona as an optical illusion, but photography proved its actual existence in the nineteenth century. Its spectrum was first observed in 1869 by the American astronomers Harkness and Young. The corona extends millions of miles above the photosphere and emits half as much light as

the full moon. Its invisibility, under ordinary circumstances, is due to the overpowering brilliance of the photosphere. Like the chromosphere, the corona can now be photographed, with the coronagraph and other instruments, under other than eclipse conditions. The first successful photograph made with a coronagraph was taken by the inventor of the instrument, Bernard Lyot, at Pic du Midi, France, in 1930.

The emission lines in the corona were not identified with known chemical elements until 1942, when they were recognized as lines of calcium, iron, and nickel that can only be produced in gases of very low density and high temperature.

The coronal gas density is extremely low. Despite its large size, the corona is very tenuous and cannot contribute appreciably to the total mass of the sun. Even more interesting, the lines of iron arise from atoms that have lost 13 or more electrons, and, very rarely, there are lines of calcium from atoms that have lost 14 or more electrons. Lines of highly ionized iron, nickel, argon, and calcium are frequently observed; much energy is required to ionize these atoms so highly (Section 7.5e). Such a high degree of ionization requires a temperature of at least 1 million degrees Kelvin. The corona, in other words, is many times hotter than the photosphere. It must be noted, however, that because of the low density of the corona, it does not contain much actual heat, despite its high temperature. One heating mechanism of the corona is believed to be shock waves originating from convective currents in the photosphere. Because of its high temperature, the bulk of the spectral lines emitted by the corona lie in the far ultraviolet. In fact, observations of the sun made from rockets show that an appreciable fraction of the far-ultraviolet solar spectrum is light from the corona and the chromosphere. Rockets and satellites have permitted observations of these lines, ranging up to those of iron 16 times ionized.

17.2 PHENOMENA OF THE SOLAR ATMOSPHERE

In its gross characteristics the sun is very stable, but the detailed features of its atmosphere are constantly changing.

(a) Photospheric Granulation

Either direct telescopic observation or photography shows that the photosphere is not perfectly smooth but has a mottled appearance resembling rice grains — this structure of the photosphere is now generally called *granulation*. Typically, granules are about 1000 km in diameter; the smallest observed are about 300 km across. They appear as bright spots surrounded by narrow darker regions.

The motions of the granules can be studied by the Doppler shifts in the spectra of gases just above them. It is found that the granules themselves are columns of hotter gases arising from

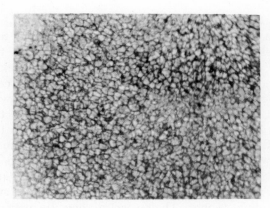

FIG. 17-3   A portion of the solar photosphere photographed from a balloon at an altitude of 80,000 feet. At this altitude, excellent resolution of the solar granulation is possible because disturbances produced by the earth's atmosphere ("seeing") are greatly reduced. *(Project Stratoscope, Princeton University.)*

below the photosphere. As the rising gas reaches the photosphere it spreads out and sinks down again. The darker intergranular regions are the cooler gases sinking back. The centers of the granules are hotter than the intergranular regions by 50 to 100°K. The vertical motions of the granules have speeds of about 2 or 3 km/sec. Individual granules persist for about 8 minutes. The granules, then, appear to be the tops of convection currents of gases seen through the photosphere.

Detailed studies of the motions of photospheric gases show that the granules themselves form part of a structure of still larger scale called *supergranulation*. Supergranules are cells about 30,000 km in diameter, within which there is a flow of gases from center to edge. In addition to the vertical currents of gases in the granules and the center-toward-edge flow in the supergranules, in each region of the sun (from 3500 to 7000 km across) the gases rhythmically pulse up and down with speeds of about ⅓ km/sec, taking about 5 minutes for a complete cycle. The surface gases of the sun, then, comprise a seething, sloshing sea.

(b) Sunspots

The most conspicuous of the photospheric features are the *sunspots*. Occasionally, spots on the sun are large enough to be visible to the naked eye, and such have been observed for many centuries. It was Galileo, however, who first showed that sunspots are actually on the surface of the sun itself, rather than being opaque patches in the earth's atmosphere or the silhouettes of planets between the sun and earth (Section 3.2b).

We know today that the spots are regions where the gases are cooler that those of the surrounding regions. Their spectra show them to be up to 1500°K cooler than the photosphere itself. Sunspots are nevertheless hotter than the surfaces of many stars. If they could be removed from the sun, they would be seen to shine brightly; they appear dark only by contrast with the hotter, brighter surrounding photosphere.

Individual sunspots have lifetimes that range from a few hours to a few months. They are first seen as small dark "pores" somewhat over 1000 mi in diameter. Most of them disappear within a day, but a few persist and may last for a week or, occasionally, much longer. If a spot lasts and develops, it is usually seen to consist of two parts: an inner darker core, the *umbra*, and a surrounding less dark region, the *penumbra*. Many spots become much larger than the earth, and a few have reached diameters of 100,000 mi. Frequently spots occur in groups of from two to twenty or more. If a group contains many spots, it is likely to include two large ones, oriented approximately east-west, and many smaller spots clustered around the two principal ones. The leading principal spot (in the direction toward which the sun is rotating) is most often the largest one of the group. The largest groups are very complex and may have over a hundred spots. Like storms on the earth, sunspots may move slowly on the surface of the sun, but their individual motions are slow when compared to the solar rotation, which carries them across the disk of the sun.

(c) The Sunspot Cycle

In 1851 a German apothecary and amateur astronomer, Heinrich Schwabe, published an important conclusion he had reached as a result of his observations of the sun over the previous decade. He found that the number of sunspots visible, on the average, varied with a period of about 10 years. Since Schwabe's work, the *sunspot cycle* has been clearly established. Although individual spots are short-lived, the total number of spots visible on the sun at any one time is likely to be very much greater during certain periods, the periods of *sunspot maximum*, than at other times, the periods of *sunspot minimum*. Sunspot maxima have occurred at an average interval of 11.1 years, but the intervals between successive maxima have ranged from as little as 8 years (from 1830 to 1838) to as long as 16 years (from 1888 to 1904). During sunspot maxima, more than 100 spots can often be seen on the sun at once. During sunspot minima, the sun sometimes has no visible spots.

FIG. 17-4   A large spot group on the sun, photographed on May 17, 1951. *(Hale Observatories.)*

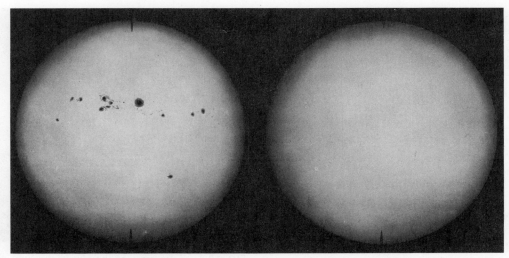

FIG. 17-5   Direct photographs of the sun near the time of sunspot maximum *(left)* and near sunspot minimum *(right). (Hale Observatories.)*

At the beginning of a cycle, just after a minimum, a few spots or groups of spots appear at latitudes of about 30° on the sun. As the cycle progresses, the successive spots occur at lower and lower latitudes, until, at the maximum of the cycle, their average latitude is about 15°. Near minimum, the last few spots of a cycle appear at about 8° latitude. About the same time, the next cycle begins with a few spots occurring simultaneously at higher latitudes. Sunspots almost never appear at latitudes greater than 40° or less than 5°. The locations of sunspots on the sun in both the Northern and Southern Hemispheres are related to the sunspot cycle in the same way; however, sunspot activity in one hemisphere may dominate for long periods.

The occurrence of the sunspot cycle, as well as of the spots themselves, is still unexplained. There is no general agreement on any of the many theories that have been proposed to explain them.

(d) Magnetic Fields on the Sun

As stated in Section 14.4b, each spectral line is split up into several components in the presence of a magnetic field, the phenomenon known as the Zeeman effect. In 1908 the American astronomer George E. Hale observed the Zeeman effect in the spectrum of sunspots and found them to possess strong magnetic fields. Magnetic fields are usually measured in terms of a unit known as the gauss. For example, the magnetic field of the earth (which aligns our compass needles) has a strength of about 1 gauss. The magnetic fields observed in sunspots range from 100 to nearly 4000 gauss. This is 10 times as great as the field of a good alnico magnet, and moreover, is spread over a region tens of thousands of miles across. The magnetic field is present in the region surrounding a spot as well as in the spot itself, and persists even after the spot has disappeared. There is also a much weaker, general magnetic field of the whole sun itself.

Whenever sunspots are observed in pairs or in groups containing two principal spots, one of the spots usually has the magnetic polarity of a north-seeking magnetic pole, and the other has the opposite polarity. Moreover, during a given cycle, the leading spots of pairs (or leading principal spots of groups) in the Northern Hemisphere all tend to have the same polarity, while those in the Southern Hemisphere all tend to have the opposite polarity. During the next sunspot cycle, however, the polarity of the leading spots is reversed in each hemisphere. For example, if during one cycle the leading spots in the Northern Hemisphere all had the polarity of a north-seeking pole, the leading spots in the Southern Hemisphere would have the polarity of a south-seeking pole; during the next cycle, the leading spots in the Northern Hemisphere would have south-seeking polarity and those of the Southern Hemisphere would have north-seeking polarity. We see, therefore, that the sunspot cycle does not repeat itself as regards magnetic polarity until *two* 11-year maxima have passed. For this reason the sunspot cycle may be said to last 22 years, rather than 11.

FIG. 17-6   Filtergram of the sun in the light of the first Balmer line of hydrogen. *(Lockheed Solar Observatory.)*

(e) Narrow-band Photographs of the Sun

In order to see regions of the sun that lie directly above the photosphere, we may observe in spectral regions to which the photospheric gases are especially opaque — at the centers of strong absorption lines such as those of hydrogen and calcium.

In 1892 Hale and, independently, Deslandres in France, invented the *spectroheliograph,* with which the sun can be photographed in light comprising only a very narrow spectral region, generally that corresponding to a small part of a particular line of some element. Today, although spectroheliograms are still used, it is more common to employ special filters that transmit light comprising only a very narrow spectral region. Most often, a small range of wavelengths is chosen that isolates light in the centers of one of the absorption lines of ionized calcium in the ultraviolet (the K line of ionized calcium) or the Hα line of hydrogen in the red. These spectral lines appear dark when viewed against the rest of the solar spectrum, but they are not completely dark;

what light remains in the centers of the lines is light that is emitted from atoms of calcium or hydrogen, respectively, in the chromosphere. Thus, these photographs reveal the appearance of the chromosphere in the light of calcium, or hydrogen. Photographs made through these filters are called *filtergrams.*

Spectacular motion pictures have been taken with spectroheliographs and through narrowband-pass filters. Time-lapse photographs, in which frames are exposed every few seconds or every few minutes and then run through a projector at normal speed, show in a dramatic way changes that occur in the solar chromosphere.

(f) Plages (Flocculi) and Faculae

Filtergrams and spectroheliograms in the light of calcium and hydrogen show bright "clouds" in the chromosphere in the magnetic-field regions around sunspots. These bright regions were formerly called *flocculi* ("tufts of wool") but are now more commonly known as *plages.* Calcium and hydrogen plages are also sometimes seen in regions where there are no visible sunspots, but these regions are generally those of higher-than-average magnetic fields.

Plages are often referred to as "clouds of calcium" or "clouds of hydrogen" (depending on the wavelength region used to make the filtergram). The description is misleading, however. All the chemical elements, including hydrogen and calcium, probably maintain uniform relative abundance distributions over the entire chromosphere. The plages, rather, are regions where calcium and hydrogen happen to be emitting more light at the observed wavelengths. These elements are partially ionized throughout most of the visible chromosphere, and some of the atoms emit light as they capture electrons and become neutral (or less ionized) or as those atoms (or ions) cascade down through the various excited energy levels. The plages, then, are not clouds of any particular element, but are regions where some of the atoms of the element observed are changing their states of ionization or excitation and are emitting more light than other regions. Plages of hydrogen and calcium usually occur in approximately the same projected regions at the same time, but they display different small-scale structure.

Light is sometimes emitted at many wavelengths and can be seen in white light, that is, in the direct image of the sun. These "white-light" plages are called *faculae* ("little torches"), and were first described by Galileo's contemporary Christopher Scheiner. Faculae are seen best near the limb of the sun where the photosphere is not so bright and the contrast is more favorable for their visibility.

(g) Spicules

The chromosphere also contains many small jet-like spikes of gas rising vertically through it. These features, called *spicules,* occur at the edges of supergranule cells, but when viewed near the limb of the sun so many are seen in projection that they give the effect of a forest of them. They show up best when the chromosphere is viewed in the light of hydrogen. They consist of gas jets moving upward at about 30 km/sec and rising to heights of from 500 to 20,000 km above the photosphere. Individual spicules last only 10 minutes or so. Through the spicules matter continually flows into the corona.

(h) Prominences

Among the more spectacular of coronal phenomena are the *prominences*. Prominences have been viewed telescopically during solar eclipses for centuries. They appear as red flamelike protuberances rising above the limb of the sun. Prominences can now be viewed at any time on spectroheliograms and filtergrams. Motions of prominences are exhibited in motion pictures. The gross features of some, the *quiescent* prominences,

FIG. 17-7   A solar prominence 205,000 miles high, photographed in violet light of the calcium K line, July 2, 1957. *(Hale Observatories.)*

may remain nearly stable for many hours, or even days, and may extend to heights of tens of thousands of miles above the solar surface. Others, the more active prominences, move upward or have arches that surge slowly back and forth. The relatively rare *eruptive* prominences appear to send matter upward into the corona at speeds up to 700 km/sec, and the most active *surge* prominences may move upward at speeds up to 1300 km/sec. Some eruptive prominences have reached heights of over 1 million miles above the photosphere. When seen silhouetted on the disk of the sun, prominences have the appearance of irregular dark filaments.

Superficially, prominences appear to be material ejected upward away from the sun, but the motion pictures show that whereas a prominence may grow in size and rise higher and higher above the photosphere, the actual material in the prominence most often appears to move downward in graceful arcs, evidently along lines of magnetic force. Apparently, most prominences form from coronal material that cools and moves downward, even though the disturbance that characterizes the prominence may move upward. Prominences are cool and dense regions in the corona where atoms and ions are capturing electrons and emitting light. Their origin is unknown, but it is significant that they usually originate near regions of sunspot activity and lie on the boundary between regions of opposite magnetic polarity. Quiescent prominences are supported by coronal magnetic fields, and eruptive prominences evidently result from sudden changes in the magnetic fields. Prominences seem to be further symptoms of the same general disturbances that produce spots and plages, that is, magnetic-field regions.

FIG. 17-8   Four successive photographs (spectroheliograms) of the great explosive prominence of June 4, 1946, taken with a coronagraph. The total elapsed time between the first *(top)* and the last *(bottom)* picture was one hour. *(Harvard Observatory.)*

(i) Flares

Occasionally, the chromospheric emission lines (the Hα and ionized calcium lines in particular) in a small region of the sun brighten up to unusually high intensity. Such an occurrence is called a *flare*. A flare is usually discovered on a spectroheliogram that is being made in the light of one of the spectral lines that brighten. It appears as an intensely bright spot on the photograph a few thousand or tens of thousands of kilometers in diameter. Very rarely, the continuous spectrum of that part of the solar surface affected also brightens during a flare. These white-light flares are among the most intense observed. A flare usually reaches its maximum intensity a few minutes after its onset. It fades out more slowly, and after an interval that ranges from a few more minutes to a few hours, it disappears altogether.

During major flares, an enormous amount of energy is released. The visible light emitted is, of course, very small compared to that from the entire sun, but a flare covering only a thousandth of the solar surface can actually outshine the sun in the ultraviolet. X-rays are emitted as well and are observed from space probes and satellites. In addition, matter is thrown out at speeds of 500 to 1000 km/sec. X-rays and atomic nuclei ejected from the sun by flares not only disturb the ionosphere in the earth's atmosphere (producing radio "fadeouts") but comprise a hazard to space travelers as well. Major, potentially hazardous, flares are fortunately rare, only a few occurring each year, but smaller flares occur every day.

Flares are most frequent in the regions of complex sunspot groups with complicated magnetic field structure. Material thrown into the corona from a flare expands and cools and, as ions and electrons recombine, it may emit light and be seen as a loop prominence over the active region of the sun. Loop prominences are the hottest of all prominences and are indicative of rather violent activity in the sun's atmospheres. As is the case with almost all solar phenomena, we do not know the source of energy of flares or the mechanism that triggers them into existence, but we feel that they are somehow connected with the magnetic fields of sunspot regions.

(j) Summary

There are many solar phenomena that seem to be associated with a particular kind of disturbance in the solar atmosphere. The disturbances are accompanied by strong magnetic fields, sunspots, plages, prominences, and solar flares. All these phenomena are related to the semiregular sunspot cycle of 11 (or perhaps 22) years. We shall see that radio radiation from the sun, as well as solar-terrestrial effects, seem also to be associated with these centers of solar activity. Even the shape of the corona varies with the sunspot cycle. During sunspot maximum it is nearly spherical and is distended, while at sunspot minimum it is contracted but extends to relatively greater distances near the plane of the solar equator.

17.3 SOLAR ROTATION

Galileo first demonstrated that the sun rotates on its axis when he observed the apparent motions of the sunspots as the turning sun carried them across its disk. He found from his spot observations that the rotation period of the sun is a little less than 1 month. The period was determined more accurately by subsequent observations of others. In 1859 Richard Carrington found that in the equatorial regions the sun rotates in about 25 days, but that at a latitude of 30° the period is about 2½ days longer. The sun's rotation rate can be determined also from the difference in the Doppler shifts of the light coming from the receding and approaching limbs (Section 9.2b). The solar rotation determined from the Doppler method confirms that the sun rotates most rapidly at low

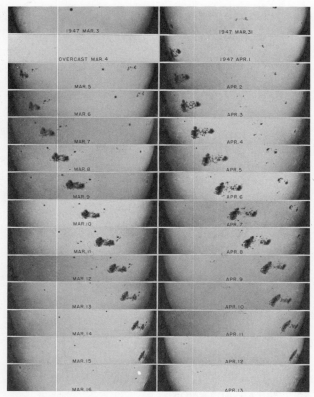

FIG. 17-9   Series of photographs showing the motions of sunspots, indicating the solar rotation. *(Hale Observatories.)*

latitudes, and with a longer period at high latitudes. At a latitude of 75°, the rotation period is about 33 days, and at the poles it may be near 35 days. Its *direction* of rotation is from west to east, like the orbital revolution of the planets and (except for Uranus and Venus) like their axial rotations. It is possible for different parts of the sun to rotate at different rates because, of course, it is fluid rather than solid like the planets. The details of its differential rotation, however, are not yet completely explained.

17.4 THE RADIO SUN

In Great Britain in 1942, unexpected noise was picked up on radar receivers. It was subsequently learned that the source of this noise was the sun, which emits electromagnetic radiation at radio as well as optical wavelengths. Since World War II, radio observations of the sun have been made regularly at many radio astronomical observatories. Shortwave radio energy (near 1 cm wavelength) can escape the sun from the lower chromosphere. The corona is more and more opaque, however, to longer and longer radio wavelengths. Those of 15 m escape the sun only if they originate high in the corona. Thus, by observing the sun at different radio wavelengths, we observe to different depths in the corona and chromosphere and can determine the heights in the solar atmosphere at which various disturbances giving rise to radio emission occur. Solar flares are often accompanied by exceptionally strong surges of radio energy called *bursts,* that rise through the corona with an outward velocity that may reach 1500 km/sec.

If the sun is typical among stars as a radio emitter, we would expect radio emission from other stars, because of their distances, to be too feeble to detect. The various cosmic radio sources known are not ordinary stars; we shall discuss them in later chapters.

17.5 SOLAR-TERRESTRIAL EFFECTS

In 1852, 1 year after Schwabe announced his discovery of the solar sunspot cycle, three investigators, Edward Sabine in England, and Rudolf Wolf and Alfred Gautier in Switzerland, independently discovered that sunspot activity was correlated with magnetic storms on the earth. During geomagnetic storms, the earth's magnetic field is disturbed, and the direction of the compass needle shows fluctuations. Today we know that long-range shortwave radio interference and displays of the aurora are also correlated with geomagnetic storms and the sunspot cycle. Some investigators are of the opinion that even changes in climate that occur with time may be connected with solar activity.

(a) Solar Corpuscular Radiation

In addition to electromagnetic radiation (X-rays, ultraviolet radiation, visible light, infrared light, and radio waves), the sun emits *corpuscular radiation*. The corpuscular radiation is in the form of charged atomic particles, mostly protons and electrons. There is a more or less continuous emission of these particles from the sun, in the form of what is called the *solar wind*. Particles in the solar wind have been recorded on instruments carried on artificial earth satellites and cosmic space probes. It has also been found that corpuscular radiation from the sun is far from constant in intensity. Following instances of strong solar activity, particularly solar flares, the flux of charged particles is greatly enhanced. Following a strong solar flare, almost invariably there is an extra intense rain of particles at the earth about 1 day later. The fastest moving of these are low-energy cosmic-ray particles. To reach the earth in 1 day, these particles must escape the sun with velocities of millions of miles per hour.

At the surface of the earth, we are protected from these particles by the atmosphere and by the earth's magnetic field. Some of the particles may be trapped temporarily in the earth's magnetic field. Many others are deflected toward the earth's magnetic poles. Only the most energetic of the particles strike the upper atmosphere of the earth above the equatorial regions. In any case, they interact with molecules of the air and do not penetrate directly to the earth's surface.

These charged particles, along with electromagnetic radiation from the sun interacting with the upper atmosphere of the earth, do, however, disrupt the high layers where the gas is ionized (the *ionosphere*). It is these ionized layers that reflect shortwave radio waves back to the earth, making long-range radio communication between distant stations on the earth possible. When the ionospheric layers are disturbed, they may disrupt the reflection of radio waves transmitted from the ground; this results in radio "fadeouts."

The rain of particles impinging upon the upper atmosphere and on the radiation belts of the earth is also responsible for the aurora (northern and southern lights). As the particles strike atoms and molecules in the upper atmosphere they excite them. Radiation from the ions and atoms in the atmosphere gives rise to the auroral emission of light. The most spectacular auroras occur at elevations of 50 to 100 mi. Even in the absence of conspicuous auroras, however, atoms in the earth's atmosphere are being bombarded by particles from the sun continually and emit light in the form of the *airglow* (also called the *night-sky glow* or *permanent aurora*). The airglow is strongest at heights of 60 to 120 mi and is the major source of illumination of the night sky in regions away from large cities and in the absence of moonlight. Like auroras, radio fadeouts, and other solar-terrestrial effects, the airglow is more intense at times of sunspot maximum.

Exercises

1. How might you convince an ignorant friend of yours that the sun was not hollow, cool inside, and inhabited? You should be able to provide, at the very least, three or four good arguments.

2. Draw a diagram illustrating limb darkening on the sun, and the explanation thereof.

3. If the corona has a temperature of the order of a million degrees, why do we derive a temperature of about 6,000° K for the sun when we observe it in visible light?

4. Draw a diagram showing a vertical cross section of a granule and the motions of gas in and around it.

5. Draw another diagram, analogous to the one required in Exercise 4, but showing supergranulation, and the gas motions in a supergranule.

6. Illustrate with a diagram, in which heliocentric latitude is plotted against time, the frequency of sunspots at various latitudes on the sun during various times in the sunspot cycle.

7. Consider Chapter 16, and compare and contrast the sun with the most common stars.

8. From the Doppler shifts of the spectral lines in the light coming from the east and west limbs of the sun, it is found that the radial velocities of the two limbs differ by about 4 km/sec. Find the approximate period of rotation of the sun.

CHAPTER 18

Unusual Stars

Most stars are stable and fit into the general scheme of classification discussed in the preceding chapters. A minority, however, deviate from the usual pattern.

18.1 STARS THAT VARY IN LIGHT

The standard international index of stars that vary in light is the Soviet *General Catalogue of Variable Stars.* The 1958 edition of this catalogue (the most recent at the time of writing) lists 14,711 known variable stars in our Galaxy. The first and second supplements of this catalogue bring the total number of catalogued variables to 18,791.

(a) Designation

Variable stars are designated in order of time of discovery in the constellation in which they occur. If a star that is discovered to vary in light already has a proper name or a Greek-letter designation, it retains that name; examples are Polaris, Betelgeuse (α Orionis), Algol, and δ Cephei. Otherwise, the first star to be recognized as variable in a constellation is designated by the capital letter R, followed by the possessive of the Latin name of the constellation (for example, R Coronae Borealis). Subsequently discovered variables in the same constellation are designated with the letters S, T, ..., Z, RR, RS, ..., RZ, SS, ST, ..., SZ, and so on, until ZZ is reached. Then the letters AA, AB, ..., AZ, BB, BC, ..., BZ, and so on, are used up to QZ (except that the letter J is omitted). This designation takes care of the first 334 variable stars in any one constellation. Thereafter, the letter V followed by a number is used, beginning with V 335. Examples are V 335 Herculis and V 969 Ophiuchi.

(b) The Light Curve

A variable star is studied by analyzing its spectrum and by measuring the variation of its light with lapse of time. Some stars show light variations that are apparent to the unaided eye. Generally, however, the apparent brightness of a variable star is determined by telescopic observation.

FIG. 18-1   Light curve of a typical cepheid variable.

A graph that shows how the magnitude of a variable star changes with time is called a *light curve* of that star. An example is given in Figure 18-1. The *maximum* is the point on the light curve where the maximum amount of light is received from the star; the *minimum* is the point where the least amount of light is received. If the light variations of a variable star repeat themselves periodically, the interval between successive maxima is called the *period* of the star. The *amplitude* is the difference in light (usually expressed in magnitudes) between the maximum and the minimum.

(c) Types of Variable Stars

The *General Catalogue of Variable Stars* lists three types of variable stars: (1) *pulsating variables,* (2) *eruptive variables,* and (3) *eclipsing variables.* Pulsating variables are stars that periodically expand and contract, pulsating in size as well as in light. Eruptive variables are stars that show sudden, usually unpredictable, outbursts of light, or, in some cases, diminutions of light. Eclipsing variables are binary stars whose orbits of mutual revolution lie nearly edge-on to our line of sight and which periodically eclipse each other (eclipsing binaries). Eclipsing variables are not, of course, true variable stars; they have been discussed in Chapter 15 and will not be considered further here. The known numbers of different kinds of variable stars (in 1958) are summarized in Table 18.1.

TABLE 18.1 Numbers of Variable Stars

TYPE	NUMBER
Pulsating	9855
Eruptive	959
Eclipsing	2763
Unclassified or unstudied	1134
All kinds	14711

18.2 PULSATING STARS

Pulsating variable stars are included among all spectral classes from B through M. Most of these stars are giants or supergiants. The largest group of pulsating stars consists of the red giant variables known as *Mira*-type stars; these are named for their prototype, Mira, in the constellation of Cetus. Other large groups of pulsating stars are the *RR Lyrae variables,* the *semiregular variables,* and the *irregular variables.* We shall open our discussion of pulsating stars, however, with an account of the relatively rare, but historically very important, *cepheid variables.*

(a) Cepheid Variables

The cepheid variables are yellow supergiants, named for the prototype and first known star of the group, δ Cephei. The variability of δ Cephei was discovered in 1784 by the young English astronomer John Goodricke just 2 years before his death at the age of 21. The magnitude of δ Cephei varies between 3.6 and 4.3 in a period of 5.4 days. The star rises rather rapidly to maximum light and then falls more slowly to minimum light.

More than 600 cepheid variables are known in our galaxy. Most cepheids have periods in the range 3 to 50 days and absolute magnitudes (at median light†) from −1.5 to −5. The amplitudes of cepheids range from 0.1 to 2 magnitudes. Polaris, the *north star,* is a small amplitude cepheid variable that varies betwen magnitudes 2.5 and 2.6 in a period of just under 4 days.

(b) Nature of Cepheid Variation

In the latter part of the last century, the light variations of Algol were shown to be caused by eclipses of one star by another in a binary system (Section 15.2e). At one time it was suspected that the light variations of many other stars, including the cepheids, might similarly be explained, but today the evidence is conclusive that the variability of the cepheids is intrinsic.

---

†The *median light* of a variable star is the amount of light it emits when it is halfway between its maximum and minimum brightness.

A hint of the physical nature of the variability of these stars comes from their spectra. At maximum light, their spectral classes correspond to higher surface temperatures than at minimum light. The light variations of a cepheid, therefore, are due in part to variations in the temperature of its radiating surface. Further spectroscopic evidence reveals that the temperature fluctuations are accompanied by actual pulsations in the sizes of these stars. The lines in the spectrum of a cepheid show Doppler shifts that vary in exactly the same period as that of the star's light fluctuations. Evidently, the changes in light are associated with a periodic rise and fall of the cepheid's radiating surface. During its pulsation cycle, the radius of δ Cephei changes by about 7 or 8 percent.

In a stable star the weights of the constituent layers, bearing toward the center of the star under the influence of gravity, are just balanced by the pressure of the hot gases within (Chapter 20). A pulsating star, on the other hand, is something like a spring: as the star contracts, its internal pressures build up until they surpass the weights of its outer layers. Eventually, these pressures start the star pulsing outward, but because of their inertia, the outward-moving layers overshoot the equilibrium point where their weights will just balance the internal pressures. As the star expands further, the weights of the overlying layers decrease, but the internal pressure decreases faster. Hence, the overlying layers are not supported adequately, and the star begins to contract. As it does so, it overshoots again, and this time it becomes too highly contracted. Once more the inner pressures cause the star to expand — and so the pulsation continues. The time required for a complete cycle of pulsation is greater for a giant star of low density than for a smaller compact star of higher density — just as a long piano string vibrates more slowly than a short one. It can be shown that for pulsating stars of any one type, the period of pulsation of a particular star is inversely proportional to the square root of its mean density.

(c) The Period-Luminosity Relation

Many other stars, besides cepheids, vary in light because of physical pulsations; indeed, cepheids are a comparatively rare class of pulsating variables. The great importance of cepheid variables lies in the fact that a relation exists between their periods of pulsation (or light variation) and their median luminosities, that is, their average absolute magnitudes. The relation was discovered in 1912 by Henrietta Leavitt, an astronomer of the Harvard College Observatory. Some hundreds of cepheid variables had been discovered in the Large and Small Magellanic Clouds, two great stellar systems that are actually neighboring galaxies (although they were not known to be galaxies in 1912 — see Chapter 21).

Many photographs of the Magellanic Clouds had been taken at the Southern Station of the Harvard College Observatory in South America. (These stellar systems are too far south in the sky to be observed from the United States.) On some of the photographs, Miss Leavitt identified 25 cepheids in the smaller cloud and plotted light curves for them. She found that the periods of the stars were related to their relative brightnesses, in the sense that the brighter-appearing ones always had the longer periods of light variation. Since the 25 stars are all in the same stellar system, they are all at about the same distance — that is, the distance to the Small Magellanic Cloud — and their relative apparent brightnesses, therefore, indicate their actual relative luminosities. Miss Leavitt, in other words, had found that the luminosities, or absolute magnitudes of the cepheid variables, were correlated with their periods. Subsequent investigation showed that this relation exists between all the cepheids in the Large and Small Magellanic Clouds.

Harlow Shapley was one of the astronomers who recognized the importance of cepheids as distance indicators, and he pioneered the work of determining distances to some of them in our galaxy; his work was later extended by others. Today, following the pioneering work of J. B.

Irwin, the best way of finding absolute magnitudes of cepheids is to identify them in certain open star clusters whose distances can be found by fitting the main sequences in their color-magnitude diagrams to the well-established main sequence on the Hertzsprung-Russell diagram. We shall see (Chapter 21) that the distances to other galaxies are based primarily on the apparent faintness of the cepheids visible in the nearer of them.

In a very few globular star clusters, and also outside of clusters in the galactic corona (Chapter 19), there are variable stars with periods in the range 10 to 30 days. The light curves of these stars are similar to those of the cepheid variables except that they fall from maximum light to minimum more slowly, and the stars are somewhat bluer in color. These objects are sometimes called *W Virginis* stars, after the prototype W Virginis, or sometimes *type II cepheids,* to distinguish them from ordinary or "classical" cepheids, which are called *type I cepheids*. Although only a few dozen type II cepheids are known, a period-luminosity relation also seems to hold for them, and it resembles that for type I cepheids, in that stars of

longer period are the more luminous. Type II cepheids are 1 to 2 magnitudes fainter than type I cepheids of the same periods.

(d) RR Lyrae Stars

Next to the long-period variables the most common variable stars are the *RR Lyrae* stars, named for RR Lyrae, best known member of the group. Nearly 2500 of these variables are known in our galaxy. Almost all of them are found in the nucleus or the corona of our galaxy or in globular clusters. In fact, nearly all globular clusters contain at least a few RR Lyrae variables, and some contain hundreds; these stars, therefore, are sometimes called *cluster*-type variables.

The periods of RR Lyrae stars are less than 1 day; most periods fall in the range 0.3 to 0.7 day. Their amplitudes never exceed 2 magnitudes, and most RR Lyrae stars have amplitudes less than 1 magnitude. They are of spectral classes A or F. Several subclasses of RR Lyrae stars are recognized, but the differences between these subclasses are small and need not be considered here.

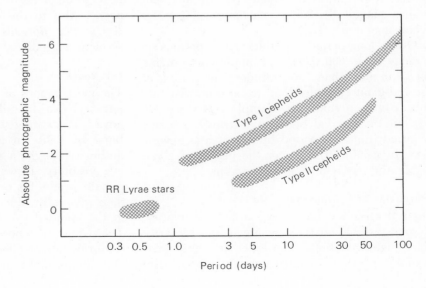

FIG. 18-2 Period-luminosity relation for type I cepheids *(upper curve)*, type II cepheids *(lower curve)*, and RR Lyrae stars *(left)*.

It is observed that the RR Lyrae stars occurring in any particular globular cluster all have about the same median apparent magnitude. Since they are all at approximately the same distance, it follows that they must also have nearly the same absolute magnitude. Because the RR Lyrae stars in different clusters are all similar to each other in observable characteristics, it has long been recognized that *all* RR Lyrae stars have about the same absolute magnitude. Recent studies show that the absolute magnitudes of RR Lyrae stars lie between 0 and +1 — that is, from 40 to 100 times as luminous as the sun.

Figure 18-2 displays the range of periods and absolute magnitudes found for the two types of cepheids and for RR Lyrae stars.

(e) Long-Period Variables

Most common among the pulsating stars are the *long-period variables*. These stars, all red giants or red supergiants, are sometimes called "red variables." A typical example, and the best known star of this group, is Mira (o Ceti). Ordinarily, Mira is at apparent magnitude 8 to 10, too faint to be visible to the unaided eye. About every 11 months, however, it brightens to naked-eye visibility and averages magnitude 2.5 at maximum. (Once it reached magnitude 1.2.)

Other long-period variables have periods in the range 80 to 600 days, and amplitudes ranging from 2.5 to more than 7 magnitudes. The periods of these red-giant variables tend to be somewhat unstable; Mira's period, for example, *averages* 330 days, but the interval between successive maxima has varied from this period by as much as several weeks. In general, the longer the period of a red variable, the less regular is its pulsation.

(f) Irregular and Semiregular Variables

Many red variables show very marked irregularity, and some exhibit no trace whatever of regularity in their light variations. The latter are designated *irregular variables*. Those stars which usually exhibit periodic variability, but which are at times unpredictably disturbed, are classed as *semiregular variables*. Generally, semiregular and irregular variables have light amplitudes that are less than 2 magnitudes — that is, a smaller range than the range of ordinary long-period variables. To the class of semiregular variables belongs the famous star Betelgeuse, which marks Orion's right shoulder.

The long-period, semiregular, and irregular red variables must be constantly watched to gather information about their light fluctuations. It would require far too much of the time of professional astronomers to keep constant vigil on all these variables. It is here that amateur astronomers provide a real service to astronomy; the society known as the *American Association of Variable Star Observers* has organized a careful surveillance of red variables and has been gathering data on them for years.

18.3 ERUPTIVE VARIABLES

There are many types of eruptive variables; they range from the *R W Aurigae* (or *T Tauri*) types, which display rapid, irregular increases in brightness, through the *novae,* to the spectacular *supernovae*. Also to the eruptive variables belong the *R Coronae Borealis variables*, stars which show sudden, unpredicted decreases in brightness.

(a) Novae

The most famous of the eruptive variables are the *novae*. Nova literally means "new." Actually, a nova is an existing star that suddenly emits an outburst of light. In ancient times, when such an outburst brought a star's luminosity up to naked-eye visibility, it seemed like a new star. Novae remain bright for only a few days or weeks and then gradually fade. They seldom remain visible to the unaided eye for more than a few months. The Chinese, whose annals record novae from centuries before Christ, called them "guest stars." Only occasionally are novae visible to the naked

FIG. 18-3   Light curve of Nova Puppis, 1942.

eye, but, on the average, two or three are found telescopically each year. Many must escape detection; altogether there may be as many as two or three dozen nova outbursts per year in our galaxy.

Novae, before and after their outbursts, are hot, subdwarf stars — stars of very high temperature but small size, so that despite their high temperatures they have lower luminosities than equally hot main-sequence stars. A typical nova, however, flares up during its outburst to thousands or even tens of thousands of times its normal luminosity, and may reach an absolute magnitude of from −6 to −9. The rise to maximum light is very rapid, often requiring less than 1 day. The subsequent decline in light is much slower; the star requires years, or even decades, to return to normal. A typical nova light curve is shown in Figure 18-3. Different types of novae, however, decline at

FIG. 18-4   Nova Herculis, as it appeared before and after its outburst in 1934. *(Yerkes Observatory.)*

different rates. Some (possibly all) show variability during their gradual fading.

At the time of a nova outburst, an outer layer of the star is ejected in the form of a shell of gas. Velocities of ejection of up to 1000 km/sec or more are found. A typical nova shell is found to contain from 10^{-5} to 10^{-4} times the solar mass. The star, therefore, ejects only a small part of its total mass. Some months after the outburst, the expanding envelope may become visible on telescopic photographs, as in the case of Nova Aquilae. (See Fig. 18-5.)

Some novae have undergone more than one outburst over the period they have been under observation; these are called *recurrent novae*. There is no way of knowing, of course, whether ordinary novae may have had previous flareups before recorded history, or whether they will produce outbursts again. It is possible, in other words, that such novae may be recurrent but show more spectacular outbursts, with longer intervals between them.

There is now rapidly accumulating evidence that novae always occur in stars that are members of binary-star systems in which the two stars, the one that undergoes the outburst and its companion, are very close together. A popular theory for the nova phenomenon involves the transfer of mass from the companion star to the nova star, and the possibility that this mass exchange triggers the outburst.

(b) Supernovae

Among the more spectacular of the cataclysms of nature is the *supernova*. In contrast to an ordinary nova, which increases in luminosity a paltry few thousands or at most tens of thousands of times, a supernova is a star that flares up to hundreds of millions of times its former brightness. At maximum light, supernovae reach absolute magnitude -14 to -18, or possibly even -20. The three most famous supernovae to have been observed during the last 10 centuries in our Galaxy are (1) the Supernova of 1054 in Taurus (described in the

FIG. 18-5   Expanding nebulosity around Nova Aquilae, Nov. 3, 1918. *Left*, July 20, 1922; *center*, September 3, 1926; *right*, August 14, 1931. *(Hale Observatories.)*

Chinese Annals), (2) Tycho's "star" of 1572 (Section 2.7a) in the constellation Cassiopeia, and (3) the supernova of 1604 in Serpens, described by both Kepler and Galileo. Supernovae are now commonly observed in other galaxies. In a typical galaxy, supernovae appear to occur at the rate of at least one every few hundred years.

The light curve of a supernova is similar to that of an ordinary nova, except for the far greater luminosity of the supernova. It now appears that there may be several different kinds of supernovae, but they all rise to maximum light extremely quickly (in a few days or less), and for a brief time one may outshine the entire galaxy in which it appears. Just after maximum, the gradual decline sets in, and the star fades in light until it disappears from telescopic visibility within a few months or years after its outburst. Bright emission lines in the spectra of supernovae indicate that they, like ordinary novae, eject material at the time of their outbursts. The velocities of ejection may be substantially greater than in ordinary novae, however, since speeds of up to 4000 km/sec have been observed. Moreover, a much larger amount of material is ejected; in fact, a large fraction of the original star may go off in the expanding envelope.

(c) The Crab Nebula

An example of the remnant of a supernova explosion is the Crab nebula in Taurus, a chaotic, expanding mass of gas, visible telescopically, and spectacular on telescopic photographs. From its rate of expansion, we calculate that the cloud of gas has required about 860 years to reach its present size, which places the event in the eleventh century. Both the location and computed time of formation of the Crab nebula are in good agreement with the occurrence of the supernova of 1054. The Crab nebula is believed, therefore, to be the material ejected during that stellar explosion.

The Crab nebula is a strong source of radio waves and X-rays. The radio spectrum (distribution

FIG. 18-6   The Crab Nebula, photographed in red light with the 200-inch telescope. *(Hale Observatories.)*

of radio energy with wavelength) resembles that expected from electrons which are spiraling at nearly the velocity of light in a magnetic field — a phenomenon known as *synchrotron radiation*. When the nature of the radio radiation from the Crab nebula was suspected, it was suggested that some of its visible light might similarly originate from the synchrotron mechanism. If so, theory predicted that the light should be polarized. The polarization of some of the light from the Crab nebula was observed at the 200-inch telescope by W. Baade in 1956. There is little doubt, therefore, that the radio energy, as well as much of the visible light from the Crab nebula, is emitted from electrons accelerated in magnetic fields; the origins of the electrons and magnetic fields remain a mystery. It is also possible to observe synchrotron radiation, incidentally, from electrons that are being accelerated in a laboratory synchrotron.

(d) Other Galactic Supernovae

Radio sources are now known that correspond to positions of other supernovae, including those observed by Tycho (1572) and Kepler (1604). If the position of a temporarily appearing bright star, as given in historical records, agrees with that of an observed radio source, and if the characteristics of the radio radiation suggest synchrotron emission, the star is generally assumed to have been a supernova. The identification is further strengthened if gas filaments or nebulosity can be observed at that same position. Provisional identifications of 14 supernovae in our galaxy within the past 2000 years have so been made. Descriptions of the temporary or "guest" stars themselves appear in the annals of the Chinese (especially), the Japanese, Koreans, Arabs, and Europeans. Several of these were observed with the naked eye for more than 1 year, and one was seen for 2 years. The latter reached an estimated apparent magnitude (at brightest) of −8, many times as bright as Venus; several were easily visible in daylight. Still other radio sources have been identified with filaments of gas that are believed to be the remnants of prehistoric supernovae. An example of these is the filamentary "loop" or Veil nebula in Cygnus.

The 14 supernovae observed in our galaxy in the past 2000 years give an average rate of one supernovae per 140 years. Many supernovae must have been missed, however, especially those in remote parts of our galaxy, hidden by interstellar dust. Shklovsky estimates that the frequency of supernovae in our galaxy may be as high as one every 30 years. This is consistent with the rate estimated for other galaxies, but the data are still so few that the figures cannot be regarded definitive.

(e) Flare Stars

Flare stars (or *UV Ceti stars*) are red main-sequence stars, mostly of spectral types M3 to M6. These stars show rare flareups in brightness. The total increase in light may be from 0.1 to several magnitudes. Their rise to maximum is very rapid — a few seconds — and the decline lasts for from a few minutes to 1 hour or so. They behave rather like miniature novae.

The color of the light associated with a flare is blue, appropriate to that from a star with a temperature of 10,000 to 20,000°K. A large area at such a temperature would radiate far more energy than is observed to come from flares. Consequently, a flare must be a localized disturbance on a small part of a star. In total energy output, a typical flare observed on a UV Ceti-type star resembles the more spectacular of solar flares (Section 17.2i).

Flares cause proportionately larger increases in the light of intrinsically faint stars than of brighter ones. Suppose a flare causes an M2 star to brighten by 1 magnitude — a factor of 2½ in light. The flare itself, then, gives off (during the

FIG. 18-7   Filamentary nebula in Cygnus, photographed with the 48-inch Schmidt telescope. The nebula is believed to be the remnant of an ancient supernova. *(Hale Observatories.)*

brief period of its existence) 1½ times as much light per second as does its parent star. A flare of exactly the same energy would increase the light of a star 2 magnitudes brighter than the M2 star by only about 20 percent, and would increase the sun's light by only 1 percent, an amount too little to be noticed if the sun were as distant as the other stars. We cannot be sure, in other words, that flares do not occur on intrinsically bright, as well as intrinsically faint, stars; they are obvious only on the latter.

18.4 SUMMARY OF VARIABLE STARS

It is neither feasible nor appropriate to describe in detail the many other kinds of variable stars in a survey of astronomy such as this book. A tabular summary of the more important types of variables is presented in Tables 18.2 and 18.3.

18.5 STARS WITH EXTENDED ATMOSPHERES

We have already seen that novae, supernovae, and novalike stars eject gas shells at the times of their outbursts. In addition, many other stars, for which no outbursts have been observed, are known to be surrounded by extended atmospheres of expanding gas shells. The extended atmosphere is usually revealed by the presence of emission lines super-posed on the continuous spectrum of the star, formed by the gas in the shell absorbing ultra-violet starlight, thus becoming ionized, and then reemitting this light as emission lines when the ions recapture electrons.

One of many classes of stars that are ejecting matter is that of the Wolf-Rayet stars. Nearly 200 are known. They are very hot O-type stars of moderately high luminosity that are ejecting gas shells at speeds of 1000 to 3000 km/sec.

Even some binary stars are found to be eject-ing matter. Certain close binaries are believed to be exchanging matter; part of this matter may flow from one star to the other, but some may flow into a large shell surrounding the system, or even be ejected into space.

The most important class of stars with ex-tended atmospheres, however, are probably the planetary nebulae.

(a) Planetary Nebulae

Planetary nebulae are large shells of gas ejected from and expanding about certain extremely hot stars. They derive their name from the fact that a few bear a superficial telescopic resemblance to planets; actually they are thousands of times larger than the entire solar system, and have nothing whatever to do with planets.

Planetary nebulae often appear large enough to see or to photograph with a telescope. The most famous example is the ring nebula, in Lyra. It is typical of many planetaries in that, although actually a hollow shell of material emitting light, it appears as a ring. The explanation is that we are looking through the *thin* dimensions of the front and rear parts of the shell, while along its periphery our line of sight encounters a long path through the glowing material. Similarly, a soap bubble often appears to be a thin ring. Another way in which planetary nebulae are identified is by their spectra. Those that are very distant from us are unresolved and appear stellar, but their spectra show emission lines which indicate the existence of luminous shells of gas surrounding stars, as do spectra of nova shells. Altogether, there are about 1000 planetary nebulae known. Doubtless there are many distant ones that have escaped detection, so there might be some thou-sands or even tens of thousands in the Galaxy.

Planetary nebulae differ from the other kinds of stars that eject gas shells in an important re-spect: An appreciable amount of material is ejected in the shell of a planetary nebula. From the light emitted by the shells, we calculate that they must have masses of 10 to 20 percent that of the sun. The shells, typically, expand about their parent stars at speeds of 20 to 30 km/sec.

TABLE 18.2 Pulsating Variables

TYPE OF VARIABLE	SPECTRA	PERIOD (DAYS)	MEDIAN MAGNITUDE (ABSOLUTE)	AMPLITUDE (MAGNITUDES)	DESCRIPTION	EXAMPLE	NUMBER KNOWN IN GALAXY[†]
Cepheids (type I)	F to G supergiants	3 to 50	-1.5 to -5	0.1 to 2	Regular pulsation; period-luminosity relation exists	δ Cep	610
Cepheids (type II)	F to G supergiants	5 to 30	0 to -3.5	0.1 to 2	Regular pulsation; period-luminosity relation exists	W Vir	(About 50; included with type I)
RV Tauri	G to K yellow and red bright giants	30 to 150	-2 to -3	Up to 3	Alternate large and small maxima	RV Tau	92
Long-period (Mira-type)	M red giants	80 to 600	$+2$ to -2	>2.5	Brighten more or less periodically	o Cet (Mira)	3657
Semiregular	M giants and supergiants	30 to 2000	0 to -3	1 to 2	Periodicity not dependable; often interrupted	α Ori	1675
Irregular	All types	Irregular	<0	Up to several magnitudes	No known periodicity; many may be semiregular, but too little data exist to classify them as such	π Gru	1370
RR Lyrae or cluster-type	A to F blue giants	<1	0 to $+1$	<1 to 2	Very regular pulsations	RR Lyr	2426
β Cephei or β Canis Majoris	B blue giants	0.1 to 0.3	-2 to -4	0.1	Maximum light occurs at time of highest compression	β Cep	11
δ Scuti	F subgiants	<1	0 to $+2$	<0.25	Similar to, and possibly related to, RR Lyrae variables	δ Sct	5
Spectrum variables	A main sequence	1 to 25	0 to $+1$	0.1	Anomalously intense lines of Si, Sr, and Cr vary in intensity with same period as light; most have strong variable magnetic fields	α^2C Vn	9

[†]According to the 1958 edition of the Soviet *General Catalogue of Variable Stars*.

TABLE 18.3 Eruptive Variables

TYPE OF VARIABLE	SPECTRA	DURATION OF INCREASED BRIGHTNESS	NORMAL ABSOLUTE MAGNITUDE	AMPLITUDE (MAGNITUDES)	DESCRIPTION	EXAMPLE	NUMBER KNOWN IN GALAXY[†]
Novae	O to A hot subdwarfs	Months to years	>0	7 to 16	Rapid rise to maximum; slow decline; ejection of gas shell	GK Per	146
Novalike variables or P Cygni stars	Hot B stars	Erratic	-3 to -6	Several magnitudes	Slow, erratic, and novalike variations in light; may be unrelated to novae. Gas shell ejected	P Cyg	35
Supernovae	?	Months to years	?	15 or more	Sudden, violent flareup, followed by decline and ejection of gas shell	CM Tau (Crab nebula)	7
R Coronae Borealis	F to K supergiant	10 to several hundred days	-5	1 to 9	Sudden and irregular drops in brightness. Low in hydrogen abundance, but high in carbon abundance	R CrB	39
T Tauri or RW Aurigae	B to M main sequence and subgiants	Rapid and erratic	0 to $+8$	Up to a few magnitudes	Rapid and irregular light variations. Generally associated with interstellar material. Subtypes from G to M are called T Tauri variables	RW Aur, T Tau	590
U Geminorum or SS Cygni or "dwarf novae"	A to F hot subdwarfs	Few days to few weeks	>0	2 to 6	Novalike outbursts at mean intervals which range from 20 to 600 days. Those with longer intervals between outbursts tend to have greater amplitudes. Many, if not all, are members of binary-star systems	SS Cyg, U Gem	112
Flare stars	M main sequence	Few minutes	>8	Up to 6	Sudden flareups in light; probably localized flares on surface of star	UV Cet	15
Z Camelopardalis variables	A to F hot subdwarfs	Few days	>0	2 to 5	Similar to U Geminorium, except that variations are sometimes interrupted by constant light for several cycles. Intervals between outbursts normally range from 10 to 40 days	Z Cam	15

†According to the 1958 edition of the Soviet *General Catalogue of Variable Stars.*

A typical planetary appears to have a diameter of about ½ LY to 1 LY. If it is assumed that the gas shell has always expanded at the speed with which it is now enlarging about its parent star, its age can be calculated. Most of the gas shells have been ejected within the past 50,000 years; an age of 20,000 years is more or less typical. After about 100,000 years, the shell is so enlarged that it is too thin and tenuous to be seen. Planetary nebulae, therefore, are temporary phenomena. When we take account of the relatively short time over which a planetary nebula exists as such, we find that they are actually very common; indeed, an appreciable fraction of all stars must sometime evolve through the planetary nebula phase.

The gas shells of planetary nebulae shine by the process of fluorescence. They absorb ultraviolet radiation from their central stars and reemit this energy as visible light. Thus, a measure of the visible light emitted by a shell enables us to calculate how much ultraviolet radiation must be leaving the star. The ultraviolet luminosities of the central stars are found to be very high, showing that the stars are very hot. Nearly all these stars are hotter than 20,000°K, and some have temperatures well in excess of 100,000° K, which makes them the hottest known stars.

Despite their high temperatures, the central stars of planetary nebulae do not have exceedingly high luminosities — some emit little more total energy than does the sun. They must, therefore, be stars of small size; some, in fact, appear to have the dimensions of white dwarfs. It does not follow, however, that the stars were small and hot when the gas shells were ejected. Analysis shows that it is more likely that the nebulae were ejected from their parent stars when the latter were, at an earlier stage in their evolution, large red giants.

18.6 X-RAY STARS

X-rays are electromagnetic radiation of wavelengths in the range 0.1 to 100 Å, to which the earth's atmosphere is opaque. Since 1962, however, we have been scanning the sky with X-ray detectors that are carried by rockets above practically all the atmosphere.

The first object from which X-rays were detected is the sun. However, if the sun is typical of other stars as an X-ray emitter, with existing equipment we could not observe X-rays from any other single star, because the other stars are so much farther away. Nevertheless, an effort was made to scan the sky with X-ray detectors of improved design carried by later rockets. On June 18, 1962, the first such systematic scan was made. A strong source of X-rays was detected in the constellation Scorpio. That first X-ray source discovered in Scorpio is now called Sco X-1.

There are now (mid-1972) about 50 X-ray sources known. Sco X-1 is now identified with an old nova, a star of magnitude 13 that emits most of its energy as X-rays. Some additional X-ray sources may be old novae. Another is the Crab nebula, an old supernova. In addition to X-rays from the entire visible Crab, there is a small source of pulsed X-rays in the middle of the nebula that is associated with the pulsar there (next section).

All X-ray sources are not peculiar stars; in addition to old novae, supernovae, and pulsars, X-ray sources have been identified with certain individual galaxies. These include the Large Cloud of Magellan, a quasar (3C273), and certain peculiar galaxies — especially Seyfert galaxies (see Chapter 21). Some clusters of galaxies are also sources of X-rays. Moreover, there is a faint more or less uniform background of X-rays from all directions in space. At the time of writing, the origin of this background is not definitely known; some astronomers believe that it may be associated with hot sparse material in intergalactic space.

18.7 PULSARS

Pulsars, or pulsating radio stars, are radio sources emitting sharp, intense, rapid, and extremely regular pulses. The first to be discovered, in mid-1967, is the constellation Vulpecula. It emits a radio pulse every 1.33728 seconds. Detailed studies show each pulse to be complex and to consist of at least three much shorter pulses, each lasting only a few milliseconds, and the entire pulse sequence only a few tens of milliseconds. Although extremely regular in period, the pulses vary considerably from one to the next in intensity. The radiation in all pulses is polarized, and appears to be synchrotron radiation.

By 1972 more than 60 other pulsars had been discovered, with periods ranging from $\frac{1}{30}$ to more than 3 seconds. Rough estimates of their distances suggest that most are from a few hundred to a few thousand parsecs away — all well inside our own galaxy.

We know the distance to the pulsar in the Crab nebula; that nebula is just over 2000 parsecs away. It is this pulsar that has the shortest known period (0.033 sec), and, as stated above, it pulses in X-rays. It is also the only pulsar yet to be observed in visble light. A star of magnitude 16 has long been known near the middle of the Crab nebula, but only after it was identified with the position of the pulsar was it observed with special photographic equipment, and found to be emitting its light in pulses with the same period as the radio and X-ray pulses. Actually, the pulses of this object are double — a main pulse and a secondary one that follows half a cycle later. Both pulses, however, last only a millisecond. Because light travels only about 186 miles in this time, the emitting source cannot be larger than 200 mi in diameter, or the pulses would appear smeared out.

We would expect pulses of such strength and regularity to be associated either with the pulsation or rotation of some kind of star. Now, if a star is unstable and can pulsate, the time it takes for one cycle (its period) is approximately in inverse proportion to the square root of its mean density. If the sun were to pulsate, its period would be something less than half an hour. Large, tenuous red giants take months or even years to pulsate (these are the long-period variables); a white dwarf, if unstable, would pulsate in a few seconds. The limit to how fast a star can rotate is set by the speed at which its equatorial regions would have the circular satellite velocity (see Section 4.6). This rotation speed is very high for a small dense star, and very low for a giant; it turns out that the shortest possible rotation period for a star is about the same as the period at which it would pulsate.

A white dwarf could pulsate or rotate in a few seconds, but even this is too long for most pulsars. Hence, it is thought that pulsars are stars far more dense than even white dwarfs, namely, *neutron stars* (see Chapter 20). Neutron stars would pulsate in a few milliseconds, and could rotate with any greater period. Most experts now consider it almost certain that pulsars are rotating neutron stars.

According to theory, the outburst that accounts for one kind of supernova originates in a nuclear explosion in a shell intermediate in position between the star's center and its surface. The explosion blows the outer part of the star out in a nebula (like the Crab nebula), and implodes the central regions into an extremely dense neutron star. Thus, a pulsar at the center of the Crab nebula, a supernova remnant, is consistent with theory.

The model for the pulses themselves is less well understood. It is usually assumed, however, that the implosion producing the neutron star also contracts the magnetic field of the original star into a field of extremely high intensity. As the neutron star rotates, ions (electrons and atomic nuclei) from its surface are thought to be spewn out into that field, perhaps from the magnetic poles of the star. Each time the star rotates, carrying a magnetic pole (or other beacon of ions) around with respect to the earth, those charged particles, interacting with the magnetic lines of force, give rise to strong synchrotron radiation in the form of pulses. The same electrons, moving on out into the nebula and later interacting with its magnetic field, can account for the synchrotron radiation observed from the Crab nebula. The positive ions (atomic nuclei) may be one source of cosmic rays.

If the theory outlined in the last paragraph is correct, we would expect that the interaction of the ions from the star with its magnetic field would slowly rob the star of energy of rotation. Thus, the period of rotation of the neutron star should gradually be increasing. Observations show that the periods of pulsars do, indeed, increase at

a very gradual rate. It is now believed that the pulsar in the Crab nebula is a relatively young one, formed by the supernova explosion in 1054, and that the other pulsars formed with prehistoric supernovae thousands of years ago, and have now slowed down to their present longer periods. The picture seems to fit together well, but many details remain to be worked out, and we warn the reader to regard the theory of pulsars described here as tentative.

Exercises

1. Refer to Figure 18-2, and estimate the distance to a type I cepheid variable with a period of about 50 days, and an apparent magnitude of +5.

2. Compare the luminosities of novae, at maximum, with the luminosities of the brightest normal stars.

3. Sketch a Hertzsprung-Russell diagram, showing the positions of the main sequence, the giant sequence, and the supergiants, and indicate on the diagram the positions of as many different kinds of variable stars as you can.

4. Sketch the spectral energy distribution of Mira, both at maximum and at minimum light, and show how a greater fraction of its total energy is in the visible spectral region when it is at maximum than when it is at minimum.

5. In a globular cluster several RR Lyrae stars are observed. At median light they appear to have apparent magnitudes between 15 and 16. Estimate the distance to the globular cluster.

6. Compare and contrast: nova shells; supernova shells; planetary nebulae.

7. The gas shell of a particular planetary nebula is expanding at the rate of 10 mi/sec. Its diameter is 1 LY. Find its age, on the assumption that it has always been expanding at the same rate. For this exercise, assume that there are 3×10^7 sec/yr, and 6×10^{12} mi/LY.

8. If the gas shell described in Exercise 7 were forcibly ejected from its parent star, do you think that it can always have expanded at exactly the same rate? Explain.

9. Assume that a pulsar is 1000 pc away. Suppose that no star brighter than apparent magnitude 23 shows up in that position of the sky. What is the brighter limit to the absolute magnitude that a star associated with the pulsar could have?

CHAPTER 19

Our Galaxy of Stars, Gas, and Dust

In 1610 Galileo described his telescopic observations of the Milky Way, which showed it to be a multitude of individual stars. In 1750 Thomas Wright published a speculative explanation, which turned out to be substantially correct — that the sun is part of a disk-shaped system of stars, and that the Milky Way is the light from the surrounding stars that lie more or less in the *plane* of the disk. The disk-shape of the stellar system to which the sun belongs — the *Galaxy* — was demonstrated quantitatively in 1785 by Herschel's "star gauging." We have already described the results of Herschel's investigations (Section 16.3c), as well as our modern view of the Galaxy (Figure 19-2).

It was the second decade of the twentieth century before astronomers had deduced, approximately, the true size of the Galaxy and the fact that the sun is located eccentrically in the disk-

FIG. 19-1   Mosaic of the Milky Way in Cygnus, photographed with the 48-inch Schmidt telescope. *(National Geographic Society-Palomar Observatory Sky Survey.)*

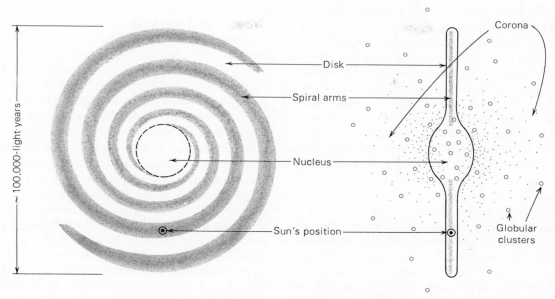

FIG. 19-2   Schematic representation of the Galaxy.

shaped system. In 1924 the existence of other, similar, stellar systems was proved; the "pinwheel" nature of many of these other galaxies suggested that our own system might also contain spiral arms winding outward from a central nucleus. Only within the last two decades, however, has substantial progress been made in identifying the spiral arms and in mapping them. This chapter deals with twentieth-century investigations of the structure of our galaxy.

19.1 SIZE OF THE GALAXY, AND OUR POSITION IN IT

Until early in the twentieth century, the Galaxy was generally believed to be centered approximately at the sun and to extend only a few thousand light years from it. The shift from the "heliocentric" to the "galactocentric" view of our system, as well as the first knowledge of its true size, came about largely through the efforts of Harlow Shapley and his investigation of the distribution of the globular clusters. Globular clusters (Section 16.3c) are great symmetrical star clusters, containing tens of thousands to hundreds of thousands of member stars each. Because of their brilliance, and the fact that they are not confined to the central plane of the Galaxy where they would otherwise be largely obscured by interstellar dust, they can be observed (with telescopes) to very large distances.

(a) Distribution of the Globular Clusters

Most globular clusters contain at least a few RR Lyrae variable stars (or cluster-variables — Section 18.2d), whose absolute magnitudes are known to be between 0 and +1. The distance to an RR Lyrae star in a globular cluster, and hence to the cluster itself, can therefore be calculated from its observed apparent magnitude (Section 14.2b). Shapley was able to determine the distances to the closer globular clusters which contained RR Lyrae stars that could be observed individually. Distance estimates to globular clusters that do not contain RR Lyrae stars, or which are too far away to permit resolution of the variables, were obtained indirectly. Shapley measured angular diameters of globular clusters of known distance, thus obtaining their true diameters. Assuming a statistical average for the true diameter of the clusters, he was then able to obtain distance estimates for the remote ones from their observed angular diameters.

From their directions and derived distances, Shapley, in 1917, mapped out the three-dimensional distribution in space of the 93 globular clusters then known. He found that the clusters formed a spheroidal system with the highest concentration of clusters at its own center. That center was not at the sun, however, but at a point in the middle of the Milky Way in the direction of Sagittarius, and at a distance of some 25,000 to 30,000 LY. Shapley then made the bold — and correct — assumption that the system of globular clusters represented the "bony frame" of the entire Galaxy; that not only is the distribution of clusters centered upon the center of the Galaxy, but, moreover, that the extent of the galactic system is indicated by the cluster distribution. Today his assumption has been roughly verified. Although the sun lies far from the galactic center, the main disk of the Galaxy probably extends a nearly equal distance beyond the sun and comprises a gigantic system 100,000 LY (more or less) across. The exact size of the disk is not known, however, nor is the exact distance of the sun from its center. Today, the center of the galactic nucleus is believed to be about 10,000 pc from the sun (or about 30,000 LY), with an uncertainty of 2000 pc.

(b) The Galactic Corona

It is to be noted that the main body of the Galaxy is confined to a relatively flat disk, while the globular clusters define a more or less spheroidal system superimposed upon the disk. A sparse "haze" of individual stars — not members of clusters — is now known to exist also in the region outlined by the globular clusters. This haze of stars and clusters forms the galactic *corona*, or "halo," a region whose volume exceeds that of the main disk of the Galaxy by many times. The presence of stars in the corona was first suspected when RR Lyrae stars not belonging to clusters were found lying in such directions and at such distances as to place them far from the galactic plane. These variables, of course, represent only an extremely minute fraction of all coronal stars, but they serve as markers, indicating the distribution of stars in the corona, not only because their distances can be determined easily, but because they have fairly high luminosities and can be seen to relatively large distances. Individual RR Lyrae stars have been found as far away as 30,000 to 50,000 LY on either side of the galactic plane, which shows that the corona must have an over-all thickness of up to 100,000 LY.

19.2 REVOLUTION OF THE SUN IN THE GALAXY

Like a gigantic "solar system," the entire Galaxy is rotating. The sun, partaking of the galactic rotation, moves with a speed of 200 to 300 km/sec in a nearly circular path about the nucleus.

(a) The Sun's Galactic Orbit

The motion of the sun in the Galaxy is deduced from the apparent motions of objects surrounding us that do not share in the general galactic rotation. The globular clusters are the most convenient of such objects. These clusters are moving, to be sure, but the fact that they are found in a spheroidal distribution, rather than being confined to the flat plane of the Galaxy, is evidence that the system of globular clusters as a whole is not rotating with the disk of the Galaxy. By analyzing the radial velocities of the globular clusters in various directions,

The ring nebula in Lyra, M57; 200-inch telescope photograph. *(Hale Observatories.)*

NGC 7293 in Aquarius; 200-inch telescope photograph. *(Hale Observatories.)*

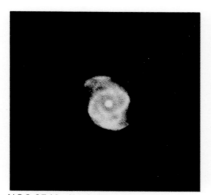

NGC 6543. *(Lick Observatory.)*

NGC 6781 in Aquila; 48-inch Schmidt telescope photograph. *(Hale Observatories.)*

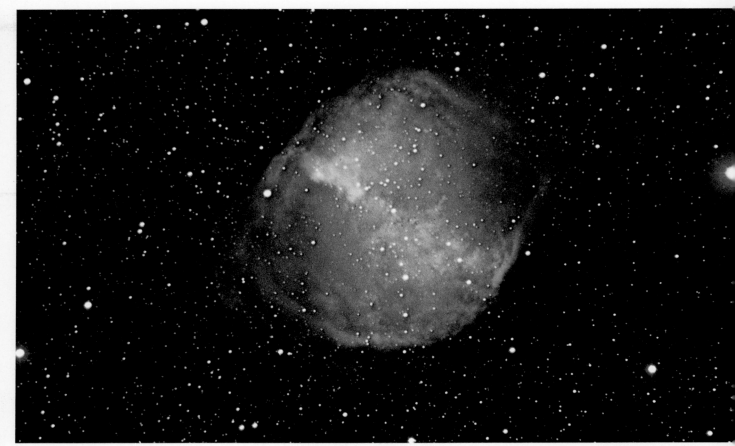

The Dumbell nebula, NGC 6853, M27; photographed by the 200-inch telescope. *(Hale Observatory.)*

The Trifid nebula, M20, in Sagittarius. The entire region is immersed in gas and dust. The blue region on the left is a reflection nebula. The region on the right is an emission nebula; 200-inch telescope photograph. *(Hale Observatories.)*

Omega nebula Sagittarius; 48-inch Schmidt telescope photograph. *(Hale Observatories.)*

Central region of the Lagoon nebula (M8). *(Lick Observatory.)*

Rosette nebula in Monoceros. NGC 2237; photographed by the 48-inch
Schmidt telescope, *(Hale Observatories.)*

we can determine the motion of the sun with respect to them. In one direction, the globular clusters, on the average, seem to be approaching us, while in the opposite direction they seem to recede from us.

When the data from various sources are combined, they indicate that the sun is moving in the direction of the constellation Cygnus, with a speed that, although somewhat uncertain, probably lies in the range 200 to 300 km/sec. This direction lies in the Milky Way, and is about 90° from the direction of the galactic center, which shows that the sun's orbit is probably nearly circular and lies, approximately, in the main plane of the Galaxy. As viewed from the north side of the galactic plane, the orbital motion of the sun is in a clockwise direction. The period of the sun's revolution about the nucleus, the *galactic year,* can be found by dividing the circumference of the sun's orbit by its speed; it comes out *roughly* 200 million (2×10^8) of our terrestrial years. We can observe, therefore, only a "snapshot" of the Galaxy in rotation; we do not actually see stars traverse appreciable portions of their orbits.

(b) High- and Low-Velocity Stars

As already mentioned, the majority of the stars near the sun move nearly parallel to the sun's path about the galactic nucleus, and their speeds with respect to the sun are generally less than 40 or 50 km/sec. These are said to be *low-velocity stars.* The radial velocities of nearby gas clouds are low and show that they, like the sun, move in roughly circular orbits about the galactic nucleus. In other words, the interstellar material in our part of the Galaxy also belongs to the class of low-velocity objects.

Some stars, on the other hand, have speeds relative to the sun in excess of 80 km/sec, and are called *high-velocity stars.* They move along orbits of rather high eccentricity that cross the sun's orbit in the plane of the Galaxy at rather large angles (Figure 19-3). Nearby stars moving on such orbits are passing through the solar neighborhood and are only temporarily near us. Stars in the galactic corona and globular clusters also have orbits very different from the sun's and are high-velocity objects.

It should be noted that the term "high velocity" or "low velocity" refers to the speed of an object *with respect to the sun* and has nothing to do with its motion in the Galaxy. Most high-velocity stars actually revolve about the Galaxy with speeds less than those of the low-velocity stars near the sun.

Galactic orbits

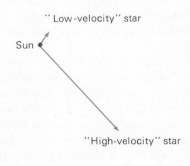

Relative velocities

FIG. 19-3   Orbits of high- and low-velocity stars.

High-velocity stars tend to be less strongly concentrated to the plane of the Galaxy than low-velocity stars. Stars in the corona are extreme examples. It is, of course, the high velocities of coronal objects perpendicular to the galactic plane that accounts for their spheroidal distribution in the Galaxy. Globular clusters, in particular, are believed to revolve about the nucleus of the Galaxy in orbits of high eccentricity and inclination to the galactic plane, perhaps rather like the comets revolving about the sun in the solar system. A globular cluster must pass through the plane of the Galaxy twice during each revolution. The large distances between stars, both within the cluster and within the Galaxy itself, make stellar collisions during the cluster's penetration of the galactic disk exceedingly improbable.

The different kinds of motion of stars in the Galaxy may be related to the times and places of formation of those stars; the high-velocity objects are believed to be among the older members of the Galaxy (see Chapter 20).

19.3 THE MASS OF THE GALAXY

The mass of the Galaxy can be calculated from the gravitational influence of the system as a whole upon the sun. For purposes of illustration, a few simplifying assumptions will be made which enable us to estimate the mass with the use of Kepler's third law.

The greatest spatial density of stars occurs in the region of the nucleus of the Galaxy. To a first approximation, therefore, we can assume that it acts, gravitationally, as if its mass were concentrated at its center. Let us assume further that the sun moves on a strictly circular orbit of radius 10,000 pc and that it completes an orbital revolution about the galactic center in 2×10^8 years. Finally, we ignore the contribution of matter farther from the galactic nucleus than the orbit of the sun and assume that all of the Galaxy's mass

is interior to the sun's orbit. With these assumptions, we can regard the sun and the Galaxy as a whole as two mutually revolving bodies.

Since there are 2×10^5 AU per parsec (Section 13.3c), the radius of the sun's orbit is 2×10^9 AU. Applying Kepler's third law (as corrected by Newton) and ignoring the mass of the sun in comparison to that of the Galaxy, we find

$$M_{\text{Galaxy}} = \frac{(2 \times 10^9)^3}{(2 \times 10^8)^2}$$

$$= \frac{8 \times 10^{27}}{4 \times 10^{16}},$$

or that the mass of the Galaxy is 2×10^{11} (200 billion) times that of the sun.

The above calculation is crude because the Galaxy cannot be expected to act, gravitationally, like a point mass. On the other hand, more sophisticated calculations that take into account the actual distribution of matter in the Galaxy (as well as it is known) give very nearly the same result. It appears, therefore, that the result of the above simple calculation is about as good a value for the mass of the Galaxy as can be obtained with our present knowledge; it is probably correct to within a factor of 2 or 3.

Presumably, most of the material of the Galaxy is in the form of stars. If the mass of the sun is taken as average, we find that the Galaxy contains some hundreds of billions of stars.

19.4 THE INTERSTELLAR MEDIUM

By earthly standards the space between the stars is empty, for in no laboratory on earth can so complete a vacuum be produced. Yet, throughout large regions of space this "emptiness" consists of vast clouds of gas and tiny solid particles. Most of this interstellar material is found between the stars in the spiral arms of our own and other galaxies. It exists, for example, in the neighborhood of the sun.

The gas and dust is not distributed uniformly, however, but has a patchy, irregular distribution, being denser in some areas than in others, hence forming "clouds."

Sometimes, these tenuous clouds are visible, or partially so, in the form of *nebulae* (Latin for "clouds"). More often, they are invisible, and their presence must be deduced. In the spiral arms, on the average, there is about one atom of gas per cubic centimeter in interstellar space, and about 25 or 50 tiny particles or "grains," each less than a thousandth of a millimeter in diameter, per cubic kilometer. In some of the denser clouds, the densities of gas and dust may exceed the average by as much as a thousand times, but even this is more nearly a vacuum than any attainable on earth. In air, for contrast, the number of molecules per cubic centimeter at sea level is of the order 10^{19}.

(a) Absorption of Starlight by Cosmic "Dust"

The tiny solid grains in interstellar space, commonly called interstellar dust, are manifested in the following ways: (1) dark nebulae, (2) general obscuration, (3) reddening of starlight, (4) reflection of starlight, and (5) polarization of starlight.

Relatively dense clouds of the solid grains produce the *dark nebulae,* the opaque-appearing clouds that are conspicuous on any photograph of the Milky Way. Even in the densest clouds, the particles are very sparse, but the clouds extend over such vast regions (measured in parsecs) that they absorb or scatter a considerable portion of the starlight passing through them. Such concentrations of dust often have the appearance of dark curtains, greatly dimming or completely obscuring the light of stars behind them.

The "dark rift," running lengthwise down a long part of the Milky Way and appearing to split it in two, is an excellent example of such an obscuring cloud. The obstruction of light from the stars located behind it is so great that less than a century ago astronomers thought that it was a

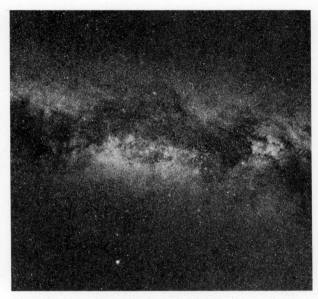

FIG. 19-4   A portion of the Milky Way in Cygnus, showing the "dark rift." *(Hale Observatories.)*

sort of "tunnel" through which they could see beyond the Milky Way, into extragalactic space.

In addition to the large dark clouds, many very small dark patches can be seen on Milky Way photographs, silhouetted against bright backgrounds of star fields or glowing gas clouds. Many of these patches, called *globules,* are round or oval and have angular diameters of only a few seconds of arc. The sharp contrast of their dark boundaries shows that they cannot be more distant than about 1000 pc, which allows us to calculate upper limits for their linear diameters. They probably range from a few thousand to a hundred thousand astronomical units across. The high opacity of the globules (they dim background objects by 5 magnitudes or more) implies that they must be very dense compared to the usual interstellar material. It has been suggested that the globules may be condensations of matter that may ultimately form into stars.

Although the distribution of the interstellar dust is spotty, and dense clouds produce conspicuous dark nebulae, some of the dust is thinly scattered more or less generally throughout the spiral arms of the Galaxy. As a result, some absorption of starlight occurs even in regions where dark clouds are not apparent. Unfortunately, the presence of such sparse absorbing matter is not obvious, and it has been the cause of considerable difficulty in the determination of stellar distances.

It has been shown (Section 14.2b) that the distance to a star can be calculated from comparison of its apparent and absolute magnitudes. If light from a star has had to pass through interstellar dust to reach us, however, it is dimmed, much as a traffic light is dimmed by fog. We therefore underestimate the apparent brightness of the star — that is, we assign to it too large (that is, faint) an apparent magnitude, and the distance we calculate for the star, corresponding to its known (or assumed) absolute magnitude, is too large. Analogously, a motorist is likely to overestimate the distance to a stoplight that he views through fog.

Astronomers once tried to calculate the extent of the Galaxy by counting the numbers of visible stars in various directions and calculating the distances required to account for their observed apparent magnitudes. Because of the general obscuration, however, they overestimated the distances of the stars, and the error increased with increasing actual distances (and hence increasing obscuration). If stars are placed too far away, they are spread out too thinly in space; in other words, interstellar absorption of light produces an *apparent* thinning out of stars with distance. The early investigators arrived at the erroneous conclusion that the Galaxy was centered on the sun, and thinned out (that is, its "edge" was reached) at a distance of only about 10,000 LY. In actual fact, we do not even see (in visible light) as far as the Galaxy's brilliant central nucleus. Were it not for the obscuring dust in space, we would be able to read at night by the light of the Milky Way.

It is a fortunate circumstance for observational astronomy that light of short wavelengths is obscured more readily than that of long wavelengths. Fifty years ago, astronomers were puzzled by the existence of stars whose spectra indicate that they are intrinsically hot and blue, of spectral type B, although they actually appear as red as cool stars of spectral type G. We know today that the light from these stars has been reddened by the interstellar absorbing material; most of their violet, blue, and green light has been obscured, leaving a greater percentage of their orange and red light, of longer wavelengths, which penetrates through the obscuring dust. This *reddening* of starlight by interstellar dust not only shows that the stars are dimmed, but also provides a means of estimating the amount of obscuration they have suffered.

The manner in which the absorption depends on wavelength can be evaluated by comparing, at various wavelengths, the relative brightnesses of two appropriate stars. Stars are chosen whose spectra show them to be approximately identical. One, however, is dimmed and reddened by interstellar dust, while the other is not, being in a direction in the sky that is relatively free of interstellar obscuration. From studies of several pairs of such obscured and unobscured stars, the absorbing power of the interstellar material at various wavelengths has been determined. Over the visible spectral region, it turns out that the absorption, expressed in magnitudes, is roughly inversely proportional to wavelength. In other words, the interstellar dust is twice as effective at obscuring starlight of 5000-Å wavelength as it is at obscuring starlight of 10,000-Å wavelength.

The law of interstellar obscuration has been found to be usually about the same in different directions in the Galaxy. Thus it is possible to estimate the total amount by which a star is dimmed from the amount that it is reddened. The reddening of the light from a star increases its apparent color index (the redder the star, the greater the color index — see Section 14.3b). The difference between the *observed* color index and the color index that the star *would have* in the absence of obscuration and reddening is called the *color excess*. The $m_{pg} - m_{pv}$ color excess, for example, is the amount by which the difference between the photographic and visual magnitudes of a star is increased by reddening. In most directions in the Galaxy, the total absorption, in visual magnitudes, is found empirically to be about three times the $m_{pg} - m_{pv}$ color excess.†

(b) Reflection Nebulae

Until now, the term "absorption" has been used loosely in this discussion. The tiny interstellar grains actually absorb some of the starlight they intercept, but most of it they merely scatter — that is, they redirect it in helter skelter directions. Since the starlight that is scattered, as well as that which is truly absorbed, does not reach us, the effect is the same as if the loss were all due to actual absorption. The whole process is more correctly termed *interstellar extinction*. The scattered or reflected light, on the other hand, illuminates the dust itself. Consequently, even the darkest dark nebulae are not completely dark but are illuminated by a faint glow of scattered starlight that can actually be measured. It is estimated that about one third of the light of the Milky Way is diffused starlight, scattered by interstellar dust.

---

†The distance modulus of a star dimmed by interstellar obscuration is given by

$$m_v - M_v = 5 \log \frac{r}{10} + 3CE,$$

where m_v and M_v are its apparent and absolute photovisual magnitudes, r its distance in parsecs, and CE its photographic minus photovisual color excess.

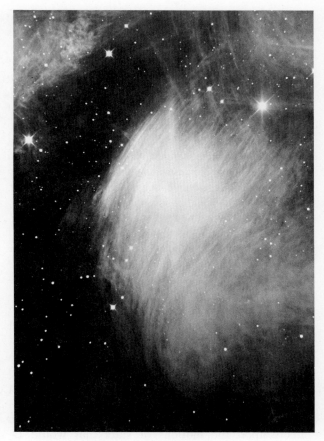

FIG. 19-5   Reflection nebula about the star Merope in the Pleiades.
(Hale Observatories.)

The light scattered by a particularly dense cloud of dust around a luminous star may be bright enough to be seen or photographed telescopically. Such a cloud of dust, illuminated by starlight, is called a *reflection nebula*. One of the best known examples is the nebulosity around each of the brightest stars in the Pleiades cluster.

Blue light is scattered more than red by the dust. A reflection nebula, therefore, always appears bluer than its illuminating star. A reflection nebula could be red only if the star which is the source of its light were very red. However, it takes a bright star to illuminate a dust cloud sufficiently for it to be visible to us, and in the regions of the Galaxy where dust clouds are found, the brightest stars are usually blue main-sequence stars. Most of the reflection nebulae that are conspicuous, therefore, are blue since they are illuminated by blue stars. Sometimes, of course, an intrinsically blue star illuminating a reflection nebula may appear much redder than it actually is because of interstellar absorption and reddening.

(c) Nature of the Interstellar Grains

The absorption of light is accomplished by *solid particles*, and not by interstellar gas. Atomic or molecular gas is almost transparent. Consider the earth's atmosphere; despite its incredibly high density, as compared to interstellar gas, it is so transparent as to be practically invisible. The absorbing power of the interstellar medium exceeds that of an equal mass of gas by more than 100,000 times. The quantity of gas that would be required to produce the observed absorption in space would have to be many thousands of times the amount that can possibly exist. The gravitational attraction of so great a mass of gas would produce effects upon the motions of stars that would be easily detected; such effects, however, are not observed.

Whereas gas can contribute only negligibly to absorption of light, we know from our everyday experience that tiny particles can be very efficient absorbers. Water vapor in the gaseous state in the air is quite invisible. When some of that vapor condenses into tiny water droplets, however, the resulting cloud is opaque. Dust storms, smoke, and smog furnish other familiar examples of the opacity of solid particles.

Calculations confirm that widely scattered particles in interstellar space must be responsible for the observed dimming of starlight. With the help of the complicated theory of the scattering of light by solid particles, and from the properties of the interstellar absorption and reflection nebulae, we can learn something of the nature of the solid grains themselves.

It is found that metallic grains, for example, iron, could produce the observed effects only if the grains were all close to a critical size. It seems more likely, however, that a variety of sizes of particles exists. Moreover, since hydrogen has been found to be by far the most abundant chemical element in the universe, one might expect that the interstellar particles are composed mostly of frozen compounds of hydrogen. Calculations show that such particles can exist in interstellar space. It is, therefore, quite probable that the interstellar "dust" is a mixture of different-sized frozen particles of nonmetallic substances, those most responsible for the spatial extinction having diameters in the range 0.0002 to 0.0005 mm.

(d) Interstellar Gas

Although interstellar gas is possibly one hundred times as dense, on the average, as the dust, because of its high transparency it is not visible by reflected starlight, nor does it contribute to the general interstellar absorption. It does manifest itself, however, in several other ways. Because of the process of *fluorescence*, clouds of gas near hot stars often shine brightly. The gas also produces narrow absorption lines superposed upon the spectra of stars whose light passes through the gas. Finally, the gas is responsible for the emission of radio waves over a broad range of wavelengths and for certain emission and absorption lines at radio wavelengths, including the important line at 21 cm.

Interstellar gas is distributed generally throughout the regions of the spiral arms of the galaxy. Hydrogen comprises about three quarters of the gas, and hydrogen and helium together comprise from 96 to 99 percent of it (by mass). Most of the gas is cold and nonluminous. Near very hot stars, however, it is ionized by the ultraviolet radiation from those stars. Since hydrogen is the main constituent of the gas, we often characterize a region of interstellar space according to whether its hydrogen is neutral — an "H I region" — or ionized — an "H II region."

The gas in the H II regions glows by the process of fluorescence. The light emitted from these regions of ionized gas consists largely of emission lines, so they are also called *emission nebulae.* Those emission nebulae in which the gas happens to be much denser than average (it occasionally reaches densities of 10^3 or 10^4 atoms per cubic centimeter — still an extremely high vacuum on earth) are especially conspicuous. The best-known example is the Orion nebula, which is barely visible to the unaided eye, but easily seen with binoculars, in the middle of the sword of the hunter. Other famous emission nebulae are the North America nebula in Cygnus and the Lagoon nebula in Sagittarius.

The cold gases in H I regions are invisible on direct photographs, but they can be observed by their radio emission. Also, some of the gases in H I regions produce absorption lines on stellar spectra.

All ultraviolet radiation of wavelength 912 Å or less can be absorbed by neutral hydrogen, and in the process the hydrogen is ionized (Section 7.5). An appreciable fraction of the energy emitted by the hottest stars lies at wavelengths shorter than 912 Å. If such a star is embedded in a cloud of interstellar gas, the ultraviolet radiation from that star ionizes the hydrogen in the gas, converting it into positive hydrogen ions (protons) and free electrons. These detached protons and electrons are then a part of the gas, each of them acting like an individual molecule, that is, a free particle. Protons in the gas are continually colliding with electrons and capturing them, becoming neutral hydrogen again. As the electrons cascade down through the various energy levels of the hydrogen atoms on their way

FIG. 19-6   The Lagoon Nebula in Sagittarius, photographed in red light with the 200-inch telescope. *(Hale Observatories.)*

to the ground states, they emit light in the form of emission lines. Lines belonging to all the series of hydrogen (Section 7.5b) are emitted — the Lyman series, Balmer series, Paschen series, and so on — but only the lines of the Balmer series are easily observed from the surface of the earth because of the opacity of our atmosphere to most wavelengths. Part of the invisible ultraviolet light from the star is thus transformed into visible light in the Balmer emission lines of hydrogen. After an atom has captured an electron and emitted light, it loses that electron again almost immediately by the subsequent absorption of another ultraviolet photon from the star. Thus, although neutral hydrogen absorbs and emits light in H II regions, almost all the hydrogen, at any given time, is in the ionized state.

The interstellar gas, of course, contains other elements besides hydrogen. Many of them are also ionized in the vicinity of hot stars and are capturing electrons and emitting light, just as the hydrogen does. Especially important is the emission of light by ionized oxygen and nitrogen and by helium. The light emitted by ionized oxygen and nitrogen arises from transitions to the lowest energy states of these ions from levels to which they have been excited by collisions with free electrons. These emission lines are observed only from a very great amount of gas that is at such a low density that it is transparent, and hence are not observed in the laboratory. For years their presence in gaseous nebulae was a mystery; they were, in fact, attributed to an unknown element, which was named *nebulium* (for the nebulae). The correct explanation was provided by the physicist I. S. Bowen in 1927.

Gas and dust are generally intermixed in space, although the proportion of one to the other is not everywhere exactly the same. The presence of dust is apparent on many photographs of emission nebulae. Clouds of dark material can be seen silhouetted on the Orion nebula, actually hiding a large part of the H II region from our view. Foreground dust clouds produce the lagoon in the Lagoon nebula, and the Atlantic Ocean and Atlantic coastline in the North America nebula. Although the dust is most conspicuous when it is in front of an emission nebula and is silhouetted against it, the dust is also intermixed with the gas. Spectra of H II regions often reveal the faint continuous spectrum (with absorption lines) of the central star, whose light is reflected to us by the dust associated with the gas. In other words, emission nebulae are generally superimposed upon *reflection nebulae*.

The gas in H II regions is heated in the ionization process, and usually has a temperature near 10,000°K. The gas in the surrounding H I regions is at about 100°K (below −270°F), and some clouds of even very much colder gas have been found. The hot gas in an H II region tries to expand, and as it pushes into the surrounding cold material of the H I region gas densities build up at the boundary. These higher-than-average densities often appear as bright "edges" in photographs of H II regions. Sometimes the hot gas expands around tongues of colder material, which then have the appearance of dark jets intruding into the bright nebula. These relatively dense regions of the interstellar medium may be sites of star formation (Chapter 20).

(e) Interstellar Absorption Lines

The cold interstellar gas — that in the H I regions — is not visible by reflected or emitted light, nor does it appreciably dim the light of stars shining through it. Yet it often reveals its presence by leaving dark absorption lines superposed upon the spectra of stars that lie beyond it.

Interstellar lines have been found of most of those elements for which observable lines would be expected. The most conspicuous interstellar lines are produced by sodium and calcium. Lines are also observed of titanium, iron, and potassium, as well as bands of the radicals CN and CH^+. The strengths of interstellar lines lead to estimates of the relative abundances of the elements that produce them. Such estimates do not differ markedly from the relative abundances of the same elements in stellar photospheres.

(f) Radio Emission From Interstellar Gas

We receive from the Milky Way a large amount of radio energy in the frequency range 10 to 300 Mc (or wavelength range 1 to 30 m). This emission from the Milky Way was, in fact, the first radio radiation of astronomical origin to be observed (Section 8.8a). Similar radiation at radio frequencies is received from nearby galaxies. The radio energy from the Milky Way does not appear to originate in stars. The sun, to be sure, is an apparently strong source of radio waves, but the sun is very close to us. If its radio emission is typical of that of the other stars, all the stars in the Galaxy would not emit a billionth of the radio energy actually observed. It is concluded, therefore, that radio energy originates in the interstellar gas.

(g) Radio Lines: The 21-cm Line of Hydrogen

A number of emission (and absorption) lines at radio wavelengths can now be observed. The first radio line to be detected is the 21-cm line of neutral hydrogen.

Some of the most important radio observations carried out in the interstellar medium are of the spectral line of hydrogen at the radio wavelength of 21.11 cm. A hydrogen atom possesses a tiny amount of angular momentum by virtue of the axial spin of its electron and the electron's orbital motion about the nucleus (proton). In addition, the proton has an axial spin of its own, and an angular momentum associated with this spin. In its usual state, the angular momenta of the electron and proton of a hydrogen atom are aligned in opposite directions. They can be aligned, however, by collisions between the hydrogen atom and other particles, in which case the atom is in a slightly excited state. Eventually such an excited hydrogen atom de-excites itself and returns to its normal state with the emission of a photon of wavelength 21 cm.

In 1944 the Dutch astronomer H. C. van de Hulst predicted that enough atoms of interstellar hydrogen would be radiating photons of 21-cm wavelength to make this radio emission line observable. Equipment sensitive enough to detect the line was not available until 1951. Since that time, "21-cm astronomy" has been a very active field of astronomical research.

Observations at 21 cm show that the neutral hydrogen in the Galaxy is confined to an extremely flat layer, most of it being in a sheet less than 100 pc thick, extending throughout the plane of the Milky Way. The strength of the line indicates that neutral interstellar hydrogen makes up from 1 to 2 percent of the mass of the Galaxy. One important use of 21-cm radiation is the measurement of its Doppler shifts in various directions, which (as we shall see) helps us to map out the spiral structure of our galactic system.

More recently, many other interstellar lines have been observed at radio wavelengths, including those originating from transitions between high energy levels of hydrogen, and also lines due to the OH radical, water, ammonia, and a growing list of organic molecules including formaldehyde and even more complex molecules. The fact that relatively complicated organic molecules can form, evidently spontaneously, in the cold and high vacuum conditions of interstellar space is highly interesting, for it thus becomes more credible that the even more complex organic molecules that are the basis of living organisms might form spontaneously in the atmosphere of the earth, and possibly other planets.

19.5 SPIRAL STRUCTURE OF THE GALAXY

Observations have been made of many other spiral galaxies which contain interstellar matter and in which individual stars can be resolved (Chapter 21). In those systems both the interstellar matter and the most luminous resolved stars are generally confined to the spiral arms. The association of the brightest stars and the gas and dust is not mysterious; highly luminous stars are almost certainly relatively young objects and, in all probability, have recently formed from clouds of interstellar gas and dust (Chapter 20). In mapping out the spiral structure of our own galaxy, therefore, we can use the gaseous nebulae and the very luminous main-sequence O and B stars as "tracers" to identify spiral arms.

Because of our position in the disk of the Galaxy, and because we are surrounded by interstellar dust, it is difficult for us to identify even the nearby

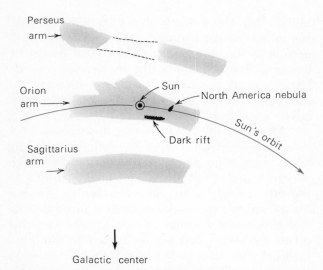

FIG. 19-7   Spiral structure of the Galaxy as deduced from optical observations.

parts of the disk have high velocities compared to the sun; the actual *radial velocity* of a gas cloud depends on its distance along a given line of sight through the galactic plane.

Now because different gas clouds along such a line of sight have different radial velocities, we observe the 21-cm radiation from the hydrogen Doppler-shifted by various amounts; that is, we do not see a single emission line at 21.11 cm, but several emission lines, all at slightly different wavelengths, each arising from a different concentration of hydrogen. The amounts of those different Doppler shifts enable us to calculate where along the line of sight each hydrogen cloud is. By observing in different directions in the galactic plane, we map out the distribution of neutral hydrogen in the galaxy, and hence detect the spiral arms. Thus maps of the spiral structure, like Figure 19-8, are obtained.

spiral arms optically. Nevertheless, short pieces of three nearby arms have been detected from the observed directions and distances of a large number of O and B stars and from the distribution of emission nebulae. The fragmentary data on the spiral structure of the Galaxy, as deduced from such optical observations, are summarized in Figure 19-7.

The sun appears to be near the inner edge of an arm which contains such conspicuous features as the North America nebula, the Coalsack (near the Southern Cross), the Cygnus Rift (great dark nebula in the summer Milky Way), and the Orion nebula. More distant, and therefore less conspicuous, emission nebulae can be identified in the *Sagittarius* and *Perseus arms,* located, respectively, about 3000 pc inside and outside the sun's position with respect to the galactic nucleus.

A more powerful method of mapping the spiral structure of the Galaxy is from radio observations of the 21-cm line of neutral hydrogen. The radio waves are not absorbed by interstellar dust and thus traverse the entire disk of the Galaxy. Because the Galaxy is rotating, the gas clouds in remote

FIG. 19-8   A drawing of the spiral structure of the Galaxy, as deduced from 21-cm observations made at Leiden (in The Netherlands) and at Sydney, Australia. The large and small dots show the location of the galactic center and the sun, respectively. *(Courtesy of Gart Westerhout.)*

19.6 DIFFERENT STELLAR POPULATIONS IN THE GALAXY

Striking correlations are found between the characteristics of stars and other objects and their locations in the Galaxy. Some classes of objects, for example, are found only in regions of interstellar matter, that is, in the spiral arms of the Galaxy. Examples are bright supergiants, main-sequence stars of high luminosity (spectral classes O and B), Wolf-Rayet stars, type I cepheid variables, and young open star clusters.

The distributions of some other classes of objects show no correlation with the location of spiral arms. These objects are found throughout the disk of the Galaxy, with greatest concentration toward the nucleus. They also extend into the sparse galactic corona. Examples are planetary nebulae, novae, type II cepheids, RR Lyrae variables, and Mira type variables of periods less than 250 days. The globular clusters, which also belong to this group of objects, are found almost entirely in the corona and nucleus of the Galaxy.

Main-sequence stars of spectral types F through M exist in all parts of the Galaxy, as do red giants and, probably, white dwarfs (observations of white dwarfs, of course, are limited to our immediate neighborhood because of their low luminosities).

There are also differences in the chemical compositions of stars in different parts of the Galaxy. Nearly all stars appear to be composed mostly of hydrogen and helium, but the residual abundance of the heavier elements seems to spread over a large range for different stars. In the sun and in other stars associated with the interstellar matter of the spiral arms, the heavy elements (elements other than hydrogen and helium) account for about 1 to 4 percent of the total stellar mass. Stars in the galactic corona and in globular clusters, however, have much lower abundances of the heavy elements — often less than one tenth, or even one hundredth that of the sun. There is also some evidence that many stars in the disk of the Galaxy, whose galactic orbits show that they are not associated with interstellar matter, may have lower heavy-element abundances than the sun.

The stars associated with the spiral arms are sometimes said to belong to *population I*, while those found elsewhere in the Galaxy are said to belong to *population II*. The terms "population I" and "population II" were first applied to different classes of stars by W. Baade, late astronomer at the Mount Wilson and Palomar Observatories. During World War II, Baade was impressed by the similarity of the stars in the nucleus of M31 — the Andromeda galaxy — to those in the globular clusters in the corona of our own galaxy and concluded that the stars situated in spiral arms must, collectively, display different properties from those located elsewhere in the Galaxy.

Today we can interpret the phenomenon of different stellar populations in the light of stellar evolution. As has been mentioned, and as will be discussed further (Chapter 20), only in the interstellar matter of spiral arms is star formation expected to take place. Thus, population I is comprised of stars of many different ages, including some that were recently formed or are still forming. Population II, on the other hand, consists entirely of old stars, formed, probably, early in the history of the Galaxy.

It is clear today that two stellar populations are insufficient to account completely for the distribution of all the different kinds of stars in the Galaxy. Modern investigators often speak of several stellar populations, ranging from "extreme population I" (spiral-arm objects), through "disk population" objects, to "extreme population II" (corona objects). The concept of two stellar populations is still a useful one, however, in interpreting many of the general characteristics of our own and other galaxies. In this book we shall use the term "population I" to apply to spiral-arm objects, and "population II" to apply to all others.

19.7 STAR CLUSTERS

A star cluster is a congregation of stars that have a stronger gravitational attraction for each other (because of their proximity) than do stars of the general field (that is, those *not* in clusters). Clusters range from rich aggregates of many thousands of stars to loose associations of only a few stars. The mutual gravitation of the stars in the larger clusters may hold them together more or less permanently; the small clusters may be held together so weakly that they must be gradually dissipating into the field.

A number of clusters bear proper names. Some of these are names of mythological characters (the Pleiades); other clusters bear the names of the constellations in which they appear (the Double Cluster in Perseus). Most of the conspicuous clusters are listed in the early catalogues of star clusters and nebulae, especially in Sir John Herschel's *General Catalogue* (1864). The catalogue designations of clusters most often referred to today are those in Messier's (1781) catalogue (for example, M13 — the famous globular cluster in Hercules), and in Dreyer's revisions of Herschel's catalogue, the *New General Catalogue* (*NGC*) and the *Index Catalogue* (*IC*), published between 1888 and 1908. For example, in Dreyer's catalogue, M13 is known as NGC 6205.

(a) Globular Clusters

As we have already mentioned, about a hundred globular clusters are known, most of them in the corona and nucleus of our Galaxy. All are very far from the sun, and some are found at distances of 60,000 LY or more from the galactic plane. A few, nevertheless, are bright enough to be seen with the naked eye; they appear as faint, fuzzy stars. One of the most famous naked-eye globular clusters is M13, in the constellation of Hercules, which passes nearly overhead on a summer evening at most places in the United States. Through a good pair of binoculars the more conspicuous globular

FIG. 19-9   The globular cluster, M3.
(Lick Observatory.)

clusters resemble tiny moth balls. A 4- to 6-in. telescope reveals their brightest stars, while a large telescope shows them to be beautiful, globe-shaped systems of stars. Visual observation, however, even through the largest telescope, does not reveal the multitude of fainter stars in globular clusters that can be recorded on telescopic photographs of long exposure.

A good photograph of a typical globular cluster shows it to be a nearly circularly-symmetrical system of stars, with the highest concentration of stars near its own center (a few globular clusters, such as Omega Centauri, appear slightly flattened). Most of the stars in the central regions of the cluster are not resolved as individual points of light but appear as a nebulous glow. Two photographs of a globular cluster made on emulsions sensitive to two different colors of light, say red and blue, show that the brightest stars are red. These stars are 2 or 3 magnitudes brighter than the RR Lyrae variable stars that are almost always found in globular clusters; since RR Lyrae stars average about absolute magnitude 0, the brighter red stars must be red giants.

Distances to globular clusters are usually calculated from the apparent magnitudes of the RR Lyrae stars they contain (Section 18.2d). From their angular sizes (typically a few minutes of arc) their actual linear diameters are found to be from 20 to 100 pc or more. In one of the nearer globular clusters more than 30,000 stars have been counted, but if those stars too faint to be observed are considered, most clusters must contain hundreds of thousands of member stars. The combined light from all these stars gives a typical globular cluster an absolute magnitude somewhere in the range −5 to −9.

The average star density in a globular cluster is about 0.4 star per cubic parsec. In the dense center of the cluster the star density may be as high as a 100 or even 1000 per cubic parsec. There is plenty of space between the stars, however, even in the center of a cluster. The "solid" photographic appearance of the central regions of a globular cluster results from the finite resolution of the telescope, seeing effects of the earth's atmosphere, and the scattering of light in the photographic

FIG. 19-10   The open star cluster M67 (NGC 2682) in Cancer. Photographed with the 200-inch telescope. *(Hale Observatories.)*

emulsion. If the earth revolved not about the sun, but about a star in the densest part of a globular cluster, the nearest neighboring stars, light-months away, would appear as points of light. Thousands of stars, however, would be scattered uniformly over the sky. The Milky Way would be hard, if not impossible, to see, and even on the darkest of nights the brightness of the sky would be comparable to what it is on earth in bright moonlight.

Globular clusters do not partake of the general galactic rotation. They are believed to revolve about the nucleus of the Galaxy on orbits of high eccentricity and high inclination to the galactic plane (rather like the orbits of comets in the solar system). Obeying Kepler's second law, a cluster spends most of its time far from the nucleus; a typical cluster probably has a period of revolution of the order of 10^8 years.

(b) Open Clusters

In contrast to the rich, partially unresolved globular clusters, *open clusters* appear comparatively loose and "open" (hence their name). They contain far fewer stars than globular clusters and show little or no strong concentration of stars toward their own centers. Although open clusters are usually more or less round in appearance, they lack the high degree of spherical symmetry that characterizes a globular cluster; some open clusters actually appear irregular. On photographs made with long-focus telescopes, the stars in these clusters are usually fully resolved, even in the central regions.

Open clusters, as mentioned previously, are found in the disk of the Galaxy, often associated with interstellar matter. Because of their locations, they are sometimes called *galactic clusters* rather than open clusters (this term should not be confused, however, with *clusters of galaxies* — Chapters 21 and 22). Nearly 900 open clusters are catalogued (at the time of writing), but probably well over 1000 are identifiable on good search photographs such as those of the Palomar Sky

Survey. Yet only the nearest open clusters can be observed, because of interstellar obscuration in the Milky Way plane. We conclude, therefore, that we see only a small fraction of the open clusters that actually exist in the Galaxy; possible tens or even hundreds of thousands of them escape detection.

Several open clusters are visible to the unaided eye. Most famous among them is the *Pleiades,* which appears as a tiny group of six stars (some people see more than six) arranged like a tiny dipper in the constellation of *Taurus;* a good pair of binoculars shows dozens of stars in the cluster, and a telescope reveals hundreds. (The Pleiades is *not* the Little Dipper; the latter is part of the constellation of *Ursa Minor,* which also contains the north star.) The *Hyades* is another famous open cluster in Taurus. To the naked eye, the cluster appears as a V-shaped group of faint stars, marking the face of the bull. Telescopes show that the Hyades actually contains more than 200 stars. The naked-eye appearance of the *Praesepe,* in Cancer, is that of a barely distinguishable patch of light; this group is often called the "Beehive" cluster, because its many stars, when viewed through a telescope, appear like a swarm of bees.

Typical open clusters contain several dozen to several hundred member stars, although a few, such as M67, contain more than a thousand. Compared to globular clusters, open clusters are small, usually having diameters of less than 10 pc. Bright supergiant stars of high luminosity in some open clusters, however, may cause them to outshine the far richer globular clusters. The RR Lyrae stars are never found in open clusters, but other kinds of variable stars, such as type I cepheids, are sometimes present.

(c) Associations

For more than 40 years it has been known that the most luminous main sequence stars — those of spectral types O and B — are not distributed at random in the sky but tend to be grouped into what are now called *associations,* lying along the spiral arms of our Galaxy. In the decade following World War II, interest was greatly revived in these groups of hot stars, especially by the Soviet astronomer V. A. Ambartsumian, who called attention to many of them and pointed out that they must be very young

FIG. 19-11 A wide angle photograph of the constellation Orion. An association of young stars is centered in the "sword" (lower part of the picture). *(Hale Observatories.)*

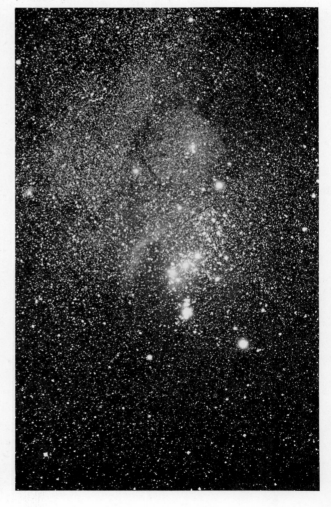

groups of stars. Because the stars of an association lie in the galactic plane and are spread over tens of parsecs, each will revolve about the galactic center with a slightly different orbital speed; Ambartsumian showed that the different orbital speeds of the different members of an association would completely disrupt the group after a few million years. The fact that associations exist at all, therefore, shows that they must be very young objects on the astronomical time scale.

An association appears as a group of several (say, 5 to 50) young stars, scattered over a region of space some 30 to 200 pc in diameter. Because these stars are rare, it would be very unlikely for so many of them to exist by chance in so relatively small a volume of space. It is assumed, therefore, that the stars in an association are either physically associated or at least have had a common origin. Stars of other spectral types may also belong to associations, but these more common stars are not conspicuous against the general star field and do not attract attention to themselves as belonging to any particular group.

Often a small open cluster is found near the center of an association. It is presumed, in such cases, that the stars in the association are the outlying members of the cluster, and that they probably had a common origin with the cluster stars. It is probable that some associations may be expanding — their member stars moving radially outward from the association center. An example of such an association is one in Perseus, known as the *Zeta Persei association.* More than a dozen of its member stars are bright enough so that their positions have been measured accurately for more than a century. Most of the stars seem to be moving outward, from the center of the group. If we extrapolate their motions backward, we find an age for the expansion of about 1½ million years. Another famous association is in *Orion,* centered on a group of stars in the middle of the Orion Nebula. Rather far from the main association are three luminous O and B stars that seem to be moving away from it at speeds of from 70 to 130 km/sec. It is tempting to postulate that these stars originated with those of the association, but there is as yet no satisfactory hypothesis to explain their extremely high speeds. Analysis of the statistics of associations and O and B stars, however, have led some investigators to conclude that all such stars have originated in associations.

About 80 associations are now catalogued. Like ordinary open clusters, however, they lie in regions occupied by interstellar matter, and many others

TABLE 19.1 Characteristics of Star Clusters

	GLOBULAR CLUSTERS	OPEN CLUSTERS	ASSOCIATIONS
Number known	119	867	82
Location in Galaxy	Corona and nucleus	Disk (especially spiral arms)	Spiral arms
Diameter (pc)	20 to 100	<10	30 to 200
Mass (solar masses)	10^4 to 10^5	10^2 to 10^3	10^2 to 10^3 ?
Number of stars	10^4 to 10^5	50 to 10^3	10 to 100 ?
Color of brightest stars	Red	Red or blue	Blue
Integrated absolute visual magnitude of cluster	-5 to -9	0 to -10	-6 to -10
Density of stars (solar masses per cubic parsec)	0.5 to 1000	0.1 to 10	<0.01
Examples	Hercules Cluster (M13)	Hyades, Pleiades, h and χ Persei, Praesepe	Zeta Persei, Orion

must be obscured. There are probably several thousand undiscovered associations in our galaxy.

(d) Summary of Clusters

Star clusters are summarized in Table 19.1. The numbers of globular clusters, open clusters, and associations are taken from the Czechoslovakian *Catalogue of Star Clusters and Associations,* which includes all objects reported up to October 7, 1961, and is the most comprehensive star cluster catalogue available. The sizes, absolute magnitudes, and numbers of stars listed for each type of cluster are approximate only and are intended as representative values.

Exercises

1. Sketch the distribution of globular clusters about the Galaxy, and show the sun's position. Show how they would appear on a Mercator-type map of the sky, with the central line of the Milky Way chosen as the "equator."

2. The globular clusters probably have highly eccentric orbits, and either oscillate back and forth through the plane of the Galaxy or revolve about its nucleus. Suppose the latter is the case; where would the clusters spend most of their time? (Think about Kepler's second law.) At any given time, would you expect most globular clusters to be moving at high or low speeds with respect to the center of the Galaxy? Why?

3. What would the mass of the Galaxy be if the sun's distance from the center were 10,000 pc but its period of revolution about the nucleus were only 100,000,000 years?

4. Suppose the mean mass of a star in the Galaxy were only 1.3 solar mass. Using the value for the mass of the Galaxy found in the text, find how many stars the system contains. What did you assume about the total mass of interstellar matter in finding your answer?

5. Identify several dark nebulae on photographs of the Milky Way in this book. Give the figure numbers of the photographs, and specify where on them the dark nebulae are to be found.

6. The red color of the sun, when seen close to the horizon, and the blue color of the daytime sky provide analogies to the reddening of starlight and the blue color of reflection nebulae. Discuss this analogy more fully, and also explain how it breaks down.

7. Where in the Galaxy do you suppose undiscovered globular clusters may exist?

8. In Table 19.1 it is indicated that stellar associations can emit even more light than a globular cluster. How is this possible if the associations have so few stars?

9. What color would a globular cluster appear? Why?

10. It is often possible to observe fainter main-sequence stars in open clusters than in globular clusters. Why do you suppose this is the case?

11. Describe the details of an experiment in which you witness the formation of "spiral arms" of cream in a cup of coffee that you have stirred vigorously before putting in the cream.

12. Distinguish clearly between the orbital motion of the sun about the galactic center and the *solar motion,* toward the *solar apex,* which was described in Chapter 13.

Orion Nebula. *(Lick Observatory.)*

CHAPTER 20

Stellar Structure and Evolution

Except for the sun, the stars appear as unresolved points of light, even when viewed with the world's greatest telescopes. It might seem, therefore, that a knowledge of the internal structure of a star, let alone how it changes in time, would be hopelessly inaccessible. Yet, in many respects we have a better understanding of the interiors and evolution of stars than we do of the interior of our own earth. This chapter will summarize how we study the structure of stars, the energy sources that keep them shining, and how stars form, age, and die.

20.1 EQUILIBRIUM IN STARS

The sun and the vast majority of other stars are *stable;* that is, they are in a *steady state* — neither expanding nor contracting. Such a star is said to be in a condition of *equilibrium;* all the forces within it are balanced, so that at each point within the star the temperature, pressure, density, and so on, are maintained at constant values. We shall see that even these stable stars, including the sun, are changing as they evolve, but such evolutionary changes are so gradual that to all intents and purposes the stars are still in a state of equilibrium.

(a) The Equilibrium Between Gravitation and Pressure

The mutual gravitational attraction between the masses of various regions within a star produces tremendous forces that tend to collapse the star toward its center. Yet, since stars like the sun have remained more or less unchanged for billions of years, the gravitational force that tends to collapse a star must be exactly balanced by a force from within.

Even at the surfaces of stars, the temperatures are so high that all chemical elements are vaporized to gases. The internal temperatures of stars are many times higher than their surface temperatures; thus, stars are gaseous throughout. Most of the internal force that supports the outer layers of a star against the downward pull of gravitation upon them is the pressure of the gases themselves. (In some very luminous stars the pressure of radiation also contributes appreciably to support of the outer layers.)

If the internal gas pressure in a star were not great enough to balance the weight of its outer parts, the star would collapse somewhat, contracting until the gas pressure inside built up to the point where it could support the star. On the other hand, if the gas pressure in a star were greater than the weight of the overlying layers, the star would expand, thus decreasing the internal pressure. Expansion would stop when the pressure at every point within the star again equaled the weight of the stellar layers above that point. This condition is called *hydrostatic equilibrium.* Stable stars are all in hydrostatic equilibrium; so are the oceans of the earth, as well as the earth's atmosphere. (Recall that the pressure of the air keeps the air from falling to the ground.)

The particles that comprise a gas (molecules or atoms) are in rapid motion, frequently colliding with each other and with the walls of the container of the gas. This constant bombardment is the pressure of the gas. A mathematical relation between the pressure and the density and temperature of a gas is given by the *perfect gas law* (also called the *equation of state* for a perfect gas).

The law states that for a perfect, or ideal, gas (one in which intermolecular or interatomic forces can be ignored), the pressure is proportional to the product of the density and temperature. The gases in most stars closely approximate an ideal gas; thus, they must obey this law.

The condition of hydrostatic equilibrium and the gas law enable us to derive minimum values for the internal pressures and temperatures of a star. It is found, for example, that for the internal gases in the sun to support their own weight, the mean pressure must be at least 500 million times the sea-level pressure of the earth's atmosphere, the central pressure must be at least 1.3 billion times that of the earth's atmosphere, and the mean temperature must be at least 2.3 million degrees Kelvin. These pressures and temperatures are only lower limits; the actual values are much higher. Under such conditions all elements are in the gaseous form, and the atoms cannot be combined into molecules. Moreover, most of the atoms are almost completely ionized — that is, stripped of their electrons (Section 7.5e). These electrons, freed from their parent atoms, become part of the gas itself, moving about as individual particles.

(b) Thermal Equilibrium

From observation we know that energy (electromagnetic energy) flows from the surfaces of the sun and stars. According to the second law of thermodynamics, heat always tries to flow from hotter to cooler regions. Therefore, as energy filters outward toward the surface of a star, it must be flowing from inner hotter regions. The temperature can never *decrease* inward in a star, or energy would flow in and heat up those regions until they were at least as hot as the outer ones. We conclude that the highest temperature occurs at the center of a star and that temperatures drop to successively lower values toward the stellar surface.† The outward flow of energy through a star, however, robs it of its internal heat and would

result in a cooling of the interior gases were that energy not replaced. There must therefore be a source of energy within each star.

If a star is in a steady state (that is, in hydrostatic equilibrium and shining with a steady luminosity), the temperature and pressure at each point within it must remain approximately constant. Energy must be supplied, therefore, to each layer in the star at just the right rate to balance the loss of heat in that layer as it passes energy outward toward the surface. Moreover, the rate at which energy is supplied to the star as a whole must, at least on the average, exactly balance the rate at which the whole star loses energy by radiating it into space; that is, the rate of energy production in a star is equal to the luminosity. We call this balance of heat gain and heat loss for the star as a whole and at each point within it the condition of *thermal equilibrium*.

20.2 STELLAR ENERGY

The rate at which the sun emits electromagnetic radiation into space, and thus the rate at which energy must be generated within it, is about 4×10^{33} ergs/sec (Section 14.3d). Moreover, the power output of the sun has been about the same throughout recorded history and, according to geological evidence, not very different since the formation of the earth billions of years ago. Our problem is now to find what sources can provide the gigantic amounts of energy required to keep the stars like the sun shining for so long.

(a) Thermal and Gravitational Energy

Two large stores of energy in a star are its internal

---

†The high temperature of the sun's corona may therefore appear to be a paradox. The actual heat energy in the sun's corona is relatively small because the corona is a highly rarefied gas. Its high temperature is believed to be maintained by shock waves or by some other process that would not exist for a gas in thermodynamic equilibrium.

heat, or *thermal energy,* and its *gravitational energy.* The heat stored in a gas is simply the energy of motion (kinetic energy) of the particles that comprise it. If the speeds of these particles can decrease, the loss in kinetic energy can be radiated away as heat and light. This is how a hot iron cools after it is withdrawn from a fire (except that the atoms in a solid vibrate within a crystalline structure, rather than moving freely, as in a gas).

Because a star is bound together by gravity, it has gravitational potential energy, as does a stone held above the ground. If the various parts of a star fall closer together, that is, if the star contracts, it converts part of its potential energy into heat, some of which can be radiated away. About the middle of the nineteenth century, the physicists Helmholtz and Kelvin postulated that the source of the sun's luminosity was indeed the conversion of part of its gravitational potential energy into radiant energy.

They showed that because of its enormous mass, the sun need contract only extremely slowly to release enough gravitational potential energy to account for its present luminosity. In fact, over the time span of recorded history, the decrease in the sun's size resulting from its contraction would be so negligible as to escape detection. It seemed to these researchers, therefore, that the sun's gravitational and thermal energies were sufficient to keep it shining for an extremely long time, and were certainly the source of its power.

The amount of potential energy that has been released since the presolar cloud (from which the sun originally formed) began to contract is of the order of 10^{49} ergs. This is the amount, according to the Helmholtz and Kelvin theory, that it could have converted to thermal energy and luminosity. Since the present luminosity of the sun is 4×10^{33} ergs/sec, or about 10^{41} ergs/year, its contraction can have kept it shining at its present rate for a period of the order of 100 million years. It is only within the present century that it was learned that the earth, and hence the sun, has an age of at least several billion years, and therefore that the sun's gravitational energy is grossly inadequate to account for the luminosity it has generated over its lifetime.

(b) Nuclear Energy

As stated in Section 7.5a the nuclei of atoms can be regarded as composed of still smaller particles called *nucleons.* The most common nucleons are protons, which carry a positive electric charge, and neutrons, which are electrically neutral. Although neutrons are slightly more massive than protons, both have masses that are approximately 1 atomic mass unit (amu); 1 amu = 1.66×10^{-24} gm.

The number of protons in the nucleus of a particular kind of atom determines the total positive charge on the nucleus, and the number of protons plus the number of neutrons is equal to the atomic mass number of the nucleus (see Chapter 7). The mass of an ordinary hydrogen nucleus is exactly that of a proton (1.00813 amu). The total mass of every other kind of nucleus, however, is slightly less than the sum of the masses of the nucleons that would be required to build it. This slight deficiency in mass, always only a tiny fraction of 1 amu, is called the *mass defect* of the nucleus. The mass defect is greatest for the nucleus of the iron atom, and is less for both more massive and less massive nuclei.

A *nuclear transformation* is a buildup of a heavier nucleus from lighter ones, or a breakup of a heavier nucleus into lighter ones. Now the total mass of all the nuclei entering a nuclear transformation (or reaction) is, in general, slightly different from the total mass of the nuclei that result from the transformation. This difference in mass is simply the difference between the sums of the mass defects of the initial and final nuclei involved in the reaction. It might seem at first sight that the principle of conservation of mass is violated in nuclear transformations. However, Einstein showed, as part of his special theory of relativity, that there is an equivalence between mass and energy, given by the famous equation

$$E = mc^2,$$

where E is the energy equivalent to the mass, m, and c is the velocity of light. The mass unaccounted for in a nuclear transformation is always balanced by the absorption or release of energy in the amount given by the above formula. In this case m is the mass discrepancy involved in the nuclear reaction. In point of fact, the equivalence of mass and energy enables us to think of the mass defect of a single nucleus as the energy binding that nucleus together.

The energy released when nuclei of uranium or plutonium break into smaller nuclei (as in some kinds of nuclear bombs and reactors) is an example of energy produced by the breakup of heavy nuclei. In this case, when the nuclear products of the breakup are of comparable mass, the process is called *nuclear fission.*

The opposite of nuclear fission is nuclear *fusion,* in which lighter nuclei combine to form heavier ones. It was suggested about 1928 that the energy source in stars might be fusion of light elements into heavier ones. Since hydrogen and helium account for about 98 or 99 percent of the mass of most stars, we logically look first to these elements as the probable reactants in any such fusion reaction. Helium nuclei are about four times as massive as hydrogen nuclei, so it would take four nuclei of hydrogen to produce one of helium. The masses of hydrogen and helium nuclei are 1.00813 and 4.00389 amu, respectively. Let us compute the difference in initial and final mass:

$$4 \times 1.00813 = \quad 4.03252 \text{ amu (mass of initial hydrogen)}$$

$$- 4.00389 \text{ amu (mass of final helium)}$$

$$\overline{ 0.02862 \text{ amu (mass lost in the transaction)}.}$$

The mass lost, 0.02862 amu, is 0.71 percent of the mass of the initial hydrogen. Thus if 1 gm of hydrogen turns into helium, 0.0071 gm of material is converted into energy. The velocity of light is 3×10^{10} cm/sec, so the actual energy released is

$$E = 0.0071 \times (3 \times 10^{10})^2$$

$$= 6.4 \times 10^{18} \text{ ergs.}$$

This 6×10^{18} ergs is enough energy to raise the 200-inch telescope nearly 100 mi above the ground.

To produce the sun's luminosity of 4×10^{33} ergs/sec, some 600 million tons of hydrogen must be converted to helium each second, with the simultaneous conversion of about 4 million tons of matter into energy. As large as these numbers are, the store of nuclear energy in the sun is still enormous. Suppose half of the sun's mass of 2×10^{33} gm is hydrogen that can ultimately be converted into helium; then the total store of nuclear energy would be 6×10^{51} ergs. Even at the sun's current rate of energy expenditure, 10^{41} ergs/year, the sun could survive for more than 10 billion years.

There is little doubt today that the principal source of energy in stars stems from thermonuclear reactions. Deep in the interiors of stars, where the temperatures range up to many millions of degrees, nuclei of atoms are changing from one kind to another with an accompanying release of energy. The most important of these changes is the conversion of hydrogen to helium. At least two processes have been discovered by which collisions between atomic nuclei, and subsequent nuclear transformations can result in the fusion of hydrogen into helium under conditions that prevail in stellar interiors. The steps in these processes (the *carbon cycle* and the *proton-proton chain*) are listed in Appendix 8. The discovery that some of the energy locked up in the nuclei of atoms can be released in the interiors of stars is perhaps the most significant contribution of astronomy in the twentieth century.

Most of the electromagnetic radiation released in these nuclear reactions is at very short wavelengths — in the form of X-rays and gamma rays. Nuclear reactions are important, however, only deep in the interior of a star. Before this released energy reaches the stellar surface, it is absorbed and reemitted by atoms a very great number of times. Photons of high energy (short wavelength) that are absorbed by atoms are often reemitted as two or more photons, each of lower energy. By the time the energy filters out to the surface of the star, therefore, it has been converted from a relatively small number of photons, each of very high energy, to a very much larger number of photons of lower energy and longer wavelength, which constitute the radiation we actually observe leaving the star.

20.3 MODEL STARS

To determine the internal structure of a star, we must now combine the principles we have described — hydrostatic equilibrium, the perfect gas law, thermal equilibrium, and the rate of energy generation from nuclear processes — with theoretical calculations of the opacity of gases (how they absorb radiation and impede the flow of radiation through them) and considerations of various modes of energy transport. These physical ideas are for-

FIG. 20-1   Modern computing center.
(Control Data Corporation.)

mulated into mathematical equations which are solved to determine the march of temperature, pressure, density, and other physical variables throughout the stellar interior. The set of solutions so obtained, based upon a specific set of physical assumptions, is called a theoretical model for the interior of the star in question.

TABLE 20.1   Model for the Structure of the Sun†

FRACTION OF RADIUS	FRACTION OF MASS	FRACTION OF LUMINOSITY	TEMPERATURE (MILLIONS OF DEGREES K)	DENSITY (GM/CM³)	FRACTION HYDROGEN (BY WEIGHT)
0.00	0.000	0.00	15.4	158	0.38
0.05	0.014	0.14	14.8	128	0.47
0.10	0.095	0.55	12.6	84	0.59
0.15	0.20	0.79	11.0	59	0.67
0.20	0.35	0.94	9.4	36	0.71
0.30	0.62	1.00	6.6	13	0.73
0.40	0.80	1.00	5.0	3	0.73
0.60	0.95	1.00	3.1	2	0.73
0.80	0.99	1.00	1.5	0.2	0.73
1.00	1.00	1.00	0.0	0.0	0.73

†Adapted from Torres-Peimbert, Ulrich, and Simpson.

The solution of the equations of structure to obtain a stellar model is a difficult and tedious business. Until the 1950's it often took as long as a year to compute a stellar model, and such a computation was a satisfactory topic for the dissertation of a student earning his Ph.D. degree. Now, however, high-speed electronic computers enable the calculation of a model in a few minutes or, in some cases, even in a few seconds.

An example of a model star is the model for the sun given in Table 20.1. This model, by Torres-Peimbert, Ulrich, and Simpson at the University of California, Berkeley, assumes that the sun was originally 73 percent hydrogen, 24.5 percent helium, and 2.5 percent heavier elements. However, the conversion of hydrogen to helium in the 4.5 billion years that the sun has been shining has reduced the hydrogen content at the center to only 38 percent.

It must be borne in mind that all models of a stellar interior, such as this one for the sun, are only as accurate as the state of our knowledge of the relevant physical processes at the times the models are constructed. New models, reflecting refinements in theory, are continually being calculated. Table 20.1, nevertheless, serves to illustrate the general nature of the internal structure of the sun.

(a) The Russell-Vogt Theorem and the Interpretation of the Main Sequence

A certain minimum amount of information, then, is required to determine the structure of a star, if it is assumed that all pertinent physical laws are perfectly understood and that infinitely accurate calculations can be performed.

It can be shown that if a star is in hydrostatic and thermal equilibrium, and if it derives all its energy from nuclear reactions, then its structure is completely and uniquely determined by its total mass and by the distribution of the various chemical elements throughout its interior. This is not to imply that we necessarily have the ability to compute perfect models for such stars; the stars themselves, nevertheless, must conform to the physical laws that govern their material and will adjust themselves to unique configurations. It should not, of course, surprise us that the very properties that a star is "born" with, and over which it "has no control," are just those that determine its structure. This important principle is known as the *Russell-Vogt theorem*.

Suppose a cluster of stars were to form from a cloud of interstellar material whose chemical composition was similar to the sun's. All condensations that become stars would then begin with the same chemical composition and would differ from each other only in mass. Suppose now that we were to compute a model for each of these stars for the time at which it became stable, and derived its energy from nuclear reactions, but before it had had time to alter its composition appreciably as a result of these reactions. The admixture of chemical elements would be the same, then, at all points within the star; such a composition is *homogeneous*.

The models we would calculate for these stars would indicate, among other things, their luminosities and radii. From Stefan's law we know that the luminosity of a star is proportional to the product of its surface area and the fourth power of its effective temperature (Section 7.4d). We can, therefore, calculate the temperature for each of the stars and plot it on the Hertzsprung-Russell diagram (Section 16.2). We would find that the most massive stars were the hottest and most luminous and would lie at the upper left corner of the diagram, while the least massive were coolest and least luminous and would lie at the lower right. The other stars would all lie along a line running diagonally across the diagram — the *main sequence*. The main sequence, then, is the locus of points on the H-R diagram representing stars of similar chemical composition but different mass. The observed fact that most stars in the Galaxy do lie along the main sequence is evidence that they have compositions similar to the sun's and are nearly chemically homogeneous. The observed scatter about the main sequence represents slight differences in the chemical compositions of actual stars. This explanation of the main sequence was first presented in the 1930s by the Danish astrophysicist B. Strömgren. At the same time, the *mass-luminosity* relation (Section 15.3) is explained; for all main-sequence stars, the masses are correlated with the radii.

Those stars that do not lie on the main sequence in the Hertzsprung-Russell diagram (for example, red giants and white dwarfs) must differ somehow from the majority in their chemical compositions, or else they are not stable or are not shining by nuclear energy alone. We have seen, however, that as stars age they convert hydrogen to helium, and so change their compositions, especially near their centers. As a consequence, they must alter their entire structures. This brings us to the topic of stellar evolution.

There are two approaches to the study of stellar evolution: the theoretical and the observational. In the theoretical approach, we calculate from the theory of stellar structure how stars should change as they contract gravitationally or age through changes in their chemical composition produced by nuclear reactions. In the observational approach, we observe stars or groups of stars that are at different stages in their evolution, and we check whether they actually exhibit the characteristics expected of them from the theoretical predictions.

The difficulty in the observational approach is that a star ages extremely slowly by human standards, and except for those brief periods in its existence where it may be variable or explosive (and the evolutionary significance of these phenomena is not yet well understood), we do not actually "see" a star evolving. A typical lifetime for a star might be something like 10 or 20 billion years, while the modern study of stellar evolution is barely a few decades old. Our problem of studying the evolution of a star by observing it for so short a time is like that of studying the evolution (or aging) of a man by observing him for only about 10 seconds. Effectively, therefore, we obtain only "snapshots" of stars of different ages.

Star clusters are the most useful objects to study in the observational approach to stellar evolution, because the stars within a cluster can usually be presumed to have a common origin and age, and to have all had, originally, similar chemical composition. From observations of cluster stars, we obtain their colors (or temperatures) and their relative brightnesses; if the distance to the cluster is known, we obtain the luminosities of the stars as well. For a cluster, therefore, we can plot a color-magnitude or Hertzsprung-Russell diagram. Consequently, it is most convenient to compare theory and observations if we consider the tracks of evolution of stars on the H-R diagram. The reader must remember that the "position" of a star, or its "evolution," on the H-R diagram does not refer to its position or motion in space. Rather, these terms refer to the position and motion on a diagram of a point that represents the star's luminosity and surface temperature — they indicate changes in the structure of the star.

20.4 EARLY STAGES OF STELLAR EVOLUTION

No star that is shining today can be infinitely old, for eventually it exhausts its sources of energy. Moreover, the stars of highest known luminosity (100 thousand to a million times that of the sun) can continue to exist at the rate they are now expending energy for only a few million years. Had they been formed when the sun was formed, billions of years ago, they would long since have burned themselves out. At least some stars, therefore, have formed recently (in the cosmic time scale), and there is every reason to expect that stars are still forming today. The "birthplaces" of stars must be the clouds of interstellar material (gas and dust). That period in a star's existence during which it condenses from interstellar matter, and contracts into an "adult" star, may be considered as its "early stages."

(a) Theoretical Studies of Early Stages of Stellar Evolution

Here and there, in comparatively dense regions of interstellar matter, small condensations begin to form — atoms of gas and particles of dust slowly begin to collect under the influence of their mutual gravitation. As a condensing region grows, so does its gravitational influence, and more and more material is attracted to it, until material over a large region of space falls toward the central condensation.

Eventually, the material in the condensation has a high enough density to become opaque to radiation. Thereafter the radiation can filter through the opaque matter only slowly. As the condensing mass continues to fall together, more and more of its potential energy is released, and since this energy is trapped in the interior rather than being radiated away at once, the interior heats up. Any solid grains vaporize and the gas becomes largely ionized. The pressures, especially, build up because of the increasing density and temperature. Finally the pressures are high enough to stop the further infall of material. Hydrostatic equilibrium has been achieved and a stable young star is formed.

It is not yet, however, a main-sequence star. Its internal temperature is not high enough to support thermonuclear reactions, however, and its only sources of energy, so far, are gravitational and thermal. Thus the star is gradually radiating energy into space and is deriving this energy from a very slow shrinking while its internal temperature and pressure continue to rise. The whole process is so gradual that hydrostatic equilibrium is never upset. This is exactly the process whereby Helmholtz and Kelvin attempted to explain the

sun's source of energy more than a century ago.

To follow theoretically the evolution of a star we compute a series of models, each successive one representing a slightly later point in time. Given one (say, initial) model, we calculate the rate at which changes should occur due to, say, gravitational contraction or changing chemical composition. Then the total change that would occur during the time interval in question is found, after which a new model for the star is computed, appropriate to the new (changed) conditions that pertain at the end of the given time interval. With each new model, the luminosity and effective temperature of the hypothetical star are calculated so that its progress on the Hertzsprung-Russell (H-R) diagram is followed as it evolves.

Now we have seen that as a star begins its existence it shines by virtue of its slow gravitational contraction. We now consider its theoretically predicted track of evolution on the H-R diagram (Figure 20-2). The Japanese astrophysicist C. Hayashi was the first to show that such young stars must lie in a zone on the H-R diagram extending nearly vertically from the lower main sequence to the right extreme of the regions occupied by red giants and red supergiants (cross-hatched region in the figure). There can be no stable star such that the point representing it on the H-R diagram lies to the right of this zone. As stars contract in the initial stages of their evolution, they move (on the H-R diagram) downward in the zone. More or less representative tracks for stars of several masses and of chemical composition more or less like the sun's are shown in Figure 20-2. We see that during this early stage a star decreases in luminosity without changing much in temperature.

Except for a star of low mass, after a period of some thousands or millions of years the star turns sharply on the H-R diagram and moves left, almost horizontally, toward the main sequence. Eventually, as the release of gravitational energy con-

FIG. 20-2 Theoretical evolutionary tracks of contracting stars on the Hertzsprung-Russell diagram. Numbers labeling the tracks are the masses of the contracting stars in units of the sun's mass.

tinues to heat up its interior, its central temperature becomes high enough to support nuclear reactions. Soon this new source of energy supplies heat to the interior of the star as fast as energy is radiated away. The central pressures and temperatures are thus maintained and the contraction of the star ceases; it is now on the main sequence. The small hooks in the evolutionary tracks of the stars shown in Figure 20-2, just before they reach the main sequence, are the points (according to theory) where the onset of nuclear-energy release occurs.

Stars of extreme low mass, on the other hand, never achieve high enough central temperature to ignite nuclear reactions. They continue to con-

tract until (after an extremely long time) they are so dense that their matter becomes degenerate and they reach the white dwarf stage (Section 20.6). The lower end of the main sequence is considered to be that point at which stars have a mass just barely great enough to sustain nuclear reactions at a sufficient rate to stop gravitational contraction; this critical mass is believed to be near $\frac{1}{12}$ that of the sun.

At the other extreme the upper end of the main sequence terminates at the point where the mass of a star would be so high and the internal temperature so great that the pressure of radiation would dominate. The radiation produced from nuclear reactions would be so extreme that when absorbed by the stellar material it would impart to it a force greater than that produced by gravitation; hence, such a star could not be stable. The upper limit to stellar mass is believed to be in the range 60 to 100 solar masses.

In general, the pre-main-sequence evolution of a star slows down with time; the points on each evolution track in Figure 20-2 are intended to divide the track into segments over which the star

FIG. 20-3　Nebulosity in Monoceros, situated in the south outer region of the young cluster NGC 2264. Photographed in red light with the 200-inch telescope. *(Hale Observatories.)*

spends roughly equal intervals of time. The time for the whole evolutionary process, however, is highly mass-dependent. Stars of mass much higher than the sun's reach the main sequence in a few million years or less; the sun must have required tens of millions of years; for stars to evolve to the lower main sequence requires hundreds of millions of years. For all stars, however, we should distinguish three evolutionary time scales:

(1) The initial gravitational collapse from interstellar matter is relatively quick. Once the condensation is, say, 1000 AU in diameter, the time for it to reach hydrostatic equilibrium is measured in thousands of years.

(2) Pre-main-sequence gravitational contraction is much more gradual; from the onset of hydrostatic equilibrium to the main sequence requires, typically, millions of years.

(3) Subsequent evolution on the main sequence is very slow, for a star changes only as thermonuclear reactions alter its chemical composition. For a star of a solar mass, this gradual process requires billions of years. All evolutionary stages are relatively faster in stars of high mass and slower in those of low mass.

So much for the theoretical predictions. Now let us examine some of the observational data.

(b) Observations of Very Young Star Clusters
Because star formation is constantly going on, among the hundreds of known star clusters we might expect to find a few that are still in the process of formation, that is, with some of their stars still in the contraction phase. The more massive and luminous cluster stars might be expected to have reached the main sequence while their less massive companions would still be "on their way in." Within the last decade, several clusters have been observed that fit this description.

The first such cluster to be studied was NGC 2264, a small open cluster embedded in a cloud of gas and dust in the constellation Monoceros. The

FIG. 20-4 Color-magnitude diagram of NGC 2264. The solid line is the position of the main sequence for "normal" stars. Stars indicated with crosses are T Tauri stars. *(Adapted from data by M. Walker.)*

H-R diagram for this cluster (by M. Walker at the Lick Observatory) is shown in Figure 20-4. The brighter stars in the cluster are on the main sequence (solid line in the figure), while the less luminous ones (presumably also less massive) are off to the right. They are interpreted as being young stars, still contracting from the interstellar material associated with the cluster.

Today, several other such clusters, evidently very young ones, are known, all of which are associated with interstellar matter. One of them is the cluster in the central part of the Orion nebula. The probability is very high that in these clusters we are witnessing the early evolution of stars.

Some of the stars in these young clusters are variable stars of the T Tauri type (Table 18.3). They show rapid and irregular light variations but are usually among the less luminous stars observed in the clusters. This type of variability may be associated with the gravitational contraction of stars of relatively low mass. The T Tauri stars in NGC 2264 are shown as crosses in Figure 20-4.

20.5 EVOLUTION FROM THE MAIN SEQUENCE TO GIANTS

As soon as a star has reached the main sequence, it derives its energy almost entirely from the thermonuclear conversion of hydrogen to helium. It remains on the main sequence for most of its "life." Since only 0.7 percent of the hydrogen used up is converted to energy, the star does not change its mass appreciably, but in its central regions, where the nuclear reactions occur, the chemical composition gradually changes as hydrogen is depleted and helium accumulates. This change of composition forces the star to change its structure, including its luminosity and size. Eventually, the point that represents it on the H-R diagram evolves away from the main sequence. The original main sequence, corresponding to stars of homogeneous chemical composition, is called the *age-zero main sequence.*

(a) Evolution on the Main Sequence

As helium accumulates at the expense of hydrogen

in the center of a star, calculations show that the temperature and density in that region must increase. Consequently, the rate of nuclear-energy generation increases, and the luminosity of the star slowly rises. A star, therefore, does not remain indefinitely *exactly* on the original age-zero main sequence. However, it increases its luminosity only by a moderate amount — probably less than a magnitude — before subsequent more rapid changes alter its structure enormously.

When the hydrogen has been depleted completely in the central part of a star, a core develops containing only helium, "contaminated" by whatever small percentage of heavier elements the star had to begin with. The energy source from hydrogen burning† is now used up, and with nothing more to supply heat to the helium core, it begins again to contract gravitationally. The star's energy is now partially supplied by potential energy released from the contracting core; the rest of its energy comes from hydrogen burning in the region immediately surrounding the core. These changes result in a substantial and rather rapid readjustment of the star's entire structure, so that the star "leaves" the vicinity of the main sequence altogether. Because the total rate of energy production in a star must be equal to its luminosity, the central hydrogen is used up first in the very luminous stars. The most massive stars spend less than 1 million years on the main sequence; a star of 1 solar mass may remain there for more than 10 billion years.

(b) Evolution to Red Giants

Detailed calculations show that the central core of a star, where the hydrogen is exhausted, contracts, and as it does so the core increases in temperature, density, and pressure. The gravitational potential

---

†The term "burning" is often used to describe the depletion of an element by nuclear reactions. This "nuclear burning" is not, of course, burning in the literal chemical sense.

energy released from the contracting central regions of the star forces the outer parts of the star to greatly distend themselves. The star as a whole, therefore, expands to enormous proportions; all but its central parts acquire a very low density. The expansion of the outer layers causes them to cool, and the star becomes red. Meanwhile, some of the potential energy released from the contracting core heats up the hydrogen surrounding it to ever higher temperatures. In these hot regions the conversion of hydrogen to helium accelerates; most stars actually increase in total luminosity. After leaving the main sequence, then, stars may be expected to move to the upper right portion of the H-R diagram; they become red giants.

(c) Observations of Star Clusters of Different Ages

The results of theoretical studies, outlined in the last two subsections, predict certain characteristics of the H-R diagrams of star clusters of different ages. In each cluster there should be some point along the main sequence where stars have just now reached the critical age where they rapidly evolve away from it; this point will be the upper termination of the main sequence for the cluster. In a young cluster, the main sequence should extend to stars of high luminosity; in successively older clusters, it will terminate at successively lower luminosities. As a cluster ages, its main sequence "burns down" like a candle.

Figure 20-5 shows a composite H-R diagram for several star clusters of different ages. On the left side is shown the absolute visual magnitude scale. On the right side is a scale that gives the approximate ages of star clusters corresponding to the points where their main sequences terminate. We see that the clusters shown range in age from only about 1 million years to several billion years.

As expected, most clusters have red giants. Those few which do not may just not happen to have any stars of the proper mass to be entering that stage of their evolution. Note the large gap

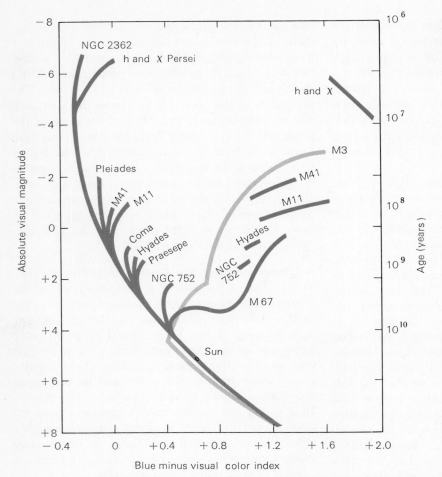

FIG. 20-5 Composite Hertzsprung-Russell diagram for several star clusters of different ages. *(Adapted from a diagram by Sandage.)*

in the center of the diagram where no cluster stars (actually, very few) appear. There are theoretical reasons to expect stars of relatively high mass to become unstable as they leave the main sequence; they would be expected, therefore, to readjust themselves to the more stable red giant configuration very rapidly. The observed absence of stars in the gap does indeed suggest that stars of high mass evolve very quickly from the main sequence to the red giant domain.

In any case, the red giant stage must be a relatively brief part of a star's life. In the youngest

clusters these stars are red *supergiants* of high luminosity. In the older clusters, stars *increase* their luminosities as they become red giants. In this stage of evolution, therefore, a star's nuclear fuel is consumed relatively quickly, and further evolutionary changes soon follow.

The oldest assemblages of stars in the Galaxy are believed to be the globular clusters. The lighter line in Figure 20-5 outlines the H-R diagram for M3, a typical globular cluster; note that the main sequence terminates and branches into the giant sequence at a lower luminosity than for any of the

Orion nebula.
(Lick Observatory.)

Central portion of Orion nebula. *(Lick Observatory.)*

The North American nebula; photographed by the
48-inch Schmidt telescope. *(Hale Observatories.)*

The Horsehead nebula in Orion; 48-inch Schmidt telescope photograph. *(Hale Observatories.)*

The 300-foot Radio Telescope. *(National Radio Astronomy Observatory.)*

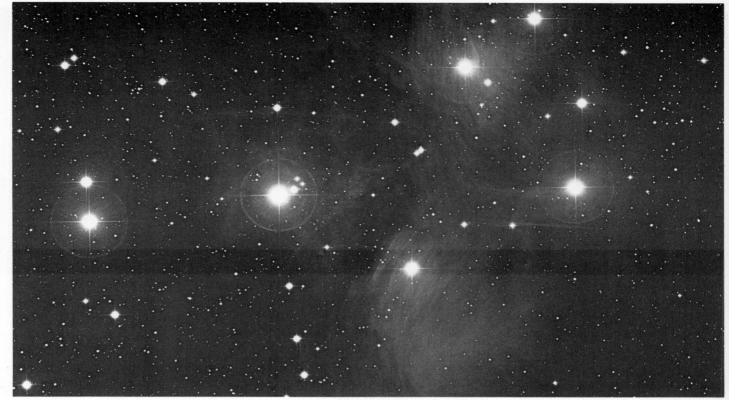

The Pleiades, and associated
nebulosity in Taurus; 48-inch
Schmidt telescope photograph.
(Hale Observatories.)

Open cluster, M16, and nebulosity in Serpens;
200-inch telescope photograph. *(Hale
Observatories.)*

The globular cluster
M13 in Hercules. *(U.S. Naval
Observatory; Flagstaff Station.)*

Veil nebula, or Cygnus "Loop," NGC 6992; photographed by the 48-inch Schmidt telescope. *(Hale Observatories.)*

The Dumbbell nebula (M27) in Valpecula. *(U.S. Naval Observatory; Flagstaff Station.)*

The Crab nebula; 200-inch telescope photograph. *(Hale Observatories.)*

other clusters. Ages for globular clusters are now variously estimated at from 5 to 20 billion years. A few open clusters (for example, NGC 188 and M67) may approach the globular clusters in age. In no cluster, however, does the main sequence terminate below the luminosity of the sun. Apparently, stars of solar mass in the Galaxy have not yet had time to evolve away from the main sequence.

Much study, both observational and theoretical, has gone into old star clusters. Figure 20-7 shows the appearance of the H-R diagram of a typical globular cluster, and Figure 20-8 shows the same for an old open cluster. The stars now evolving off the main sequences of these old clusters appear to have masses of about 1.1 or 1.2 times that of the sun.

Today, the observed giant sequences in the H-R diagrams of star clusters are quite well understood theoretically. We summarize by noting that globular clusters (and, presumably, other population II systems) have main sequences that terminate at about absolute magnitude 3.5; these systems are evidently old and have not experienced recent star formation. Open clusters, on the other hand, are often located in regions of interstellar matter, where star formation can still take place. Indeed, we find open clusters of all ages from less than 1 million to several billion years. The H-R diagrams of the oldest of them resemble those of globular clusters (at least as far as main-sequence and giant stars are concerned), except for differences that are now understood in terms of small differences in chemical composition of the two kinds of clusters.

FIG. 20-6   The globular cluster M13 in Hercules. *(Hale Observatories.)*

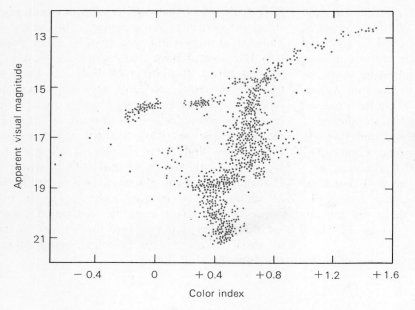

FIG. 20-7 Color-magnitude diagram for the globular cluster M3. *(Adapted from data by Arp, Baum, and Sandage.)*

FIG. 20-8 Color-magnitude diagram for the old, open cluster M67. *(Adapted from data by Sandage and Johnson.)*

20.6 LATER EVOLUTION STAGES

Our knowledge of the evolution of stars subsequent to their being red giants is less complete. We think we know, however, how stars end, and we can make some educated guesses about some of the intermediate steps.

(a) Termination of the Red Giant Stage

Part of the gravitational potential energy that is released in the contracting cores of red giants heats up those cores. By the time a star reaches the end of the red giant sequence in the H-R diagram, the temperature of its central regions must exceed 100 million degrees Kelvin.

At such a high temperature, nuclear processes other than the carbon cycle and proton-proton chain are possible. The most important of these is the formation of a carbon nucleus by three helium nuclei (the *triple-alpha process* — so named because the nucleus of a helium atom is also called an *alpha particle*). Successive bombardments of a carbon nucleus by helium nuclei can build up other, still heavier nuclei. One group of physicists and astronomers has found mechanisms whereby virtually all the chemical elements can be built up in the centers of red giant stars, in approximately the relative abundances that they occur in nature. It now seems quite probable that a gradual buildup of the heavy elements (heavier than helium) is continually going on in the hot centers of these stars.

The new energy released expands the core rather quickly and reverses the growth of the outer parts of the red giant. The star then shrinks rapidly and increases in surface temperature. Although the calculations cannot be regarded as definitive, they indicate that the point representing the star on the H-R diagram moves more or less horizontally to the left. As soon as the helium is exhausted in a central region, however, the energy release from the triple-alpha process is shut off, and we have a situation analogous to a main-sequence star when its central hydrogen is used up and the hydrogen burning ceases in its center. Now, however, we have a core of carbon (or perhaps heavier elements) surrounded by a shell where helium is still burning; farther out in the star we come to another shell where hydrogen is left and is still burning. The star now moves back to the right in the H-R diagram and returns to the red giant domain. The calculations indicate that a star may actually move first to the left across the H-R diagram, and then back to be a red giant several times, each time as a consequence of the onset of new nuclear reactions or of nuclear energy being released in new parts of the star. All these evolutionary stages occur in tens or hundreds of millions of years or less — a brief time compared to the stars' main-sequence lives. Some observational evidence supporting the theoretical calculations is the presence of the "horizontal branch" of stars on the H-R diagram of globular clusters and possibly in some open clusters. How the red giant phase of a star's evolution is finally terminated is not known. There is some thought that many, if not all, stars of about 1.2 solar masses eject planetary-nebula shells to end their careers as red giants (see below).

One stage of evolution of horizontal branch stars is through the "gap" occupied by the RR Lyrae variable stars, where no star is stable. As stars evolve across the horizontal branch they must some time pass through this stage of instability — as RR Lyrae variables (Section 18.2d). Other types of stellar variability must also represent stages of evolution of stars of other masses (compositions, or both), but ideas on how they fit into the picture are still rather speculative.

(b) White Dwarfs — One Final Stage of Evolution

Sooner or later a star must exhaust its store of nuclear energy. Then it can only contract and release more of its potential energy. Eventually, the shrinking star will obtain an enormous density. We observe such stars — the extremely compact white dwarf stars (Section 16.2d), whose mean densities range up to over 1 million times that of water.

White dwarf stars are far more dense, of course, than any solid substance. The high density of a white dwarf is possible because the atoms that comprise the gases in a stellar interior are almost completely ionized, that is, stripped of virtually all their electrons. Most of a neutral atom is empty space; once an atom is completely ionized, however, it and its freed electrons can occupy a volume many times smaller than when the electrons are still revolving about the nucleus.

It might seem that a star could collapse until the nuclei and freed electrons of its atoms were packed together like sardines. The contraction must cease, however, long before that. There is a theoretical limit to which certain kinds of particles, including electrons, can be packed. According to the quantum theory, no two electrons with almost the same velocity can exist in the same very small volume. There is a minimum volume, therefore, into which a given number of electrons can be crowded. How large that smallest possible volume is depends on the total range of velocities of the electrons; the *slower* they are moving, on the average, the *larger* is that minimum volume.

A star has a very great many electrons, each acting as a free particle in the stellar gas. Since some of the electrons in a white dwarf may have speeds near that of light, the total range of speeds of the electrons in the star is very large, but still it is not infinite. There exists, therefore, a minimum volume into which the electrons of a white dwarf can be compressed. As a star contracts, eventually the point is reached where its electrons have been packed into their smallest possible volume. They then exert an enormous pressure, which prevents the star from contracting any more. The electrons cannot slow down (that is, "cool") and allow further contraction, because if they lower their speeds they must occupy a *larger* volume, and there is no energy available to expand the star again.

A white dwarf is a star of such high density that its electrons are certainly packed to this limit, except in an extremely thin layer at the very surface of the star. In this state, the electrons are said to comprise a *degenerate* gas. White dwarfs are thus said to be degenerate stars. They have now reached their final size. There is still, however, a large amount of empty space between the gas particles; they are very far from "touching."

The other particles that make up the gas in a white dwarf (the atomic nuclei) are still free to move as molecules of a perfect gas, and they exert their usual gas pressure. The pressure exerted by the degenerate electrons, however, is computed to far exceed that of the nuclei. The structure of white dwarfs, therefore, is dominated by this degenerate electron pressure.

The structure of white dwarfs was first investigated by A. Eddington, the great British physicist and theoretical astronomer who was largely responsible for laying the foundations of the theory of stellar structure. Some of the electrons in some white dwarfs, however, have speeds near that of light, and Newtonian mechanics is not applicable to their behavior. The first person to apply the theory of relativity to the construction of models of white dwarfs was the Indian (now American) astrophysicist, S. Chandrasekhar.

It turns out that the entire structure of a white dwarf, including its radius, is specified uniquely by its *mass*. There is, therefore, a unique relation between the mass and radius of a white dwarf star. Chandrasekhar's mass-radius relation (with some small modifications that modern theory predicts) is shown in Figure 20-9. Note that the larger the mass of the star, the smaller is its radius. A white dwarf with a mass of about 1.2 solar masses would have a radius of zero! According to the theory, therefore, no white dwarf can have a mass quite as large as 1.2 times that of the sun.

The white dwarf upper limit of about 1.2 solar masses means that no stable electron-degenerate

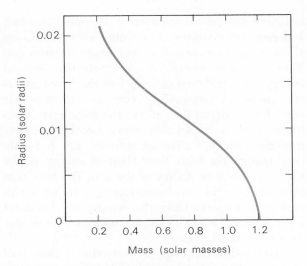

FIG. 20-9   Theoretical relation between the masses and radii of white dwarf stars.

configuration of greater mass can exist. However, those stars that have had time to exhaust their nuclear fuel supply and evolve to the white dwarf stage must have had original masses greater than 1.2 solar masses, for those more massive stars are the very ones that use up their energy store most rapidly. We know, at present, of two paths of evolution that may be open to stars of mass too great to become white dwarfs. If a star has a mass less than about two solar masses — only a little greater than the white-dwarf limit — theory predicts that it may be able to contract to a neutron star (which we shall return to in Section 20.6f). For a still more massive star, there is the interesting possibility that as it contracts its electrons gain enough speed to avoid ever becoming degenerate; such a star eventually contracts through what is called the *event horizon*, where its surface gravity is so great that light cannot escape. It is then called a *black hole* (Section 20.6g).

On the other hand, white dwarf stars are plentiful, and they must have come from somewhere. Moreover, the number of white dwarfs is high enough, as nearly as can be determined, to account for all evolved stars of original mass greater than 1.2 solar masses. It is widely accepted, therefore, that most stars eventually do, in fact, become white dwarfs. Consequently they must lower their masses somehow, before reaching that stage, by ejecting matter into space.

(c) Ejection of Matter by Stars

Shell stars, Wolf-Rayet stars, emission-line B stars, novae, supernovae, and various kinds of variable stars are all observed to be ejecting matter into space. It has even been shown that many red giants as well are steadily ejecting matter from their outer layers.

For stars of 1.2 or so solar masses — the most massive stars among red giants in globular clusters and in the galactic corona, nucleus, and disk — one of the most important mass-ejection mechanisms may be the planetary nebula phenomenon. Only about 1000 planetary nebulae are known, but when account is taken of the fact that they are identifiable as planetaries for less than 10^5 years, it is found that the number of planetaries is sufficient to be compatible with the hypothesis that all these red giants eventually undergo that form of mass loss. Moreover, there are compelling (although not conclusive) arguments that the planetary nebula shells are ejected from red giants and that their hot central stars are the inner parts of the red giants that are left after the mass ejection. The mass of a typical nebular shell is about 0.2 solar mass. Many of their central stars already have the dimensions of white dwarfs, and may well be evolving to that final state.

Stars of different mass probably undergo different means of mass loss.

(d) Increase of Heavy Elements in the Universe

We have seen that stars convert hydrogen into helium and, moreover, that at least some stars in some stages of their evolution must be building up helium into carbon and heavier elements. Thus, inside stars, some of the lighter elements of the universe are gradually being converted into heavier ones. As these stars eject matter into the interstellar medium, that matter is richer in heavy elements than was the material from which the stars were formed. There is taking place, in other words, a gradual enrichment of the heavy-element abundance in interstellar matter. The heavy-element abundance in stars that are forming now should be higher than in those that formed in the past. The fact that the oldest known stars (those in globular clusters) are the stars with the lowest known abundance of heavy elements seems to support this idea.

Some astronomers have gone so far as to speculate that originally *all* stars in the Galaxy were formed of pure hydrogen, the lightest and simplest element, and that all other elements were formed in the hot centers of stars at advanced stages of evolution. Stars such as the sun, in whose outer layers heavy elements are observed spectroscopically, would then be of the second or third (or even higher) "generation," that is, have been formed of matter that was once part of other stars. The "original" pure hydrogen stars, much less massive than the sun, would not, however, be expected to leave the main sequence for many billions of years; while they remain on the main sequence, these stars would convert hydrogen to helium, but heavier elements would not yet be formed in them. Unless the Galaxy were at least 50 to 100 billion years old, we should see many of these "first-generation" stars without heavy elements; we do not. Moreover, there is now reason to believe that a substantial abundance of helium existed in the material from which the Galaxy formed — helium built up in the original "fireball" or "big bang" with which the universe began its present expansion (Chapter 22).

(e) Evolution of White Dwarfs

A white dwarf is presumed to have exhausted its available nuclear-energy sources. It cannot contract and release gravitational potential energy because of the great pressure of the degenerate electron gas. Thus the only source of energy is the thermal energy (that is, kinetic energy) of the nondegenerate nuclei of atoms, behaving as ordinary gas particles, scattered throughout the degenerate electrons. As these nuclei slow down (cool), the electron gas (which, unlike an ordinary gas, is highly conducting) conducts their thermal energy to the surface. At the boundary of the star, the very thin skinlike layer of nondegenerate gas radiates this energy into space. Only the opacity of this outer layer keeps the nuclei in the interior of the star from cooling off at once.

Gradually, however, a white dwarf does cool off, much like a hot iron when it is removed from a stove. The cooling is relatively rapid at first, but as the star's internal temperature drops, so does its cooling rate. Eventually, a white dwarf will cease to shine at all. It will then be a *black dwarf*, a cold mass of degenerate gas, floating through space. An extremely long time may be required, however, for a star to cool off to the black dwarf stage — perhaps trillions of years.

(f) Neutron Stars

Another possible end product of stellar evolution is a neutron star. Neutron stars are hypothetical configurations composed entirely of neutrons. Ordinarily, a free neutron (one not bound in an atomic nucleus) survives only about 15 minutes before decaying into a proton and an electron. Under extremely high pressures, however, a neutron is stable. Suppose, somehow, that all the electrons in a star could be forced, under tremendous pressure, into the atomic nuclei. Since stars are electrically neutral there are just as many electrons as there are protons in the nuclei. Thus all the matter would become neutrons.

Neutrons behave as do electrons in the sense that they become degenerate if crowded into a sufficiently small volume for a given velocity range. Thus, the structure of neutron stars is very analogous to that of white dwarfs. They, too, follow a mass-radius relation, similar to that for white dwarfs except that they are very much smaller. A neutron star of 1 solar mass would have a radius of about 10 km (6 mi) and a density near 10^{15}gm/cm³ — ten billion tmes as great as that of a typical white dwarf. The upper mass limit for a neutron star is not very accurately determined (at the time of writing), but is believed to be about 2 solar masses.

One kind of star that might become a neutron star is one whose mass exceeds the white-dwarf limit, but not the limit for neutron stars — that is, which lies in the range 1.2 to 2 solar masses. When it exhausts its store of nuclear fuel and gravitationally contracts, it would miss the white dwarf configuration and keep contracting until the pressures built up by its tremendous internal gravitational field force the electrons into the nuclei, producing a sea of neutrons. The contraction ceases when the neutrons themselves become degenerate, keeping the star at a size dictated by the mass-radius relation for neutron stars — in any case not over a few miles in diameter.

FIG. 20-10   Summary of evolutionary track on the Hertzsprung-Russell diagram for a star of 1.2 solar masses.

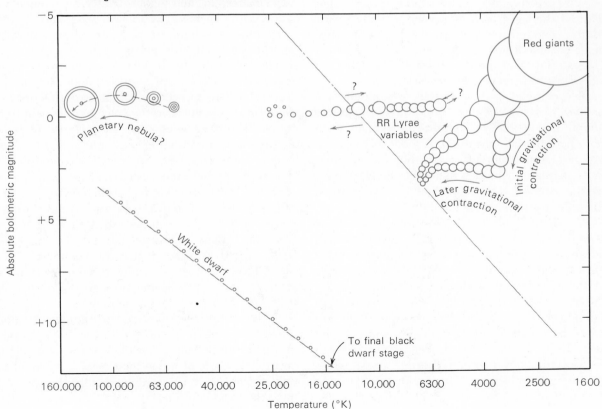

Another possibility for building a neutron star is an implosion in the center of a supernova. We have already described this kind of event in our consideration of a current theory for pulsars (Sections 18.7). At present, pulsars provide our only observational evidence that neutron stars do actually exist.

(g) Black Holes

We turn now to some of the strangest objects of nature — the *black holes*. Perhaps black holes do not exist at all; there is as yet no definite observational evidence for them. But theoretically they could exist, and many theoreticians consider it almost certain that they do.

We have seen that a star of mass greater than 1.2 times the sun's cannot become a white dwarf, and that one greater than about twice the sun's cannot become a neutron star either. However, we have also seen that most such stars must somehow shed mass and become white dwarfs, because white dwarfs are very common. Still, we might ask whether a massive star could avoid ejecting material and possibly contract forever. (Let us ignore here the complications introduced by stellar rotation, or else assume that somehow a star transfers its angular momentum to another body, thereby lowering its own rotation.) Could a star contract until its density becomes infinite?

We might think ourselves out on a limb here, for we have no understanding of the laws of physics that apply to matter of extremely high density. However, the General Theory of Relativity predicts that the actual state of the matter is not relevant to what we observe from the outside world. We recall (Section 3.7a) that the paths of light are affected by the gravitational fields around material bodies. Light rays grazing the surface of the sun are deflected about a second of arc. A dense white dwarf deflects light more than a minute of arc. At the surface of a neutron star, light is deflected 30°.

Consider light radiated from the surface of a neutron star. That which emerges normal to the surface (that is, perpendicular to it) flows out radially from the star. That emitted at an angle of, say, 30° to the normal (hence, 60° to the surface) leaves the star at an angle somewhat greater than 30° to the normal, because of the gravitational deflection. That light which is emitted 60° or more from the normal, however, is deflected so much that it never escapes the star. Thus, from each point on the surface of the neutron star, that portion of the light which can pass on out into space is emitted into a cone about the normal to the surface of half angle about 60°.

Now imagine a more massive star, that shrinks to a smaller size, and higher density, than a neutron star. As the surface gravity increases, the angular cone into which light can be emitted from any point on the star and still escape into space becomes smaller. Eventually, as the star becomes small enough and its gravity strong enough, a point is reached where the cone of light escape shrinks to zero; this happens when the velocity of escape from the star's surface is equal to the velocity of light. As the star contracts still more, light and everything else is trapped inside, unable to escape through that surface where the escape velocity is the speed of light. That surface is the *event horizon,* and its radius is the *Schwarzschild radius,* named for Karl Schwarzschild, who first described the situation a few years after Einstein introduced General Relativity. It is this surface that is the boundary of the black hole. All that is inside it is hidden forever from us; as the star shrinks through the black hole it literally disappears from the universe.

The size of the Schwarzschild radius is proportional to the mass of the star. For a star of 1 solar mass, the black hole is about 3 km in radius; the entire black hole of a one-solar mass star would be only about 4 mi across — about a third the size of a one-solar mass neutron star.

(h) Detection of Black Holes

If a star falling into a black hole disappears how can we know of its existence? As measured by a wristwatch worn by an observer on the surface of such a star, collapse through the black hole occurs fairly rapidly (except that the atoms of the observer and his watch are crushed). But relativity predicts that the passage of time for an outside observer will be slower. As the star approaches the Schwarzchild radius, its time (to us outside) would appear to slow so much that the star's surface would never quite cross the event horizon. However, the light leaving the shrinking star would become so feeble that we would cease to see it. On the other hand, the gravitational field set up by the infalling matter would remain and build up outside the star. The black hole would thus act as an invisible source of gravitation.

We might be able to detect a black hole if it were a member of a binary star system. We recall how the companion of Sirius was discovered before it was seen (Section 15.2a), and how a companion of Barnard's star of planetary mass is inferred from its gravitational influence (Section 15.2c). If the motion of a star were such as to indicate that it must have a companion of such high mass that it should be visible were it a normal star, a black hole might be indicated. There are two or three binary systems that are possible candidates, but none is so well investigated that we can claim the "discovery" of a black hole. We might also infer the existence of black holes in a cluster of objects that are moving so fast as to suggest a far stronger gravitational field for the cluster than would be expected from the visible members. The galaxies in some clusters of galaxies seem to be moving at surprisingly high speeds in those clusters, and some investigators suggest that black holes might account for the "missing mass," giving those clusters greater than expected gravitational fields. Others suggest that black holes throughout the universe might give rise to gravitational fields produced by unseen matter. Observational evidence for black holes, however, is very weak at best. Moreover, we should remember that, at least in our own galaxy, there are enough white dwarfs to account for most if not all evolved stars of high mass.

(i) Gravitational Collapse and Gravity Waves

We remind the reader that a black hole's size is proportional to its mass. A one-solar mass black hole is about 4 mi in diameter; a 10-solar mass black hole would be 40 mi across. If an entire galaxy of 10^{11} solar masses could collapse, its corresponding black hole would be a little under 0.1 LY across. Perhaps, within galaxies, large masses could go into gravitational collapse. The potential energy released could give rise to the immense localized energy sources observed, for example, in Seyfert galaxies and quasars (Chapter 21).

If a large mass of matter falls toward a black hole, theory predicts that gravitational waves should also be emitted. The transport of gravitational waves is thought to be by means of hypothetical *gravitons*, analogous to the transport of electromagnetic energy by photons. Over the past several years, Joseph Weber, at the University of Maryland, has been conducting experiments to detect gravitational waves. His apparatus consists of pairs of large metal cylinders in widely separated places (Chicago and Maryland), which start vibrating when exposed to gravitational waves. Local disturbances, such as trucks passing by, also start vibrations in the cylinders, but if both members of a pair separated by a thousand or so miles respond at the same time, there is the possibility that they were exposed to gravitational waves from space. So far, a large number of such simultaneous events have been observed. If these are actually due to gravitational waves from space, they probably originate near the center of our galaxy. Preliminary results (still in need of verification) suggest that something like 1000 solar masses of matter are gravitationally collapsing into black holes each year.

20.7 PAST AND FUTURE OF THE SUN AND SOLAR SYSTEM

Since the sun is a typical star, much of the theory of stellar evolution that we have discussed applies to it. From theoretical calculations we can now form a fairly clear picture of the past history of the sun, and can predict a good deal about its future.

(a) Formation of the Solar System

A great many hypotheses have been proposed for the origin of the solar system. Most of these ideas fall into two general categories. First, Chamberlin, Moulton, Jeans, and Jeffreys are among those who have suggested that the planets were formed from material pulled from the sun by the tidal force of a passing star. Second, it was proposed in the mid-eighteenth century by Kant, and later by Laplace, that the planets and sun formed together from the same primeval matter — a large cloud of gas or dust or both. The facts that the planets all revolve in the same direction, and that their orbits lie in nearly the same plane, as well as the expected rarity of a close stellar encounter, have combined to lead modern investigators to reject the first mechanism.

We can speculate that the original cloud of matter that formed the solar system must have been much larger than the orbits of the most distant planets. It must also have had at least some slight rotation (perhaps caused by the differential rotation of the Galaxy). As the cloud contracted, it rotated faster and faster to conserve its angular momentum. Most of the mass of the cloud found its way to the denser central condensation, ultimately to become the sun. A fraction of the material, however, was left behind and, because of its rotation, flattened into a disk. The planets and their satellites were formed from this disk.

The precise mechanism by which the planets formed is more speculative, as is the origin of the minor planets, meteoroids, and comets. One hypothesis, suggested first by C. F. Von Weizsäcker in the mid-1940s, has been modified by Kuiper and is now known as the protoplanet hypothesis. The protoplanets, according to the theory, were condensations that developed in the rotating disk. The planets themselves formed in the denser, central parts of these condensations and their satellites from smaller subcondensations. Only a small fraction of the material of these protoplanets, however, ended up in the planets and satellites; most of it was "blown" from the solar system by the force of electromagnetic and corpuscular radiation from the sun. The smaller planets (such as the earth), in fact, were not even able to retain the most abundant of the lighter elements.

Theories of the formation of the solar system, unfortunately, are less subject to observational verification than those of stellar evolution, and are less well founded. The reader who wishes to study this rather speculative subject will find references in Appendix 1.

(b) Early Solar Evolution

Several calculations of the sun's early evolution have been made. The exact track of evolution of the sun in the H-R diagram depends on its assumed initial chemical composition and on rather uncertain data involving the opacity of the outer layers of the young sun. The time required for it to contract to the main sequence is probably a few tens of millions of years.

Since it reached the main sequence, the sun has increased somewhat in luminosity, probably by about 30 to 50 percent. During that interval of from 4 to 5 billion years, it has depleted much of the hydrogen at its very center, but a pure helium core has not yet had time to form. It is not certain how much more time the sun has before starting to evolve to the red giant stage, but a good guess is that it has lived out about half of its main-sequence life. We can probably look forward to at least another 5 billion years before the sun's structure undergoes large changes.

(c) The Future of the Solar System

All available evidence leads us to expect that sometime in the future the sun will leave the main sequence and evolve to a red giant. According to theoretical calculations by UCLA's Roger Ulrich, the sun will grow until, about 5 billion years from now, its photosphere reaches the orbit of Mars. The earth, then well inside the sun, exposed to temperatures of thousands of degrees, will gradually vaporize. The gases in the greatly distended outer layers of the sun will, of course, be very tenuous, but they should still offer enough resistance to the partially vaporized earth to slow it in its orbital motion. Ulrich believes that the earth will spiral inward toward the very hot interior of the sun, reaching its final end about ten thousand years after being swallowed by the sun.

If, in that remote time, man could leave the doomed earth but remain in the solar system, he would find that most of the naked-eye stars in our Twentieth-century sky would long since have exhausted their nuclear fuel and evolved to white dwarfs (or perhaps to some other end.) Main-sequence stars less massive than the sun would still be shining, but only those few of them passing temporarily through the solar neighborhood would be near enough to see with the unaided eye. It is doubtful, however, if by then *all* star formation in the Galaxy would have ceased. Luminous young stars might be shining in remote clouds of interstellar matter; a Milky Way might still stretch around the sky.

(d) Is the Solar System Unique?

In the solar neighborhood, at least, more than half the stars appear to be members of double- or multiple-star systems. Some of these systems contain invisible members whose presence is revealed only by their gravitational effects on the motions of the other stars in the systems. In a few cases, these unseen stars have very small masses; a companion of the star 61 Cygni, for example, has a mass that is only $\frac{1}{60}$ that of the sun, and Barnard's star (Section 15.2c) has a companion with a mass not much greater than Jupiter's. Some writers describe such objects as "planetlike." Actually, however, most cannot be solid bodies. Jupiter is nearly as massive as a solid object can be; if it had only a few times greater mass, it would be a feebly shining gaseous star.

Some astronomers have speculated that stars may form either in binary or multiple systems or else have planets. In a way, the sun and Jupiter can be considered a sort of "degenerate binary star." If the formation of the solar system were along the lines outlined above, it would not be surprising if there were many stars with planetary systems. Unfortunately, our most powerful telescopes could not reveal an object as insignificant as the earth revolving about even the nearest other star. We may never know, then, whether worlds similar to the earth exist beyond the solar system.

At least we have no reason to believe that other planetary systems do *not* exist. They may be common, but at present we can only speculate on this possibility. Whether any of these unknown worlds possesses intelligent life, or even life at all, is a subject of further speculation which belongs in the field of biology, not astronomy. When we consider the many billions of other stars that could have planets, as does the sun, we might be tempted to believe that intelligent life on at least some of those hypothetical planets would be almost inevitable. Unfortunately, however, we have no basis whatsoever for guessing upon what fraction of those planets life may have developed, nor do we know the probability that simple life forms will ultimately develop into intelligent beings. If, for example, the chances of the latter are one in a million, intelligent life may abound elsewhere, but if they are only one in a trillion, we may be unique as "reasoning" beings.

Exercises

1. Compare hydrostatic equilibrium in a star to the state of equilibrium in an inflated balloon or tire. Describe what happens if air is let out of the balloon, or if more is blown in. What if the air inside is heated or cooled?

2. Discuss the hydrostatic equilibrium at various levels in a lake.

3. If the atmospheric pressure were the same on two different days, but if one day were much hotter than the other, what could you say about the relative density of air on the two days?

4. Verify that some 600 million tons of hydrogen are converted to helium in the sun each second.

5. Stars exist that are as much as 10^5 times as luminous as the sun. Consider a star with a mass of 2×10^{34} gm (10 solar masses) and a luminosity of 4×10^{38} erg/sec. Assume that the star is 100 percent hydrogen, all of which can be converted to helium, and calculate how long it can shine at its present luminosity. There are about 3×10^7 sec/yr.

6. Perform a similar calculation for a typical star less massive than the sun — say with a mass of 1×10^{33} gm and a luminosity of 4×10^{32} ergs/sec.

7. Why do you suppose so great a fraction of the sun's energy comes from its centralmost regions? Within what fraction of the sun's radius does practically all of the sun's luminosity originate? (See Table 20.1.) Within what radius of the sun has its original hydrogen been partially used up? Discuss what relation the answers to these questions bear to each other.

8. The H-R diagram for field stars (that is, stars all around us in the sky) shows very luminous main-sequence stars and also various kinds of red giants and supergiants. We also find white dwarfs among the field stars. Explain these features, and interpret the H-R diagram for field stars in terms of our current ideas about stellar evolution.

9. In the H-R diagrams for some young clusters, stars of very low and very high luminosity are off to the right of the main sequence, while those of intermediate luminosity are on the main sequence. Can you offer an explanation? Sketch an H-R diagram for such a cluster.

10. By the time the sun becomes a white dwarf, the constellations familiar to us now could not be seen, even if their stars had never changed in luminosity. Why?

11. Prepare a chart, or diagram, which exhibits the relative sizes of a typical red giant, the sun, a typical white dwarf, and a (hypothetical) neutron star. You may have to be clever to devise such a diagram.

CHAPTER 21

Galaxies

The "analogy [of the nebulae] with the system of stars in which we find ourselves . . . is in perfect agreement with the concept that these elliptical objects are just [island] universes — in other words, Milky Ways . . . ."

So wrote Immanuel Kant (1724–1804) in 1755† concerning the faint patches of light which telescopes revealed in large numbers. Unlike the true gaseous nebulae that populate the Milky Way (Chapter 19), the nebulous-appearing luminous objects referred to by Kant are found in all directions in the sky *except* where obscuring clouds of interstellar dust intervene. Despite Kant's (and others') speculation that these patches of light are actually systems like our own Milky Way Galaxy, the weight of astronomical opinion rejected the hypothesis, and their true nature remained a subject of controversy until 1924. The realization, only a few decades ago, that our Galaxy is not unique and central in the universe ranks with the acceptance of the Copernican system as one of the great advances in cosmological thought.

21.1 GALACTIC OR EXTRAGALACTIC?

The discovery and cataloguing of nebulae had reached full swing by the close of the eighteenth century. A very significant contribution to our knowledge of these objects was provided by the work of the great German-English astronomer William Herschel (1738–1822) and his only son, John (1792–1871). William surveyed the northern sky by scanning it visually with the world's first

large reflecting telescope, instruments of his own design and manufacture. John took his father's telescopes to the Southern Hemisphere and extended the survey to the rest of the sky.

For a while, the elder Herschel himself, who had discovered thousands of "nebulae,"* regarded these objects as galaxies, like the Milky Way system; he was known to remark once that he had discovered more than 1500 "universes." He found, however, that many of the nebulae appeared as individual (although rather indistinct) "stars," surrounded by hazy glows of light. The fact that it is difficult to reconcile the appearance of such an object with that of a remote stellar system led Herschel, in 1791, to abandon the island-universe hypothesis. Still, the concept of the possibility of other galaxies never quite disappeared from astronomical thought.

(a) Catalogues of Nebulae

One of the earliest catalogues of nebulous-appearing objects was prepared in 1781 by the French astronomer Charles Messier (1730–1817). Messier was a comet hunter, and as an aid to himself and others in his field he placed on record 103 objects that might be mistaken for comets.

---

†*Universal Natural History and Theory of the Heavens.*

---

*Nebula (plural *nebulae*) literally means "cloud." Faint star clusters, glowing gas clouds, dust clouds reflecting starlight, and galaxies all appear as faint, unresolved luminous patches when viewed visually with telescopes of only moderate size. Since the true natures of these various objects were not known to the early observers, all of them were called "nebulae." Today, we usually reserve the word "nebula" for the true gas or dust clouds (Chapter 19), but some astronomers still refer to galaxies as nebulae or *extragalactic nebulae.*

Because Messier's list contains some of the most conspicuous star clusters, nebulae, and galaxies in the sky, these objects are often referred to by their numbers in his catalogue — for example, M31, the great galaxy in Andromeda (see also Chapter 19).

In the years from 1786 to 1802, William Herschel presented to the Royal Society three catalogues, containing a total of 2500 nebulae. The *General Catalogue of Nebulae,* published by John Herschel in 1864, contains 5079 objects, of which 4630 had been discovered by him and his father. The *General Catalogue* was revised and enlarged into a list of 7840 nebulae and clusters by J. L. E. Dreyer in 1888. Today most bright galaxies are known by their numbers in Dreyer's *New General Catalogue* — for example, NGC 224 = M31. Two supplements to the *New General Catalogue,* known as the first and second *Index Catalogues* (abbreviated "IC"), were published in 1895 and in 1908.

By 1908 nearly 15,000 nebulae had been catalogued and described. Some had been correctly identified as star clusters and others as gaseous nebulae (such as the Orion nebula). The nature of most of them, however, still remained unexplained. If they were nearby, with distances comparable to those of observable stars, they would have to be luminous clouds — probably of gas, and possibly with intermixed stars — within our Galaxy. If, on the other hand, they were very remote, far beyond the foreground stars of the Galaxy, they could be unresolved *systems* of billions of stars, galaxies in their own right — or as Kant had described them, "island universes." The resolution of the problem required the determination of the distances to at least some of the nebulae.

(b) Arguments For and Against the Island-Universe Hypothesis

By the early twentieth century the nebulae could be divided into two distinct groups: those largely irregular and amorphous, which are concentrated near the Milky Way, and those (either elliptical or wheel-shaped) showing symmetry, which are most numerous in parts of the sky far from the Milky Way. Those of the former type have bright-line spectra, and were clearly recognized as gaseous nebulae. The few available spectra of the brighter nebulae in the latter category, on the other hand, showed absorption lines like the spectra of stars.

In 1908 construction was completed of the 60-inch telescope on Mount Wilson, and within a decade the 100-inch telescope was also in operation. Photographs obtained with these instruments clearly resolved the brightest stars in some of the nearer "nebulae." To calculate their distances, however, it was necessary to know what kind of stars they were — that is, what their absolute magnitudes were. Moreover, not all astronomers agreed that real stars had actually been resolved in any of the nebulae. Even the Mount Wilson astronomers who had photographed those stars were not convinced of their true stellar nature and had described them as "nebulous stars."

Two of the major protagonists in the controversy over the nature of the nebulae were Harlow Shapley, of the Mount Wilson Observatory, and H. D. Curtis, of the Lick Observatory. Their opposing views culminated on April 26, 1920, in the famous Shapley-Curtis debate before the National Academy of Sciences. Curtis supported the island-universe theory, and Shapley opposed it. Of course, the controversy was not settled by the debate; according to A. R. Sandage, "Perhaps the fairest statement that can be made is that Shapley used many of the correct arguments but came to the wrong conclusion. Curtis, whose intuition was better in this case, gave rather weak and sometimes incorrect arguments from the facts, but reached the correct conclusion."[†]

[†]A. R. Sandage, *The Hubble Atlas of Galaxies.* Carnegie Institution, Washington, D.C., 1961, p. 3.

FIG. 21-1   A field of variable stars in the Andromeda Galaxy, with two variables marked. Photographed with the 200-inch telescope. *(Hale Observatories.)*

(c) The Resolution of the Controversy

The final resolution of the controversy was brought about by the discovery of variable stars in some of the nearer "nebulae" in 1923 and 1924. Edwin Hubble, at the Mount Wilson Observatory, analyzed the light curves (Section 18.1b) of variables he had discovered in M31, M33, and NGC 6822 and found that they were cepheids (Section 18.2a). Although cepheid variables are supergiant stars, the ones studied by Hubble appeared very faint — near magnitude 18. Those stars, therefore, and the systems in which they were found, must be very remote; the "nebulae" had been established as galaxies. Hubble's exciting results were presented to the American Astronomical Society at its thirty-third meeting, which began on December 30, 1924.

21.2 DISTANCES TO THE GALAXIES

When we realize how recently in the history of astronomy it was learned that galaxies are remote stellar systems, it may not seem strange to us that today, less than half a century later, their distances are known only roughly. Distances to even the nearest galaxies are not yet known with an accuracy of 1 part in 10. Most visible galaxies in the universe lie at distances that are uncertain by a factor of 2. The geometrical triangulation techniques that can be applied to yield many-place accuracy in the solar system do not work at all on galaxies; after many lifetimes of observation, the sun's motion in our Galaxy is still too little to provide us with a sufficient base line. We must, therefore, resort to indirect methods to "survey" the galaxies. The most important of these are based upon the photometry of individual stars and other objects contained within galaxies and upon the light of the entire galaxies themselves.

(a) Luminosity Methods of Estimating Distances of Galaxies

The most direct and reliable method of finding the distance to a galaxy is to identify in it an object or objects that are similar to known objects in our own Galaxy and which, presumably, have the same intrinsic luminosities (or absolute magnitudes). By comparing the apparent magnitude of such an object with its known absolute magnitude, we find its distance modulus, and hence its distance — which is, of course, the distance of the galaxy itself — by the method outlined in Section 14.2b. This was the procedure followed by Hubble when he derived distances to several nearby galaxies from the cepheid variables in them.

Cepheid variables are still our most important link to galactic distances, which is why these relatively rare stars are so important. By measuring the light from cepheids, we can determine, with fair precision, distances to several neighboring galaxies in which they can be observed. Unfortunately, however, there are only about 30 galaxies containing cepheids that are near enough so that we can observe those variables, even with the world's largest telescope. The nearest of these (the Clouds of Magellan) are only about 150,000 to 200,000 LY away from the sun, and the most distant are within 20 million LY.

The brightest cepheids have an absolute magnitude of about −6. The most luminous stars in some kinds of galaxies are even brighter, and can be seen, therefore, to a greater distance. The brightest stars found in our own and other galaxies (usually blue supergiant stars) have absolute magnitudes of about −9. Bright novae, at maximum, reach about the same luminosity, and this is also about the absolute magnitude of the most luminous globular clusters. Thus the brightest supergiant stars in a galaxy, occasional nova outbursts that occur in it, and its brightest globular clusters all are distance indicators that can be observed to as far away as 80 million LY. Finally, distances to some kinds of galaxies even somewhat farther away can be estimated from the angular sizes of the largest H II regions (emission nebulae) that can be resolved in them.

If all galaxies were identical, they would all emit the same total amount of light, and the magnitude of a galaxy as a whole would indicate its distance. We shall see that galaxies range enormously in total luminosity. Many galaxies, probably the majority, do not have distinguishing characteristics that enable us to estimate their absolute magnitudes; we can only tell which ones are highly luminous and which ones are not if we see a collection of them, of various brightnesses, side by side in a cluster (see below).

(b) Distribution of Galaxies

The distances that separate galaxies are hundreds of thousands to millions of light-years. There are less than 20 known galaxies within 2½ million LY, but there are many thousands within 50 million LY; galaxies extend in all directions as far as we can see. The more distant galaxies, of course, appear less conspicuous than the nearer ones. Tiny images of extremely remote galaxies are actually more numerous than images of the foreground stars of our own Galaxy on photographs obtained with large telescopes directed away from the Milky Way. The images of galaxies at the limit of detection can be distinguished from those of faint stars only because the galaxian images are rather fuzzy or elongated and lack the sharp, point-like appearance of the star images.

In the years following his announcement of the true nature of the "nebulae," Hubble made an extensive study of galaxies and their distribution in space. He realized that it would take thousands of years to photograph the entire sky with the 100-inch telescope and to count all the galaxies that could be observed with that instrument. Instead, he photographed 1283 sample regions of the sky with the 100-inch telescope. The results of Hubble's survey are shown in Figure 21-2. Each symbol in the figure shows one of the sample regions, and the size and type of the symbol shows how many galaxies were visible on the photograph of each region. From the 44,000-odd galaxies Hubble counted in these selected regions, he calculated that nearly 100 million galaxies must exist within the range of the 100-inch telescope. There may be 1 billion galaxies within the reach of the 200-inch instrument at Palomar.

Hubble found the largest number of galaxies in the regions near the galactic poles (90° from the great circle running through the Milky Way) and determined that, on the average, fewer and

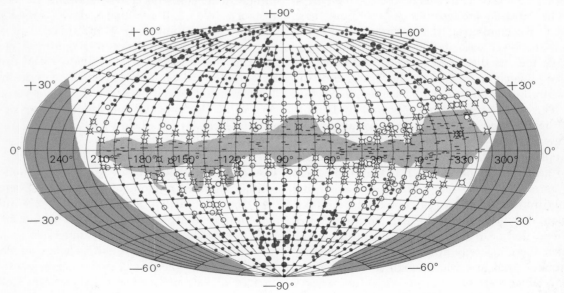

FIG. 21-2 Distribution of galaxies. The size of each dot plotted on the map of the sky indicates how many galaxies were counted on a photograph taken of that location with the 100-inch telescope. Dashes and open circles indicate few or no galaxies, and the irregular region across the center of the map where almost no galaxies are to be seen is the zone of avoidance. This zone coincides with the visible Milky Way, where interstellar dust obscures the exterior galaxies. The empty regions on the plot are those of the sky too far south to survey from Mount Wilson.

fewer galaxies could be seen in directions successively closer to the Milky Way. This, of course, is an effect of absorption of light by interstellar dust in our galaxy (see Section 19.4a). The obscuration of light by interstellar dust is so complete in a region along the Milky Way that no galaxy shows through (or nearly none); Hubble called this region the "zone of avoidance".

Hubble found that when allowance is made for interstellar absorption, the distribution of galaxies, on the large scale, is the same in all directions, and that the mean density of galaxies in space seems to be about the same at all distances. On the small scale, however, galaxies tend to be grouped into *clusters of galaxies*. A discussion of these clusters is given below, but a description of how relative distances to them can be estimated is relevant here.

(c) Relative Distances to Clusters of Galaxies

All the galaxies in a cluster are at about the same distance — the distance to the cluster itself. We can intercompare those galaxies, therefore, and pick out those which are intrinsically bright and those which are faint. If a cluster is rich, that is, if it has many member galaxies, it is probable that it will contain a few highly luminous galaxies. We can estimate relative distances to different clusters, therefore, by comparing the apparent magnitudes of their brightest members. If the light received from the brightest galaxies in cluster *A*, for example, is one-fourth as much as from the brightest galaxies in cluster *B*, cluster *A* is probably twice as distant as *B*.

Using the magnitudes of the brightest cluster members in this way works only for rich clusters, which can be expected to have similar galaxies for their brightest members. Unfortunately, there is no very rich cluster close enough to us so that we can determine its distance by independent means (say, from resolved stars in some of its galaxies). We are not certain, therefore, of the actual luminosities of the brightest galaxies in rich clusters. *Relative* distances to clusters are thus known more reliably than absolute distances.

(d) Summary

Distances to the nearest galaxies are found from the apparent faintness of stars and other objects of known luminosity that can be recognized in them. Although the distances found for these neighboring galaxies are very great compared to the largest astronomical distances known a half century ago, they are still very small compared to those of the overwhelming majority of galaxies that can be observed with large telescopes. Distances can be obtained for remote galaxies only when they happen to be members of rich clusters, and then only with considerable uncertainty.

We have not referred in this section to the well-known correlation between the distances of galaxies and their radial velocities, as indicated by the Doppler shifts of the lines in their spectra. Although distances are often estimated for galaxies from their radial velocities, the method is not really independent, for it depends on the calibration of the velocity-distance relation, which, in turn, must be based on distances measured by other means to selected galaxies with various observed Doppler shifts. The accurate calibration of this relation is one of the major problems of extragalactic astronomy. The velocity-distance relation, and its calibration, is so fundamental to the study of modern *cosmology* that discussion of it is postponed to Chapter 22.

21.3 DETERMINATION OF GROSS PROPERTIES OF GALAXIES

The linear size of that part of a galaxy that corresponds to an observed angular size can be calculated once the distance to the galaxy is known, just as we calculate the diameter of the sun or of a planet. Also, if we know its distance, we can apply the inverse-square law of light and calculate the total luminosity, or the absolute visual magnitude of a galaxy from the amount of light flux we receive from it or from its apparent visual magnitude. Thus a knowledge of galaxian radii and luminosities is provided by a knowledge of galaxian distances.

The determination of masses of galaxies, however, is more difficult, and, in fact, is possible (by present techniques) for only a small fraction of them. There are several techniques for measuring galaxian masses.

(a) Mass of Galaxies

We determine the masses of galaxies, like those of other astronomical bodies, by measuring their gravitational influences on other objects or on the stars within them. Internal motions in galaxies provide the most reliable methods of measuring their masses but are observable only in compara-

FIG. 21-3   The Andromeda Galaxy, M31, and its rotation curve. *(Yerkes Observatory.)*

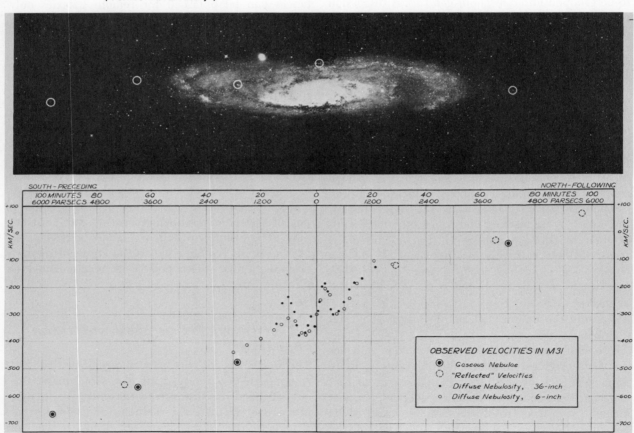

tively nearby galaxies. The most important procedure is to observe the rotation of a galaxy and then to compute its mass with the help of Kepler's third law.

As an illustration, we shall consider the rotation of M31 — the Andromeda galaxy (Figure 21-3). Galaxy M31 is probably quite similar to our own galaxy; it has a brilliant central nucleus and conspicuous spiral arms that wind through a presumably circular disk. The galaxy is inclined at an angle of only about 15° to our line of sight, so we see it highly foreshortened. Figure 21-4 is a plot of the radial velocities measured at various points along the major axis of the projected image of M31. The radial velocity of the brilliant nucleus shows that the galaxy as a whole approaches us at nearly 300 km/sec. The still more negative radial velocities of regions southwest of the nucleus indicate that that side of the galaxy is turning toward us and the northeast side away from us. We see that the maximum rotational speeds in M31 occur at about 70′ from its center. Parts farther from the center of the galaxy have smaller orbital speeds. They approach those speeds that would be pre-

dicted at various distances from the nucleus by Kepler's third law, on the assumption that those parts are well outside the bulk of the mass, or that the whole mass acts as if it were concentrated at the center (like the sun in the solar system).

From Figure 21-4 we see that a star 2° from the center of the Andromeda galaxy would have a speed of about 200 km/sec relative to the center of the galaxy. Since the distance to M31 is about 2.2×10^6 LY, that star would be about 2.4×10^4 pc, or 5×10^9 AU, from the nucleus, and the circumference of its orbit would be 3×10^{10} AU. A speed of 200 km/sec corresponds to 42 AU/year, so the period of the star would be 7×10^8 years. If we apply Kepler's third law to the mutual revolution of M31 and this hypothetical star, we find that the mass of the galaxy is given approximately by

$$\text{mass} = \frac{(5 \times 10^9)^3}{(7 \times 10^8)^2} = 2.5 \times 10^{11} \text{ solar masses,}$$

which is of the same order as the mass of our own Galaxy. The masses have now been determined

FIG. 21-4   Rotation curve of M31, based on measures of radial velocities of emission nebulae in the spiral arms by N. U. Mayall.

of more than two dozen nearby galaxies from comparable analyses of their rotations. Some galaxies contain enough neutral hydrogen to be observed in 21-cm radiation (Section 19.4g). For such galaxies the rotation curves can sometimes be obtained from the Doppler shifts of the 21-cm line.

Many galaxies are not highly flattened and rotating rapidly. Nevertheless, the velocities of the stars in such a galaxy depend upon its gravitational attraction for them, and hence upon its mass. The spectrum of a galaxy is a composite of the spectra of its many stars, whose different motions produce different Doppler shifts. The lines in the composite spectrum, therefore, are broadened, and the amount by which they are broadened indicates the range of speeds with which the stars are moving with respect to the center of mass of the galaxy, and so allows us to estimate its mass.

Like stars, galaxies are often observed in close pairs. Double galaxies are so common that no more than a small fraction can be "optical doubles," that is, two objects at different distances, merely lined up in projection. Most, in other words, are true binary systems of mutually revolving galaxies. However, we cannot "see" the motion of one galaxy about the other, as we can observe the mutual revolution of the members of a binary star system. Nevertheless, from a study of the differences between the radial velocities of the members of double-galaxy systems, and from the separation of their galaxies, we can learn something about average galaxian masses. The masses of clusters of galaxies can be calculated by the same technique used to "weigh" star clusters. The radial velocities of many galaxies in a cluster are first measured. The average of these velocities is that of the center of mass of the cluster, and the differences between the velocities of individual galaxies and this mean value tells us how fast they are moving within the cluster. These speeds depend on the total gravitational potential energy

FIG. 21-5   NGC 4565, a spiral galaxy in Coma Berenices, seen edge on. Photographed in red light with the 200-inch telescope. *(Hale Observatories.)*

of the cluster, and a knowledge of them thus enables us to estimate the cluster mass.

21.4 TYPES OF GALAXIES

We have seen how the fundamental properties of galaxies — their luminosities, radii, and masses — are estimated. Now let us consider briefly the different kinds of galaxies that are found. The vast majority of observed galaxies fall into two general classes: spirals and ellipticals. A small minority are classed as irregular.

(a) Spiral Galaxies

Our own galaxy and M31, which is believed to be much like it, are typical spiral galaxies. Like our galaxy (Chapter 19), a spiral consists of a nucleus, a disk, a corona, and spiral arms. Interstellar material is usually observed in the arms of spiral galaxies. Bright emission nebulae are present, and absorption of light by dust is also often apparent, especially in those systems turned almost edge on to our line of sight (see Figure 21-5). The spiral arms contain the young stars, which include luminous supergiants. These bright stars and the emission nebulae make the arms of spirals stand out like the arms of a fourth-of-July pinwheel. The individual interarm stars are usually not observable at all, save in the nearest galaxies, al-

FIG. 21-7   NGC 3031 (M81), spiral galaxy in Ursa Major. Photographed with the 200-inch telescope. *(Hale Observatories.)*

FIG. 21-6   The Sc galaxy NGC 5194 (M51) and its irregular II companion, NGC 5195. Photographed with the 200-inch telescope. *(Hale Observatories.)*

though their collective light may be appreciable as a uniform glow. Open star clusters can be seen in the arms of nearer spirals, and globular clusters are often visible in their coronas; in M31, for example, more than 200 globular clusters have been identified. The spiral arms of a galaxy are composed of population I stars, while the stars in the nucleus, corona, and disk (except for the arms themselves) are population II. Spiral galaxies, in other words, have mixed stellar population types.

FIG. 21-8   NGC 1300, barred spiral galaxy in
Eridanus, photographed with the 200-inch telescope.
(Hale Observatories.)

A large minority (perhaps a fourth or more) of spiral galaxies display "bars" running through their nuclei; the spiral arms of such a system usually begin from the ends of the bar, rather than winding out directly from the nucleus. These are called *barred spirals*. A famous example is NGC 1300 (Figure 21-8). The bar in a barred spiral is in a sense a straight portion of spiral arm and usually contains interstellar matter and population I stars. Studies of the rotations of some barred spirals show that their inner parts (out to the ends of the bars) are rotating approximately as solid wheels. In the absence of differential shearing rotation, the straight bar can persist, rather than winding up; the detailed structures and dynamics of barred spirals, however, are not yet understood.

In both normal and barred spirals we observe a gradual transition of morphological types. At one extreme, the nucleus is large and luminous, the arms are small and tightly coiled, and bright emission nebulae and supergiant stars are inconspicuous. At the other extreme are spirals in which the nuclei are small — almost lacking — and the arms are loosely wound, or even wide open. In these latter galaxies, there is a high degree of resolution of the arms into luminous stars, star clusters, and emission nebulae. Our Galaxy and M31 are both intermediate between these two extremes. So far as is known, all spirals and barred spirals rotate so that their arms trail, as does our galaxy (Chapter 19).

Perhaps about one percent of the spiral galaxies belong to a special class known as *Seyfert galaxies* (after the astronomer who first described them). Seyfert galaxies usually appear normal on direct photographs, but their spectra display strong emission lines originating from hot gas at the centers of the galaxies' nuclei. The emitting region is usually rather localized, but often enough energy is involved to suggest an explosive or other cataclysmic event taking place there.

(b) Elliptical Galaxies

More than two thirds of the thousand most conspicuous galaxies in the sky are spirals. For this reason it is often said that most galaxies are spirals. Actually, however, the most numerous galaxies in any given volume of space are those of relatively low luminosity, which cannot be seen at large distances, and which, therefore, are not among the brightest-appearing galaxies. (Similarly, the most numerous stars are faint main-sequence stars, very few of which can be seen with the unaided eye — see Section 16.1.) Most of these dwarf galaxies fall into the class of *elliptical* galaxies. Moreover, the rich clusters, which contain a good fraction of all galaxies, contain mostly ellipticals. They are really far more numerous than spirals.

Elliptical galaxies are spherical or ellipsoidal systems which consist entirely of population II stars and contain no trace of spiral arms. They resemble the nucleus and corona components of spiral galaxies. Although dust and conspicuous emission nebulae are not easily observed in elliptical galaxies, some do show evidence of sparse interstellar gas in their spectra. In the larger (nearby) ones, many globular clusters can be identified. The elliptical galaxies show various degrees of flattening and range from systems that are approximately spherical to those that approach the flatness of spirals (Figure 21-9). The distribution of light in a typical elliptical galaxy shows that while it has many stars concentrated toward its center, a sparse scattering of stars extends for very great distances and merges imperceptibly into the void of intergalactic space (compare with the corona of our own galaxy — Section 19.1b). For this reason it is nearly impossible to define the total size of an elliptical galaxy. Similarly, it is not obvious how far the corona of a spiral galaxy extends.

The fact that elliptical galaxies are not disk-shaped shows that they are not rotating as rapidly as the spirals. It is hypothesized that they are systems that formed from pregalaxian material that had little angular momentum per unit mass —

FIG. 21-9 Types of elliptical galaxies. *(Hale Observatories.)*

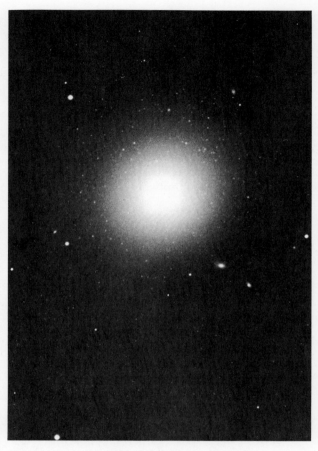

FIG. 21-10 NGC 4486 (M87), giant elliptical galaxy in Virgo, photographed with the 200-inch telescope. Note the many visible globular clusters in the galaxy. *(Hale Observatories.)*

that is, that their original material had low net rotation. Consequently, as such a cloud of primeval material contracted, it did not flatten into a disk, and the density of the material was high enough that it completely (or nearly completely) condensed into stars. In a spiral, on the other hand, a considerable amount of gas (and/or dust) in the flat, rapidly rotating disk was not able to condense into stars at once. This material was presumably formed into spiral arms by the rotation of the galaxy, where it now still slowly condenses into stars at a gradual rate. Elliptical galaxies must consist entirely of old stars (at least if those galaxies are as old as our Galaxy), which would explain why they are pure population II systems.

Elliptical galaxies have a much greater range in size, mass, and luminosity than do the spirals. The rare giant ellipticals (for example, M87 — see Figure 21-10) are more luminous than any known spiral. The brightest ellipticals in some rich clusters

FIG. 21-12   NGC 6822, an irregular galaxy in the Local Group. *(Hale Observatories.)*

FIG. 21-11   Leo II, a dwarf elliptical galaxy (negative print). Photographed with the 200-inch telescope. *(Hale Observatories.)*

probably reach absolute magnitudes that are brighter than -24, more than 100 billion times the luminosity of the sun, and more than 10 times the luminosity of the Andromeda galaxy. The mass data available for giant ellipticals suggest that some may have masses in excess of a trillion times that of the sun. While, as stated above, the diameters of these large galaxies are difficult to define, they certainly extend over at least several hundred thousand light-years. The often-made statement that the largest of the galaxies are spirals like our own is a misconception.

Elliptical galaxies range all the way from the giants, just described, to dwarfs, which are believed to be the most common kind of galaxy. An example is the Leo II system, shown in Figure 21-11. There are so few bright stars in this galaxy that even its central regions are transparent. The total number of stars, however (most too faint to show in Figure 21-11), is probably at least several million. The absolute magnitude of this typical dwarf is about −10; its luminosity is about 1 million times that of the sun. Intermediate between the giant and dwarf elliptical galaxies are systems such as M32 and NGC 205, two near companions to M31. They can be seen in the photograph of M31 (see page 384); NGC 205 is the one that is farther from M31.

(c) Irregular Galaxies

About 3 percent of the brightest appearing galaxies in the northern sky are classed as irregular. They show no trace of circular or rotational symmetry, but have an irregular or chaotic appearance. The irregular galaxies divide into two groups. The first group, denoted *Irr I* galaxies, consists of objects showing high resolution into O and B stars and emission nebulae. The best-known examples are the Large and Small Clouds of Magellan (Figure 21-13), our nearest galaxian neighbors. We find many star clusters in these galaxies, as well as variable stars, supergiants, and gaseous nebulae; they contain both population I and population II stars. The Small Cloud, however, is apparently lacking in dust, although it does contain interstellar gas. The lack of conspicuous dust clouds is common in this first kind of irregular galaxy.

Galaxies of the second irregular type (*Irr II*) resemble the Irr I objects in their lack of symmetry. These objects, however, display no resolution into stars or clusters, but are completely amorphous in texture. Their spectra are continuous with absorption lines and resemble the spectra of type A5 stars, showing that stars not luminous enough to be resolved must exist in these galaxies. The Irr II galaxies generally also show conspicuous dark lanes of absorbing interstellar dust. Examples are M82 (Figure 21-14) and the companion to the spiral galaxy, M51 (Figure 21-6).

(d) Summary

The gross features of the different kinds of galaxies are summarized in Table 21.1. Many of the figures given, especially for mass, luminosity, and diameter, are very rough and are intended only to illustrate orders of magnitude, not precise values.

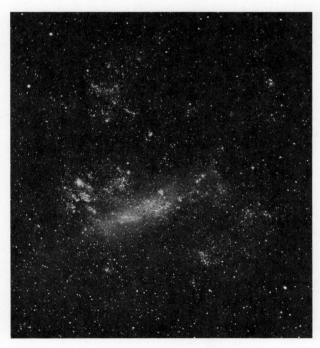

FIG. 21-13   The Large Magellanic Cloud. *(Lick Observatory.)*

21.5 CLUSTERS OF GALAXIES

Until World War II, only a few dozen clusters of galaxies were known; most of these had been discovered by accident on telescopic photographs. Recent photographic surveys of the sky have shown, however, that clusters are far more common than had previously been thought. It now appears that all or nearly all galaxies may be members of groups or clusters. Clusters of galaxies, in other words, are now regarded as fundamental condensations of matter in the universe.

FIG. 21-14   NGC 3034 (M82), an irregular II galaxy in Ursa Major.
(Hale Observatories.)

(a) Types of Clusters

Clusters of galaxies can be roughly classified into two categories: *regular clusters* and *irregular clusters*. The regular clusters have spherical symmetry and show marked central concentration. They tend to be very rich clusters, and most of them probably contain at least 1000 members brighter than absolute magnitude −16. The regular clusters have structures resembling those of globular star clusters; some investigators refer to them as *globular clusters of galaxies*. Regular clusters consist entirely, or nearly entirely, of elliptical galaxies.

TABLE 21.1 Gross Features of Galaxies of Different Types

	SPIRALS	ELLIPTICALS	IRREGULARS
Mass (solar masses)	10^9 to 2×10^{11}	10^6 to 10^{13}	10^8 to 3×10^{10}
Diameter (thousands of light-years)	20 to 150	2 to 500 (?)	5 to 30
Luminosity (solar units)	10^8 to 10^{10}	10^6 to 10^{11}	10^7 to 2×10^9
Absolute visual magnitude	−15 to −20	−9 to −23	−13 to −18
Population type	I and II	II	I and II
Composite spectral type	A to K	G to K	A to F
Interstellar matter	Both gas and dust	Almost no dust; little gas	Much gas; much dust in Irr II; little or no dust in Irr I

FIG. 21-15   Cluster of galaxies in Corona Borealis, photographed wtih the 200-inch telescope. *(Hale Observatories.)*

The irregular clusters, sometimes called *open clusters,* have a more nearly amorphous appearance and possess little or no spherical symmetry nor central concentration. The irregular clusters sometimes, however, have several small subcondensations, and resemble loose swarms of small clusters. They contain all kinds of galaxies — spirals, ellipticals, and irregulars. Irregular clusters are more numerous than the regular clusters and range from rather rich aggregates of more than 1000 galaxies (for example, the Virgo cluster) to small groups of a few dozen members or less. An example of the latter is the *Local Group,* the small cluster of galaxies to which our Galaxy belongs.

(b) The Local Group

The Local Group contains 17 known members, spread over a region about 3 million LY in diameter.

FIG. 21-16   Irregular cluster of galaxies in Hercules, photographed with the 200-inch telescope. *(Hale Observatories.)*

TABLE 21.2 The Local Group

GALAXY	TYPE	RIGHT ASCENSION 1950	DECLINATION 1950	VISUAL MAGNITUDE (m_V)	DISTANCE (KPC)	DISTANCE (1000 LY)	DIAMETER (KPC)	DIAMETER (1000 LY)	ABSOLUTE MAGNITUDE (M_V)	RADIAL VELOCITY (KM/SEC)	MASS (SOLAR MASSES)
Our Galaxy	Spiral	——	——	—	—	——	30	100	(−21)	—	2×10^{11}
Large Magellanic Cloud	Irr I	5^h 26^m	−69°	0.9	48	160	10	30	−17.7	+276	2.5×10^{10}
Small Magellanic Cloud	Irr I	0 50	−73	2.5	56	180	8	25	−16.5	+168	
Ursa Minor system	dwarf ell.	15 8.2	+67 18′		70	220	1	3	(−9)		
Sculptor system	dwarf ell.	0 57.5	−33 58	8.0	83	270	2.2	7	−11.8		(2 to 4 × 10^6)
Draco system	dwarf ell.	17 19.4	+57 58		100	330	1.4	4.5	(−10)		
Fornax system	dwarf ell.	2 37.7	−34 44	8.4	250	800	4.5	15	−13.6	+39	(1.2 to 2 × 10^7)
Leo II system	dwarf ell.	11 10.8	+22 26	12.04	230	750	1.6	5.2	−10.0		(1.1 × 10^6)
Leo I system	dwarf ell.	10 5.8	+12 33	12.0	280	900	1.5	5	−10.4		
NGC 6822	Irr I	19 42.1	−14 54	8.9	460	1500	2.7	9	−14.8	−32	
NGC 147	elliptical	0 30.4	+48 13	9.73	570	1900	3	10	−14.5		
NGC 185	elliptical	0 36.1	+48 4	9.43	570	1900	2.3	8	−14.8	−305	
NGC 205	elliptical	0 37.6	+41 25	8.17	680	2200	5	16	−16.5	−239	
NGC 221 (M32)	elliptical	0 40.0	+40 36	8.16	680	2200	2.4	8	−16.5	−214	
IC 1613	Irr I	1 00.6	+ 1 41	9.61	680	2200	5	16	−14.7	−238	
Andromeda galaxy (NGC 224; M31)	Spiral	0 40.0	+41 0	3.47	680	2200	40	130	−21.2	−266	3×10^{11}
NGC 598 (M33)	Spiral	1 31.0	+30 24	5.79	720	2300	17	60	−18.9	−189	8×10^{9}

There are 3 large spirals (our own, M31, and M33), 4 irregulars, and at least 10 ellipticals, 6 of which are dwarfs. Among the outlying galaxies that could also be members are a large elliptical and another spiral discovered by Paolo Maffei on infrared photographs of a region where dust clouds of the Milky Way dim visible light (however, spectra of the Maffei galaxies show them to be moving too fast to be bound to the local Group). Van den Bergh has recently announced discovery of several dwarfs near M31, and there are doubtless other undiscovered dwarf ellipticals.

Table 21.2 lists the known members of the Local Group that all investigators agree are galaxies. Some of the data are very uncertain and are given in parentheses. Figure 21-17 is a plot of the Local Group; the galaxies have been projected onto an arbitrary plane centered on our Galaxy; then their distances from the center of the plot have been increased so that they are shown at the correct relative distances from us.

The radial velocities of some of the Local Group galaxies have been measured. The Andromeda galaxy, for example, approaches us at about 260 km/sec, while the Large Cloud of Magellan recedes from us at about 276 km/sec. Most of the apparent radial velocity of each of these objects is due to the revolution of the sun about the center of the Galaxy and the motion of the Galaxy in the Local Group. From an analysis of the radial velocities of all the other galaxies we find that the sun's orbital velocity is about 250 km/sec, and that our Galaxy is moving at a speed of from 100 to 150 km/sec in the Local Group. The average of the motions of all the galaxies in the Group indicates that its total mass is near 5×10^{11} solar masses.

(c) The Neighboring Groups and Clusters and the Local Supercluster

Beyond the Local Group, at distances of a few times its diameter, we find other similar small

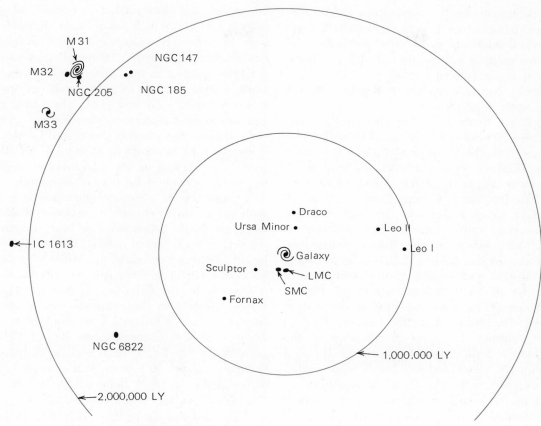

FIG. 21-17   Plot of the Local Group.

groups of galaxies. Distances to the nearer of these can be determined from cepheid variables that are observed in them. Examples are groups of galaxies centered about each of the bright spirals, M51 and M81 (the galaxies pictured in Figures 21-6 and 21-7, respectively).

The nearest rich cluster of galaxies is the Virgo cluster (so named because it is in the direction of the constellation Virgo). Its distance is probably between 30 and 50 million LY — too remote to observe the cepheid variables in it. The distance to that cluster is estimated from the apparent faint-

ness of the O and B supergiant stars that are barely observable in some of its galaxies, and from the apparent brightness of globular clusters in M87, a bright elliptical galaxy in the cluster. The Virgo cluster appears as a great cloud covering a 10° by 12° region of the sky, within which at least 1000 galaxies can be seen. The brightest of these are giant elliptical galaxies and large spirals; the faintest (observable) are dwarf ellipticals more or less like the Sculptor and Fornax galaxies in the Local Group (see Table 21.2). The linear diameter of the cluster is about 7 million LY.

The Virgo cloud is an example of a rich irregular cluster. Its structure is very complicated, there being several relatively dense subcondensations of galaxies within it, as well as many double and triple galaxies. It is possible that it may not even be a single cluster, but two clusters of slightly different distances, seen in projection.

Altogether, there are several thousand galaxies within about 70 million LY. Many of these galaxies are grouped into the Virgo cluster and a few other somewhat smaller irregular clusters, and most of the rest of them are in relatively small clusters or groups like the Local Group. Beyond 50 to 70 million LY, galaxies seem to thin out and are sparse in space until a much larger distance is reached. Many experts in this field of astronomy, therefore, are of the opinion that this large collection of clusters and smaller groups comprises a *second-order cluster* — that is, a *cluster of clusters*. G. de Vaucouleurs, who has studied the system extensively, has termed it the *local supergalaxy,* or *local supercluster*. The over-all diameter of the supercluster is believed to be between 100 and 150 million LY. Some evidence has been presented which suggests that the local supercluster is flattened and is rotating, but more observational data are needed to verify this suspicion. The total mass of the local supercluster is probably of the order 10^{15} solar masses.

(d) More Distant Systems of Galaxies

There must be small groups of galaxies that lie beyond the local supercluster, but which are not conspicuous to us. In a typical system like the Local Group, for example, we would see only the one or two brightest galaxies, and would not identify the group as a cluster. Rich clusters of galaxies, on the other hand, especially the *regular clusters,* stand out conspicuously and are recognized to very great distances.

The nearest rich regular cluster is the Coma cluster (in the constellation Coma Berenices), which lies at a distance that is probably about 300 million LY, and which has a linear diameter of at least 10 million LY. Despite its great distance, we can observe more than 1000 member galaxies. The brightest galaxies in the cluster are two giant ellipticals, whose absolute visual magnitudes are likely to be between -23 and -24. The Coma cluster contains more and more member galaxies at magnitudes that are successively fainter and fainter. There is every reason to expect there to be dwarf elliptical galaxies present; if so, the total number of galaxies in the cluster might be tens of thousands.

From the Sky Survey photographs obtained with the 48-inch Schmidt telescope at Palomar, a fairly complete catalogue has been prepared of the very rich clusters of galaxies that are not hidden by the obscuring clouds of the Milky Way and that lie in the northern three quarters of the sky that is covered by the Palomar Sky Survey. The most remote clusters in the catalogue are at distances between 2 and 3 billion LY — about 10 times the distance of the Coma cluster. In all, 2712 rich clusters are listed. The total volume of space they occupy is about 10^{14} times that occupied by our own Galaxy — itself believed to be the entire "universe" less than half a century ago.

The distribution of these rich clusters (Figure 21-18) provides our best estimate of the general distribution of matter in the universe. We find that on the large scale the clusters are spread uniformly throughout space in all directions, out to the greatest distance for which our survey of them is complete. The universe, in other words, seems to be homogeneous on the large scale.

On the small scale, however, the clusters are *not* distributed uniformly. Just as we find a tendency for galaxies to be found in clusters, so the clusters themselves sometimes are grouped into "clumps" or *second-order clusters*. Rigorous statistical tests appear to verify the reality of some clustering tendency of clusters of galaxies. The average size of the second-order clusters is similar to the

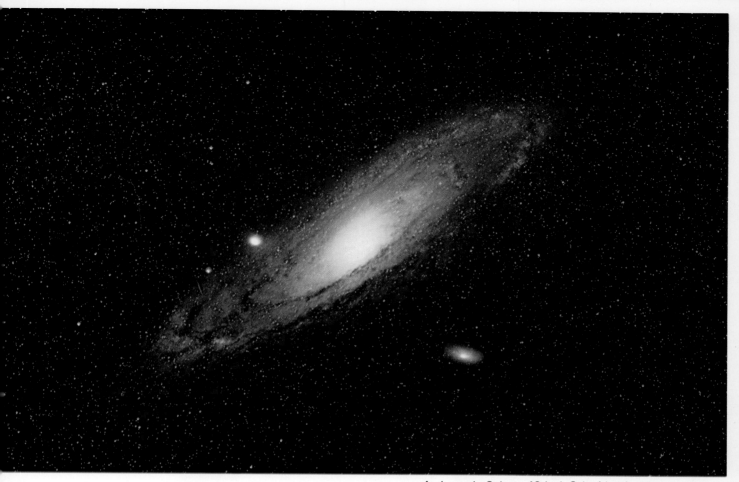

Andromeda Galaxy; 48-inch Schmidt telescope photograph.
(Hale Observatories.)

NGC 7331; 200-inch telescope
photograph. *(Hale
Observatories.)*

NGC 3034, M82; 200-inch telescope photograph.
(Hale Observatories.)

NGC 253; photographed with the 48-inch
Schmidt telescope. *(Hale Observatories.)*

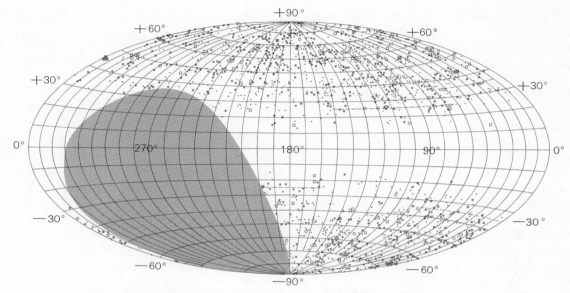

FIG. 21-18   Distribution of rich clusters of galaxies. Each symbol on the plot (sky map) represents a cluster of galaxies; the larger the symbol, the closer the cluster. The most distant are probably more than 2 billion LY away. The empty region along the middle of the plot is the zone of avoidance, as in Figure 21-2. The large, oval, empty region on the left is part of the sky too far south to be surveyed from Palomar Mountain, where the study of rich clusters was carried out by the author.

size of the local supercluster, which suggests that they may be similar supersystems. Evidence is lacking for or against the existence of *third-order clusters.*

21.6 GALAXIES AS RADIO SOURCES

The investigations that followed Jansky's discovery of cosmic radio waves in 1931 (Section 8.8) revealed that continuous radio energy is emitted from the disk and corona of our Galaxy. In addition, since World War II, some thousands of discrete radio sources, each occupying a small region in the sky, have been discovered and catalogued. There is growing evidence that many or most of these discrete sources are extragalactic. There are now about 100 radio sources that have been identified with individual galaxies.

These "radio galaxies" fall into two groups: "normal" and "peculiar" galaxies. The normal radio galaxies are simply ordinary galaxies that, like our Galaxy, emit some of their radiation at radio wavelengths. Radio energy with a continuous spectrum, for example, has been observed to emanate from the disks and coronas of all nearby spirals. Their radiation at radio wavelengths amounts, on the average, to only about 10^{38} ergs/sec, or 10^{28} kilowatts, a tiny fraction of their energy output at visible wavelengths. In addition, radiation at the 21-cm line of neutral hydrogen (Section 19.4g) has been observed in some galaxies. The Doppler shift of this emission line indicates the same radial velocity for a galaxy as does the shift of a line in its visible spectrum.

The peculiar radio galaxies are galaxies which for some reason emit unusually large amounts of radio energy — from 10^{39} to 10^{44} ergs/sec. Some peculiar radio galaxies resemble ordinary galaxies in all their visual aspects.

Other peculiar galaxies have unique properties that may be associated with their unusual radio "luminosities." For example, M87, a giant elliptical galaxy in the Virgo cluster, emits thousands of times as much radio energy as is typical of bright galaxies. This galaxy looks normal on photographs of full exposure. On photographs of short exposure, however, the outer parts of the galaxy are not "burned out," and a luminous feature resembling a jet is seen emanating from the center of M87 (Figure 21-19). The visible light from this jet is

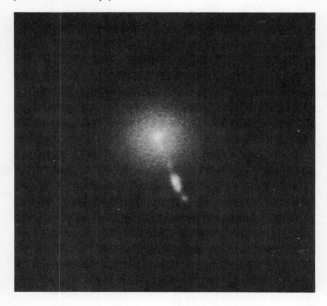

FIG. 21-19   The elliptical galaxy NGC 4486 (M87). A 120-inch telescope photograph of very short exposure to show the nuclear "jet." (Lick Observatory.)

FIG. 21-20   Location of radio sources associated with a typical peculiar radio galaxy.

highly polarized and is therefore believed to be synchrotron radiation, produced by electrons moving with speeds near that of light and spiraling around lines of force in a magnetic field.

Some radio galaxies are very complex. An example is the Cygnus A source, which is one of the strongest *observed* radio sources despite the fact that the cluster with which it is identified has an estimated distance of about 700 million LY. The power emitted by the Cygnus A source at radio frequencies is about 10^{44} ergs/sec — many times the visible light output of the galaxy or galaxies involved. A remarkable feature of these strong complex sources is that the radio energy from such a galaxy often appears to come from *two* regions on opposite sides of, and about 10^{5} LY from, its center (Figure 21-20).

It is very difficult to account for the radio emission from some peculiar radio galaxies. In the case of M87, the radio energy is probably associated with the presence of the jet. Among the speculations on the origin of the radio energy emitted by other galaxies are hypotheses that involve chain reactions between supernovae, the fission of galactic nuclei, and potential energy released when a galaxy condenses from pregalaxian material. It is not clear, however, that any of the mechanisms suggested can produce the power required; if the radio energy is, indeed, synchrotron radiation, the amount of energy tied up in the magnetic fields and moving electrons in some galaxies must be near 10^{60} ergs — an appreciable fraction of the total nuclear energy available in a galaxy! It is difficult to understand what the source of so great an amount of energy could be. Still more remarkable are the *quasi-stellar sources*.

21.7 QUASARS

If the sun, as has been stated before, were typical among stars as a radio emitter, we would not expect to be able to observe a single other star at radio wavelengths; the radio emission from the stars would be too feeble to detect with existing instruments. It was with considerable surprise, therefore, that in 1960 two radio sources were identified with what appeared to be stars. There seemed to be no chance that the identifications were in error, because the precise positions of the radio sources were pinned down by noting the exact instants they were occulted by the moon. By 1963 the number of such "radio stars" had increased to four. They were especially perplexing objects because their optical spectra showed emission lines that at first could not be identified with known chemical elements.

The breakthrough came in 1963 when M. Schmidt, an astronomer at the Hale Observatories, recognized the emission lines in one of the objects to be the Balmer lines of hydrogen (Section 7.5b) shifted far to the red from their normal wavelengths. If the redshift is a Doppler shift, the object must be receding from us at about 15 percent the speed of light! With this hint, the emission lines in the other objects were reexamined to see if they too might be well-known lines with large redshifts. Such proved, indeed, to be the case, but the other objects were found to be receding from us at even greater speeds. Evidently, they could not be neighboring stars; their stellar appearance must be due to the fact that they are very distant. They are called, therefore, *quasi-stellar-radio sources*, or simply *quasi-stellar sources* (abbreviated QSS). In modern usage, the term has become shortened to *quasar*.

The discovery of these peculiar objects prompted a search for others. The procedure now followed is to look for stellar-appearing objects at the positions of all unidentified radio sources. By 1972, over two hundred quasi-stellar sources had been discovered. It is estimated that about a third of all radio sources are QSSs. All have spectra that show enormous redshifts. The relative shifts of wavelength range up to $\Delta\lambda/\lambda = 2.7$, and for the majority $\Delta\lambda/\lambda$ is greater than 1.0. If we apply the exact formula for the Doppler shift (from relativity theory — given in Section 7.3b) we find that $\Delta\lambda/\lambda = 2.2$ corresponds to a velocity of recession of 82 percent of the speed of light.

We shall see in Chapter 22 that remote galaxies are receding from us also, and that there is a correlation between their radial velocities and distances. Most investigators (at the time of writing) regard the redshifts of the quasars as indicative that they are at very great extragalactic distances and that they conform to the same relation between radial velocity and distance that ordinary galaxies do; the vast majority of quasars, however, have much higher speeds than any known galaxy, and must, therefore, be even more distant. In our description of quasars, we assume that this interpretation of their redshifts is correct, but we shall also consider some alternatives.

(a) Characteristics of Quasars

The quasars are all unresolved optically — that is, they appear stellar. A few, however, are associated with tiny wisps or filaments of nebulous-appearing matter. Some are resolved at radio wavelengths, which indicates that the radio energy (at least for some) comes from larger regions than the visble photographic images — perhaps not especially surprising, for radio radiation from galaxies, as we have seen, often originates from outside their optical images. At least some of the radio radiation appears to be synchroton.

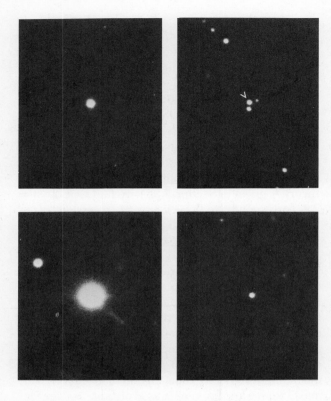

FIG. 21-21   Quasi-stellar radio sources 3C-48, 3C-147, 3C-273, and 3C-196, photographed with the 200-inch telescope. *(Hale Observatories.)*

Although they differ considerably from each other in luminosity, the QSSs are nevertheless extremely luminous at all wavelengths. In radio energy, they are as bright as the brightest peculiar radio galaxies, and in visible light most are far more luminous than the brightest elliptical galaxies — their absolute magnitudes range down to −26 or −27. They are very blue in color — in fact, one of their recognizable characteristics is their excess amount of ultraviolet radiation, as compared to normal stars and galaxies. Most surprising of all is that almost all of them are variable, both in radio emission and visible light. Their variation is irregular, evidently at random, by a few tenths of a magnitude or so, but sometimes flareups of more than a magnitude are observed in an interval of a few weeks. Now quasars are highly luminous, so a change in brightness by a magnitude (a factor of 2.5 in light) means an extremely great amount of energy is released rather suddenly. Moreover, since the fluctuations occur in such short times, the part of a quasar responsible for the light (and radio) variations must be smaller than the distance light travels in a month or so; otherwise light emitted at one time from different parts of the object would reach earth at different times (because of the range of distances light would have to travel to reach us), and we would see the increase spread over a longer time.

We have then the perplexing picture of a quasi-stellar source: an extremely luminous object of small size displaying enormous changes in energy output over intervals of months or less from regions less than a few light-months across; 100 times the luminosity of our entire Galaxy is released from a volume more than 10^{17} times smaller than the Galaxy.

(b) Quasi-Stellar Galaxies

The quasars are stellar-appearing objects of extragalactic distance that are also radio sources. A natural question is whether there exist similar objects that are not radio sources or that radiate too weakly at radio wavelengths to be detected. In an effort to answer this question, A. R. Sandage, an astronomer at the Hale Observatories, initiated the investigation, in 1965, of faint, very blue objects far from the plane of the Milky Way that had been presumed to be stars in the corona of our Galaxy. Spectra revealed that most of such objects are, in fact, stars, but some showed large redshifts, as did the QSSs. Thus, radio-quiet quasi-stellar objects do exist. Sandage has dubbed them *quasi-stellar galaxies* (QSGs).

It now appears that the QSGs probably outnumber the QSSs by many times, although they are harder to find, because radio emission does not call attention to them. Many investigators believe that the radio-emitting QSSs are temporary evolutionary stages of longer-lived QSGs. In any event, another remarkable class of objects has been shown to exist.

(c) The Significance of the Redshifts

Until now we have assumed that the redshifts of quasars and QSGs are Doppler shifts, and that their distances are those of hypothetical galaxies of similar redshifts. As we have seen, however, such enormous distances require that these objects have truly astounding luminosities, and, in particular, the variations involve fantastic changes in luminosity over short periods and in relatively small regions. Some investigators, therefore, look to other interpretations for the redshifts.

The only known cause of redshift, other than Doppler, is an exceedingly strong gravitational field. It has not been absolutely ruled out that the redshifts of the quasars are entirely or partially gravitational, but most experts regard it as very unlikely. It is difficult, even theoretically, to construct a configuration of matter that fits the general characteristics of a QSS and has so large a gravitational redshift. The weight of opinon, therefore, is that the observed redshifts imply that the quasars do, in fact, have extremely large radial velocities.

There is less agreement, on the other hand, that the quasars obey the same velocity-distance relation observed for ordinary galaxies. The luminosities need not be so extreme and the small variable regions are not so difficult to explain, if the QSSs are presumed to be relatively nearby. They still, however, must be extragalactic. Objects inside our Galaxy moving as fast as the QSSs would certainly have observable proper motions; the QSSs do not. We are left with the possibility, therefore, that they are at extragalactic distance, but not so extremely far away as galaxies of comparable radial velocities. But then where did they come from, and what gave them their great speeds? It has been suggested that they are objects that have been ejected from other nearby galaxies, or even from our own Galaxy. But if they are ejected from other galaxies, why are they all receding from us, rather than some approaching us? And a disturbance violent enough to eject them from our own Galaxy would be expected to produce other observed effects, for which no evidence exists. Moreover, to explain the energy required to eject such objects from galaxies at such speeds is even harder than to explain the high luminosities required if they are at very great distances.

(d) Theories of Quasars

It is, perhaps, unfortunate that we feel obliged to say anything about the various theories that have been advanced to explain the quasi-stellar sources, for they are all almost certain to be out of date, if not completely wrong, by the time this is read. We shall, in any case, be brief.

One model is that QSSs are large masses of gas, probably hundreds or thousands of solar masses, in gravitational collapse, the material freely falling together under its mutual gravitation. Such bodies, too massive to ever reach hydrostatic equilibrium, would continue to collapse indefinitely. The energy they radiate comes, of course, from the gravitational potential energy they release as they contract (see Section 20.2a). Such collapsing masses are presumed to be either isolated or perhaps within galaxies; the galaxies themselves (in the latter case) would be too faint, because of their distances, to be observed.

Another theory is that the QSSs are galaxies in the process of formation. Some elliptical galaxies are nearly spherical, and thus have relatively low angular momentum in relation to their masses. Galaxies presumably form initially as condensations from pregalaxian gas drawn together by its mutual gravitation, much as we imagine star formation to start. Stars within the galaxies are thus subcondensations in the main gas mass. According to this theory of quasars, when the collapsing mass of gas achieves a critical density, the stellar subcondensations form within it. Now the stars occupy so little volume compared to the galaxy as a whole that they do not collide with each other but rather fall on through the new galaxy, interact gravitationally with each other, and finally, as a result of such gravitational interactions, reach the equilibrium configuration of the finished elliptical galaxy. At one point in its evolution, just after the star formation begins, the galaxy would be much smaller than when its equilibrium is reached. At the same time, there are many young stars of high luminosity — main-sequence O and B stars. It is mainly these young stars, according to the theory, that give rise to the great luminosity of the QSS, and coming from such a small volume of space. These high-mass stars evolve very rapidly (Chapter 20) and in the course of their evolution produce supernova outbursts. Such supernovae, going off at random, are presumed (in this theory) to account for the light variations of the quasars.

Still another suggestion is that the QSSs are old galaxies in which the central nuclear regions have evolved to very high stellar densities, as a result of stellar gravitational encounters. If the stars are crowded together close enough, they can actually collide with each other. The colliding stars would have very high relative speeds, and their collisions would rip them apart. The high-intensity, fluctuating luminosity of the quasars in this model, then, is the energy emitted as stars collide in the nuclei of distant galaxies — the nuclei being the only parts of the galaxies bright enough to observe.

There are also other theories for quasi-stellar sources, many very esoteric. No model yet proposed, however, satisfactorily accounts for all observed phenomena associated with these mysterious objects, and probably none is correct.

21.8 EXTENT OF THE OBSERVABLE UNIVERSE

As far as we can see in all directions we find galaxies and clusters of galaxies. The more intrinsically luminous a galaxy is, the greater is the distance to which it can be observed. At the farthest depth of observable space, we see only the greatest giants among galaxies — that is, the brightest members of individual rich clusters — or perhaps quasi-stellar sources. If we knew the upper limit to the luminosities of galaxies or quasars, we could calculate the distance of those most remote observed objects. At present, we can only guess that they are between 5 and 15 billion LY away. It is possible that some or many of the unidentified sources of radio waves may be galaxies, or clusters, or quasars at still greater distances — beyond the range of our largest optical telescopes.

The size and extent of the universe, its motions, and its evolution, are the subjects of Chapter 22.

Exercises

1. The brightest-appearing stars in the Andromeda galaxy (M31) are at apparent magnitude 15. If early investigators assumed that these stars were about like the sun (absolute magnitude +5), what distance would they have calculated for M31?

2. What would the answer be to Exercise 1 if those brightest stars were recognized as supergiants, and their absolute magnitude had been assumed to be −10? (Actually, −9 is more nearly correct, but assume −10 to make the arithmetic easier.)

3. Why is the term "island universe" a misnomer?

4. Starting with the determination of the size of the earth, outline all of the steps that one has to go through to obtain the distance to a remote cluster of galaxies.

5. Why do we use the *brightest* galaxies in a cluster as indicators of its distance, rather than average galaxies in it? (Hint: There are two reasons, one having to do with the definition of "average," the other involving the distances to typical clusters.)

6. Compare and contrast the kinds of stellar evolution going on in spiral and elliptical galaxies. Sketch what an H-R diagram would look like that included a representative sample of stars in each kind of galaxy.

7. Suggest some ways by which you might be able to check whether certain galaxies seen near a distant cluster in the sky are actually members of that cluster, or whether they are galaxies in the foreground or background with respect to the cluster.

8. The tenth brightest galaxy in cluster *A* is at apparent magnitude 10, while the tenth brightest galaxy in cluster *B* is at apparent magnitude 15. Which cluster is more distant and by how many times?

9. How can we determine the inclination of M31 (The Andromeda galaxy) to our line of sight?

10. Where might the gas and dust (if any) in an elliptical galaxy come from?

The peculiar elliptical galaxy NGC 5128 in Centaurus;
a strong radio source. Photographed with the 200-inch telescope.
(Hale Observatories.)

CHAPTER 22

Cosmology

In its modern usage by astronomers, the term *cosmology* embraces not only *cosmogony* — the origin and evolution of the universe — but also the content and organization of the universe. It is the problem of the modern cosmologist to piece together those properties of the universe that are revealed by observation into a self-consistent hypothesis that describes its structure and evolution.

In one respect the study of cosmology differs from research in other realms of physical science; the cosmological problem is unique. We can check the *generality of* gravitational theory, for example, by noting that it applies to a very great many different physical phenomena. By its very nature, however, a cosmological theory — a theory of the universe — can be applied only to one system — the universe itself. For this reason, cosmology is one of the most difficult, and most speculative, fields of science.

The observational basis for cosmology comes from two sources: the large-scale distribution of matter in space, which was discussed in Chapter 21, and the large-scale motions in the universe, which will be considered now.

22.1 THE "EXPANDING UNIVERSE"

Within the first decade after the true nature of the galaxies was established, it was found that the universe may be expanding. The evidence comes from the radial velocities of the galaxies.

(a) The Law of the Redshifts

The first radial velocities of galaxies were measured by V. M. Slipher at the Lowell Observatory. In

FIG. 22-1   Hydra cluster of galaxies, showing the galaxy observed for Doppler shift, and the spectrum showing the large redshift (indicated by the length of the arrow) equivalent to a radial velocity of about 60,000 km/sec. 200-inch telescope photographs. *(Hale Observatories.)*

the years between 1912 and 1925 he obtained spectra from which he measured the velocities of more than 40 "nebulae." Slipher did not, of course, know the true nature of the objects he investigated, but he found very high velocities for those that were later shown to be galaxies. A very few of them showed velocities of approach (negative radial velocities) — these are some of the Local Group galaxies — but Slipher found the vast majority to be moving away from us, with speeds up to 1125 mi/sec.

In the mid-1920s there seemed to be some evidence that the velocities of the nebulae might be correlated with their distances, but the distances were too poorly known at that time to verify the suspicion. By 1929, however, Edwin Hubble, at the Mount Wilson Observatory, had determined new estimates for the distances to many of the galaxies whose velocities had been measured, and he found that the velocities of recession of those galaxies were, in fact, *proportional* to their distances.

Meanwhile, M. L. Humason, also at Mount Wilson, had begun to photograph spectra of fainter galaxies and galaxies in clusters with the 60- and 100-inch telescopes. He and Hubble collaborated, and had definitely established, by the early 1930s, that the more distant a galaxy, the greater, in direct proportion, is its speed of recession, as determined by the shift of its spectral lines to the longer (or red) wavelengths. This relation is now known as the *law of redshifts,* or sometimes the *Hubble law.*

The study of the radial velocities of distant galaxies and clusters of galaxies — that is of the redshifts of their spectral lines — has been continued, most recently with the 200-inch telescope at Palomar. As the investigation progressed, more and more remote galaxies of greater and greater speed of recession have been discovered. The largest speed yet measured (at the time of writing) is of a very distant cluster, known as 3C295, that is moving away from us at 41 percent of the speed of light, or about 76,000 mi/sec. The actual distances to the galaxies

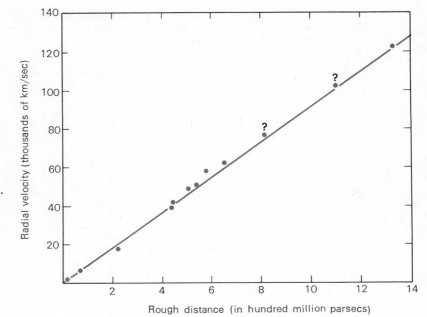

FIG. 22-2   Velocity-distance relation for clusters of galaxies.

whose velocities have been measured cannot be determined accurately, but the *relative* distances to clusters of known radial velocity (or redshift) are fairly well established (Section 21.2c). To the accuracy of the present observations, these clusters have radial velocities that are proportional to their distances. The velocity, *V*, of a cluster, in other words, is represented by the equation

$$V = Hr,$$

where *r* is the distance to the cluster, and *H* is a constant of proportionality, called the *Hubble constant,* that specifies the rate of recession of galaxies or clusters of various distances. The Hubble constant is now believed to lie in the range 30 to 100 km/sec per million parsecs, a recent determination giving 50 km/sec per million parsecs. In other words, a cluster moves away from us at a speed of 50 km/sec for every million parsecs of its distance. This velocity-distance relation for clusters of galaxies in shown in Figure 22-2.

(b) Interpretation of the Law of Redshifts

The fact that distant galaxies all seem to rush away from us may seem to imply that we must, somehow, be at the "center" of the universe, but such is not the case. A simple and familiar analogy is that of a balloon whose surface is covered with dots of ink. As the balloon is inflated, the dots all separate, and a tiny insect on any one dot would see all other dots moving away from it; yet none of the dots is at a "center."

For a three-dimensional analogy, consider the loaf of raisin bread shown in Figure 22-3. Before being placed in the oven, the loaf is 12 in. in diameter (a). The cook, however, mistakenly put far too much yeast in the dough, and within 1 hour the loaf has grown in size to a diameter of 24 in. (b). During the expansion of the bread, the raisins in it all separated from each other. Suppose, now, that we are microbes, sitting on raisin *A* (intentionally *not* at the center of the loaf), and that we measure the speeds with which the other raisins

FIG. 22-3   Separation of raisins in a hypothetical loaf of raisin bread that is 12 inches in diameter when it is placed in the oven (a), and grows to 24 inches in diameter at the end of 1 hour (b).

(a)

(b)

TABLE 22.1

RAISIN	PRESENT DISTANCE (IN.)	SPEED OF RECESSION (IN./HR)
B	4	2
C	6	3
D	12	6
E	16	8

recede from us. At the end of the hour, raisin B, originally 2 in. away, has increased its distance to 4 in.; evidently, its speed of recession has been 2 in./hour. Raisin C increased its distance from 3 to 6 in., and so moved at 3 in./hour, and so on for the other raisins; we obtain the data in Table 22.1. If we plot the data in the table, we will find a linear (straight-line) velocity-distance relation for the raisins analogous to that observed for galaxies. The same result (or plot) would be obtained irrespective of which raisin were chosen for "us." We would, in fact, observe an exactly linear velocity-distance relation for the raisins if we could measure their instantaneous velocities and distances, for *any* uniform expansion or contraction of the bread, whether or not it expands (or contracts) at a constant rate.

If the universe were now changing scale — either uniformly expanding or contracting — we should expect to observe a linear relation, or proportionality, between the speeds and distances of galaxies and clusters, just as in the raisin bread analogy. In an expanding universe, galaxies would move away from us at speeds proportional to their distances; in a contracting universe, they would *approach* us at speeds proportional to their distances. The law of redshifts, therefore, is generally interpreted as observational evidence that the universe is now expanding; in this expansion the galaxies and clusters are being carried farther and farther from each other.

Galaxies and clusters do, of course, have individual motions of their own, superimposed upon the general expansion. Galaxies in pairs, for ex-

ample, revolve about each other, and those in clusters move about within the clusters. In fact, a few galaxies in nearby groups and clusters move fast enough within those systems so that they are actually *approaching* us even though the clusters of which they are a part are moving, as units, away.

(c) Are the Redshifts Really Doppler Shifts?

Not all scientists accept the interpretation of the redshift law as an indication of an expanding universe. Some argue that the observed redshifts in the spectra of distant galaxies may not be Doppler shifts at all but may be caused by some unknown effect on light as it travels over large distances. It has been suggested, for example, that photons may lose energy, or "tire," as they traverse space; since the energy of a photon is inversely proportional to its wavelength, the "tired-light" hypothesis could explain the redshift of the lines in spectra of remote objects. However, there is not a shred of evidence, either observational or theoretical, to support the hypothesis.

The Doppler shift, on the other hand, is an effect of speed that is well confirmed in laboratory experiments, in observations in the solar system, and in observations of double stars. There we can see the light sources, or planets, or stars, moving. The Doppler effect is the only *known* cause that can account for the redshifts observed in the spectra of galaxies.

Many people, apparently, are troubled by the concept of the extremely high speeds with which galaxies are receding. On the other hand, it requires very special assumptions about the nature of natural laws to understand how the galaxies could be fixed in space in a *static* universe. Their mutual gravitation should pull them together, causing the universe to *contract*. It would be easy to understand their mutual separation, however, if an expansion were once started — say, by a primeval "explosion."

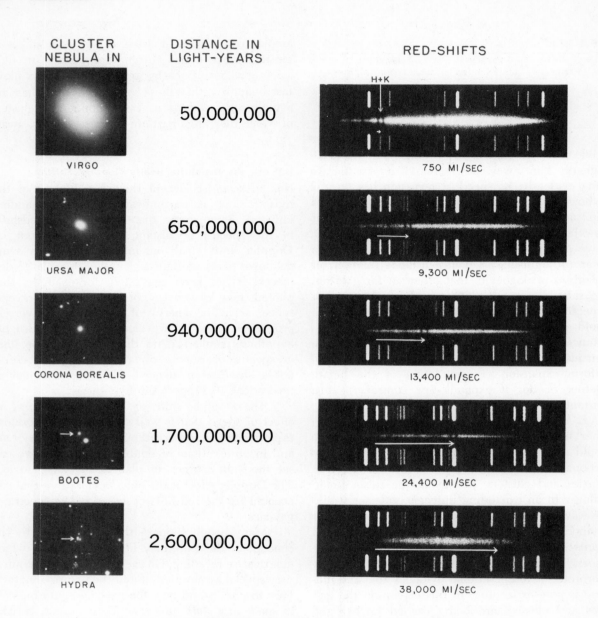

CLUSTER NEBULA IN	DISTANCE IN LIGHT-YEARS	RED-SHIFTS
VIRGO	50,000,000	750 MI/SEC
URSA MAJOR	650,000,000	9,300 MI/SEC
CORONA BOREALIS	940,000,000	13,400 MI/SEC
BOOTES	1,700,000,000	24,400 MI/SEC
HYDRA	2,600,000,000	38,000 MI/SEC

FIG. 22-4 *(Left)* Photographs of individual galaxies in successively more distant clusters; *(right)* the spectra of those galaxies, showing the Doppler shift of two strong absorption lines due to ionized calcium. The distances are estimated from the faintness of galaxies in each cluster, and are provisional. *(Hale Observatories.)*

Einstein's general theory of relativity provides another basis for understanding an expanding universe. The solution of Einstein's equations, as they apply to a field of gravitating bodies, allows for the possibility of a *repulsion* between distant objects in the universe. The magnitude of this repulsion is represented in the equations by a term called the *cosmological constant*. Unfortunately, the theory of relativity does not specify what value the cosmological constant must have, but leaves it arbitrary. Einstein originally chose the constant so that the repulsion between distant masses would exactly balance their gravitational attraction, to maintain a static universe.

According to the theory, however, there is no a priori reason why the universe must be static. Larger or smaller values of the cosmological constant (including zero) are possible; an expanding, a static, or a contracting universe are all compatible with general relativity.

Most investigators, therefore, tentatively accept the expanding universe interpretation of the redshift law. They are ready, however, to revise any conclusions that are based upon this interpretation, when and if new evidence is brought to light that would support alternate explanations of the redshifts.

22.2 COSMOLOGICAL MODELS

Having examined the observational basis for cosmology, we may investigate some of the hypotheses that have been suggested to describe its structure and evolution. Such a hypothesis is called a *cosmological model*.

All cosmological models are based on certain assumptions, without which no hypothesis could be formed. For example, an assumption common to most models, and to all discussed here, is that the redshifts of galaxies are really Doppler shifts and that the universe is really expanding. Another even more basic postulate is that known as the cosmological principle.

(a) The Cosmological Principle

It would be impossible even to begin to formulate a theory of the universe if we did not assume that our observations give us information that applies to the whole universe, not just to our own part of it. In other words, we must assume that the part of the universe that we actually observe is representative of the entire cosmos, and that we are not located in some very unusual place, fundamentally different from the rest of the universe. Stated more generally, we must assume that whereas there will be local variations in the exact details of galaxies and clusters, all observers, everywhere in space, would view the same large-scale picture of the universe. This assumption is known as the *cosmological principle*.

(b) Evolutionary Cosmologies

As the universe expands, it thins out. If no new matter is being created — that is, if the total mass of the universe is constant — and if no previous contraction preceded the present expansion, it follows that all the matter of the universe must have once been close together. This suggests that some original "explosion" may have started galaxies moving away from each other. The fastest-moving galaxies would have gone the farthest since the explosion, and the slow-moving ones less far. From any one place in the exploding material, all other places would appear to move away at speeds proportional to their distances — that is, the redshift law would apply. The idea that such an explosion took place is called the "big-bang" theory.

The big-bang hypothesis implies that the universe has a finite age — at least since the explosion. Since no galaxy could have a greater age, all galaxies would be aging, and evolving, together. Models based on some version of the big-bang theory, therefore, are usually called *evolutionary models* or *evolutionary cosmologies*.

We can estimate an upper limit to the age of the universe since the explosion from the present rate of recession of the galaxies. If it is assumed that the speeds of galaxies have not changed since then, a galaxy at a distance r, moving away with a radial velocity V, must have been receding from us for a time r/V. But r/V is simply the reciprocal of the Hubble constant; the maximum "age" of the universe, depending upon the value of the Hubble constant, would lie in the range 10 to 25 billion years. Actually, if there is no force of repulsion present, the mutual gravitation of the galaxies would have slowed them down; in the past, in other words, they would have to have been moving faster than they now are and would have required a somewhat shorter time to reach their present distances.

The Belgian cosmologist Abbe Lemaitre proposed a version of the big-bang theory in the 1920s which he called the theory of the "primeval atom," referring to an original single chunk of matter that exploded. A group of American physicists, led by the late George Gamow, have used the idea in an attempt to explain how the chemical elements were formed in the relative abundances we observe today. Gamow and his collaborators considered what nuclear reactions could go on in all the matter of the universe during the big bang. Their idea was that electrons, protons, and neutrons would combine to form hydrogen, helium, and heavier atoms. As the material expanded and the hot gases cooled, condensations formed that later became galaxies, and smaller condensations within these became stars.

There are difficulties in the calculations of element formation, and today there are alternative hypotheses that some of the heavy elements are formed within stars (Chapter 20). Otherwise, however, the general scheme of the big-bang theory provides a basis for understanding the universe as we see it today. The start of the explosion some billions of years ago takes on special significance; if all matter was then in the form of subatomic particles at a high temperature, we could never expect to discover by observation today what occurred before then. That is, the beginning of the explosion was a "beginning" in a broader sense; it can be considered that the universe was "created" then.

(c) General Relativity in Cosmology

Concepts like "straight" and "curved" lines can be defined only in terms of the paths followed by light. According to Einstein's general theory of relativity, the path of light, that is, the geometry of space, is affected by the presence of matter in it. The curvature is locally greater, for example, near a massive body than it is far from all objects. The general curvature of space depends on the average density of matter in the universe.

It is unlikely that any of us can visualize curved space, but an analogy may make it less mysterious. Consider the curved surface of the earth, which has a finite *area,* but is unbounded, and upon which the geometry is not Euclidean. In a very large triangle on the earth's surface, for example, the angles do not add up to 180°. There is no "center" to the surface area of the earth, just as there is no center to the finite volume of curved three-dimensional space. Moreover, there is a "radius" of curvature of space that is connected with its volume much in the way that the radius of the earth is connected with its surface area. An important difference between the curved surface of the earth and the curved volume of space, however, is that the space curvature may be changing — in fact, *must* be changing — if the density of matter in space is changing.

Without repulsion (that is, with no cosmological constant) the motions of galaxies after an initial explosion would be slowed down by gravitation in the same way that a rocket fired away from the earth is slowed down by the earth's gravity. If the rocket has an initial speed of about

11 km/sec (the "escape velocity"), it will move away from the earth forever. The galaxies, if they were started fast enough in the big bang, could likewise move away from each other forever; if they started slower than some critical escape velocity, however, they would eventually fall back together again. In the latter case, possibly a future explosion would send the galaxies out again; it is even conceivable that alternate expansions and contractions might lead to an "oscillating" universe.

The speeds with which galaxies must recede to escape from each other depends on the mean density of matter in the universe. The present mean density is estimated to lie in the range 10^{-29} to 10^{-31} gm/cm^3 but is not known accurately enough to predict the future motions of the galaxies. Thus the hypotheses that the universe will expand forever, and that it will eventually stop expanding and begin to contract, are both possible evolutionary cosmological models (for which the cosmological constant is assumed to be zero). These models require ages for the universe, however, that are probably less than 20 billion years, and may be less than 10 billion years. These ages are comparable to, or even less than, those we estimate for the oldest star clusters in our galaxy.

We emphasize that the models considered so far are among the simplest ones. We have assumed: (1) the cosmological principle; (2) that the mass and energy of the universe are jointly conserved; (3) that general relativity, the most accurate description of nature we know today, is, in fact, absolutely correct; (4) that the cosmological constant is zero; and (5) that there are no complications, such as rotation of the universe as a whole. It is appropriate that we try the simplest models first, and turn to more complex ones only as observations show that the simple ones must be rejected.

We now describe an even simpler model — the *steady-state theory*. Recent observations, as we shall see, have shown that the steady state is almost certainly wrong, but it is worth discussion anyway, because it has received a great deal of interest since the late 1940s, because it illustrates how an alternative cosmological model can be constructed, and because we shall be able to see how observational tests can be applied to force the rejection of a model.

(d) The Steady-State Theory

The problems of the finite age implied by the evolving cosmologies, and of beginnings and endings, led several astronomers (Bondi and Gold, and later Hoyle and others) to speculate on the possibilities of another cosmological hypothesis. Theirs was the simplest of all possible cosmological models, for it rested upon only a single assumption. This assumption, an extension of the cosmological principle, is called the *perfect cosmological principle;* it states that the universe is not only the same *everywhere* (except for local small-scale irregularities), but at *all times.* In other words, if we could return to life billions of years hence, we would find the universe, on the whole, as it is now. Since it was assumed that nothing changes in time, the hypothesis was known as the *steady-state cosmology.*

By starting with this premise, Bondi and Gold ruled out creation of the universe at a definite time in the past. Moreover, they explained how our view of the universe can remain the same when galaxies are moving apart — how the average density of matter can stay constant when galaxies are leaving our region of space at high speeds on all sides. They explained this by the bold assumption that matter is being *created continuously* in empty space, at a rate just sufficient to replace what is leaving due to the observed recession of galaxies. The theory, therefore, was sometimes called the "continuous-creation theory." Moreover, to satisfy the perfect cosmological principle, there must always be the same proportion of young and old galaxies in any given volume of space. This con-

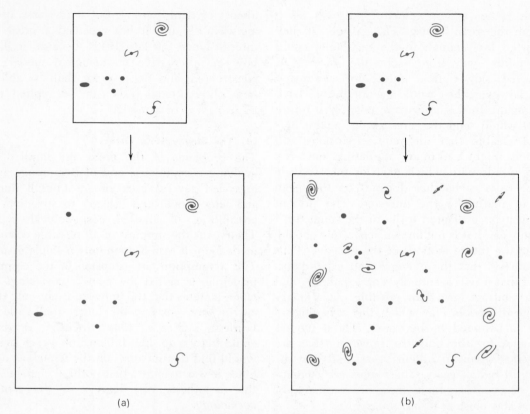

FIG. 22-5 In the evolving cosmologies (a), the density of matter in space thins out as the universe expands; in the steady-state cosmology (b), new matter is created spontaneously, and the density of the universe remains constant.

dition requires that the rate of creation of new matter be always proportional to the present amount of matter, and that the rate of increase of distance between any two galaxies (or clusters of galaxies) be proportional to their current separation. The new matter that comes into being (in this theory) was usually presumed to be in the form of gases composed of individual atoms (possibly pure hydrogen), which later condense into galaxies and stars.

In the steady-state cosmology, *individual* galaxies were assumed to form, age, grow old, and,

as their stars evolve to the black dwarf state, finally die. Only when a large volume of space, containing many galaxies, is considered are things unchanging — there would always be young and old galaxies, side by side, and always in the same proportions. The formation of new matter would need to occur only at a very slow rate, and we would not have necessarily expected to find newly formed galaxies in our immediate vicinity in space. The universe itself, according to the theory, is infinitely large and infinitely old — with no beginning and no end.

To the objection that matter cannot be created spontaneously from nothing, the proponents of the theory answered that it is no more difficult to imagine matter being created gradually and steadily than all at once in or prior to some "big bang" in the past.

The steady-state cosmology was particularly appealing philosophically because of its symmetry and simplicity. It is contrasted with the evolutionary cosmologies in Figure 22-5.

We shall see in Section 22.3 how we can subject these models to observational tests. At the time of writing (mid-1972) the observations appear to be inconsistent with the steady state. However, it must be emphasized that the two types of models described here are not *the* two opposing views of the universe, but are merely two of the simplest of an infinite number of possibilities. The rejection of one *does not* imply the correctness of the other.

22.3 TESTS FOR COSMOLOGICAL MODELS

It is clear that evidence favoring one hypothesis over another will depend on *differences* between them. Since all the cosmological models described here refer to a universe expanding at the rate indicated by the Hubble-Humason (redshift) law, the differences between them must reflect the different ways in which the theories predict that the universe will change in time. In other words, for testing cosmological theories we need observations of the universe as it was long ago to compare with present conditions. Fortunately, we can and do look into the past. Light travels at a finite speed, so we see remote objects as they were when light left them — up to billions of years ago for galaxies at distances of billions of light-years.

One of the possibly observable differences between the steady-state and evolutionary cosmologies is in the predicted mixture of galaxies of different ages. According to the steady-state theory, individual galaxies age and die, but new ones are always forming, so that old and young galaxies have always existed side by side. No matter how far out in space we look, therefore, we should not find galaxies, on the average, any different from those nearby us. Evolutionary cosmology, on the other hand, assumes a distinct "beginning," so that all galaxies have about the same age. Thus, if we look out to a great distance, and hence far back into time, we should find galaxies that are systematically younger than those near us — that is, remote galaxies should be at an earlier stage of evolution.

Another test for cosmology is the way the density of matter in space changes in time. Evolutionary cosmologies require that matter gradually thins out in space; remote galaxies, therefore, should appear relatively closer together (as they were in the past) than nearby ones. The steady-state theory, however, predicts that the average distance between galaxies never changes and so should be the same for galaxies at all distances.

A third, less obvious test involves the speeds of receding galaxies at great distances. The evolutionary models predict that for no repulsion (cosmological constant zero), the expansion should slow down, because of the gravitational forces between galaxies. In the steady-state theory, on the other hand, the rate of the separation of any two galaxies is ever increasing. Because of these changes in the rate of expansion of the universe, the radial velocities of remote galaxies should deviate from a relation exactly proportional to their observed distances; that is, a graph of radial velocity versus distance should not necessarily be a straight line for very distant galaxies. The exact form of the redshift-distance relation is different as predicted by each cosmological theory.

The differences between the observable quantities predicted by various models are small until very large distances are reached — that is, until we look back through an appreciable interval of time.

Unfortunately, precise distances of remote objects, as well as their luminosities and other characteristics, are very difficult to determine, and critically accurate observations are required. Therefore, the tests described above are not easy to apply. At present we may be on the verge of being able to observe some quantities with enough precision to choose between some of the evolutionary models (or perhaps eliminate them all), but we have not yet succeeded.

Whereas we cannot yet discriminate between the various evolutionary models, there are at least two significant observations that do, if our interpretations of them are correct, appear to rule out the steady state. We shall describe them briefly next.

(a) Observational Evidence Against the Steady State

One clue that the steady-state theory is probably wrong comes from the quasi-stellar sources (Section 21.7). If we assume that they obey the same velocity-distance relation that galaxies do (in the professional parlance, that their redshifts are "cosmological") the distribution of quasars in space would have to be uniform to be consistent with the steady state. M. Schmidt, an astronomer at the Hale Observatories, has recently studied a statistically homogeneous sample of about 30 quasi-stellar sources. He found that the overwhelming majority of them—far more than would be expected in a uniform distribution—are at very great distances. In other words, the quasars lie at such distances (or at least, have such redshifts)

FIG. 22-6   Central part of the great cluster of galaxies in Coma Berenices, which has an estimated distance of about 300 million light years. Photographed with the 200-inch telescope. *(Hale Observatories.)*

that nearly all of them represent phenomena occurring long ago in the history of the universe. It is as if the quasars are things that existed early in the evolution of the universe — perhaps things happening to or in galaxies far in the past — and that occur only rarely today. If this analysis is correct, it clearly violates the perfect cosmological principle, and hence the steady state cosmology.

The other evidence against the steady state is perhaps one of the most exciting discoveries in recent decades. In 1965, A. Penzias and R. Wilson, of the Bell Telephone Laboratories, attempted to calibrate the random noise in the Laboratories' very sensitive radio receiver. They found, however, that the antenna, operating at 7.35-cm wavelength, picked up a weak radio signal that they could not account for unless they assumed that it was extraterrestrial in origin. It seemed to be coming from all directions, as if the entire celestial sphere were emitting the radiation. The radio radiation was of about the intensity that would be received if the whole universe were emitting radiation like a very cold black body (perfect radiator) with a temperature of about 3°K (3 centigrade degrees above absolute zero).

Now both Gamow and the Princeton physicist R. H. Dicke had predicted that if there were a "big bang" to start the expansion of the universe, it should now be possible to observe some of the radiation originating in that explosion. Just after the "bang," according to the hypothesis, the matter in the universe was extremely dense, hot, and opaque. Very intense radiation flowed from atom to atom in this *primeval fireball*. As the universe rapidly expanded, however, the matter thinned out and soon became transparent. From that point, most of the radiation was no longer absorbed and flowed freely throughout the universe; only those photons that chanced to strike material bodies (such as stars, gas clouds, and planets) would be absorbed. Some of that radiation should strike the earth and we should be able to detect it.

Calculations indicate that the primeval fireball would become transparent when its density dropped to about 1000 atoms/cm³. At that time, its temperature would have been about 3000°K. Thus, if we had observed its radiation then it would have looked like that from a black body of about 3000°K. However, to see this radiation now we must look far back into time, and hence off into space where the redshift is extremely large.

To see how large the redshift would be, note that for the density of the universe to drop from 1000 atoms/cm³ to its present average value of about one atom per million cubic centimeters, it would have to have increased in volume by a factor of 10⁹. The average separation of atoms now is thus 1000 times as great as when the fireball became transparent, and waves of electromagnetic radiation would have been stretched out in wavelength by a thousand times. In other words, radiation which at the time of the fireball would have been visible light would now appear as radio waves, characteristic of that emitted by a cold black body only a few degrees above absolute zero.

When Penzias and Wilson announced their discovery of radiation at 7.35 cm, it occurred to Dicke and his colleagues that we might actually be looking at the "glowing embers" of the "big bang." Hence they designed delicate equipment to see if radiation corresponding to a 3°K black body could be detected at other wavelengths as well. The observations that followed confirmed the Bell Laboratories' findings. By mid-1968, isotropic radiation (that is, radiation equally intense from all directions) had been observed at 0.86, 1.58, 3.2, 7.35, and 20.7 cm. The radiation at all these wavelengths fits very well that from a black body with a temperature of 2.7°K. An independent check comes from observations of interstellar absorption lines (Sections 7.5 and 19.4e) of cyanogen (CN) in the spectra of the stars Zeta Ophiuchi and Zeta Persei. The observed lines showed that a significant fraction of the CN molecules are in a level of excitation that would be expected if they were bathed in radio-wavelength radiation of 2.6 cm. The implied strength of this radiation is also compatible with a general background radiation like that from a black body of temperature between 2 and 3°K.

Most cosmologists are of the opinion that the background radio radiation is good evidence for a primeval fireball and hence for an evolving cosmology. The observations are not yet definitive enough, however, for us to discriminate between the open or closed models of the universe — that is, whether the expansion will continue forever, or eventually give way to contraction (or oscillate).

Note that in all directions in space, as we look out to great distances, we are also looking back in time to the same (hypothetical) "big bang." Thus the observed radiation comes equally from all directions and gives no direction to a "center" of the universe; the universe, its "center," and its origin are all around us.

(b) Conclusion

Present-day evidence suggests that the universe is not a steady state but is evolving from a "big bang" billions of years ago. We should not, however, take our ideas too seriously. Our think-

ing in the realm of cosmology may be as rudimentary as that of the Greek philosophers 2500 years ago, who believed that all heavenly motions must occur in perfect circles. It will be very surprising if, after the next century, scientists finally settle on one or another of the specific theories we have described here.

The following words, written by Edwin Hubble in 1936, are still appropriate†:

Thus the explorations of space end on a note of uncertainty. And necessarily so. We are, by definition, in the very center of the observable region. We know our immediate neighborhood rather intimately. With increasing distance, our knowledge fades, and fades rapidly. Eventually, we reach the dim boundary — the utmost limits of our telescopes. There, we measure shadows, and we search among ghostly errors of measurement for landmarks that are scarcely more substantial.

The search will continue. Not until the empirical resources are exhausted, need we pass on to the dreamy realms of speculation.

---

†*The Realm of the Nebulae,* Yale University Press, 1936, pp. 201-202. Quoted by permission of the publisher.

Exercises

1. Consider the analogy (to the expanding universe) of the expanding rubber balloon covered with ink dots. Sketch such a balloon. Select an ink dot, and estimate its distance, measuring along the surface of the rubber, to various other dots. Now imagine the balloon to expand considerably, and again consider the distances to those same other dots. Show that a velocity-distance relation results, as is done in the text for an expanding loaf of raisin bread. Discuss the limitation of only two dimensions (along the rubber surface) in the analogy.

2. Suppose (in Exercise 1) you had not measured the distances along the rubber surface, but the actual distances between the dots in three-dimensional space. Now suppose light were constrained to move only along the rubber surface. Discuss the analogy to "curved space."

3. Is a systematic evolution of galaxies consistent with the cosmological principle? With the perfect cosmological principle? Can a stationary universe (one which is not expanding) be compatible with either or both?

4. Suppose everything in the universe were expanding together, right on down to atoms. Would we then be aware of the expansion? Why or why not?

5. Explain why the redshifts in the spectra of galaxies cannot be produced by the absorption and reddening of the light from the galaxies by dust in intergalactic space.

6. Suppose we had the ability and time to measure the radial velocities of all galaxies visible with the 200-inch telescope. If we were then to plot those measured redshifts against the apparent brightnesses of those galaxies (that is, their apparent magnitudes) would all of the points lie along a smooth line? Why or why not?

7. Suppose we could obtain accurate distances to clusters of galaxies. If we now picked out just one galaxy in each of a number of clusters, and measured its radial velocity, and then plotted those measured radial velocities of the individual galaxies in different clusters against the cluster distances, would the points lie exactly on a straight line, or would there be some scatter of the points? Why?

8. Do you suppose that the Hubble constant can be evaluated very well (or at all) from the ratio of the radial velocity of the Andromeda galaxy to its distance? Why or why not?

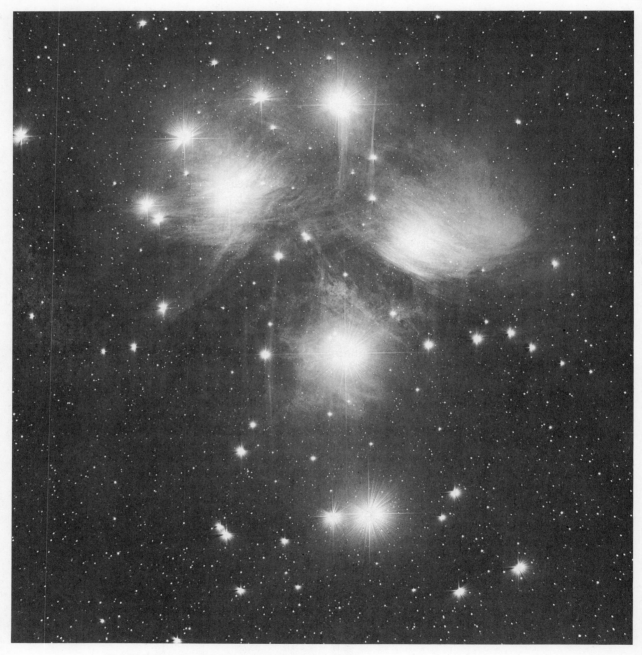

The Pleiades. North is to the left. *(Lick Observatory.)*

CHAPTER 23

Epilogue

The ancients, as we have seen, thought the earth to be in the center of the universe, occupying a very special place in space. At least partly they were right, for the earth is not located typically in the cosmos. Let us imagine that we could view the universe from a really typical place in it — say from within an imaginary space ship with which we could travel at will with the speed of light. (Let us ignore here the limitations placed on such travel by practical considerations, and also the relativistic effects such a speed would have on distance and time.)

Around us, from this typical place, we would find virtually complete blackness — empty space. Perhaps a few faint, indistinct blurs of light might be seen with the naked eye. With a telescope we could make out many more tiny specks of light — all apparently very remote — most, or perhaps all, in small groups or clusters. We pick out one such group for study, and to see it better, race our space ship toward it at the speed of light.

After, say, 2.1 billion years, we are close enough to see that group of indistinct specks of light more easily. They are not points, we find, but they present finite angular sizes to our view, and they are separated by distances, on the average, equal to at least 10 times their own sizes. We plunge our ship directly toward the cluster.

In another 6 or 7 million years we are within it. Now we are surrounded by large centers of light — some spheroidal in shape, others flat and resembling great pinwheels in rapid rotation. These centers of light are not brilliant — in fact, they are rather low in surface brightness, and are so far apart that we still feel ourselves in incredibly empty space. Yet we see them clearly — at least with the space ship's telescope. We select one of these patches of light to investigate more closely, and race our ship toward it.

Two million years later we are well inside it, but *still* we find ourselves in empty space. Now, however, there are tiny points of light — thousands of them visible to our eyes, and millions to our telescope — in all directions. In some directions they merge, in projection, into nearly solid faint patches of light, and an irregular pattern of these luminous patches seems to form a complete circle surrounding us. Some of the points of light are in small clusters, and others appear to be emersed in faint nebulous glowing clouds. At random we choose one of those insignificant points of light. Still at the speed of light, we fly toward it.

After 123 years we are near enough to see that that point is not a point at all, but an intensely hot sphere, pouring blinding amounts of light into space. Our instruments reveal that it is nearly a million miles in diameter and that it has a temperature at its surface of many thousands of degrees. If we approach it any closer, our ship, and we, will surely vaporize. Thus we veer to one side to escape its heat.

Only 8 minutes later we are comfortably far away from that sphere of heat, although it is still blinding to look at. Thus we look around us in other directions, and suddenly find that we are approaching a most remarkable object — a tiny sphere, merely 8000 miles across, mostly blue and green in color. It spins rapidly, turning different portions of its surface into illumination by the brilliant globe that almost destroyed us. But the heat and light falling on this small world is quite gentle, and makes it a place of warmth and beauty — an abode of life — truly a most atypical place in the universe.

Bibliography

General Texts

Struve, O., Lynds, B., and Pillans, H., *Elementary Astronomy*. New York: Oxford University Press, 1959.

Russell, H. N., Dugan, R. S., and Stewart, J. Q., *Astronomy* (2 volumes). Boston: Ginn & Company, 1938, 1945.

Motz, L., and Duveen, A., *Essentials of Astronomy*. Belmont, Calif.: Wadsworth Publishing Co., Inc., 1966.

Wyatt, S. P., *Principles of Astronomy,* (2nd Ed.) Boston: Allyn and Bacon, Inc., 1971.

Baker, R. H., and Fredrick, L. W., *Astronomy*. Princeton, N. J.: D. Van Nostrand Company, Inc., 1971.

More Elementary Texts

Inglis, S. J., *Planets, Stars, and Galaxies* (2nd ed.). New York: John Wiley & Sons, Inc., 1967.

Brandt, J. C., and Maran, S. P., *New Horizons in Astronomy*. San Francisco: W. H. Freeman, 1972.

Hynek, J. A., and Apfel, N. H., *Astronomy One*. Menlo Park, Calif.: W. A. Benjamin, Inc., 1972.

Baker, R. H., and Fredrick, L. W., *An Introduction to Astronomy*. Princeton, N. J.: D. Van Nostrand Company, Inc., 1968.

Alter, D., Cleminshaw, C. H., and Phillips, J., *Pictorial Astronomy*. New York: Thomas Y. Crowell Company, 1963.

Histories of Astronomy

Pannekoek, A., *A History of Astronomy*. New York: Interscience Publishers, 1961.

Berry, A., *A Short History of Astronomy*. New York: Dover Publications, Inc., 1960.

Hoyle, F., *Astronomy*. New York: Doubleday & Company, Inc., 1962.

Vaucouleurs, G. de, *Discovery of the Universe*. New York: The Macmillan Company, 1957.

Shapley, H., and Howarth, H. E., *Source Book in Astronomy*. New York: McGraw-Hill, Inc., 1929.

Struve, O., and Zebergs, V., *Astronomy of the Twentieth Century*. New York: The Macmillan Company, 1962.

King, H. C., *Exploration of the Universe*. New York: New American Library, Inc., 1964.

Koestler, A., *The Watershed* (A Biography of Johannes Kepler). New York: Doubleday & Company, Inc., 1960.

Koestler, A., *The Sleepwalkers*. New York: The Macmillan Company, 1959.

Celestial Mechanics

Ryabov, Y., *An Elementary Survey of Celestial Mechanics*. New York: Dover Publications, Inc., 1961.

Ahrendt, M. H., *The Mathematics of Space Exploration*. New York: Holt, Rinehart and Winston, 1965.

Space Science

Newell, H. E., *Express to the Stars*. New York: McGraw-Hill, Inc., 1961.

Lundquist, C. A., *Space Science*. New York: McGraw-Hill, Inc., 1966.

Telescopes and Optics

Miczaika, G., and Sinton, W., *Tools of the Astronomer*. Cambridge, Mass.: Harvard University Press, 1961.

Page, T., and Page, L. W., *Telescopes; How to Make Them and Use Them*. New York: The Macmillan Company, 1966.

Steinberg, J. L., and Lequeux, J., *Radio Astronomy*. New York: McGraw-Hill, Inc., 1963.

Earth and Solar System

Whipple, F., *Earth, Moon, and Planets*. Cambridge, Mass. Harvard University Press, 1963.

Watson, F., *Between the Planets*. Cambridge, Mass.: Harvard University Press, 1956.

Menzel, D., *Our Sun*. Cambridge, Mass.: Harvard University Press, 1959.

Baldwin, R. B., *A Fundamental Survey of the*

Moon. New York: McGraw-Hill, Inc., 1965.

Hawkins, G. S., *Meteors, Comets, and Meteorites.* New York: McGraw-Hill, Inc., 1964.

Stellar Astronomy

Brandt, J. C., *The Sun and Stars.* New York: McGraw-Hill, Inc., 1966.

Aller, L. H., *Atoms, Stars, and Nebulae* (Rev. Ed.). Cambridge, Mass.: Harvard University Press, 1971.

Bok, B. J., and Bok, P. F., *The Milky Way.* Cambridge, Mass.: Harvard University Press, 1957.

Campbell, L., and Jacchia, L., *The Story of Variable Stars.* New York: McGraw-Hill, Inc., 1941.

Galaxies and Cosmology

Shapley, H., *Galaxies.* Cambridge, Mass.: Harvard University Press, 1961.

Hubble, E., *The Realm of the Nebulae.* New Haven, Conn.: Yale University Press, 1936.

Shapley, H., *The Inner Metagalaxy.* New Haven, Conn.: Yale University Press, 1957.

Sandage, A. R., *The Hubble Atlas of Galaxies.* Washington, D.C.: Carnegie Institution, 1961.

Baade, W., *Evolution of Stars and Galaxies.* Cambridge, Mass.: Harvard University Press, 1963.

Sciama, D. W., *The Unity of the Universe.* New York: Doubleday & Company, Inc., 1959.

Hodge, P. W., *Galaxies and Cosmology.* New York: McGraw-Hill, Inc., 1966.

Journals and Periodicals

Sky and Telescope, published monthly by the Sky Publishing Corporation, Harvard College Observatory, Cambridge, Massachusetts.

The Griffith Observer, published monthly by the Griffith Observatory, Los Angeles, California.

The Review of Popular Astronomy, published bimonthly by the Sky Map Publications, St. Louis, Missouri.

Scientific American, published monthly by Scientific American, New York, New York.

Mercury, journal of the Astronomical Society of the Pacific, published bimonthly by the Society, care of the California Academy of Sciences, Golden Gate Park, San Francisco, California.

APPENDIX 2

Glossary

aberration (of starlight). Apparent displacement in the direction of a star due to the earth's orbital motion.

absolute magnitude. Apparent magnitude a star would have at a distance of 10 pc.

absolute zero. A temperature of $-273°C$ (or $0°K$) where all molecular motion stops.

absorption spectrum. Dark lines superimposed on a continuous spectrum.

accelerate. To change velocity; either to speed up, slow down, or change direction.

acceleration of gravity. Numerical value of the acceleration produced by the gravitational attraction on an object at the surface of a planet or star.

active sun. The sun during times of unusual solar activity — spots, flares, and associated phenomena.

age-zero main sequence. Main sequence for a system of stars that have completed their contraction from interstellar matter, are now deriving all their energy from nuclear reactions, but whose chemical composition has not yet been altered by nuclear reactions.

airglow. Fluorescence in the atmosphere.

albedo. The fraction of incident sunlight that a planet or minor planet reflects.

almanac. A book or table listing astronomical events.

alpha particle. The nucleus of a helium atom, consisting of two protons and two neutrons.

altitude. Angular distance above or below the horizon, measured along a vertical circle, to a celestial object.

amplitude. The range in variability, as in the light from a variable star.

angstrom (Å). A unit of length equal to 10^{-8} cm.

angular diameter. Angle subtended by the diameter of an object.

angular momentum. A measure of the momentum associated with motion about an axis or fixed point.

annular eclipse. An eclipse of the sun in which the moon is too distant to appear to cover the sun completely, so that a ring of sunlight shows around the moon.

anomalistic month. The period of revolution of the moon about the earth with respect to its line of apsides, or to the perigee point.

anomalistic year. The period of revolution of the earth about the sun with respect to its line of apsides, or to the perihelion point.

antarctic circle. Parallel of latitude 66½° S; at this latitude the noon altitude of the sun is 0° on the date of the summer solstice.

aphelion. Point in its orbit where a planet is farthest from the sun.

apogee. Point in its orbit where an earth satellite is farthest from the earth.

apparent magnitude. A measure of the observed light flux received from a star or other object at the earth.

apparent relative orbit. The projection onto a plane perpendicular to the line of sight of the relative orbit of the fainter of the two components of a visual binary star about the brighter.

apparent solar day. The interval between two successive transits of the sun's center across the meridian.

apparent solar time. The hour angle of the sun's center *plus* 12 hours.

arctic circle. Parallel of latitude 66½° N; at this latitude the noon altitude of the sun is 0° on the date of the winter solstice.

argument of perifocus (or of perihelion or perigee). Angle at the focus of the orbit of a body (or at the center of the sun or earth), measured in the orbital plane and in the direction of motion of the body, from the ascending node to the perifocus (or perihelion or perigee) point.

artificial satellite. A manmade object put into a closed orbit about the earth.

ascendant (astrological term). The point on the zodiac that is on the eastern horizon, just rising at the moment of birth.

ascending node. The point along the orbit of a body where it crosses from the south to the north of some reference plane, usually the plane of the celestial equator or of the ecliptic.

aspect. The situation of the sun, moon, or planets with respect to one another.

association. A loose cluster of stars whose spectral types, motions, or positions in the sky indicate that they have probably had a common origin.

asteroid. A synonym for "minor planet."

astigmatism. A defect in an optical system whereby pairs of light rays in different planes do not focus at the same place.

astrology. The pseudoscience that treats with supposed influences of the configurations and locations in the sky of the sun, moon, and planets on human destiny; a primitive religion having its origin in ancient Babylonia.

astrometric binary. A binary star in which one component is not observed, but whose presence is deduced from the orbital motion of the visible component.

astrometry. That branch of astronomy that deals with the determination of precise positions and motions of celestial bodies.

astronautics. The science of the laws and methods of space flight.

astronomical unit (AU). Originally meant to be the semimajor axis of the orbit of the earth; now defined as the semimajor axis of the orbit of a hypothetical body with the mass and period that Gauss assumed for the earth. The semimajor axis of the orbit of the earth is 1.000 000 230 AU.

astronomy. The branch of science that treats of the physics and morphology of that part of the universe which lies beyond the earth's atmosphere.

astrophysics. That part of astronomy which deals principally with the physics of stars, stellar systems, and interstellar material. Astrophysics also deals, however, with the structures and atmospheres of the sun and planets.

atmospheric refraction. The bending, or refraction, of light rays from celestial objects by the earth's atmosphere.

atom. The smallest particle of an element that retains the properties which characterize that element.

atomic mass unit. *Chemical:* one sixteenth of the mean mass of an oxygen atom. *Physical:* one twelfth of the mass of an atom of the most common isotope of carbon. The atomic mass unit is approximately the mass of a hydrogen atom, 1.67×10^{-24} gm.

atomic number. The number of protons in each atom of a particular element.

atomic transition. A change in the state of energy of an atom; the atom may gain or lose energy by collision with another particle or by the emission or absorption of a photon.

atomic weight. The mean mass of an atom of a particular element in atomic mass units.

aurora. Light radiated by atoms and ions in the ionosphere, mostly in the polar regions.

autumnal equinox. The intersection of the ecliptic and celestial equator where the sun crosses the equator from north to south.

azimuth. The angle along the celestial horizon, measured eastward from the north point, to the intersection of the horizon with the vertical circle passing through an object.

Baily's beads. Small "beads" of sunlight seen passing through valleys along the limb of the moon in the instant preceding totality in a solar eclipse.

ballistic missile. A missile or rocket that is given its entire thrust during a brief period at the beginning of its flight, and that subsequently "coasts" to its target along an orbit.

Balmer lines. Emission or absorption lines in the spectrum of hydrogen that arise from transitions between the second (or first excited) and higher energy states of the hydrogen atoms.

bands (in spectra). Emission or absorption lines, usually in the spectra of chemical compounds or radicals, so numerous and closely spaced that they coalesce into broad emission or absorption bands.

barred spiral galaxy. Spiral galaxy in which the spiral arms begin from the ends of a "bar" running through the nucleus rather than from the nucleus itself.

barycenter. The center of mass of two mutually revolving bodies.

base line. That side of a triangle used in triangulation or surveying whose length is known (or can be measured), and which is included between two angles that are known (or can be measured).

"big-bang" theory. A theory of cosmology in which the expansion of the universe is presumed to have begun with a primeval explosion.

binary star. A double star; two stars revolving about each other.

black body. A hypothetical perfect radiator, which absorbs and reemits all radiation incident upon it.

black dwarf. The presumed final state of evolution for a star, in which all of its energy sources are exhausted and it no longer emits radiation.

black hole. A configuration of matter from which the velocity of escape equals or exceeds the speed of light, and hence, which cannot be observed.

blink microscope. A microscope in which the user's view is shifted rapidly back and forth between the corresponding portions of two different photographs of the same region of the sky.

Bode's law. A scheme by which a sequence of numbers can be obtained that give the approximate distances of the planets from the sun in astronomical units.

Bohr atom. A particular model of an atom, invented by Niels Bohr, in which the electrons are described as revolving about the nucleus in circular orbits.

bolide. A very bright fireball or meteor; sometimes defined as a fireball accompanied by sound.

bolometer. An instrument for measuring radiation received from a luminous body over the entire electromagnetic spectrum.

bolometric correction. The difference between the visual (or photovisual) and bolometric magnitudes of a star.

bolometric magnitude. A measure of the flux of radiation from a star or other object received just outside the earth's atmosphere, as it would be detected by a device sensitive to *all* forms of electromagnetic energy.

burnout. The instant when a rocket stops firing.

calculus. A branch of mathematics that permits computations involving rates of change (*differential* calculus) or of the contribution of an infinite number of infinitesimal quantities (*integral* calculus).

carbon cycle. A series of nuclear reactions involving carbon as a catalyst, by which hydrogen is transformed to helium.

cardinal points. The four principal points of the compass: North, East, South, and West.

celestial equator. A great circle on the celestial sphere 90° from the celestial poles; the circle of intersection of the celestial sphere with the plane of the earth's equator.

celestial mechanics. That branch of astronomy which deals with the motions and gravitational influences of the members of the solar system.

celestial navigation. The art of navigation at sea or in the air from sightings of the sun, moon, planets, and stars.

celestial poles. Points about which the celestial sphere appears to rotate; intersections of the celestial sphere with the earth's polar axis.

celestial sphere. Apparent sphere of the sky; a sphere of large radius centered on the observer. Directions of objects in the sky can be denoted by the positions of those objects on the celestial sphere.

center of gravity. Center of mass.

center of mass. The mean position of the various mass elements of a body or system, weighted according to their distances from that center of mass; that point in an isolated system which moves with constant velocity, according to Newton's first law of motion.

centrifugal force (or acceleration). An imaginary force (or acceleration) that is often introduced to account for the illusion that a body moving on a curved path tends to accelerate radially from the center of curvature. The actual force present is the one that diverts the body's motion from a straight line and is directed *toward* the center of curvature. It is, however, legitimate to introduce a fictitious centrifugal force field in a rotating (and hence noninertial) coordinate system.

centripetal force (or acceleration). The force required to divert a body from a straight path into a curved path (or the acceleration experienced by the body); it is directed toward the center of curvature.

Cepheid variable. A star that belongs to one of two classes (type I and type II) of yellow supergiant pulsating stars.

Ceres. Largest of the minor planets and first to be discovered.

chromatic aberration. A defect of optical systems whereby light of different colors is focused at different places.

chromosphere. That part of the solar atmosphere that lies immediately above the photospheric layers.

chronograph. A device for recording and measuring the times of events.

chronometer. An accurate clock.

circular velocity. The critical speed with which a revolving body can have a circular orbit.

circumpolar regions. Portions of the celestial sphere near the celestial poles that are either always above or always below the horizon.

Clouds of Magellan. Two neighboring galaxies visible to the naked eye from southern latitudes.

cluster of galaxies. A system of galaxies containing from several to thousands of member galaxies.

cluster variable (RR Lyrae variable). A member of a certain large class of pulsating variable stars, all with periods less than 1 day. These stars are often present in globular star clusters.

color excess. The amount by which the color index of a star is increased when its light is reddened in passing through interstellar absorbing material.

color index. Difference between the magnitudes of a star or other object measured in light of two different spectral regions, for example, photographic *minus* photovisual magnitudes.

color-magnitude diagram. Plot of the magnitudes (apparent or absolute) of the stars in a cluster against their color indices.

coma. A defect in an optical system in which off-axis rays of light striking different parts of the objective do not focus in the same place.

coma (of comet). The diffuse gaseous component of the head of a comet.

comet. A swarm of solid particles and gases, which revolves about the sun, usually in an orbit of high eccentricity.

comparison spectrum. The spectrum of a vaporized element (such as iron) photographed beside the image of a stellar spectrum, and with the same camera, for purposes of comparison of wavelengths.

compound. A substance composed of two or more chemical elements.

compound nucleus. An excited nucleus, usually temporary, formed by nuclei of two or more simpler atoms.

conduction. The transfer of energy by the direct passing of energy or electrons from atom to atom.

configuration. Any one of several particular orientations in the sky of the moon or a planet with respect to the sun.

conic section. The curve of intersection between a circular cone and a plane; these curves can be ellipses, circles, parabolas, or hyperbolas.

conjunction. The configuration of a planet when it has the same celestial longitude as the sun, or the configuration when any two celestial bodies have the same celestial longitude or right ascension.

conservation of angular momentum. The law that angular momentum is conserved in the absence of any force not directed toward or away from the point or axis about which the angular momentum is referred — that is, in the absence of a torque.

constellation. A configuration of stars named for a particular object, person, or animal; or the area of the sky assigned to a particular configuration.

continuous spectrum. A spectrum of light comprised of radiation of a continuous range of wavelengths or colors rather than only certain discrete wavelengths.

convection. The transfer of energy by moving currents of a fluid containing that energy.

core (of earth). The central part of the earth, believed to be a liquid of high density.

coriolis effect. The deflection (with respect to the ground) of projectiles moving across the surface of the rotating earth.

corona. Outer atmosphere of the sun.

corona (or halo) of Galaxy. The outer portions of the Galaxy, especially on either side of the plane of the Milky Way.

coronagraph. An instrument for photographing the chromosphere and corona of the sun outside of eclipse.

corpuscular radiation. Charged particles, mostly atomic nuclei and electrons, emitted into space by the sun and possibly other objects.

cosine. One of the trigonometric functions of an angle. If the angle is in a right triangle, the cosine is the ratio of the lengths of the shorter side adjacent to the angle to the hypotenuse.

cosmic rays. Atomic nuclei (mostly protons) that are observed to strike the earth's atmosphere with exceedingly high energies.

cosmogony. The study of the origin of the world or universe.

cosmological constant. A term that arises in the development of the field equations of general relativity, which represents a repulsive force in the universe. The cosmological constant is often assumed to be zero.

cosmological model. A specific model, or theory, of the organization and evolution of the universe.

cosmological principle. The assumption that, on the large scale, the universe at any given time is the same everywhere.

cosmology. The study of the organization and evolution of the universe.

coudé focus. An optical arrangement in a reflecting telescope whereby light is reflected by two or more secondary mirrors down the polar axis of the telescope to a focus at a place separate from the moving parts of the telescope.

crater (lunar). A more or less circular depression in the surface of the moon.

crater (meteoritic). A crater on the earth caused by the collision of a meteoroid with the earth, and a subsequent explosion.

crescent moon. One of the phases of the moon when its elongation is less than 90° from the sun and it appears less than half full.

crust (of earth). The outer layer of the earth.

cyclonic motion. A counterclockwise circular circulation of winds (in the Northern Hemisphere) that results from the coriolis effect.

dark nebula. A cloud of interstellar dust that obscures the light of more distant stars, and appears as an opaque curtain.

daylight saving time. A time 1 hour more advanced than standard time, usually adopted in spring and summer to take advantage of long evening twilights.

declination. Angular distance north or south of the celestial equator to some object, measured along an hour circle passing through that object.

deferent. A stationary circle in the Ptolemaic system along which moves the center of another circle (epicycle), along which moves an object or another epicycle.

degenerate gas. A gas in which the allowable states for the electrons have been filled; it behaves according to different laws from those that apply to "perfect" gases.

density. The ratio of the mass of an object to its volume.

descending node. The point along the orbit of a body where it crosses from the north to the south of some reference plane, usually the plane of the celestial equator or of the ecliptic.

deuterium. A "heavy" form of hydrogen, in which the nucleus of each atom consists of one proton and one neutron.

differential equation. An equation that involves derivatives or rates of change.

differential galactic rotation. The state of rotation of the galaxy whereby it does not rotate as a solid wheel, so that parts adjacent to each other do not always stay close together.

differential gravitational force. The difference between the respective gravitational forces exerted on two bodies near each other by a third, more distant body.

diffraction. The spreading out of light in passing the edge of an opaque body.

diffraction grating. A system of closely spaced equidistant slits or reflecting strips which, by diffraction and interference, produce a spectrum.

diffraction pattern. A pattern of bright and dark fringes produced by the interference of light rays, diffracted by different amounts, with each other.

diffuse nebula. A reflection or emission nebula produced by interstellar matter (not a planetary nebula).

disk (of planet or other object). The apparent disk that a planet (or the sun, or moon, or a star) displays when seen in the sky or viewed telescopically.

disk of Galaxy. The central disk or "wheel" of our Galaxy, superimposed on the spiral structure.

dispersion. Separation, from white light, of different wavelengths being refracted by different amounts.

distance modulus. Difference between the apparent and absolute magnitudes of an object.

diurnal. Daily.

diurnal circle. Apparent path of a star in the sky during a complete day due to the earth's rotation.

diurnal libration. The phenomenon whereby slightly different hemispheres of the moon can be observed during a day because of the motion of the observer caused by the earth's rotation.

diurnal motion. Motion during one day.

diurnal parallax. Apparent change in direction of an object caused by a displacement of the observer due to the earth's rotation.

Doppler shift. Apparent change in wavelength of the radiation from a source due to its relative motion in the line of sight.

draconic month. The period of revolution of the moon about the earth with respect to the nodes of the moon's orbit.

dwarf (star). A main-sequence star (as opposed to a giant or supergiant).

dyne. The metric unit of force; the force required to accelerate a mass of 1 gram in the amount 1 centimeter per second per second.

east point. The point on the horizon 90° from the north point (measured clockwise as seen from the zenith).

eccentric. A point, about which an object revolves on a circular orbit, that is not at the center of the circle.

eccentricity (of ellipse). Ratio of the distance between the foci to the major axis.

eclipse. The cutting off of all or part of the light of one body by another passing in front of it.

eclipse path. The track along the earth's surface swept out by the tip of the shadow of the moon (or the extension of its shadow) during a total (or annular) solar eclipse.

eclipse season. A period during the year when an eclipse of the sun or moon is possible.

eclipsing binary star. A binary star in which the plane of revolution of the two stars is nearly edge on to our line of sight, so that the light of one star is periodically diminished by the other passing in front of it.

ecliptic. The apparent annual path of the sun on the celestial sphere.

ecliptic limit. The maximum angular distance from a node where the moon can be for an eclipse to take place.

electromagnetic radiation. Radiation consisting of waves propagated through the building up and breaking down of electric and magnetic fields; these include radio, infrared, light, ultraviolet, X-rays, and gamma rays.

electromagnetic spectrum. The whole array or family of electromagnetic waves.

electron. A negatively charged subatomic particle that normally moves about the nucleus of an atom.

electron volt. The kinetic energy acquired by an electron that is accelerated through an electric potential of 1 volt; 1 electron volt is 1.60207×10^{-12} erg.

element. A substance that cannot be decomposed, by chemical means, into simpler substances.

elements (of orbit). Any of several quantities that describe the size, shape, and orientation of the orbit of a body.

ellipse. A conic section: the curve of intersection of a circular cone and a plane cutting completely through the cone.

elliptical galaxy. A galaxy whose apparent photometric contours are ellipses and which contains no conspicuous interstellar material.

ellipticity. The ratio (in an ellipse) of the major axis minus the minor axis to the major axis.

elongation. The difference between the celestial longitudes of a planet and the sun.

emission line. A discrete bright spectral line.

emission nebula. A gaseous nebula that derives its visible light from the fluorescence of ultraviolet light from a star in or near the nebula.

emission spectrum. A spectrum consisting of emission lines.

encounter (gravitational). A near passing, on hyperbolic orbits, of two objects that influence each other gravitationally.

energy. The ability to do work.

energy level (in an atom or ion). A particular level, or amount, of energy possessed by an atom or ion above the energy it possesses in its least energetic state.

energy spectrum. A table or plot showing the relative numbers of particles (in cosmic rays or corpuscular radiation) of various energies.

ephemeris. A table that gives the positions of a celestial body at various times, or other astronomical data.

ephemeris time. A kind of time that passes at a strictly uniform rate; used to compute the instants of various astronomical events.

epicycle. A circular orbit of a body in the Ptolemaic system, the center of which revolves about another circle (the deferent).

equant. A stationary point in the Ptolemaic system not at the center of a circular orbit about which a body (or the center of an epicycle) revolves with uniform angular velocity.

equation of state. An equation relating the pressure, temperature, and density of a substance (usually a gas).

equation of time. The difference between apparent and mean solar time.

equator. A great circle on the earth, 90° from its poles.

equatorial mount. A mounting for a telescope, one axis of which is parallel to the earth's axis, so that a motion of the telescope about that axis can compensate for the earth's rotation.

equinox. One of the intersections of the ecliptic and celestial equator.

equivalent width. A measure of the strength of a spectral line; the width of an absorption line of rectangular profile and zero intensity at its center.

erg. The metric unit of energy; the work done by a force of 1 dyne moving through a distance of one centimeter.

eruptive variable. A variable star whose changes in light are erratic or explosive.

establishment of port. The time interval, at a particular port, between the meridian passage of the moon and high tide.

evolutionary cosmology. A theory of cosmology that assumes that all parts of the universe have a common age and evolve together.

eyepiece. A magnifying lens used to view the image produced by the objective of a telescope.

excitation. The process of imparting to an atom or an ion an amount of energy greater than that it has in its normal or least-energy state.

extinction. Attenuation of light from a celestial body produced by the earth's atmosphere, or by interstellar absorption.

extragalactic. Beyond the Galaxy.

faculus (p. faculae). Bright regions near the limb of the sun.

filar micrometer. A device used with a telescope to measure the angular separations of closely separated pairs of stars or the diameters of extended objects.

filtergram. A photograph of the sun (or part of it) taken through a special narrow-band-pass filter.

fireball. A spectacular meteor.

fission. The breakup of a heavy atomic nucleus into two or more lighter ones.

flare. A sudden and temporary outburst of light from an extended region of the solar surface.

flare star. A member of a class of stars that show occasional, sudden, unpredicted increases in light.

flash spectrum. The spectrum of the very limb of the sun obtained in the instant before totality in a solar eclipse.

flocculus (p. flocculi). Bright regions of the solar surface observed in the monochromatic light of some spectral line; flocculi are now usually called *plages*.

fluctions. Name given by Newton to the calculus.

fluorescence. The absorption of light of one wavelength and reemission of it at another wavelength; especially the conversion of ultraviolet into visible light.

focal length. The distance from a lens or mirror to the point where light converged by it comes to a focus.

focal ratio (speed). Ratio of the focal length of a lens or mirror to its aperture.

focus. Point where the rays of light converged by a mirror or lens meet.

focus of a conic section. Mathematical point associated with a conic section, whose distance to any point on the conic bears a constant ratio to the distance from that point to a straight line known as the *directrix*.

forbidden lines. Spectral lines that are not usually observed under laboratory conditions because they result from atomic transitions that are highly improbable.

force. That which can change the momentum of a body; numerically, the rate at which the body's momentum changes.

Foucault pendulum. An experiment first conducted by Jean Foucault in 1851 to demonstrate the rotation of the earth.

Fraunhofer line. An absorption line in the spectrum of the sun or of a star.

Fraunhofer spectrum. The array of absorption lines in the spectrum of the sun or of a star.

free-free transition. An atomic transition in which the energy associated with an atom or ion and passing electron changes during the encounter, but without capture of the electron by the atom or ion.

frequency. Number of vibrations per unit time; number of waves that cross a given point per unit time (in radiation).

fringes (interference). Successive dark and light fringes, or lines, caused by interference of light waves with each other before they strike a screen or detecting device and are observed.

full moon. That phase of the moon when it is at opposition (180° from the sun) and its full daylight hemisphere is visible from the earth.

fusion. The building up of heavier atomic nuclei from lighter ones.

galactic cluster. An "open" cluster of stars located in the spiral arms or disk of the Galaxy.

galactic equator. Intersection of the principal plane of the Milky Way with the celestial sphere.

galactic latitude. Angular distance north or south of the galactic equator to an object, measured along a great circle passing through that object and the galactic poles.

galactic longitude. Angular distance, measured eastward along the galactic equator from the galactic center, to the intersection of the galactic equator with a great circle passing through the galactic poles and an object.

galactic poles. The poles of the galactic equator; the intersections with the celestial sphere of a line through the observer that is perpendicular to the plane of the galactic equator.

galactic rotation. Rotation of the Galaxy.

galaxy. A large assemblage of stars; a typical galaxy contains millions to hundreds of billions of stars.

Galaxy. The galaxy to which the sun and our neighboring stars belong; the Milky Way is light from remote stars in the Galaxy.

gamma rays. Photons (of electromagnetic radiation) of energy higher than those of X-rays; the most energetic form of electromagnetic radiation.

gauss. A unit of magnetic flux density.

gegenschein (counterglow). A very faint, diffuse glow of light opposite the sun in the sky, believed to be caused by sunlight reflected from interplanetary particles.

geod. The mean figure of the earth.

geomagnetic. Referring to the geometrical center of the earth's magnetic field.

geomagnetic poles. The poles of a hypothetical

bar magnet whose magnetic field most nearly matches that of the earth.

giant (star). A star of large luminosity and radius.

gibbous moon. One of the phases of the moon in which more than half, but not all, of the moon's daylight hemisphere is visible from the earth.

globular cluster. One of about 120 large star clusters that form a system of clusters centered on the center of the Galaxy.

globule. A small, dense, dark nebula; believed to be a possible protostar.

granulation. The "rice-grain"-like structure of the solar photosphere.

gravitation. The tendency of matter to attract itself together.

gravitational constant, G. The constant of proportionality in Newton's law of gravitation; in metric units G has the value 6.668×10^{-8} dyne $\cdot$ cm^2/gm^2.

gravitational energy. Energy that can be released by the gravitational collapse, or partial collapse, of a system.

great circle. Circle on the surface of a sphere that is the curve of intersection of the sphere with a plane passing through its center.

greatest elongation (east or west). The largest separation in celestial longitude (to the east or west) that an inferior planet can have from the sun.

Greenwich meridian. The meridian of longitude passing through the site of the old Royal Greenwich Observatory, near London; origin of longitude on the earth.

Gregorian calendar. A calendar (now in common use) introduced by Pope Gregory XIII in 1582.

H I region. Region of neutral hydrogen in interstellar space.

H II region. Region of ionized hydrogen in interstellar space.

half-life. The time required for half of the radioactive atoms in a sample to disintegrate.

halo (around sun or moon). A ring of light around the sun or moon caused by refraction by the ice crystals of cirrus clouds.

halo (of galaxy). See corona.

harmonic law. Kepler's third law of planetary motion: the cubes of the semimajor axes of the planetary orbits are in proportion to the squares of the sidereal periods of the planets.

harvest moon. The full moon nearest the time of the autumnal equinox.

head (of comet). The main part of a comet, consisting of its nucleus and coma.

"heavy" elements. In astronomy, usually those elements of greater atomic number than helium.

heliacal rising. The rising of a star or planet the first time that it can be observed in the morning before sunrise.

helio-. Prefix referring to the sun.

heliocentric. Centered on the sun.

helium flash. The nearly explosive ignition of helium in the triple-alpha process in the dense core of a red giant star.

Hertzsprung-Russell (H-R) diagram. A plot of absolute magnitude against temperature (or spectral class or color index) for a group of stars.

high-velocity star (or object). A star (or object) with high space motion; generally an object that does not share the high orbital velocity of the sun about the galactic nucleus.

homogeneous star (or stellar model). A star (or theoretical model of a star) whose chemical composition is the same throughout its interior.

horizon (astronomical). A great circle on the celestial sphere 90° from the zenith.

horizon system. A system of celestial coordinates (altitude and azimuth) based on the astronomical horizon and the north point.

horizontal branch. A sequence of stars on the Hertzsprung-Russell diagram of a typical globular cluster of approximately constant absolute magnitude (near $M_v = 0$).

horizontal parallax. The angle by which an object appears displaced (after correction for atmospheric refraction) when viewed on the horizon from a place on the earth's equator, as compared to its direction if it were viewed from the center of the earth.

horoscope (astrological term). A chart showing the positions along the zodiac and in the sky of the sun, moon, and planets at some given instant and place on earth — generally corresponding to the time and place of a person's birth.

hour angle. The angle measured westward along the celestial equator from the local meridian to the hour circle passing through an object.

hour circle. A great circle on the celestial sphere passing through the celestial poles.

Hubble constant. Constant of proportionality in the relation between the velocities of remote galaxies and their distances. The Hubble constant is approximately 75 km/sec · 10^6 pc.

hydrostatic equilibrium. A balance between the weights of various layers, as in a star or the earth's atmosphere, and the pressures that support them.

hyperbola. A conic section of eccentricity greater than 1.0; the curve of intersection between a circular cone and a plane that is at too small an angle with the axis of the cone to cut all of the way through it, and is not parallel to a line in the face of the cone.

hyperfine transition. A change in the energy state of an atom that involves a change in the spin of its nucleus.

hypothesis. A tentative theory or supposition, advanced to explain certain facts or phenomena, and which is subject to further tests and verification.

image. The optical representation of an object produced by light rays from the object being refracted or reflected by a lens or mirror.

image tube. A device in which electrons, emitted from a photocathode surface exposed to light, are focused electronically.

inclination (of an orbit). The angle between the orbital plane of a revolving body and some fundamental plane — usually the plane of the celestial equator or of the ecliptic.

Index Catalogue, IC. The supplement to Dreyer's *New General Catalogue* of star clusters and nebulae.

index of refraction. A measure of the refracting power of a transparent substance; specifically, the ratio of the speed of light in a vacuum to its speed in the substance.

inertia. The property of matter that requires a force to act on it to change its state of motion; momentum is a measure of inertia.

inferior conjunction. The configuration of an inferior planet when it has the same longitude as the sun, and is between the sun and earth.

inferior planet. A planet whose distance from the sun is less than the earth's.

infrared radiation. Electromagnetic radiation of wavelength longer than the longest (red) wavelengths that can be perceived by the eye, but shorter than radio wavelengths.

insolation. The rate at which all radiation from the sun is received per unit area on the ground.

intercalate. To insert, as a day in a calendar.

interferometer (stellar). An optical device, making use of the principle of interference of light waves, with which small angles can be measured.

International Date Line. An arbitrary line on the surface of the earth near longitude 180° across which the date changes by one day.

international magnitude system. The system of photographic and photovisual magnitudes, referring to the blue and yellow spectral regions, at one time adopted by international agreement, but now largely superseded by the *U, B, V* system.

interplanetary medium. The sparse distribution of gas and solid particles in the interplanetary space.

interstellar dust. Microscopic solid grains, believed to be mostly dielectric compounds of hydrogen and other common elements, in interstellar space.

interstellar gas. Sparse gas in interstellar space.

interstellar lines. Absorption lines superimposed on stellar spectra, produced by the interstellar gas.

interstellar matter. Interstellar gas and dust.

ion. An atom that has become electrically charged by the addition or loss of one or more electrons.

ionization. The process by which an atom gains or loses electrons.

ionization potential. The energy required to remove an electron from an atom.

ionosphere. The upper region of the earth's atmosphere in which many of the atoms are ionized.

irregular galaxy. A galaxy without rotational symmetry; neither a spiral or elliptical galaxy.

irregular variable. A variable star whose light variations do not repeat with a regular period.

island universe. Historical synonym for galaxy.

isotope. Any of two or more forms of the same element, whose atoms all have the same number but different masses.

isotropic. The same in all directions.

iteration. The "closing in " on the solution to a mathematical problem by repetitive calculations.

Jovian planet. Any of the planets Jupiter, Saturn, Uranus, and Neptune.

Julian calendar. A calendar introduced by Julius Caesar in 45 B.C.

Julian day. The number of the day in a running sequence beginning January 1, 4713 B.C.

Jupiter. The fifth planet from the sun in the solar system.

Kepler's laws. Three laws, discovered by J. Kepler, that describe the motions of the planets.

kinetic energy. Energy associated with motion; the kinetic energy of a body is one half the product of its mass and the square of its velocity.

kinetic theory (of gases). The science that treats the motions of the molecules that compose gases.

Kirkwood's gaps. Gaps in the spacing of the minor planets that arise from perturbations produced by the major planets.

latitude. A north-south coordinate on the surface of the earth; the angular distance north or south of the equator measured along a meridian passing through a place.

law. A statement of order or relation between phenomena that, under given conditions, is presumed to be invariable.

law of areas. Kepler's second law: the radius vector from the sun to any planet sweeps out equal areas in the planet's orbital plane in equal intervals of time.

law of the redshifts. The relation between the radial velocity and distance of a remote galaxy: the radial velocity is proportional to the distance of the galaxy.

lead sulfide cell. A device used to measure infrared radiation.

leap year. A calendar year with 366 days, intercalated approximately every 4 years to make the average length of the calendar year as nearly equal as possible to the tropical year.

libration. Any of several phenomena by which an observer on earth, over a period of time, can see more than one hemisphere of the moon.

libration in latitude. Libration caused by the fact that the moon's axis of rotation is not perpendicular to its plane of revolution.

libration in longitude. Libration caused by the regularity in the moon's rotation but irregularity in its orbital speed.

light. Electromagnetic radiation that is visible to the eye.

light curve. A graph that displays the variation in light or magnitude of a variable or eclipsing binary star.

light-year. The distance light travels in a vacuum in 1 year; 1 LY = 9.46×10^{17} cm, or about 6×10^{12} mi.

limb (of sun or moon). Apparent edge of the sun or moon as seen in the sky.

limb darkening. The phenomenon whereby the sun is less bright near its limb than near the center of its disk.

limiting magnitude. The faintest magnitude that can be observed with a given instrument or under given conditions.

line broadening. The phenomenon by which spectral lines are not precisely sharp but have finite widths.

line of apsides. The line connecting the apses of an orbit (the perifocus and farthest from focus points); or the line along the major axis of the orbit.

line of nodes. The line connecting the nodes of an orbit.

line of position. A part of a circle of position, so small that it can be considered a straight line.

line profile. A plot of the intensity of light versus wavelength across a spectral line.

linear diameter. Actual diameter in units of length.

Local Group. The cluster of galaxies to which our Galaxy belongs.

local supercluster (or **supergalaxy**). A proposed cluster of clusters of galaxies, to which the Local Group belongs.

longitude. An east-west coordinate on the earth's surface; the angular distance, measured east or west along the equator from the Greenwich meridian, to the meridian passing through a place.

longitude of the ascending node. The angle measured eastward from a reference direction (usually the vernal equinox) in a fundamental plane (usually the plane of the celestial equator or of the ecliptic) to the ascending node of the orbit of a body.

low-velocity star (or **object**). A star (or object) that has low space velocity; generally an object that shares the sun's high orbital speed about the galactic center.

luminosity. The rate of radiation of electromagnetic energy into space by a star or other object.

luminosity class. A classification of a star according to its luminosity for a given spectral class.

luminosity function. The relative numbers of stars (or other objects) of various luminosities or absolute magnitudes.

luminous energy. Light.

lunar. Referring to the moon.

lunar eclipse. An eclipse of the moon.

Lyman lines. A series of absorption or emission lines in the spectrum of hydrogen that arise from transitions to and from the lowest energy states of the hydrogen atoms.

magnetic field. The region of space near a magnetized body within which magnetic forces can be detected.

magnetic pole. One of two points on a magnet (or the earth) at which the greatest density of lines of force emerge. A compass needle aligns itself along the local lines of force on the earth and points more or less toward the magnetic poles of the earth.

magnifying power. The number of times larger (in angular diameter) an object appears through a telescope than with the naked eye.

magnitude. A measure of the amount of light flux received from a star or other luminous object.

main sequence. A sequence of stars on the Hertzsprung-Russell diagram, containing the majority of stars, that runs diagonally from the upper left to the lower right.

major axis (of ellipse). The maximum diameter of an ellipse.

major planet. A Jovian planet.

mantle (of earth). The greatest part of the earth's interior, lying between the crust and the core.

mare. Latin for "sea"; name applied to many of the "sealike" features on the moon or Mars.

Mars. Fourth planet from the sun in the solar system.

mass. A measure of the total amount of material in a body; defined either by the inertial properties of the body or by its gravitational influence on other bodies.

mass defect. The amount by which the mass of an atomic nucleus is less than the sum of the masses of the individual nucleons that compose it.

mass function. A numerical relation between the masses of the components of a binary star and the inclination of their plane of mutual revolution to the plane of the sky; the mass function is determined from an analysis of the radial-velocity curve of a spectroscopic binary when the spectral lines of only one of the stars are visible.

mass-luminosity relation. An empirical relation between the masses and luminosities of many (principally main-sequence) stars.

mass-radius relation (for white dwarfs). A theoretical relation between the masses and radii of white dwarf stars.

mean solar day. Interval between successive meridian passages of the mean sun; average length of the apparent solar day.

mean solar time. Local hour angle of the mean sun *plus* 12 hours.

mean sun. A fictitious body that moves eastward with uniform angular velocity along the celestial equator, completing one circuit of the sky with respect to the vernal equinox in a tropical year.

mechanics. That branch of physics which deals with the behavior of material bodies under the influence of, or in the absence of, forces.

Mercury. Nearest planet to the sun in the solar system.

meridian (celestial). The great circle on the celestial sphere that passes through an observer's zenith and the north (or south) celestial pole.

meridian (terrestrial). The great circle on the surface of the earth that passes through a particular place and the north and south poles of the earth.

mesosphere. The layer of the ionosphere immediately above the stratosphere.

Messier catalogue. A catalogue of nonstellar objects compiled by Charles Messier in 1787.

meteor. The luminous phenomenon observed when a meteoroid enters the earth's atmosphere and burns up; popularly called a "shooting star."

meteor shower. Many meteors appearing to radiate from a common point in the sky caused by the collision of the earth with a swarm of meteoritic particles.

meteorite. A portion of a meteoroid that survives passage through the atmosphere and strikes the ground.

meteorite fall. The occurrence of a meteorite striking the ground.

meteoroid. A meteoritic particle in space before any encounter with the earth.

micrometeorite. A meteoroid so small that, on entering the atmosphere of the earth, it is slowed quickly enough that it does not burn up or ablate but filters through the air to the ground.

Milky Way. The band of light encircling the sky, which is due to the many stars and diffuse nebulae lying near the plane of the Galaxy.

minor axis (of ellipse). The smallest or least diameter of an ellipse.

minor planet. One of several tens of thousands of small planets, ranging in size from a few hundred miles to less than 1 mile in diameter.

Mira Ceti–type variable star. Any of a large class of red-giant long-period or irregular pulsating variable stars, of which the star Mira is a prototype.

missile. A projectile, especially a rocket.

model atmosphere (or photosphere). The result of a theoretical calculation of the run of temperature, pressure, density, and so on, through the outer layers of the sun or a star.

molecule. A combination of two or more atoms bound together; the smallest particle of a chemical compound or substance that exhibits the chemical properties of that substance.

momentum. A measure of the inertia or state of motion of a body; the momentum of a body is the product of its mass and velocity. In the absence of a force, momentum is conserved.

monochromatic. Of one wavelength or color.

n-**body problem.** The problem of determining the positions and motions of more than two bodies in a system in which the bodies interact under the influence of their mutual gravitation.

nadir. The point on the celestial sphere 180° from the zenith.

nautical mile. The mean length of 1 minute of arc on the earth's surface along a meridian.

navigation. The art of finding one's position and course at sea or in the air.

neap tide. The lowest tides in the month which occur when the moon is near first or third quarter.

nebula. Cloud of interstellar gas or dust.

Neptune. Eighth planet from the sun in the solar system.

neutrino. A particle that has no mass or charge but that carries energy away in the course of certain nuclear transformations.

neutron. A subatomic particle with no charge and with mass approximately equal to that of the proton.

neutron star. A hypothetical star of extremely high density composed entirely of neutrons.

New General Catalogue (NGC). A catalogue of star clusters, nebulae, and galaxies compiled by J. L. E. Dreyer in 1888.

new moon. Phase of the moon when its longitude is the same as that of the sun.

Newtonian focus. An optical arrangement in a reflecting telescope where the light is reflected by a flat mirror to a focus at the side of the telescope tube just before it reaches the focus of the objective.

Newton's laws. The laws of mechanics and gravitation formulated by Isaac Newton.

night sky light. The faint illumination of the night sky; the main source is usually fluorescence by atoms high in the atmosphere.

node. The intersection of the orbit of a body with a fundamental plane — usually the plane of the celestial equator or of the ecliptic.

nodical month. The period of revolution of the

moon about the earth with respect to the line of nodes of the moon's orbit.

nodical (eclipse) year. Period of revolution of the earth about the sun with respect to the line of nodes of the moon's orbit.

north point. That intersection of the celestial meridian and astronomical horizon lying nearest the north celestial sphere.

north polar sequence. A group of stars in the vicinity of the north celestial pole whose magnitudes serve as standards for the international magnitude system.

nova. A star that experiences a sudden outburst of radiant energy, temporarily increasing its luminosity by hundreds to thousands of times.

nuclear. Referring to the nucleus of the atom.

nuclear transformation. Transformation of one atomic nucleus into another.

nucleon. Any one of the subatomic particles that compose a nucleus.

nucleus (of atom). The heavy part of an atom, composed mostly of protons and neutrons, and about which the electrons revolve.

nucleus (of comet). A swarm of solid particles in the head of a comet.

nucleus (of galaxy). Central concentration of stars, and possibly gas, at the center of a galaxy.

O-association. A stellar association in which the stars are predominately of types O and B.

objective. The principal image-forming component of a telescope or other optical instrument.

objective prism. A prismatic lens that can be placed in front of a telescope objective to transform each star image into an image of its spectrum.

oblate spheroid. A solid formed by rotating an ellipse about its minor axis.

oblateness. A measure of the "flattening" of an oblate spheroid; numerically, the ratio of the difference between the major and minor diameters (or axes) to the major diameter (or axis).

obliquity of the ecliptic. Angle between the planes of the celestial equator and the ecliptic; about 23½°.

obscuration (interstellar). Absorption of starlight by interstellar dust.

occultation. An eclipse of a star or planet by the moon or another planet.

ocular. Eyepiece.

opacity. Absorbing power; capacity to impede the passage of light.

open cluster. A comparatively loose or "open" cluster of stars, containing from a few

dozen to a few thousand members, located in the spiral arms or disk of the Galaxy; galactic cluster.

opposition. Configuration of a planet when its elongation is 180°.

optical binary. Two stars at different distances nearly lined up in projection so that they appear close together, but which are not really dynamically associated.

optics. The branch of physics that deals with light and its properties.

orbit. The path of a body that is in revolution about another body or point.

Pallas. Second minor planet to be discovered.

parabola. A conic section of eccentricity 1.0; the curve of intersection between a circular cone and a plane parallel to a straight line in the surface of the cone.

paraboloid. A parabola of revolution; a curved surface of parabolic cross section. Especially applied to the surface of the primary mirror in a standard reflecting telescope.

parallactic ellipse. A small ellipse that a comparatively nearby star appears to trace out in the sky, which results from the orbital motion of the earth about the sun.

parallax. An apparent displacement of an object due to a motion of the observer.

parallax (stellar). An apparent displacement of a nearby star that results from the motion of the earth around the sun; numerically, the angle subtended by 1 AU at the distance of a particular star.

parallelogram of forces. A geometrical construction that permits the determination of the resultant of two different forces.

parsec. The distance of an object that would have a stellar parallax of 1 second of arc; 1 parsec = 3.26 light-years.

partial eclipse. An eclipse of the sun or moon in which the eclipsed body does not appear completely obscured.

peculiar velocity. The velocity of a star with respect to the local standard of rest; that is, its space motion, corrected for the motion of the sun with respect to our neighboring stars.

penumbra. The portion of a shadow from which only part of the light source is occulted by an opaque body.

penumbral eclipse. A lunar eclipse in which the moon passes through the penumbra, but not the umbra, of the earth's shadow.

perfect cosmological principle. The assumption that, on the large scale, the universe appears the same from every place and at all times.

perfect gas. An "ideal" gas that obeys the perfect gas laws.

perfect gas laws. Certain laws that describe the behavior of an ideal gas; Charles' law, Boyle's law, and the equation of state for a perfect gas.

perfect radiator. Black body; a body that absorbs and subsequently reemits all radiation incident upon it.

periastron. The place in the orbit of a star in a binary star system where it is closest to its companion star.

perifocus. The place on an elliptical orbit that is closest to the focus occupied by the central force.

perigee. The place in the orbit of an earth satellite where it is closest to the center of the earth.

perihelion. The place in the orbit of an object revolving about the sun where it is closest to the center of the sun.

period. A time interval; for example, the time required for one complete revolution.

period-luminosity relation. An empirical relation between the periods and luminosities of cepheid-variable stars.

periodic comet. A comet whose orbit has been determined to have an eccentricity of less than 1.0.

perturbation. The disturbing effect, when small, on the motion of a body as predicted by a simple theory, produced by a third body or other external agent.

phases of the moon. The progression of changes in the moon's appearance during the month that results from the moon's turning different portions of its illuminated hemisphere to our view.

photocell (photoelectric cell). An electron tube in which electrons are dislodged from the cathode when it is exposed to light and are accelerated to the anode, thus producing a current in the tube, whose strength serves as a measure of the light striking the cathode.

photographic magnitude. The magnitude of an object, as measured on the traditional, blue- and violet-sensitive photographic emulsions.

photometry. The measurement of light intensities.

photomultiplier. A photoelectric cell in which the electric current generated is amplified at several stages within the tube.

photon. A discrete unit of electromagnetic energy.

photosphere. The region of the solar (or a stellar) atmosphere from which radiation escapes into space.

photovisual magnitude. A magnitude corresponding to the spectral region to which the human eye is most sensitive, but measured by photographic methods with suitable green- and yellow-sensitive emulsions and filters.

plage. A bright region of the solar surface observed in the monochromatic light of some spectral line; flocculus.

Planck's constant. The constant of proportionality relating the energy of a photon to its frequency.

Planck's radiation law. A formula from which can be calculated the intensity of radiation at various wavelengths emitted by a black body.

planet. Any of nine solid, nonluminous bodies revolving about the sun.

planetarium. An optical device for projecting on a screen or domed ceiling the stars and planets and their apparent motions as they appear in the sky.

planetary nebula. A shell of gas ejected from, and enlarging about, a certain kind of extremely hot star.

planetoid. Synonym for minor planet.

Pluto. Ninth planet from the sun in the solar system.

polar axis. The axis of rotation of the earth; also, an axis in the mounting of a telescope that is parallel to the earth's axis.

polarization. A condition in which the planes of vibration (or the E vectors) of the various rays in a light beam are at least partially aligned.

polarized light. Light in which polarization is present.

Polaroid. Trade name for a transparent substance that produces polarization in light.

Population I and II. Two classes of stars (and systems of stars), classified according to their spectral characteristics, chemical compositions, radial velocities, ages, and locations in the Galaxy.

position angle. Direction in the sky of one celestial object from another; for example, the angle, measured to the east from the north, of the fainter component of a visual binary star in relation to the brighter component.

positron. An electron with a positive rather than a negative charge.

postulate. An essential prerequisite to a hypothesis or theory.

potential energy. Stored energy that can be converted into other forms; especially gravitational energy.

precession (of earth). A slow, conical motion of the earth's axis of rotation, caused principally by the gravitational torque of the moon and sun on the earth's equatorial bulge. *Luni-solar precession,* precession caused by the moon and sun only; *planetary precession,* a slow change in the orientation of the plane of the earth's orbit caused by planetary perturbations; *general precession,* the combination of both on the motion of the earth's axis with respect to the stars.

precession of the equinoxes. Slow westward motion of the equinoxes along the ecliptic that results from precession.

primary cosmic rays. The cosmic-ray particles that arrive at the earth from beyond its atmosphere as opposed to the secondary particles that are produced by collisions between primary cosmic rays and air molecules.

primary minimum (in the light curve of an eclipsing binary). The middle of the eclipse during which the most light is lost.

prime focus. The point in a telescope where the objective focuses the light.

prime meridian. The terrestrial meridian passing through the site of the old Royal Greenwich Observatory; longitude 0°.

primeval atom. A single mass whose explosion (in some cosmological theories) has been postulated to have resulted in all the matter now present in the universe.

primeval fireball. The extremely hot opaque gas that is presumed to comprise the entire mass of the universe at the time of or immediately following the "big bang"; the exploding primeval atom.

Principia. Contraction of *Philosophiae Naturalis Principia Mathematica,* the great book by Newton in which he set forth his laws of motion and gravitation in 1687.

prism. A wedge-shaped piece of glass that is used to disperse white light into a spectrum.

prolate spheroid. The solid produced by the rotation of an ellipse about its major axis.

prominence. A phenomenon in the solar corona that commonly appears like a flame above the limb of the sun.

proper motion. The angular change in direction of a star per year.

proton. A heavy subatomic particle that carries a positive charge, and one of the two principal constituents of the atomic nucleus.

proton-proton chain. A chain of thermonuclear reactions by which nuclei of hydrogen are built up into nuclei of helium.

protoplanet (or star or galaxy). The original material from which a planet (or a star or galaxy) condensed.

pulsar. One of several pulsating radio sources of small angular size that emit radio pulses in very regular short periods.

pulsating variable. A variable star that physically pulsates in size and luminosity.

quadrature. A configuration of a planet in which its elongation is 90°.

quantum mechanics. The branch of physics that deals with the structure of atoms and their interactions with each other and with radiation.

quarter moon. Either of the two phases of the moon when its longitude differs by 90° from that of the sun; the moon appears half full at these phases.

quasar. Popular term for quasi-stellar source.

quasi-stellar galaxy (QSG). A stellar-appearing object of very large red shift presumed to be extragalactic and highly luminous.

quasi-stellar source (QSS). A stellar-appearing object of very large red shift that is a strong source of radio waves; presumed to be extragalactic and highly luminous.

R Coronae Borealis variables. Eruptive variable stars that show sudden and irregular drops in brightness; the class is named for the prototype, R Coronae Borealis.

RR Lyrae variable. One of a class of giant pulsating stars with periods less than 1 day; a cluster variable.

RW Aurigae stars. Variable stars, generally associated with interstellar matter, that show rapid and irregular light variations.

radar. A technique for observing the reflection of radio waves from a distant object.

radial velocity. The component of relative velocity that lies in the line of sight.

radial velocity curve. A plot of the variation of radial velocity with time for a binary or variable star.

radiant (of meteor shower). The point in the sky from which the meteors belonging to a shower seem to radiate.

radiation. A mode of energy transport whereby energy is transmitted through a vacuum; also the transmitted energy itself, either electromagnetic or corpuscular.

radiation pressure. The transfer of momentum carried by electromagnetic radiation to a body that the radiation impinges upon.

radical. A bond of two or more atoms that does not, in itself, comprise a molecule, but that has characteristics of its own and enters into chemical reactions as if it were a single atom.

radio astronomy. The technique of making

astronomical observations in radio wavelengths.

radio telescope. A telescope designed to make observations in radio wavelengths.

radioactivity (radioactive decay). The process by which certain kinds of atomic nuclei naturally decompose with the spontaneous emission of other subatomic particles and gamma rays.

ray (lunar). Any of a system of bright elongated streaks, sometimes associated with a crater on the moon.

Rayleigh scattering. Scattering of light (photons) by molecules of a gas.

recurrent nova. A nova that has been known to erupt more than once.

red giant. A large, cool star of high luminosity; a star occupying the upper right portion of the Hertzsprung-Russell diagram.

reddening (interstellar). The reddening of starlight passing through interstellar dust, caused by the dust scattering blue light more effectively than red.

redshift. A shift to longer wavelengths of the light from remote galaxies; presumed to be produced by a Doppler shift.

reflecting telescope. A telescope in which the principal optical component (objective) is a concave mirror.

reflection. The reflection of light rays by an optical surface.

reflection nebula. A relatively dense dust cloud in interstellar space that is illuminated by starlight.

refracting telescope. A telescope in which the principal optical component (objective) is a lens or system of lenses.

refraction. The bending of light rays passing from one transparent medium (or a vacuum) to another.

regression of nodes. A consequence of certain perturbations on the orbit of a revolving body whereby the nodes of the orbit slide westward in the fundamental plane (usually the plane of the ecliptic or of the celestial equator).

relative orbit. The orbit of one of two mutually revolving bodies referred to the other body as origin.

relativity. A theory formulated by Einstein that describes the relations between measurements of physical phenomena by two different observers who are in relative motion at constant velocity (the *special theory of relativity*), or at accelerated motion (the *general theory of relativity*).

resolution. The degree to which fine details in an image are separated or resolved.

resolving power. A measure of the ability of an optical system to resolve or separate fine details in the image it produces; in astronomy, the angle in the sky that can be resolved by a telescope.

retrograde motion. An apparent westward motion of a planet on the celestial sphere or with respect to the stars.

revolution. The motion of one body around another.

right ascension. A coordinate for measuring the east-west positions of celestial bodies; the angle measured eastward along the celestial equator from the vernal equinox to the hour circle passing through a body.

rille (or rill). A crevasse or trenchlike depression in the moon's surface.

rotation. Turning of a body about an axis running through it.

Russell-Vogt theorem. The theorem that the mass and chemical composition of a star determine its entire structure if it derives its energy entirely from thermonuclear reactions.

saros. A particular cycle of similar eclipses that recur at intervals of about 18 years.

satellite. A body that revolves about a larger one; for example, a moon of a planet.

Saturn. The sixth planet from the sun in the solar system.

scale (of telescope). The linear distance in the image corresponding to a particular angular distance in the sky; say, so many centimeters per degree.

Schmidt telescope. A type of reflecting telescope invented by B. Schmidt, in which certain aberrations produced by a spherical concave mirror are compensated for by a thin objective correcting lens.

science. The attempt to find order in nature or to find laws that describe natural phenomena.

scientific method. A specific procedure in science: (1) the observation of phenomena or the results of experiments; (2) the formulation of hypotheses that describe these phenomena, and that are consistent with the body of knowledge available; (3) the testing of these hypotheses by noting whether or not they adequately predict and describe new phenomena or the results of new experiments.

scintillation counter. Device for recording primary or secondary cosmic-ray particles from flashes of light produced when these particles strike fluorescent materials.

Sculptor-type system. A dwarf elliptical galaxy,

of which the system in Sculptor is a typical example.

second-order cluster of galaxies. A cluster of clusters of galaxies.

secondary cosmic rays. Secondary particles produced by interactions between primary cosmic rays from space and the atomic nuclei in molecules of the earth's atmosphere.

secondary minimum (in an eclipsing binary light curve). The middle of the eclipse of the cooler star by the hotter, in which the light of the system diminishes less than during the eclipse of the hotter star by the cooler.

secular. Not periodic.

secular parallax. A mean parallax for a selection of stars, derived from the components of their proper motions that reflect the motion of the sun.

seeing. The unsteadiness of the earth's atmosphere, which blurs telescopic images.

seismic waves. Vibrations traveling through the earth's interior that result from earthquakes.

seismograph. An instrument used to record and measure seismic waves.

seismology. The study of earthquakes and the conditions that produce them, and of the internal structure of the earth as deduced from analyses of seismic waves.

selected areas. Small regions of the sky in which magnitudes of stars have been accurately measured and serve as standards for magnitude systems.

seleno-. Prefix referring to the moon.

semimajor axis. Half the major axis of a conic section.

semiregular variable. A variable star, usually a red giant or supergiant, whose period of pulsation is far from constant.

separation (in a visual binary). The angular separation of the two components of a visual binary star.

shadow cone. The umbra of the shadow of a spherical body (such as the earth) in sunlight.

shell star. A type of star, usually of spectral type B to F, surrounded by a gaseous ring or shell.

shower (meteor). Many meteors, all seeming to radiate from a common point in the sky, caused by the encounter by the earth of a swarm of meteoroids moving together through space.

sidereal day. The interval between two successive meridian passages of the vernal equinox.

sidereal month. The period of the moon's revolution about the earth with respect to the stars.

sidereal period. The period of revolution of one body about another with respect to the stars.

sidereal time. The local hour angle of the vernal equinox.

sidereal year. Period of the earth's revolution about the sun with respect to the stars.

sign (of zodiac) (Astrological term). Any of twelve equal sections along the ecliptic, each of length 30°. Starting at the vernal equinox, and commencing eastward, the signs are Aries, Taurus, Gemini, Cancer, Leo, Virgo, Libra, Scorpio, Sagittarius, Capricornus, Aquarius, and Pisces.

sine (of angle). One of the trigonometric functions; the sine of an angle (in a right triangle) is the ratio of the length of the side opposite the angle to that of the hypotenuse.

sine curve. A graph of the sine of an angle plotted against the angle.

small circle. Any circle on the surface of a sphere that is not a great circle.

solar activity. Phenomena of the solar atmosphere associated with sunspots, plages, and related phenomena.

solar antapex. Direction away from which the sun is moving with respect to the local standard of rest.

solar apex. The direction toward which the sun is moving with respect to the local standard of rest.

solar constant. Mean amount of solar radiation received per unit time, by a unit area, just outside the earth's atmosphere, and normal to the direction to the sun; the numerical value is 1.37×10^6 ergs/cm².sec.

solar motion. Motion of the sun, or the velocity of the sun, with respect to the local standard of rest.

solar parallax. Angle subtended by the equatorial radius of the earth at a distance of 1 AU.

solar system. The system of the sun and the planets, their satellites, the minor planets, comets, meteoroids, and other objects revolving around the sun.

solar time. A time based on the sun; usually the hour angle of the sun *plus* 12 hours.

solar wind. A radial flow of corpuscular radiation leaving the sun.

solstice. Either of two points on the celestial sphere where the sun reaches its maximum distances north and south of the celestial equator.

south point. Intersection of the celestial meridian and astronomical horizon 180° from the north point.

space motion. The velocity of a star with respect to the sun.

space probe. An unmanned interplanetary rocket carrying scientific instruments to obtain data on other planets or on the interplanetary environment.

space technology. The applied science of the immediate space environment of the earth.

specific gravity. The ratio of the density of a body or substance to that of water.

spectral class (or type). A classification of a star according to the characteristics of its spectrum.

spectral sequence. The sequence of spectral classes of stars arranged in order of decreasing temperatures of stars of those classes.

spectrogram. A photograph of a spectrum.

spectrograph. An instrument for photographing a spectrum; usually attached to a telescope to photograph the spectrum of a star.

spectroheliogram. A photograph of the sun obtained with a spectroheliograph.

spectroheliograph. An instrument for photographing the sun, or part of the sun, in the monochromatic light of a particular spectral line.

spectrophotometry. The measurement of the intensity of light from a star or other source at different wavelengths.

spectroscope. An instrument for directly viewing the spectrum of a light source.

spectroscopic binary star. A binary star in which the components are not resolved optically, but whose binary nature is indicated by periodic variations in radial velocity, indicating orbital motion.

spectroscopic parallax. A parallax (or distance) of a star that is derived by comparing the apparent magnitude of the star with its absolute magnitude as deduced from its spectral characteristics.

spectroscopy. The study of spectra.

spectrum. The array of colors or wavelengths obtained when light from a source is dispersed, as in passing it through a prism or grating.

spectrum analysis. The study and analysis of spectra, especially stellar spectra.

spectrum binary. A binary star whose binary nature is revealed by spectral characteristics that can only result from the composite of the spectra of two different stars.

spectrum (α^2 Canum Venaticorum) variable. Any of a class of main-sequence spectral type A stars that show anomalously strong lines of certain elements which vary periodically in intensity.

speed. The rate at which an object moves without regard to its direction of motion; the numerical or absolute value of velocity.

spherical aberration. A defect of optical systems whereby on-axis rays of light striking different parts of the objective do not focus at the same place.

spicule. A narrow jet of rising material in the solar chromosphere.

spiral arms. Arms of interstellar material and young stars that wind out in a plane from the central nucleus of a spiral galaxy.

spiral galaxy. A flattened, rotating galaxy with pinwheel-like arms of interstellar material and young stars winding out from its nucleus.

sporadic meteor. A meteor that does not belong to a shower.

spring tide. The highest tide of the month, produced when the sun and moon have longitudes that differ from each other by nearly 0° or 180°.

Sputnik. Russian for "satellite," or "fellow traveler"; the name given to the first Soviet artificial satellite.

stadium. A Greek unit of length, based on the Olympic Stadium; roughly, 1/10 mi.

standard time. The local mean solar time of a standard meridian, adopted over a large region to avoid the inconvenience of continuous time changes around the earth.

star. A self-luminous sphere of gas.

star cluster. An assemblage of stars held together by their mutual gravitation.

statistical parallax. The mean parallax for a selection of stars, derived from the radial velocities of the stars and the components of their proper motions that cannot be affected by the solar motion.

steady state (theory of cosmology). A theory of cosmology embracing the perfect cosmological principle, and involving the continuous creation of matter.

Stefan's law. A formula from which the rate at which a black body radiates energy can be computed; the total rate of energy emission from a unit area of a black body is proportional to the fourth power of its absolute temperature.

stellar evolution. The changes that take place in the sizes, luminosities, structures, and so on, of stars as they age.

stellar model. The result of a theoretical calculation of the run of physical conditions in a stellar interior.

stellar parallax. The angle subtended by 1 AU

at the distance of a star; usually measured in seconds of arc.

stratosphere. The layer of the earth's atmosphere above the troposphere, where most weather takes place, and below the ionosphere.

Strömgren sphere. A region of ionized gas in interstellar space surrounding a hot star; an H II region.

Stonehenge. An assemblage of upright stones in Salisbury Plain, England, believed to have been constructed by early people for astronomical observations connected with timekeeping and the calendar.

subdwarf. A star of luminosity lower than that of main-sequence stars of the same spectral type.

subgiant. A star of luminosity intermediate between those of main-sequence stars and normal giants of the same spectral type.

summer solstice. The point on the celestial sphere where the sun reaches its greatest distance north of the celestial equator.

sun. The star about which the earth and other planets revolve.

sundial. A device for keeping time by the shadow a marker (gnomon) casts in sunlight.

sunspot. A temporary cool region in the solar photosphere that appears dark by contrast against the surrounding hotter photosphere.

sunspot cycle. The semiregular 11-year period with which the frequency of sunspots fluctuates.

supergiant. A star of very high luminosity.

superior conjunction. The configuration of a planet in which it and the sun have the same longitude, with the planet being more distant than the sun.

superior planet. A planet more distant from the sun than the earth.

supernova. A stellar outburst or explosion in which a star suddenly increases its luminosity by from hundreds of thousands to hundreds of millions of times.

surface gravity. The weight of a unit mass at the surface of a body.

surveying. The technique of measuring distances and relative positions of places over the surface of the earth (or elsewhere); generally accomplished by triangulation.

synchrotron radiation. The radiation emitted by charged particles being accelerated in magnetic fields and moving at speeds near that of light.

synodic month. The period of revolution of the moon with respect to the sun, or its cycle of phases.

synodic period. The interval between successive occurrences of the same configuration of a planet; for example, between successive oppositions or successive superior conjunctions.

syzygy. A configuration of the moon in which its elongation is 0° or 180° (new or full).

T-association. A stellar association containing T Tauri stars.

T Tauri stars. Variable stars associated with interstellar matter that show rapid and erratic changes in light.

tail (of comet). Gases and solid particles ejected from the head of a comet and forced away from the sun by radiation pressure or corpuscular radiation.

tangent (of angle). One of the trigonometric functions; the tangent of an angle (in a right triangle) is the ratio of the length of the side opposite the angle to that of the shorter of the adjacent sides.

tangential (transverse) velocity. The component of a star's space velocity that lies in the plane of the sky.

tektites. Rounded glassy bodies that are suspected to be of meteoritic origin.

telescope. An optical instrument used to aid the eye in viewing or measuring, or to photograph distant objects.

telluric. Of terrestrial origin.

temperature (absolute). Temperature measured in centigrade degrees from absolute zero.

temperature (centigrade). Temperature measured on a scale where water freezes at 0° and boils at 100°.

temperature (color). The temperature of a star as estimated from the intensity of the stellar radiation at two or more colors or wavelengths.

temperature (effective). The temperature of a black body that would radiate the same total amount of energy that a particular body does.

temperature (excitation). The temperature of a star as estimated from the relative strengths of lines in its spectrum that originate from atoms in different stages of excitation.

temperature (Fahrenheit). Temperature measured on a scale where water freezes at 32° and boils at 212°.

temperature (ionization). The temperature of a star as estimated from the relative strengths of lines in its spectrum that originate from atoms in different stages of ionization.

temperature (Kelvin). Absolute temperature measured in centigrade degrees.

temperature (kinetic). A measure of the speeds or mean energy of the molecules in a substance.

temperature (radiation). The temperature of a black body that radiates the same amount of energy in a given spectral region as does a particular body.

terminator. The line of sunrise or sunset on the moon.

terrestrial planet. Any of the planets Mercury, Venus, Earth, Mars, and sometimes Pluto.

theory. A set of hypotheses and laws that have been well demonstrated as applying to a wide range of phenomena associated with a particular subject.

thermal energy. Energy associated with the motions of the molecules in a substance.

thermal equilibrium. A balance between the input and outflow of heat in a system.

thermocouple. A device for measuring the intensity of infrared radiation.

thermodynamics. The branch of physics that deals with heat and heat transfer among bodies.

thermonuclear energy. Energy associated with thermonuclear reactions or that can be released through thermonuclear reactions.

thermonuclear reaction. A nuclear reaction or transformation that results from encounters between nuclear particles that are given high velocities (by heating them).

thermopile. A device consisting of a series or pile of thermocouples, which is used to measure the intensity of infrared radiation.

tidal force. A differential gravitational force that tends to deform a body.

tide. Deformation of a body by the differential gravitational force exerted on it by another body; in the earth, the deformation of the ocean surface by the differential gravitational forces exerted by the moon and sun.

ton (American short). 2000 lbs.

ton (English long). 2240 lbs.

ton (metric). One million grams.

total eclipse. An eclipse of the sun in which the sun's photosphere is entirely hidden by the moon, or an eclipse of the moon in which it passes completely into the umbra of the earth's shadow.

train (of meteor). A temporarily luminous trail left in the wake of a meteor.

transit. An instrument for timing the exact instant a star or other object crosses the local meridian. Also, the passage of a celestial body across the meridian; or the passage of one body (say, a planet) across the disk of a larger one (say, the sun).

triangulation. The operation of measuring some of the elements of a triangle so that other ones can be calculated by the methods of trigonometry, thus determining distances to remote places without having to span them directly.

trigonometry. The branch of mathematics that deals with the analytical solutions of triangles.

trillion. In the United States — 1 thousand billions or million millions (10^{12}); in Great Britain — 1 million billions or million-million millions (10^{18}).

triple-alpha process. A series of two nuclear reactions by which three helium nuclei are built up into one carbon nucleus.

Trojan minor planet. One of several minor planets that share Jupiter's orbit around the sun, but located approximately 60° around the orbit from Jupiter.

Tropic of Cancer. Parallel of latitude 23½° N.

Tropic of Capricorn. Parallel of latitude 23½° S.

tropical year. Period of revolution of the earth about the sun with respect to the vernal equinox.

tsunami. A series of very fast waves of seismic origin traveling through the ocean; popularly called "tidal waves."

U,B,V system. A system of stellar magnitudes consisting of measures in the ultraviolet, blue, and green-yellow spectral regions.

UV Ceti stars. Main-sequence stars, mostly of spectral class M, which show occasional, unpredicted flare-ups in light; also called "flare stars."

ultraviolet radiation. Electromagnetic radiation of wavelengths shorter than the shortest (violet) wavelengths to which the eye is sensitive; radiation of wavelengths in the approximate range 100 to 4000 angstroms.

umbra. The central, completely dark part of a shadow.

universal time. The local mean time of the prime meridian.

universe. The totality of all matter and radiation and the space occupied by same.

Uranus. Seventh planet from the sun in the solar system.

Van Allen layer. Doughnut-shaped region surrounding the earth where many rapidly moving charged particles are trapped in its magnetic field.

variable star. A star that varies in luminosity.

variation of latitude. A slight semiperiodic

change in the latitudes of places on the earth that results from a slight shifting of the body of the earth with respect to its axis of rotation.

vector. A quantity that has both magnitude and direction.

velocity. A vector that denotes both the speed and direction a body is moving.

velocity of escape. The speed with which an object must move in order to enter a parabolic orbit about another body (such as the earth), and hence move permanently away from the vicinity of that body.

Venus. The second planet from the sun in the solar system.

vernal equinox. The point on the celestial sphere where the sun crosses the celestial equator passing from south to north.

vertical circle. Any great circle passing through the zenith.

visual binary star. A binary star in which the two components are telescopically resolved.

visual photometer. An instrument used with a telescope for visually measuring the light flux from a star.

volume. A measure of the total space occupied by a body.

von Jolly balance. A balance invented by von Jolly in 1881 to measure the mass of the earth.

Vulcan. A hypothetical planet once believed to exist and have an orbit between that of Mercury and the sun; the existence of Vulcan is now generally discredited.

W Ursae Majoris star. Any of a class of eclipsing binaries whose components are nearly in contact and hence suffer tidal distortion and loss or transfer of matter.

W Virginis star (type II Cepheid). A variable star belonging to the relatively rare class of population II Cepheids.

walled plain. A large lunar crater.

wandering of the poles. A semiperiodic shift of the body of the earth relative to its axis of rotation; responsible for variation of latitude.

watt. A unit of power; 10 million ergs expended per second.

wavelength. The spacing of the crests or troughs in a wave train.

weight. A measure of the force due to gravitational attraction.

west point. The point on the horizon 270° around the horizon from the north point, measured in a clockwise direction as seen from the zenith.

whistles (of meteors). Brief interruptions in high-frequency radio broadcasting reception due to the ionization of the air by meteors.

white dwarf. A star that has exhausted most or all of its nuclear fuel and has collapsed to a very small size; believed to be a star near its final stage of evolution.

Widmanstätten figures. Crystalline structure that can be observed in cut and polished meteorites.

Wien's law. Formula that relates the temperature of a black body to the wavelength at which it emits the greatest intensity of radiation.

winter solstice. Point on the celestial sphere where the sun reaches its greatest distance south of the celestial equator.

Wolf-Rayet star. One of a class of very hot stars that eject shells of gas at very high velocity.

World Calendar. A suggested calendar reform in which each of the four quarters would be identical and the same date of the month would always fall on the same day of the week.

X-rays. Photons of wavelengths intermediate between those of ultraviolet radiation and gamma rays.

X-ray stars. Stars (other than the sun) that emit observable amounts of radiation at X-ray frequencies.

year. The period of revolution of the earth around the sun.

Zeeman effect. A splitting or broadening of spectral lines due to magnetic fields.

zenith. The point on the celestial sphere opposite to the direction of gravity; or the direction opposite to that indicated by a plumb bob.

zenith distance. Arc distance of a point on the celestial sphere from the zenith; 90° minus the altitude of the object.

zodiac. A belt around the sky centered on the ecliptic.

zodiacal light. A faint illumination along the zodiac, believed to be sunlight reflected and scattered by interplanetary dust.

zone of avoidance. A region near the Milky Way where obscuration by interstellar dust is so heavy that few or no exterior galaxies can be seen.

zone time. The time, kept in a zone 15° wide in longitude, that is the local mean time of the central meridian of that zone. Zone time is used at sea, but over land the boundaries are irregular to conform to political boundaries, and it is called standard time.

Some Principles of Arithmetic, Algebra, and Geometry

A3.1 POWERS AND ROOTS

If a number is raised to a power, it is multiplied by itself the number of times indicated by the power or exponent:

$$2^3 = 2 \times 2 \times 2 = 8;$$

$$a^5 = a \times a \times a \times a \times a.$$

Numbers can be raised to fractional as well as integral powers; thus $3^{2-1/2}$ and $7^{3.86}$ are numbers that exist. However, nonintegral powers cannot be evaluated with simple arithmetic. Any number raised to the zero power is unity:

$$a^0 = 1; \quad 2^0 = 1; \quad (8.7621)^0 = 1.$$

The nth root of a number is a quantity which, when raised to the nth power, is equal to the number. Examples:

$$\text{square root of } 64 = \sqrt{64} = 64^{1/2} = 8;$$

$$4\text{th root of } 16 = \sqrt[4]{16} = 16^{1/4} = 2;$$

$$n\text{th root of } a = \sqrt[n]{a} = a^{1/n}.$$

Usually roots are not integral numbers:

$$2^{1/2} = 1.414214 \ldots;$$

$$17^{1/3} = 2.571282 \ldots.$$

Nonintegral roots, such as $^{3.47}\sqrt{33}$ and $21^{1/1.73}$, exist but cannot be evaluated with simple arithmetic.

The nth power or root of a fraction is evaluated by calculating the ratio of the nth power or root of the numerator to the nth power or root of the denominator:

$$\left(\frac{a}{b}\right)^n = \frac{a^n}{b^n}; \qquad \left(\frac{a}{b}\right)^{1/n} = \frac{a^{1/n}}{b^{1/n}};$$

$$\left(\frac{4}{2}\right)^2 = \frac{16}{4} = 4; \qquad \left(\frac{8}{27}\right)^{1/3} = \frac{8^{1/3}}{27^{1/3}} = \frac{2}{3}.$$

The negative power or root of a number is the reciprocal of the same positive power or root of the number:

$$a^{-n} = \frac{1}{a^n}; \quad 2^{-2} = \frac{1}{2^2} = \frac{1}{4}; \quad 3^{-1/3} = \frac{1}{3^{1/3}}.$$

Further properties of powers and roots are:

$$(ab)^n = a^n b^n; \qquad (abc)^n = a^n b^n c^n;$$

$$(ab)^{1/n} = a^{1/n} b^{1/n}; \qquad (abc)^{1/n} = a^{1/n} b^{1/n} c^{1/n}.$$

Examples:

$$(2 \times 3)^3 = 2^3 \times 3^3 = 8 \times 27 = 216;$$

$$(4 \times 9)^{1/2} = 4^{1/2} \times 9^{1/2} = 2 \times 3 = 6.$$

$$a^n a^m = a^{n+m}; \quad \frac{a^n}{a^m} = a^n a^{-m} = a^{n-m}.$$

Examples:

$$2^2 \times 2^3 = 2^5 = 32; \quad \frac{2^3}{2^2} = 2^{3-2} = 2.$$

$$(a^n)^m = a^{nm}; \qquad (a^n)^{1/m} = a^{n/m};$$

$$(a^{1/n})^m = a^{m/n}; \qquad (a^{1/n})^{1/m} = a^{1/nm}.$$

A3.2 POWERS-OF-10 NOTATION

It is often necessary to deal with very large or very small numbers. For example, the earth is 93,000,000 miles from the sun and the mass of the hydrogen atom is 0.000 000 000 000 000 000 000 000 06 oz. Instead of writing and carrying so many zeros, the numbers are usually written as figures between 1 and 10 multiplied by the appropriate power of 10. For example, 93,000,000 is $9.3 \times 10,000,000$, or 9.3×10^7. Similarly, 0.000 000 000 000 000 000 000 06 is 6/100 000 000 000 000 000 000 000 or $6/10^{26} = 6 \times 10^{-26}$. The rule in reading numbers written in this notation is that the exponent of 10 is the number of places the decimal point is to be moved to the right (if the exponent is positive) or to the left (if the exponent is negative).

Multiplication, division, and exponentiation of numbers are facilitated in powers-of-10 notation. Examples:

$$6,000,000 \times 400 = 6 \times 10^6 \times 4 \times 10^2$$
$$= (6 \times 4) \times (10^6 \times 10^2) = 24 \times 10^8$$
$$= 2.4 \times 10^9.$$

$$\frac{6 \times 10^{-26}}{9.3 \times 10^7} = \frac{6}{9.3} \times \frac{10^{-26}}{10^7} = \frac{6}{9.3} \times 10^{-26-7}$$
$$= 0.645 \times 10^{-33} = 6.45 \times 10^{-34}.$$

$$(4000)^3 = (4 \times 10^3)^3 = 4^3 \times (10^3)^3 = 64 \times 10^9$$
$$= 6.4 \times 10^{10}.$$

$$(64,000,000)^{1/2} = (64 \times 10^6)^{1/2} = 64^{1/2} \times (10^6)^{1/2}$$
$$= 8 \times 10^3.$$

A3.3 ANGULAR MEASURE

The most common units of angular measure used in astronomy are the following:
(1) arc measure:
   one circle contains 360 degrees = 360°;
   1° contains 60 minutes of arc = 60′;
   1′ contains 60 seconds of arc = 60″.
(2) time measure:
   one circle contains 24 hours = 24^h;
   1^h contains 60 minutes of time = 60^m;
   1^m contains 60 seconds of time = 60^s.
(3) radian measure:
   one circle contains 2π radians.

A radian is the angle at the center of a circle subtended by a length along the circumference of the circle equal to its radius. Since the circumference of a circle is 2π times its radius, there are 2π radians in a circle.

Relations between these different units of angular measure are given in the following table.

ARC MEASURE	TIME MEASURE	RADIANS	SECONDS OF ARC
57°.2958	3^h.820	1.0	206,264.806
15°	1^h	0.2618	54,000
1°	4^m	1.745×10^{-2}	3,600
15′	1^m	4.363×10^{-3}	900
1′	4^s	2.909×10^{-4}	60
15″	1^s	7.27×10^{-5}	15
1″	0^s.0667	4.85×10^{-6}	1

A3.4 PROPERTIES OF CIRCLES AND SPHERES

The ratio of the circumference of a circle to its diameter is always the same, regardless of the size of the circle. This ratio is universally symbolized by the Greek letter pi (π). Because π is a type of number called an *irrational number*, it cannot be expressed as a simple ratio of two integers, and so its value can never be specified exactly. It can, however, be approximated to any desired degree of accuracy by methods of mathematical analysis. Even the Greeks had determined the value of π by geometrical means to considerable accuracy. For many purposes, a sufficient approximation to π is

$$\pi = 3\tfrac{1}{7} = \tfrac{22}{7}.$$

The value of π has been evaluated to hundreds of decimal places; however, it is seldom needed to greater accuracy than

$$\pi = 3.14159265.$$

In a circle of circumference C, diameter D, and radius R $(R = \tfrac{1}{2}D)$, we have, from the definition of π,

$$\pi = \frac{C}{D} \quad \text{or} \quad C = \pi D = 2\pi R.$$

The circumference of a sphere of radius R is the circumference of any circle on the surface of the sphere whose center coincides with the center of the sphere (a *great circle*), and hence which also has a radius R. Thus, also on a sphere

$$C = 2\pi R.$$

For example, the radius of the earth is about 4000 mi. Its circumference is thus

$$C = 2\pi(4000) = 25,000 \text{ mi.}$$

The *area* of a circle of radius R and diameter D is

$$A = \pi R^2 \quad \text{or} \quad A = \frac{\pi D^2}{4}.$$

The *surface area* of a sphere of radius R is

$$A = 4\pi R^2.$$

This is the total area over the outside surface and must not be confused with the volume. For example, the surface area of the earth is approximately:

$$A = 4\pi (4000)^2 = 2 \times 10^8 \text{ mi}^2.$$

The *volume* of a sphere of radius R is

$$V = \frac{4}{3}\pi R^3.$$

A3.5 PROPORTIONS

If a is to b as c is to d, the ratios a/b and c/d are equal, and the equation

$$\frac{a}{b} = \frac{c}{d}$$

expresses a proportion. This same equation, of course, could also be written

$$ad = bc; \qquad a = b\frac{c}{d}; \qquad c = d\frac{a}{b}.$$

If x and y are both variable quantities and a is a fixed quantity (a *constant*), x and y are said to be proportional to each other (for example, y is proportional to x) if

$$y = ax.$$

This equation means that if x is doubled, y is doubled; if x is increased by 10 times, so is y; if x is diminished to $\frac{1}{1000}$ its former value, so is y. The two always change proportionately. The constant a is thus called a *constant of proportionality*. The above equation can also be written

$$y \propto x,$$

where the symbol $\propto$ means "proportional to."

For example, if x is the length of a board in feet and y is its length in inches, y is proportional to x and the constant of proportionality is 12. If one board has a length of twice as many inches as another, it also has a length of twice as many feet. As a further example, the circumference of a circle is proportional to its radius, the constant of proportionality being 2π.

In an equation like $y = ax$, y is said to be *directly* proportional to x. If, on the other hand, $y = a/x$, which can also be written $y \propto 1/x$, y is said to be *inversely* proportional to x (it is pro-

portional to the reciprocal of x). Again, a is the constant of proportionality. Here, if x is doubled, y is halved; if x is increased by 10 times, y is decreased by 10 times; if x is decreased by 10 times, y is increased by 10 times, and so on.

As an example, if one has a fixed amount of money and apples cost p cents each, the number of apples he can buy, N, is inversely proportional to their cost, p, for

$$N = \frac{A}{p} \qquad \text{or} \qquad N \propto \frac{1}{p};$$

here A, the constant of proportionality, is the total amount of money available in cents.

If $y = ax^2$, y is proportional to the *square* of x; a is still the constant of proportionality. For example, the area of a circle is proportional to the square of its radius (the constant of proportionality is π); a circle of 3 in. radius, in other words, has nine times the area of a circle of 1-in. radius.

In general, proportionalities hold regardless of the units used to express the values of the variables. Only the constant of proportionality changes with a change in units. Suppose the area, in square feet, is desired of a circle of radius R feet. Then $A = \pi R^2$; the constant of proportionality is π. Now, however, suppose R is still expressed in feet but that the area is desired in square inches. Then, since 1 square foot $= 12^2 = 144$ in.^{2}, $A = 144\pi R^2$; the constant of proportionality is now 144π. If R is given in feet but A is desired in square miles, since there are $(5280)^2$ square feet per square mile, we have

$$A = \frac{\pi}{(5280)^2} R^2 \text{ mi}^2.$$

Here the constant of proportionality is $\pi/(5280)^2$. In all cases, however, $A \propto R^2$; only the constant of proportionality changes as the units used to express R and A are changed.

If $y = ax^n$, y is proportional to the nth power of x; if $y = a/x^2$, y is inversely proportional to the square of x. In both cases a is the constant of proportionality.

Among objects of similar shape, no matter what that shape may be, linear distances on those objects (circumferences, diameters, and so on) are proportional to the linear dimensions of the objects. For example, the circumference of a cylinder is proportional to its diameter; the altitudes of cones, similar to each other, are in proportion to their bases, and so on.

Areas of similarly shaped objects, or parts thereof, are proportional to the *square* of their linear dimensions. The surface area of a sphere, for example, is proportional to the square of its radius, and the total outside area of a cube is proportional to the square of the length along one of its edges. The constants of proportionality are 4π and 6, respectively.

The *volumes* of similar objects are proportional to the *cubes* of their linear dimensions. For example, the volume of a sphere is proportional to the cube of its radius, and the volume of a cube is proportional to the cube of the length along one of its edges. The constants of proportionality here are $4\pi/3$ and 1, respectively.

APPENDIX 4

Metric and English Units

In the English system of measure the fundamental units of length, mass, and time are the yard, pound, and second, respectively. There are also, of course, larger and smaller units, which include the ton (2000 lb), the mile (1760 yd), the rod ($16\frac{1}{2}$ ft), the inch ($\frac{1}{36}$ yd), the ounce ($\frac{1}{16}$ lb), and so on. Such units are inconvenient for conversion and arithmetic computation.

In science, therefore, it is more usual to use the metric system, which has been adopted universally in nearly all except the English-speaking countries. The fundamental units of the metric system are:

    length: 1 meter (m)
    mass: 1 kilogram (kg)
    time: 1 second (sec)

A meter was originally intended to be 1 ten-millionth of the distance from the equator to the North Pole along the surface of the earth. It is about 1.1 yd. A kilogram is about 2.2 lb. The second is the same in metric and English units. The most commonly used quantities of length and mass of the metric system are the following:

length

1 km	= 1 kilometer	= 1000 meters	= 0.6214 mile	
1 m	= 1 meter	= 1.094 yards	= 39.37 inches	
1 cm	= 1 centimeter	= 0.01 meter	= 0.3937 inch	
1 mm	= 1 millimeter	= 0.001 meter	= 0.1 cm	= 0.03937 inch
1μ	= 1 micron	= 0.000 001 meter	= 0.0001 cm	= 3.937×10^{-5} inch

also: 1 mile = 1.6093 km
             1 inch = 2.5400 cm

mass

1 metric ton	= 10^6 grams	= 1000 kg	= 2.2046×10^3 lb
1 kg	= 1000 grams	= 2.2046 lb	
1 gm	= 1 gram	= 0.0022046 lb	= 0.0353 oz
1 mg	= 1 milligram	= 0.001 gm	= 2.2046×10^{-6} lb

also: 1 lb = 453.6 gm
          1 oz = 28.3495 gm

Temperature Scales

Three temperature scales are in general use:
(1) Fahrenheit (F); water freezes at 32°F and boils at 212°F.
(2) Centigrade (C); water freezes at 0°C and boils at 100°C.
(3) Kelvin or absolute (K); water freezes at 273°K and boils at 373°K.
All molecular motion ceases at $-459°F = -273°C = 0°K$. Thus Kelvin temperature is measured from this lowest possible temperature, called *absolute zero*. It is the temperature scale most often used in astronomy. Kelvin degrees have the same value as centigrade degrees, since the difference between the freezing and boiling points of water is 100 degrees in each.

On the Fahrenheit scale, water boils at 212 degrees and freezes at 32 degrees; the difference is 180 degrees. Thus to convert centigrade or Kelvin degrees to Fahrenheit it is necessary to multiply by $180/100 = 9/5$. To convert from Fahrenheit to centigrade or Kelvin degrees, it is necessary to multiply by $100/180 = 5/9$.

Example 1: What is 68°F in centigrade and in Kelvin?

$$68°F - 32°F = 36°F \text{ above freezing.}$$

$$\frac{5}{9} \times 36° = 20°;$$

thus,

$$68°F = 20°C = 293°K.$$

Example 2: What is 37°C in Fahrenheit and in Kelvin?

$$37°C = 273° + 37° = 310°K;$$

$$\frac{9}{5} \times 37° = 66.6°;$$

thus,

$$37°C \text{ is } 66.6°F \text{ above freezing}$$

or

$$37°C = 32° + 66.6° = 98.6°F.$$

APPENDIX 6

Mathematical, Physical, and Astronomical Constants

MATHEMATICAL CONSTANTS:

$$\pi = 3.1415926536$$
$$1 \text{ radian} = 57°2957795$$
$$= 3437\!.'74677$$
$$= 206264\!.''806$$

Number of square degrees on a sphere = 41 252.96124

PHYSICAL CONSTANTS:

velocity of light	$c = 2.99793 \times 10^{10}$ cm/sec
constant of gravitation	$G = 6.668 \times 10^{-8}$ dyne·cm²/gm²
Planck's constant	$h = 6.624 \times 10^{-27}$ erg·sec
Boltzmann's constant	$k = 1.380 \times 10^{-16}$ erg/deg
mass of hydrogen atom	$m_H = 1.673 \times 10^{-24}$ gm
mass of electron	$m_e = 9.1085 \times 10^{-28}$ gm
charge on electron	$\epsilon = 4.803 \times 10^{-10}$ electrostatic units
Stefan-Boltzmann constant	$\sigma = 5.669 \times 10^{-5}$ erg/cm²·deg⁴·sec
constant in Wien's law	$\lambda_{max} T = 0.28979$ cm·deg
Rydberg's constant	$R = 1.09737 \times 10^5$ per cm
1 electron volt	$eV = 1.60207 \times 10^{-12}$ erg
1 angstrom	$Å = 10^{-8}$ cm

ASTRONOMICAL CONSTANTS:

astronomical unit	$AU = 1.49597893 \times 10^{13}$ cm
parsec	$pc = 206265$ AU
	$= 3.262$ LY
	$= 3.086 \times 10^{18}$ cm
light-year	$LY = 9.4605 \times 10^{17}$ cm
	$= 6.324 \times 10^4$ AU
tropical year	$= 365.242199$ ephemeris days
sidereal year	$= 365.256366$ ephemeris days
	$= 3.155815 \times 10^7$ sec
mass of earth	$M_{\oplus} = 5.977 \times 10^{27}$ gm
mass of sun	$M_{\odot} = 1.991 \times 10^{33}$ gm
equatorial radius of earth	$R_{\oplus} = 6378.24$ km
radius of sun	$R_{\odot} = 6.960 \times 10^{10}$ cm
luminosity of sun	$L_{\odot} = 3.86 \times 10^{33}$ erg/sec
solar constant	$S = 1.36 \times 10^6$ erg/cm²·sec
	$= 1.97$ cal/m²·min
obliquity of ecliptic (1900)	$\epsilon = 23°27'8\!.''26$
direction of galactic center (1950)	$\alpha = 17^h42^m4$
	$\delta = -28°55'$
direction of north galactic pole (1950)	$\alpha = 12^h49^m$
	$\delta = +27°4$

Astronomical Coordinate Systems

There are several astronomical coordinate systems that are in common use. In each of these systems the position of an object in the sky, or on the celestial sphere, is denoted by two angles. These angles are referred to a *reference plane*, which contains the observer, and a *reference direction*, which is a direction from the observer to some arbitrary point lying in the reference plane. The intersection of the reference plane and the celestial sphere is a great circle, which defines the "equator" of the coordinate system. At two points, each 90° from this equator, are the "poles" of the coordinate system. Great circles passing through these poles intersect the equator of the system at right angles.

One of the two angular coordinates of each coordinate system is measured from the equator of the system to the object along the great circle passing through it and the poles. Angles on one side of the equator (or reference plane) are reckoned as positive; those on the opposite side are negative. The other angular coordinate is measured along the equator from the reference direction to the intersection of the equator with the great circle passing through the object and the poles.

The system of terrestrial latitude and longitude provides an excellent analogue. Here the plane of the terrestrial equator is the fundamental plane and the earth's equator is the equator of the system; the North and South terrestrial Poles are the poles of the system. One coordinate, the *latitude* of a place, is reckoned north (positive) or south (negative) of the equator along a meridian passing through the place. The other coordinate, *longitude*, is measured along the equator to the intersection of the equator and the meridian of the place from the intersection of the equator and the Greenwich meridian. The direction (from the center of the earth) to this latter intersection is the reference direction. Terrestrial longitude is either east or west (whichever is less), but the corresponding coordinate in celestial systems is generally reckoned in one direction from 0 to 360° (or, equivalently, from 0 to 24^h).

The following table lists the more important astronomical coordinate systems and defines how each of the angular coordinates is defined.

Astronomical Coordinate Systems

SYSTEM	REFERENCE PLANE	REFERENCE DIRECTION	"LATITUDE" COORDINATE	RANGE	"LONGITUDE" COORDINATE	RANGE
Horizon	Horizon plane	North point (formerly the south point was used by astronomers)	Altitude, h; toward the zenith $(+)$ toward the nadir $(-)$	$\pm 90°$	Azimuth, A; measured to the east along the horizon from the north point	0 to 360°
Equator	Plane of the celestial equator	Vernal equinox	Declination, δ; toward the north celestial pole $(+)$ toward the south celestial pole $(-)$	$\pm 90°$	Right ascension, α or R.A.; measured to the east along the celestial equator from the vernal equinox	0 to 24^h
Ecliptic	Plane of the earth's orbit (ecliptic)	Vernal equinox	Celestial latitude, β; toward the north ecliptic pole $(+)$ toward the south ecliptic pole $(-)$	$\pm 90°$	Celestial longitude, λ; measured to the east along the ecliptic from the vernal equinox	0 to 360°
Galactic	Mean plane of the Milky Way	Direction to the galactic center	Galactic latitude, b or b''; toward the north galactic pole $(+)$ toward the south galactic pole $(-)$	$\pm 90°$	Galactic longitude, l or l''; measured along the galactic equator to the east from the galactic center	0 to 360°

APPENDIX 8

Nuclear Reactions in Astronomy

Given here are the series of thermonuclear reactions that are most important in stellar interiors. The subscript to the left of a nuclear symbol is the atomic number; the superscript to the right is the atomic mass number. β^+ is the symbol for a positron, ν for neutrino, and γ for a photon (generally of gamma-ray energy).

1. THE PROTON-PROTON CHAIN
   (Important below $15 \times 10^{6}°K$)
   (a) $_1H^1 + {}_1H^1 \rightarrow {}_1H^2 + \beta^+ + \nu$
   (b) $_1H^2 + {}_1H^1 \rightarrow {}_2He^3 + \gamma$
   (c) $_2He^3 + {}_2He^3 \rightarrow {}_2He^4 + 2{}_1H^1$

2. THE CARBON-NITROGEN CYCLE
   (Important above $15 \times 10^{6}°K$)
   (a) $_6C^{12} + {}_1H^1 \rightarrow {}_7N^{13} + \gamma$
   (b) $\quad\quad {}_7N^{13} \rightarrow {}_6C^{13} + \beta^+ + \nu$
   (c) $_6C^{13} + {}_1H^1 \rightarrow {}_7N^{14} + \gamma$
   (d) $_7N^{14} + {}_1H^1 \rightarrow {}_8O^{15} + \gamma$
   $\quad\quad\quad {}_8O^{15} \rightarrow {}_7N^{15} + \beta^+ + \nu$
   (f) $_7N^{15} + {}_1H^1 \rightarrow {}_6C^{12} + {}_2He^4$

3. THE TRIPLE-ALPHA PROCESS
   (Important above $10^{8}°K$)
   (a) $_2He^4 + {}_2He^4 \rightarrow {}_4Be^8$
   (b) $_2He^4 + {}_4Be^8 \rightarrow {}_6C^{12}$

Orbital Data
for the Planets

| PLANET | SYMBOL | SEMIMAJOR AXIS | | SIDEREAL PERIOD | | SYNODIC PERIOD (DAYS) | MEAN ORBITAL SPEED (KM/SEC) | ORBITAL ECCENTRICITY | INCLINATION OF ORBIT TO ECLIPTIC |
		AU	10^6 KM	TROPICAL YEARS	DAYS				
Mercury	☿	0.3871	57.9	0.24085	87.97	115.88	47.8	0.20563	7°004
Venus	♀	0.7233	108.1	0.61521	224.70	583.92	35.0	0.00679	3.394
Earth	⊕	1.0000	149.5	1.000039	365.26		29.8	0.01673	0.0
Mars	♂	1.5237	227.8	1.88089	686.98	779.94	24.2	0.09337	1.850
(Ceres)	①	2.7673	414	4.604		466.6	17.9	0.0765	10.615
Jupiter	♃	5.2028	778	11.86223		398.88	13.1	0.04844	1.305
Saturn	♄	9.5388	1426	29.4577		378.09	9.7	0.05568	2.490
Uranus	♂ or ♅	19.182	2868	84.013		369.66	6.8	0.04721	0.773
Neptune	♆	30.058	4494	164.793		367.48	5.4	0.00858	1.774
Pluto	♇	39.439	5896	247.686		366.72	4.7	0.25024	17.170

For the mean equator and equinox of 1960. Source: Supplement to *The American Ephemeris and Nautical Almanac*.

Physical Data for the Planets

PLANET	DIAMETER		MASS (EARTH = 1)	MEAN DENSITY (GM/CM³)	PERIOD OF ROTATION	INCLINATION OF EQUATOR TO ORBIT	OBLATENESS	SURFACE GRAVITY (EARTH = 1)	ALBEDO	VISUAL MAGNITUDE AT MAXIMUM LIGHT	VELOCITY OF ESCAPE (KM/SEC)
	KM	EARTH = 1									
Mercury	4,880	0.38	0.056	5.1	59 days	?	0	0.39	0.06	−1.9	4.3
Venus	12,112	0.95	0.82	5.3	242.9 days	23°	0	0.91	0.76	−4.4	10.3
Earth	12,742	1.00	1.00	5.52	$23^h56^m04^s$	23°27'	1/298.2	1.00	0.39	—	11.2
Mars	6,800	0.53	0.108	3.94	$24^h37^m23^s$	24°	1/192	0.38	0.15	−2.8	5.1
Jupiter	143,000	11.19	318.0	1.33	9^h50^m to 9^h55^m	3°	1/15	2.64	0.51	−2.5	57.5
Saturn	121,000	9.47	95.2	0.69	10^h14^m to 10^h38^m	27°	1/9.5	1.13	0.50	−0.4	35.4
Uranus	47,000	3.69	14.6	1.56	10^h45^m	98°	1/14	1.07	0.66	+5.6	21.9
Neptune	45,000	3.50	17.3	2.27	16^h?	29°	1/40	1.41	0.62	+7.9	24.4
Pluto	<6,000	<0.47	0.06?	?	6.387 days		0	?	?	+14.9	?

Satellites of Planets

PLANET	SATELLITE	DISCOVERED BY	MEAN DISTANCE FROM PLANET (KM)	SIDEREAL PERIOD (DAYS)	ORBITAL ECCENTRICITY	DIAMETER OF SATELLITE (KM)†	MASS (PLANET = 1)	APPROXIMATE MAGNITUDE AT OPPOSITION
Earth	Moon	—	384,405	27.322	0.055	3476	0.0123	−12.5
Mars	Phobos	A. Hall (1877)	9,380	0.319	0.021	25		+12
	Deimos	A. Hall (1877)	23,500	1.262	0.003	13		13
Jupiter	V	Barnard (1892)	180,500	0.498	0.003	(150)		13
	I Io	Galileo (1610)	421,600	1.769	0.000	3240	4×10^{-5}	5
	II Europa	Galileo (1610)	670,800	3.551	0.000	2830	2.5×10^{-5}	6
	III Ganymede	Galileo (1610)	1,070,000	7.155	0.002	4900	8×10^{-5}	5
	IV Callisto	Galileo (1610)	1,882,000	16.689	0.008	4570	5×10^{-5}	6
	VI	Perrine (1904)	11,470,000	250.57	0.158	(120)		14
	VII	Perrine (1905)	11,800,000	259.65	0.207	(50)		18
	X	Nicholson (1938)	11,850,000	263.55	0.130	(20)		19
	XII	Nicholson (1951)	21,200,000	631.1	0.169	(20)		18
	XI	Nicholson (1938)	22,600,000	692.5	0.207	(25)		19
	VIII	Melotte (1908)	23,500,000	738.9	0.378	(50)		17
	IX	Nicholson (1914)	23,700,000	758	0.275	(22)		19
Saturn	Janus	A. Dollfus (1966)	157,500	0.749	0.020	(350)		14
	Mimas	W. Herschel (1789)	185,400	0.942	0.004	(500)	6.7×10^{-8}	12
	Enceladus	W. Herschel (1789)	237,900	1.370	0.000	(500)	1.5×10^{-7}	12
	Tethys	Cassini (1684)	294,500	1.888	0.000	(1000)	1.1×10^{-6}	11
	Dione	Cassini (1684)	377,200	2.737	0.002	(1000)	1.8×10^{-6}	11
	Rhea	Cassini (1672)	526,700	4.518	0.001	1350	4×10^{-6}	10
	Titan	Huygens (1655)	1,221,000	15.945	0.029	4950	2.5×10^{-4}	8
	Hyperion	Bond (1848)	1,479,300	21.277	0.104	(400)	2×10^{-7}	14
	Iapetus	Cassini (1671)	3,558,400	79.331	0.028	(1200)	2.7×10^{-6}	11
	Phoebe	W. Pickering (1898)	12,945,500	550.45	0.163	(300)		14
Uranus	Miranda	Kuiper (1948)	123,000	1.414	0			17
	Ariel	Lassell (1851)	191,700	2.520	0.003	(600)		15
	Umbriel	Lassell (1851)	267,000	4.144	0.004	(400)		15
	Titania	W. Herschel (1787)	438,000	8.706	0.002	(1000)		14
	Oberon	W. Herschel (1787)	585,960	13.463	0.001	(900)		14
Neptune	Triton	Lassell (1846)	353,400	5.877	0.000	4000	1.3×10^{-3}	14
	Nereid	Kuiper (1949)	5,560,000	359.881	0.749	(300)		19

†A diameter of a satellite given in parentheses is estimated from the amount of sunlight it reflects.

The Nearest Stars

STAR	RIGHT ASCENSION (1950) h m	DECLINATION (1950) ° '	DISTANCE (pc)	PROPER MOTION "	RADIAL VELOCITY (km/sec)	SPECTRA OF COMPONENTS† A	B	C	VISUAL MAGNITUDES OF COMPONENTS A	B	C	ABSOLUTE VISUAL MAGNITUDES OF COMPONENTS A	B	C
α Centauri	14 36.2	−60 38	1.31	3.68	−23	G2V	K5V	M5eV	−0.01	+1.4	+10.7	+4.4	+5.8	+15
Barnard's Star	17 55.4	+4 33	1.83	10.30	−108	M5V	+		+9.54			+13.2		
Wolf 359	10 54.1	+7 19	2.35	4.84	+13	M6eV			+13.66			+16.8		
Lalande 21185	11 00.6	+36 18	2.49	4.78	−86	M2V	+		+7.47			+10.5		
Sirius	6 42.9	−16 39	2.67	1.32	−8	A1V	wd		−1.42	+8.7		+1.4	+11.5	
Luyten 726−8	1 36.4	−18 13	2.67	3.32	+29	M5.5eV	M6eV		+12.5	+12.9		+15.4	+15.8	
Ross 154	18 46.8	−23 54	2.94	0.67	−4	M4.5eV			+10.6			+13.3		
Ross 248	23 39.4	+43 55	3.16	1.58	−81	M5.5eV			+12.24			+14.7		
ε Eridani	3 30.6	−9 38	3.30	0.97	+15	K2V			+3.73			+6.1		
Ross 128	11 45.2	+1 06	3.37	1.36	−13	M5V			+11.13			+13.5		
Luyten 789−6	22 35.8	−15 36	3.37	3.27	−60	M5.5eV			+12.58			+14.9		
61 Cygni	21 04.7	+38 30	3.40	5.22	−64	K5V	K7V		+5.19	+6.02		+7.5	+8.3	
Procyon	7 36.7	+5 21	3.47	1.25	−3	F5 IV-V	wd		+0.38	+10.7		+2.7	+13.0	
ε Indi	21 59.6	−57 00	3.51	4.67	−40	K5V			+4.73			+7.0		
Σ 2398	18 42.2	+59 33	3.60	2.29	+8	M4V	M5V		+8.90	+9.69		+11.1	+11.9	
BD+43°44	0 15.5	+43 44	3.60	2.91	+18	M2.5eV	M4eV(?)		+8.07	+11.04		+10.3	+13.2	
τ Ceti	1 41.8	−16 12	3.64	1.92	−16	G8Vp			+3.50			+5.7		
CD−36°15693	23 02.6	−36 08	3.66	6.87	+10	M2V			+7.39			+9.6		
BD+5°1668	7 24.7	+5 23	3.76	3.73	+26	M4V			+9.82			+11.9		
CD−39°14192	21 14.3	−39 04	3.92	3.46	+21	M0V			+6.72			+8.7		
Kruger 60	22 26.2	+57 27	3.95	0.87	+26	M3V	M4.5eV		+9.77	+11.43		+11.8	+13.4	
Kapteyn's Star	5 09.7	−45 00	3.98	8.79	+242	M0			+8.81			+10.8		
Ross 614	6 26.8	−2 46	4.03	0.97	+24	M4.5eV(?)	?		+11.13	+14.8		+13.1	+16.8	
BD−12°4523	16 27.5	−12 32	4.10	1.24	−13	M4.5V			+10.13			+12.0		
v. Maanen's Star	0 46.5	+5 09	4.24	2.98	+30	wd			+12.36			+14.3		
Wolf 424	12 30.8	+9 18	4.45	1.87	−5	M5.5eV	M6eV		+12.7	+12.7		+14.4	+14.4	
BD+50°1725	10 08.3	+49 42	4.51	1.45	−27	M0V			+6.59			+8.3		
CD−37°15492	0 02.5	−37 36	4.57	6.09	+24	M3V			+8.59			+10.3		
CD−46°11540	17 24.9	−46 51	4.70	1.15	—	M4V			+9.34			+11.3		
BD+20°2465	10 16.9	+20 07	4.72	0.49	+10	M4eV	+		+9.5			+11.1		
CD−44°11909	17 33.5	−44 16	4.78	1.14	—	M5V			+11.2			+12.8		
CD−49°13515	21 30.2	−49 13	4.78	0.78	—	M3V			+9			+11		
AOe 17415−6	17 36.7	+68 23	4.84	1.31	−17	M3V			+9.1			+10.7		
Ross 780	22 50.5	−14 31	4.84	1.12	+9	M5V			+10.2			+11.8		
Lalande 25372	13 43.2	+15 10	4.87	2.30	+15	M2V			+8.6			+10.2		
CC 658	11 42.7	−64 33	4.90	2.69	—	wd			+11			+12.5		
40 Eridani	4 13.0	−7 44	5.00	4.08	−42	K0V	wd	M5eV	+4.5	+9.2	+11.0	+6.0	+10.7	+12.5

†A + symbol indicates that the star is an astrometric binary and has an unseen companion. See also the Notes to Appendix 13.

The Twenty Brightest Stars

STAR	RIGHT ASCENSION (1950) h m	DECLINATION (1950) ° '	DISTANCE (pc)†	PROPER MOTION "	SPECTRA OF COMPONENTS A	B	C	VISUAL MAGNITUDES OF COMPONENTS A	B	C	ABSOLUTE VISUAL MAGNITUDES OF COMPONENTS A	B	C
Sirius	6 42.9	−16 39	2.7	1.32	A1V	wd		−1.42	+8.7		+1.4	+11.5	
Canopus	6 22.8	−52 40	30	0.02	F0Ib-II			−0.72			−3.1		
α Centauri	14 36.2	−60 38	1.3	3.68	G2V	K5V	M5eV	−0.01	+1.4	+10.7	+4.4	+5.8	+15
Arcturus	14 13.4	+19 27	11	2.28	K2IIIp			−0.06			−0.3		
Vega	18 35.2	+38 44	8.0	0.34	A0V			+0.04			+0.5		
Capella	5 13.0	+45 57	14	0.44	GIII	MIV	M5V	+0.05	+10.2	+13.7	−0.7	+9.5	+13
Rigel	5 12.1	−8 15	250	0.00	B8 Ia	B9		+0.14	+6.6		−6.8	−0.4	
Procyon	7 36.7	+5 21	3.5	1.25	F5IV-V	wd		+0.38	+10.7		+2.7	+13.0	
Betelgeuse	5 52.5	+7 24	150	0.03	M2Iab			+0.41v			−5.5		
Achernar	1 35.9	−57 29	20	0.10	B5V			+0.51			−1.0		
β Centauri	14 00.3	−60 08	90	0.04	B1III	?		+0.63	+4		−4.1	−0.8	
Altair	19 48.3	+8 44	5.1	0.66	A7IV-V			+0.77			+2.2		
α Crucis	12 23.8	−62 49	120	0.04	B1IV	B3		+1.39	+1.9		−4.0	−3.5	
Aldebaran	4 33.0	+16 25	16	0.20	K5III	M2V		+0.86	+13		−0.2	+12	
Spica	13 22.6	−10 54	80	0.05	B1V			+0.91v			−3.6		
Antares	16 26.3	−26 19	120	0.03	MIIb	B4eV		+0.92v	+5.1		−4.5	−0.3	
Pollux	7 42.3	+28 09	12	0.62	K0III			+1.16			+0.8		
Fomalhaut	22 54.9	−29 53	7.0	0.37	A3V	K4V		+1.19	+6.5		+2.0	+7.3	
Deneb	20 39.7	+45 06	430	0.00	A2Ia			+1.26			−6.9		
β Crucis	12 44.8	−59 24	150	0.05	B0.5IV			+1.28v			−4.6		

†Distances of the more remote stars have been estimated from their spectral types and apparent magnitudes, and are only approximate.

Notes: Several of the components listed are themselves spectroscopic binaries. A "v" after a magnitude denotes that the star is variable, in which case the magnitude at median light is given. A "p" after a spectral type indicates that the spectrum is peculiar. An "e" after a spectral type indicates that emission lines are present. When the luminosity classification is rather uncertain a range is given.

The Messier Catalogue of Nebulae and Star Clusters

M	NGC OR (IC)	RIGHT ASCENSION (1950) h	m	DECLI-NATION (1950) °	′	APPARENT VISUAL MAGNITUDE	DESCRIPTION
1	1952	5	31.5	+22	00	11.3	"Crab" nebula in Taurus; remains of SN 1054
2	7089	21	30.9	−1	02	6.4	Globular cluster in Aquarius
3	5272	13	39.8	+28	38	6.3	Globular cluster in Canes Venatici
4	6121	16	20.6	−26	24	6.5	Globular cluster in Scorpio
5	5904	15	16.0	+2	17	6.1	Globular cluster in Serpens
6	6405	17	36.8	−32	10		Open cluster in Scorpio
7	6475	17	50.7	−34	48		Open cluster in Scorpio
8	6523	18	00.6	−24	23		"Lagoon" nebula in Sagittarius
9	6333	17	16.3	−18	28	8.0	Globular cluster in Ophiuchus
10	6254	16	54.5	−4	02	6.7	Globular cluster in Ophiuchus
11	6705	18	48.4	−6	20		Open cluster in Scutum Sobieskii
12	6218	16	44.7	−1	52	7.1	Globular cluster in Ophiuchus
13	6205	16	39.9	+36	33	5.9	Globular cluster in Hercules
14	6402	17	35.0	−3	13	8.5	Globular cluster in Ophiuchus
15	7078	21	27.5	+11	57	6.4	Globular cluster in Pegasus
16	6611	18	16.1	−13	48		Open cluster with nebulosity in Serpens
17	6618	18	17.9	−16	12		"Swan" or "Omega" nebula in Sagittarius
18	6613	18	17.0	−17	09		Open cluster in Sagittarius
19	6273	16	59.5	−26	11	7.4	Globular cluster in Ophiuchus
20	6514	17	59.4	−23	02		"Trifid" nebula in Sagittarius
21	6531	18	01.6	−22	30		Open cluster in Sagittarius
22	6656	18	33.4	−23	57	5.6	Globular cluster in Sagittarius
23	6494	17	54.0	−19	00		Open cluster in Sagittarius
24	6603	18	15.5	−18	27		Open cluster in Sagittarius
25	(4725)	18	28.7	−19	17		Open cluster in Sagittarius
26	6694	18	42.5	−9	27		Open cluster in Scutum Sobieskii
27	6853	19	57.5	+22	35	8.2	"Dumbbell" planetary nebula in Vulpecula
28	6626	18	21.4	−24	53	7.6	Globular cluster in Sagittarius
29	6913	20	22.2	+38	21		Open cluster in Cygnus
30	7099	21	37.5	−23	24	7.7	Globular cluster in Capricornus

M	NGC	RIGHT ASCENSION (1950)		DECLINATION (1950)		APPARENT VISUAL MAGNITUDE	DESCRIPTION
		h	m	°	′		
31	224	0	40.0	+41	00	3.5	Andromeda galaxy
32	221	0	40.0	+40	36	8.2	Elliptical galaxy; companion to M31
33	598	1	31.0	+30	24	5.8	Spiral galaxy in Triangulum
34	1039	2	38.8	+42	35		Open cluster in Perseus
35	2168	6	05.7	+24	21		Open cluster in Gemini
36	1960	5	33.0	+34	04		Open cluster in Auriga
37	2099	5	49.1	+32	33		Open cluster in Auriga
38	1912	5	25.3	+35	47		Open cluster in Auriga
39	7092	21	30.4	+48	13		Open cluster in Cygnus
40		12	20	+59			Close double star in Ursa Major
41	2287	6	44.9	−20	41		Loose open cluster in Canis Major
42	1976	5	32.9	−5	25		Orion nebula
43	1982	5	33.1	−5	19		Northeast portion of Orion nebula
44	2632	8	37	+20	10		Praesepe; open cluster in Cancer
45		3	44.5	+23	57		The Pleiades; open cluster in Taurus
46	2437	7	39.5	−14	42		Open cluster in Puppis
47	2478	7	52.4	−15	17		Loose group of stars in Puppis
48		8	11	−1	40		"Cluster of very small stars"; not identifiable
49	4472	12	27.3	+8	16	8.5	Elliptical galaxy in Virgo
50	2323	7	00.6	−8	16		Loose open cluster in Monoceros
51	5194	13	27.8	+47	27	8.4	"Whirlpool" spiral galaxy in Canes Venatici
52	7654	23	22.0	+61	20		Loose open cluster in Cassiopeia
53	5024	13	10.5	+18	26	7.8	Globular cluster in Coma Berenices
54	6715	18	51.9	−30	32	7.8	Globular cluster in Sagittarius
55	6809	19	36.8	−31	03	6.2	Globular cluster in Sagittarius
56	6779	19	14.6	+30	05	8.7	Globular cluster in Lyra
57	6720	18	51.7	+32	58	9.0	"Ring" nebula; planetary nebula in Lyra
58	4579	12	35.2	+12	05	9.6	Spiral galaxy in Virgo
59	4621	12	39.5	+11	56	10.0	Spiral galaxy in Virgo
60	4649	12	41.1	+11	50	9.0	Elliptical galaxy in Virgo
61	4303	12	19.3	+4	45	9.6	Spiral galaxy in Virgo
62	6266	16	58.0	−30	02	7.3	Globular cluster in Scorpio
63	5055	13	13.5	+42	17	8.6	Spiral galaxy in Canes Venatici
64	4826	12	54.2	+21	57	8.5	Spiral galaxy in Coma Berenices
65	3623	11	16.3	+13	22	9.4	Spiral galaxy in Leo
66	3627	11	17.6	+13	16	9.0	Spiral galaxy in Leo; companion to M65
67	2682	8	48.4	+12	00		Open cluster in Cancer
68	4590	12	36.8	−26	29	8.2	Globular cluster in Hydra
69	6637	18	28.1	−32	24	8.0	Globular cluster in Sagittarius
70	6681	18	40.0	−32	20	8.1	Globular cluster in Sagittarius
71	6838	19	51.5	+18	39		Globular cluster in Sagitta
72	6981	20	50.7	−12	45	9.3	Globular cluster in Aquarius
73	6994	20	56.2	−12	50		Open cluster in Aquarius
74	628	1	34.0	+15	32	9.3	Spiral galaxy in Pisces
75	6864	20	03.1	−22	04	8.6	Globular cluster in Sagittarius
76	650	1	39.1	+51	19	11.4	Planetary nebula in Perseus
77	1068	2	40.1	−0	12	8.9	Spiral galaxy in Cetus

M	NGC	RIGHT ASCENSION (1950)		DECLI-NATION (1950)		APPARENT VISUAL MAGNITUDE	DESCRIPTION
		h	m	°	′		
78	2068	5	44.2	+0	02		Small emission nebula in Orion
79	1904	5	22.1	−24	34	7.5	Globular cluster in Lepus
80	6093	16	14.0	−22	52	7.5	Globular cluster in Scorpio
81	3031	9	51.7	+69	18	7.0	Spiral galaxy in Ursa Major
82	3034	9	51.9	+69	56	8.4	Irregular galaxy in Ursa Major
83	5236	13	34.2	−29	37	8.3	Spiral galaxy in Hydra
84	4374	12	22.6	+13	10	9.4	Elliptical galaxy in Virgo
85	4382	12	22.8	+18	28	9.3	Elliptical galaxy in Coma Berenices
86	4406	12	23.6	+13	13	9.2	Elliptical galaxy in Virgo
87	4486	12	28.2	+12	40	8.7	Elliptical galaxy in Virgo
88	4501	12	29.4	+14	42	9.5	Spiral galaxy in Coma Berenices
89	4552	12	33.1	+12	50	10.3	Elliptical galaxy in Virgo
90	4569	12	34.3	+13	26	9.6	Spiral galaxy in Virgo
91		omitted					
92	6341	17	15.6	+43	12	6.4	Globular cluster in Hercules
93	2447	7	42.4	−23	45		Open cluster in Puppis
94	4736	12	48.6	+41	24	8.3	Spiral galaxy in Canes Venatici
95	3351	10	41.3	+11	58	9.8	Barred spiral galaxy in Leo
96	3368	10	44.1	+12	05	9.3	Spiral galaxy in Leo
97	3587	11	12.0	+55	17	11.1	"Owl" nebula; planetary nebula in Ursa Major
98	4192	12	11.2	+15	11	10.2	Spiral galaxy in Coma Berenices
99	4254	12	16.3	+14	42	9.9	Spiral galaxy in Coma Berenices
100	4321	12	20.4	+16	06	9.4	Spiral galaxy in Coma Berenices
101	5457	14	01.4	+54	36	7.9	Spiral galaxy in Ursa Major
102		omitted					
103	581	1	29.9	+60	26		Open cluster in Cassiopeia
104	4594	12	37.4	−11	21	8.3	Spiral galaxy in Virgo
105	3379	10	45.2	+13	01	9.7	Elliptical galaxy in Leo
106	4258	12	16.5	+47	35	8.4	Spiral galaxy in Canes Venatici
107	6171	16	29.7	−12	57	9.2	Globular cluster in Ophiuchus

The Chemical Elements

ELEMENT	SYMBOL	ATOMIC NUMBER	ATOMIC WEIGHT† (CHEMICAL SCALE)	NUMBER OF ATOMS PER TRILLION (10^{12}) HYDROGEN ATOMS‡
Hydrogen	H	1	1.0080	1.0×10^{12}
Helium	He	2	4.003	1.6×10^{11}
Lithium	Li	3	6.940	3.2×10^{3}
Beryllium	Be	4	9.013	6.3×10^{2}
Boron	B	5	10.82	7.6×10^{2}
Carbon	C	6	12.011	4.0×10^{8}
Nitrogen	N	7	14.008	1.1×10^{8}
Oxygen	O	8	16.0000	8.9×10^{8}
Fluorine	F	9	19.00	1.0×10^{6}
Neon	Ne	10	20.183	5.0×10^{8}
Sodium	Na	11	22.991	2.8×10^{5}
Magnesium	Mg	12	24.32	2.5×10^{7}
Aluminum	Al	13	26.98	1.7×10^{6}
Silicon	Si	14	28.09	3.2×10^{7}
Phosphorus	P	15	30.975	2.5×10^{5}
Sulfur	S	16	32.066	2.2×10^{7}
Chlorine	Cl	17	35.457	1.8×10^{6}
Argon	Ar(A)	18	39.944	7.6×10^{6}
Potassium	K	19	39.100	4.6×10^{4}
Calcium	Ca	20	40.08	1.4×10^{6}
Scandium	Sc	21	44.96	6.3×10^{2}
Titanium	Ti	22	47.90	6.0×10^{4}
Vanadium	V	23	50.95	1.3×10^{4}
Chromium	Cr	24	52.01	1.1×10^{5}
Manganese	Mn	25	54.94	7.4×10^{4}
Iron	Fe	26	55.85	5.4×10^{6}
Cobalt	Co	27	58.94	2.6×10^{4}
Nickel	Ni	28	58.71	8.9×10^{5}
Copper	Cu	29	63.54	3.2×10^{3}
Zinc	Zn	30	65.38	3.3×10^{3}
Gallium	Ga	31	69.72	3.2×10^{2}
Germanium	Ge	32	72.60	3.1×10^{2}
Arsenic	As	33	74.91	1.3×10^{2}
Selenium	Se	34	78.96	2.1×10^{3}
Bromine	Br	35	79.916	4.5×10^{2}
Krypton	Kr	36	83.80	1.6×10^{3}
Rubidium	Rb	37	85.48	2.2×10^{2}
Strontium	Sr	38	87.63	5.0×10^{2}

ELEMENT	SYMBOL	ATOMIC NUMBER	ATOMIC WEIGHT† (CHEMICAL SCALE)	NUMBER OF ATOMS PER TRILLION (10^{12}) HYDROGEN ATOMS‡
Yttrium	Y	39	88.92	1.6×10^3
Zirconium	Zr	40	91.22	4.5×10^2
Niobium (Columbium)	Nb(Cb)	41	92.91	2.0×10^2
Molybdenum	Mo	42	95.95	2.0×10^2
Technetium	Tc(Ma)	43	(99)	—
Ruthenium	Ru	44	101.1	66
Rhodium	Rh	45	102.91	23
Palladium	Pd	46	106.4	19
Silver	Ag	47	107.880	11
Cadmium	Cd	48	112.41	46
Indium	In	49	114.82	19
Tin	Sn	50	118.70	1.1×10^2
Antimony	Sb	51	121.76	2.6
Tellurium	Te	52	127.61	1.1×10^2
Iodine	I(J)	53	126.91	22
Xenon	Xe(X)	54	131.30	1.1×10^2
Cesium	Cs	55	132.91	14
Barium	Ba	56	137.36	3.2×10^2
Lanthanum	La	57	138.92	13
Cerium	Ce	58	140.13	20
Praseodymium	Pr	59	140.92	4.6
Neodymium	Nd	60	144.27	23
Promethium	Pm	61	(147)	—
Samarium	Sm(Sa)	62	150.35	7.8
Europium	Eu	63	152.0	3.0
Gadolinium	Gd	64	157.26	11
Terbium	Tb	65	158.93	1.7
Dysprosium	Dy(Ds)	66	162.51	12
Holmium	Ho	67	164.94	2.5
Erbium	Er	68	167.27	6.9
Thulium	Tm(Tu)	69	168.94	1.2
Ytterbium	Yb	70	173.04	6.0
Lutecium	Lu(Cp)	71	174.99	1.1
Hafnium	Hf	72	178.50	2.5
Tantalum	Ta	73	180.95	5.6
Tungsten	W	74	183.86	4.0
Rhenium	Re	75	186.22	8.0
Osmium	Os	76	190.2	25
Iridium	Ir	77	192.2	16
Platinum	Pt	78	195.09	50
Gold	Au	79	197.0	4.6
Mercury	Hg	80	200.61	5.6
Thallium	Tl	81	204.39	3.5
Lead	Pb	82	207.21	32
Bismuth	Bi	83	209.00	3.2
Polonium	Po	84	(209)	—
Astatine	At	85	(210)	—
Radon	Rn	86	(222)	—
Francium	Fr(Fa)	87	(223)	—
Radium	Ra	88	226.05	—
Actinium	Ac	89	(227)	—
Thorium	Th	90	232.12	1.0
Protoactinium	Pa	91	(231)	—

ELEMENT	SYMBOL	ATOMIC NUMBER	ATOMIC WEIGHT† (CHEMICAL SCALE)	NUMBER OF ATOMS PER TRILLION (10^{12}) HYDROGEN ATOMS‡
Uranium	U(Ur)	92	238.07	0.5
Neptunium	Np	93	(237)	—
Plutonium	Pu	94	(244)	—
Americium	Am	95	(243)	—
Curium	Cm	96	(248)	—
Berkelium	Bk	97	(247)	—
Californium	Cf	98	(251)	—
Einsteinium	E	99	(254)	—
Fermium	Fm	100	(253)	—
Mendeleevium	Mv	101	(256)	—
Nobelium	No	102	(253)	—

†Where mean atomic weights have not been well determined, the atomic mass numbers of the most stable isotopes are given in parentheses.
‡Adapted from a compilation by L. H. Aller.

APPENDIX 16

The Constellations

CONSTELLATION (LATIN NAME)	GENITIVE CASE ENDING	ENGLISH NAME OR DESCRIPTION	ABBRE- VIA- TION	APPROXIMATE POSITION α (h)	δ (°)
Andromeda	Andromedae	Princess of Ethiopia	And	1	+40
Antlia	Antliae	Air pump	Ant	10	−35
Apus	Apodis	Bird of Paradise	Aps	16	−75
Aquarius	Aquarii	Water bearer	Aqr	23	−15
Aquila	Aquilae	Eagle	Aql	20	+5
Ara	Arae	Altar	Ara	17	−55
Aries	Arietis	Ram	Ari	3	+20
Auriga	Aurigae	Charioteer	Aur	6	+40
Boötes	Boötis	Herdsman	Boo	15	+30
Caelum	Caeli	Graving tool	Cae	5	−40
Camelopardus	Camelopardis	Giraffe	Cam	6	−70
Cancer	Cancri	Crab	Cnc	9	+20
Canes Venatici	Canum Venaticorum	Hunting dogs	CVn	13	+40
Canis Major	Canis Majoris	Big dog	CMa	7	−20
Canis Minor	Canis Minoris	Little dog	CMi	8	+5
Capricornus	Capricorni	Sea goat	Cap	21	−20
†Carina	Carinae	Keel of Argonauts' ship	Car	9	−60
Cassiopeia	Cassiopeiae	Queen of Ethiopia	Cas	1	+60
Centaurus	Centauri	Centaur	Cen	13	−50
Cepheus	Cephei	King of Ethiopia	Cep	22	+70
Cetus	Ceti	Sea monster (whale)	Cet	2	−10
Chamaeleon	Chamaeleontis	Chameleon	Cha	11	−80

CONSTELLATION (LATIN NAME)	GENITIVE CASE ENDING	ENGLISH NAME OR DESCRIPTION	ABBRE-VIA-TION	APPROXIMATE POSITION	
				α (h)	δ (°)
Circinus	Circini	Compasses	Cir	15	−60
Columba	Columbae	Dove	Col	6	−35
Coma Berenices	Comae Berenices	Berenice's hair	Com	13	+20
Corona Australis	Coronae Australis	Southern crown	CrA	19	−40
Corona Borealis	Coronae Borealis	Northern crown	CrB	16	+30
Corvus	Corvi	Crow	Crv	12	−20
Crater	Crateris	Cup	Crt	11	−15
Crux	Crucis	Cross (southern)	Cru	12	−60
Cygnus	Cygni	Swan	Cyg	21	+40
Delphinus	Delphini	Porpoise	Del	21	+10
Dorado	Doradus	Swordfish	Dor	5	−65
Draco	Draconis	Dragon	Dra	17	+65
Equuleus	Equulei	Little horse	Equ	21	+10
Eridanus	Eridani	River	Eri	3	−20
Fornax	Fornacis	Furnace	For	3	−30
Gemini	Geminorum	Twins	Gem	7	+20
Grus	Gruis	Crane	Gru	22	−45
Hercules	Herculis	Hercules, son of Zeus	Her	17	+30
Horologium	Horologii	Clock	Hor	3	−60
Hydra	Hydrae	Sea serpent	Hya	10	−20
Hydrus	Hydri	Water snake	Hyi	2	−75
Indus	Indi	Indian	Ind	21	−55
Lacerta	Lacertae	Lizard	Lac	22	+45
Leo	Leonis	Lion	Leo	11	+15
Leo Minor	Leonis Minoris	Little lion	LMi	10	+35
Lepus	Leporis	Hare	Lep	6	−20
Libra	Librae	Balance	Lib	15	−15
Lupus	Lupi	Wolf	Lup	15	−45
Lynx	Lyncis	Lynx	Lyn	8	+45
Lyra	Lyrae	Lyre or harp	Lyr	19	+40
Mensa	Mensae	Table Mountain	Men	5	−80
Microscopium	Microscopii	Microscope	Mic	21	−35
Monoceros	Monocerotis	Unicorn	Mon	7	−5
Musca	Muscae	Fly	Mus	12	−70
Norma	Normae	Carpenter's level	Nor	16	−50
Octans	Octantis	Octant	Oct	22	−85
Ophiuchus	Ophiuchi	Holder of serpent	Oph	17	0
Orion	Orionis	Orion, the hunter	Ori	5	+5
Pavo	Pavonis	Peacock	Pav	20	−65
Pegasus	Pegasi	Pegasus, the winged horse	Peg	22	+20
Perseus	Persei	Perseus, hero who saved Andromeda	Per	3	+45
Phoenix	Phoenicis	Phoenix	Phe	1	−50
Pictor	Pictoris	Easel	Pic	6	−55
Pisces	Piscium	Fishes	Psc	1	+15
Piscis Austrinus	Piscis Austrini	Southern fish	PsA	22	−30
†Puppis	Puppis	Stern of the Arognauts' ship	Pup	8	−40
†Pyxis (= Malus)	Pyxidis	Compass on the Argonauts' ship	Pyx	9	−30
Reticulum	Reticuli	Net	Ret	4	−60
Sagitta	Sagittae	Arrow	Sge	20	+10
Sagittarius	Sagittarii	Archer	Sgr	19	−25
Scorpius	Scorpii	Scorpion	Sco	17	−40

CONSTELLATION (LATIN NAME)	GENITIVE CASE ENDING	ENGLISH NAME OR DESCRIPTION	ABBRE-VIA-TION	APPROXIMATE POSITION α (h)	δ (°)
Sculptor	Sculptoris	Sculptor's tools	Scl	0	−30
Scutum	Scuti	Shield	Sct	19	−10
Serpens	Serpentis	Serpent	Ser	17	0
Sextans	Sextantis	Sextant	Sex	10	0
Taurus	Tauri	Bull	Tau	4	+15
Telescopium	Telescopii	Telescope	Tel	19	−50
Triangulum	Trianguli	Triangle	Tri	2	+30
Triangulum Australe	Trianguli Australis	Southern triangle	TrA	16	−65
Tucana	Tucanae	Toucan	Tuc	0	−65
Ursa Major	Ursae Majoris	Big bear	UMa	11	+50
Ursa Minor	Ursae Minoris	Little bear	UMi	15	+70
†Vela	Velorum	Sail of the Argonauts' ship	Vel	9	−50
Virgo	Virginis	Virgin	Vir	13	0
Volans	Volantis	Flying fish	Vol	8	−70
Vulpecula	Vulpeculae	Fox	Vul	20	+25

†The four constellations Carina, Puppis, Pyxis, and Vela originally formed the single constellation, Argo Navis.

APPENDIX 17

Star Maps

The star maps, one for each month, are printed on six pages at the back of this book, following the index. To learn the stars and constellations, the card containing the map for the current month should be taken outdoors and compared directly with the sky. The maps were designed for a latitude of about 35° N but are useful anywhere in the continental United States. Each map shows the appearance of the sky at about 9:00 P.M. (Standard Time) near the middle of the month for which it is intended; near the beginning and end of the month, it shows the sky as it appears about 10:00 P.M. and 8:00 P.M., respectively. To use a map, hold the card vertically, and turn it so that the direction you are facing is shown at the bottom.

These star maps were originally prepared by C. H. Cleminshaw for the *Griffith Observer* (published by the Griffith Observatory, P.O. Box 27787, Los Angeles, CA. 90027), and are reproduced here by the very kind permission of the Griffith Observatory, William J. Kaufmann, director.

Index

Page numbers in boldface refer to principal discussions. Page numbers in parentheses refer to figures.

NORTHERN HORIZON

DRACO

URSA MAJOR "BIG DIPPER"

URSA MINOR "LITTLE DIPPER"

POLARIS "NORTH STAR"

CEPHEUS

DENEB

CYGNUS

CASSIOPEIA

LEO

REGULUS

CANCER

CASTOR

POLLUX

GEMINI

CAPELLA

AURIGA

PERSEUS

ANDROMEDA

TRIANGULUM

PEGASUS

EASTERN HORIZON

HYDRA

ARIES

PISCES

PLEIADES

TAURUS

ALDEBARAN

WESTERN HORIZON

CANIS MINOR

PROCYON

BETELGEUSE

ORION

RIGEL

CETUS

SIRIUS

LEPUS

CANIS MAJOR

COLUMBA

CANOPUS

SOUTHERN HORIZON

THE NIGHT SKY IN JANUARY

Latitude of chart is 34°N, but it is practical throughout the continental United States.

To use: Hold chart vertically and turn it so the direction you are facing shows at the bottom.

Chart time (Local Standard):

10 p.m. First of month

9 p.m. Middle of month

8 p.m. Last of month

Star Chart from GRIFFITH OBSERVER monthly magazine

NORTHERN HORIZON

EASTERN HORIZON

WESTERN HORIZON

SOUTHERN HORIZON

THE NIGHT SKY IN FEBRUARY

Latitude of chart is 34°N, but it is practical throughout the continental United States.

To use: Hold chart vertically and turn it so the direction you are facing shows at the bottom.

Chart time (Local Standard):
10 p.m. First of month
9 p.m. Middle of month
8 p.m. Last of month

Star Chart from GRIFFITH OBSERVER monthly magazine

NORTHERN HORIZON

EASTERN HORIZON

WESTERN HORIZON

SOUTHERN HORIZON

THE NIGHT SKY IN MARCH

Latitude of chart is 34°N, but it is
practical throughout the continental
United States.

To use: Hold chart vertically and turn
it so the direction you are facing
shows at the bottom.

Chart time (Local Standard):
10 p.m. First of month
9 p.m. Middle of month
8 p.m. Last of month

Star Chart from GRIFFITH OBSERVER monthly magazine

NORTHERN HORIZON

EASTERN HORIZON

WESTERN HORIZON

SOUTHERN HORIZON

THE NIGHT SKY IN APRIL

Latitude of chart is 34°N, but it is practical throughout the continental United States.

To use: Hold chart vertically and turn it so the direction you are facing shows at the bottom.

Chart time (Local Standard):

10 p.m. First of month
9 p.m. Middle of month
8 p.m. Last of month

Star Chart from GRIFFITH OBSERVER monthly magazine

THE NIGHT SKY IN MAY

Latitude of chart is 34°N, but it is practical throughout the continental United States.

To use: Hold chart vertically and turn it so the direction you are facing shows at the bottom.

Chart time (Local Standard):

10 p.m. First of month
9 p.m. Middle of month
8 p.m. Last of month

Star Chart from GRIFFITH OBSERVER monthly magazine

NORTHERN HORIZON

EASTERN HORIZON

WESTERN HORIZON

SOUTHERN HORIZON

THE NIGHT SKY IN JUNE

Latitude of chart is 34°N, but it is practical throughout the continental United States.

To use: Hold chart vertically and turn it so the direction you are facing shows at the bottom.

Chart time (Local Standard):
10 p.m. First of month
9 p.m. Middle of month
8 p.m. Last of month

Star Chart from GRIFFITH OBSERVER monthly magazine

SOUTHERN HORIZON

THE NIGHT SKY IN JULY

Latitude of chart is 34°N, but it is
practical throughout the continental
United States.

To use: Hold chart vertically and turn
it so the direction you are facing
shows at the bottom.

Chart time (Local Standard):

10 p.m. First of month

9 p.m. Middle of month

8 p.m. Last of month

Star Chart from GRIFFITH OBSERVER monthly magazine

NORTHERN HORIZON

EASTERN HORIZON

WESTERN HORIZON

SOUTHERN HORIZON

THE NIGHT SKY IN AUGUST

Latitude of chart is 34°N, but it is
practical throughout the continental
United States.

To use: Hold chart vertically and turn
it so the direction you are facing
shows at the bottom.

Chart time (Local Standard):
10 p.m. First of month
9 p.m. Middle of month
8 p.m. Last of month

Star Chart from GRIFFITH OBSERVER monthly magazine

NORTHERN HORIZON

EASTERN HORIZON

WESTERN HORIZON

SOUTHERN HORIZON

THE NIGHT SKY IN SEPTEMBER

Latitude of chart is 34°N, but it is practical throughout the continental United States.

To use: Hold chart vertically and turn it so the direction you are facing shows at the bottom.

Chart time (Local Standard):

10 p.m. First of month
9 p.m. Middle of month
8 p.m. Last of month

Star Chart from GRIFFITH OBSERVER monthly magazine

NORTHERN HORIZON

"BIG DIPPER"
URSA MAJOR

"LITTLE DIPPER"
URSA MINOR
POLARIS "NORTH STAR"

BOÖTES

DRACO

CEPHEUS

CORONA BOREALIS

HERCULES

SERPENS

AURIGA
CAPELLA

CASSIOPEIA

PERSEUS

TAURUS
ALDEBARAN

PLEIADES

TRIANGULUM

ANDROMEDA

VEGA
LYRA

DENEB
CYGNUS
"NORTHERN CROSS"

OPHIUCHUS

ARIES

SAGITTA

EASTERN HORIZON

WESTERN HORIZON

PEGASUS

DELPHINUS

ALTAIR

SERPENS

PISCES

AQUILA

CETUS

AQUARIUS

CAPRICORNUS

SAGITTARIUS

FOMALHAUT

GRUS

SOUTHERN HORIZON

THE NIGHT SKY IN OCTOBER

Latitude of chart is 34°N, but it is
practical throughout the continental
United States.

To use: Hold chart vertically and turn
it so the direction you are facing
shows at the bottom.

Chart time (Local Standard):

10 p.m. First of month

9 p.m. Middle of month

8 p.m. Last of month

Star Chart from GRIFFITH OBSERVER monthly magazine

THE NIGHT SKY IN NOVEMBER

Latitude of chart is 34°N, but it is practical throughout the continental United States.

To use: Hold chart vertically and turn it so the direction you are facing shows at the bottom.

Chart time (Local Standard):

10 p.m. First of month
9 p.m. Middle of month
8 p.m. Last of month

Star Chart from GRIFFITH OBSERVER monthly magazine

NORTHERN HORIZON

EASTERN HORIZON

WESTERN HORIZON

SOUTHERN HORIZON

THE NIGHT SKY IN DECEMBER

Latitude of chart is 34°N, but it is
practical throughout the continental
United States.

To use: Hold chart vertically and turn
it so the direction you are facing
shows at the bottom.

Chart time (Local Standard):

10 p.m. First of month
9 p.m. Middle of month
8 p.m. Last of month

Star Chart from GRIFFITH OBSERVER monthly magazine